Cognitive Radio Technology

Cognitive Radio Technology

Edited by Bruce A. Fette

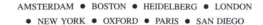

AMSTERDAM • BOSTON • HEIDELBERG • LONDON
• NEW YORK • OXFORD • PARIS • SAN DIEGO
• SAN FRANCISCO • SINGAPORE • SYDNEY • TOKYO

Newnes is an imprint of Elsevier

ELSEVIER

Newnes

Newnes is an imprint of Elsevier
30 Corporate Drive, Suite 400, Burlington, MA 01803, USA
Linacre House, Jordan Hill, Oxford OX2 8DP, UK

Recognizing the importance of preserving what has been written, Elsevier prints its books
on acid-free paper whenever possible.

Library of Congress Cataloging-in-Publication Data

Cognitive radio technology / edited by Bruce A. Fette.—1st ed.
    p. cm.—(Communications engineering series)
 Includes bibliographical references and index
 ISBN-13: 978-0-7506-7952-7 (alk. paper)
 ISBN-10: 0-7506-7952-2 (alk. paper)
 1. Software radio. 2. Artificial intelligence. 3. Wireless communication systems. I. Fette, Bruce A.
II. Series.

  TK5103.4875.C64   2006
  621.384—dc22   2006016824

British Library Cataloguing-in-Publication Data
A catalogue record for this book is available from the British Library.

ISBN 13: 978-0-7506-7952-7
ISBN 10: 0-7506-7952-2

For information on all Newnes publications visit our Web site at
www.books.elsevier.com

06 07 08 09 10   10 9 8 7 6 5 4 3 2 1

Typeset by Charon Tec Ltd, Chennai, India
www.charontec.com
Printed in the United States of America

Contents

Chapter 13: Roles of Ontologies in Cognitive Radios

List of Contributors

Kenneth Baclawski
Computer and Information Science
Northeastern University
360 Huntington Avenue
Boston, MA, 02115

Charles W. Bostian
Wireless @ Virginia Tech
Bradley Department of Electrical
    and Computer Engineering
Virginia Tech
Mail Code 0111
Blacksburg, VA, 24060-0111
Email: bostian@vt.edu

David Brady
ECE Dept
Northeastern University
360 Huntington Avenue
Boston MA, 02115
Email: brady@ece.neu.edu

Joseph P. Campbell
Senior MTS
Information Systems Technology
    Group
MIT Lincoln Laboratory
244 Wood Street, C-290A
Lexington, MA, 02420-9185
Email: j.campbell@ieee.org

William M. Campbell
Information Systems Technology
    Group
MIT Lincoln Laboratory
244 Wood Street, C-243
Lexington, MA, 02420-9185
Email: wmcampbell@ieee.org

Bruce A. Fette
Chief Scientist
Communication Networks Division
General Dynamics C4 Systems
8220 E Roosevelt
Scottsdale, AZ, 85257
Email: brucefette@yahoo.com

Mieczyslaw M. Kokar
Department of Electrical and
    Computer Engineering
Northeastern University
360 Huntington Avenue
Boston, MA, 02115
Email: mkokar@ece.neu.edu

Paul Kolodzy
Kolodzy Consulting
P.O. Box 1443
Centerville, VA, 20120
Email: pkolodzy@kolodzy.com

Vincent J. Kovarik Jr.
Harris Corporation
Mail Stop W2-11F
P.O. Box 37
Melbourne FL, 32902-0037
Email: vkovarik@acm.org

Bin Le
Center for Wireless
    Telecommunications
Wireless @ Virginia Tech
Bradley Department of Electrical
    and Computer Engineering
Virginia Tech, Mail Code 0111
Blacksburg, VA, 24061-0111
Email: binle@vt.edu

Scott M. Lewandowski
Information Systems Technology
    Group
MIT Lincoln Laboratory
244 Wood Street, C-256
Lexington, MA, 02420-9185
Email: scl@ll.mit.edu

Preston Marshall
Defense Advanced Research Projects
    Agency
Email: pmarshall@darpa.mil

Allen B. MacKenzie
Bradley Department of Electrical
    and Computer Engineering
Virginia Polytechnic Institute and
    State University, Mail Code 0111
Blacksburg, VA, 24061-0111
Email: macenab@vt.edu

Joseph Mitola III
Consulting Scientist
Tampa, FL, 33604
Email: jmitola@tampabay.rr.com

James O. Neel
Mobile and Portable Radio Research
    Group
Wireless @ Virginia Tech
Bradley Department of Electrical
    and Computer Engineering
Virginia Tech
432 Durham Hall, MS 0350
Blacksburg, VA, 24061
Email: janeel@vt.edu

John Polson
Principal Engineer
Bell Helicopter, Textron Inc.
P.O. Box 482
Fort Worth, TX, 76101
Email: jtpolson@bellhelicopter.
    textron.com

Jeffrey H. Reed
Mobile and Portable Radio Research
    Group
Wireless @ Virginia Tech
Bradley Department of Electrical
    and Computer Engineering
Virginia Tech
432 Durham Hall, MS 350
Blacksburg, VA, 24061
Email: reedjh@vt.edu

Pablo Robert
Mobile and Protable Radio Research
    Group (MPRG)
Bradley Department of Electrical
    and Computer Engineering
Virginia Tech
Blacksburg, VA, 24060-0111
Email: probert@vt.edu

Thomas W. Rondeau
Bradley Department of Electrical
    and Computer Engineering
Virginia Tech
Mail Code 0111
Blacksburg, VA, 24060-0111
Email: trondeau@vt.edu

Jonathan M. Smith
Defense Advanced Research Projects
    Agency
Email: jmsmith@darpa.mil

Clifford J. Weinstein
Group Leader
Information Systems Technology
    Group
MIT Lincoln Laboratory
244 Wood Street, C-290A
Lexington, MA, 02420-9185
Email: cjw@ll.mit.edu

Robert J. Wellington
University of Minnesota
9740 Russel Circle S.
Bloomington MN, 55431
Email: rwellington@mn.rr.com

Youping Zhao
Mobile and Portable Radio Research
    Group
Wireless @ Virginia Tech
432 Durham Hall, MS 350
Blacksburg, VA, 24061
Email:yozhao@vt.edu

Foreword

This introduction takes a visionary look at ideal cognitive radios (CRs) that integrate advanced software-defined radios (SDR) with CR techniques to arrive at radios that learn to help their user using computer vision, high-performance speech understanding, global positioning system (GPS) navigation, sophisticated adaptive networking, adaptive physical layer radio waveforms, and a wide range of machine learning processes.

CRs Know Radio Like TellMe® Knows 800 Numbers

When you dial 1-800-555-1212, a speech synthesis algorithm says "Toll Free Directory Assistance powered by TellMe® Networks (www.tellme.com, Mountain View, CA, 2005). Please say the name of the listing you want." If you mumble it says, "OK, United Airlines. If that is not what you wanted press 9, otherwise wait while I look up the number." Reportedly, some 99 percent of the time TellMe gets it right, replacing the equivalent of thousands of directory assistance operators of yore. TellMe, a speech-understanding system, achieves a high degree of success by its focus on just one task: finding a toll-free telephone number. Narrow task focus is one key to algorithm successes.

The cognitive radio architecture (CRA) is the building block from which to build cognitive wireless networks (CWNs), the wireless mobile offspring of TellMe. CRs and networks are emerging as practical, real-time, highly focused applications of computational intelligence technology. CRs differ from the more general artificial intelligence (AI) based services like intelligent agents, computer speech, and computer vision in degree of focus. Like TellMe, CRs focus on very narrow tasks. For CRs, the task is to adapt radio-enabled information services to the specific needs of a specific user. TellMe, a network service, requires

---

Note: Adapted from J. Mitola III, *Aware, Adaptive and Cognitive Radio: The Engineering Foundations of Radio XML*, Wiley, 2006.

substantial network computing resources to serve thousands of users at once. CWNs, on the other hand, may start with a radio in your purse or on your belt, a cell phone on steroids, focused on the narrow task of creating from the myriad available wireless information networks and resources just what is needed by just one user: you. Each CR fanatically serves the needs and protects the personal information of just one owner via the CRA using its audio and visual sensory perception and automated machine learning (AML).

TellMe is here and now, while CRs are emerging in global wireless research centers and industry forums like the SDR Forum and Wireless World Research Forum (WWRF). This book introduces the technologies to bootstrap CR systems, introducing technical challenges and approaches, emphasizing CR as a technology enabler for rapidly emerging commercial CWN services.

CRs See What You See, Discovering Radio Frequency Uses, Needs, and Preferences

Although the common cell phone may have a camera, it lacks vision algorithms, so it does not know what it is seeing. It can send a video clip, but it has no perception of the visual scene in the clip. If it had vision-processing algorithms, it could perceive and understand the visual scene. It could tell whether it were at home, in the car, at work, shopping, or driving up the driveway on the way home. If vision algorithms show it that you are entering your driveway in your car, a CR could learn to open the garage door for you wirelessly. Thus, you would not need to fish for the garage door opener, yet another wireless gadget. In fact, you do not need a garage door opener anymore, once CRs enter the market. To open the car door, you will not need a key fob either. As you approach your car, your personal CR perceives the common scene and, as trained, synthesizes the fob radio frequency (RF) transmission and opens the car door for you.

CRs do not attempt everything. They learn about your radio use patterns because they know a lot about radio, generic users, and legitimate uses of radio. CRs have the a priori knowledge needed to detect opportunities to assist you with your use of the radio spectrum accurately, delivering that assistance with minimum intrusion.

Products realizing the visual perception of this vignette are demonstrated on laptop computers today. Reinforcement learning (RL) and case-based reasoning (CBR) are mature AML technologies with radio network applications now being demonstrated in academic and industrial research settings as technology pathfinders for CR[1] and CWN.[2] Two or three Moore's law cycles or 3 to 5 years from now, these vision and learning algorithms will fit in your cell phone. In the interim, CWNs will begin to offer such services, presenting consumers with new tradeoffs between privacy and ultra-personalized convenience.

CRs Hear What You Hear, Augmenting Your Personal Skills

The cell phone on your waist is deaf. Although your cell phone has a microphone, it lacks embedded speech-understanding technology, so it does not perceive what it hears. It can let you talk to your daughter, but it has no perception of your daughter, nor of the content of your conversation. If it had speech-understanding technology, it could perceive your speech dialog. It could detect that you and your daughter are talking about common subjects like homework or your favorite song. With CR, speech algorithms would detect your daughter saying that your favorite song is now playing on WDUV. As an SDR, not just a cell phone, your CR then could tune to FM 105.5 so that you can hear "The Rose."

With your CR, you no longer need a transistor radio. Your CR eliminates from your pocket, purse or backpack yet another RF gadget. In fact, you may not need iPOD®, Game Boy® and similar products as high-end CRs enter the market (or iPODs or Game Boys with CR may become the single pocket pal instead: you never know how market demand will shape products toward the "killer app," do you?). Your CR could learn your radio listening and information use patterns, accessing songs, downloading games, snipping broadcast news, sports, and stock quotes as you like as the CR re-programs its internal SDR to better serve your needs and preferences. Combining vision and speech perception, as you approach your car your CR perceives this common scene and, as you had the morning before, tunes your car radio to WTOP for your favorite "Traffic and weather together on the eights."

---

[1] Mitola's reference for CR pathfinders.

[2] *Semantic Web*: Researchers formulate CRs as sufficiently speech-capable to answer questions about <Self/> and the <Self/> use of <Radio/> in support of its <Owner/>. When an ordinary concept like "owner" has been translated into a comprehensive ontological structure of Computational primitives, for example, via Semantic Web technology, the concept becomes a computational primitive for autonomous reasoning and information exchange. Radio XML, an emerging CR derivative of the eXtensible Markup Language, XML, offers to standardize such radio-scene perception primitives. They are highlighted in this brief treatment by <Angle-brackets/>. All CR have a <Self/>, a <Name/>, and an <Owner/>. The <Self/> has capabilities like <GSM/> and <SDR/>, a self-referential computing architecture, which is guaranteed to crash unless its computing ability is limited to real-time response tasks; this is appropriate for CR but may be too limiting for general-purpose computing.

For AML, CRs need to save speech, RF, and visual cues, all of which may be recalled by the user, expanding the user's ability to remember details of conversations and snapshots of scenes, augmenting the skills of the <owner/>.[3] Because of the brittleness of speech and vision technologies, CRs try to "remember everything" like a continuously running camcorder. Since CRs detect content, such as speakers' names, and keywords like "radio" and "song," they can retrieve some content asked for by the user, expanding the user's memory in a sense. CRs thus could enhance the personal skills of their users, such as memory for detail.

CRs Learn to Differentiate Speakers to Reduce Confusion

To further limit combinatorial explosion in speech, CR may form speaker models, statistical summaries of the speech patterns of speakers, particularly of the <Owner/>. Speaker modeling is particularly reliable when the <Owner/> uses the CR as a cell phone to place a phone call. Contemporary speaker recognition algorithms differentiate male from female speakers with high accuracy. With a few different speakers to be recognized (i.e., fewer than 10 in a family) and with reliable side information like the speaker's telephone number, today's state-of-the-art algorithms recognize individual speakers with better than 95 percent accuracy.

Over time, each CR learns the speech patterns of its <Owner/> in order to learn from the <Owner/> and not be confused by other speakers. CR thus leverages experience incrementally to achieve increasingly sophisticated dialog. Today, a 3 GHz laptop supports this level of speech understanding and dialog synthesis in real time, making it likely to be available in a cell phone in 3 to 5 years.

The CR must both know a lot about radio and learn a lot about you, the <Owner/>, recording and analyzing personal information and thus placing a premium on trustworthy privacy technologies. Increased autonomous customization of wireless service include secondary use of broadcast spectrum. Therefore, the CRA incorporates speech recognition to enable learning without requiring overwhelming amounts of training, allowing it to become sufficiently helpful without being a nuisance.

More Flexible Secondary Use of Radio Spectrum

In 2004, the US Federal Communications Commission (FCC) issued a Report and Order that radio spectrum allocated to TV, but unused in a particular broadcast

---

[3] Ibid.

market, such as a rural area, could be used by CR as secondary users under Part 15 rules for low-power devices—for example, to create ad hoc networks. SDR Forum member companies have demonstrated CR products with these elementary spectrum perception and use capabilities. Wireless products—both military and commercial—are realizing that the FCC vignettes already exist.

Complete visual and speech perception capabilities are not many years distant. Productization is underway. Thus, many chapters of Bruce's outstanding book emphasize CR spectrum agility, suggesting pathways toward enhanced perception technologies, with new long-term growth paths for the wireless industry. This book's contributors hope that it will help you understand and create new opportunities for CR technologies.

Dr. Joseph Mitola III
Tampa, Florida

CHAPTER 1

History and Background of Cognitive Radio Technology

Bruce A. Fette
Communications Networks Division
General Dynamics C4 Systems
Scottsdale, AZ, USA

1.1 The Vision of Cognitive Radio

Just imagine if your cellular telephone, personal digital assistant (PDA), laptop, automobile, and TV were as smart as "Radar" O'Reilly from the popular TV series M*A*S*H.[1] They would know your daily routine as well as you do. They would have things ready for you as soon as you ask, almost in anticipation of your need. They would help you find people, things, and opportunities; translate languages; and complete tasks on time. Similarly, if a radio were smart, it could learn services available in locally accessible wireless computer networks, and could interact with those networks in their preferred protocols, so you would have no confusion in finding the right wireless network for a video download or a printout. Additionally, it could use the frequencies and choose waveforms that minimize and avoid interference with existing radio communication systems. It might be like having a friend in everything that's important to your daily life, or like you were a movie director with hundreds of specialists running around to help you with each task, or like you were an executive with hundred assistants to find documents, summarize them into reports, and then synopsize the reports into an integrated picture. A cognitive radio is the convergence of the many pagers, PDAs, cell phones, and many other

---

[1] "Radar" O'Reilly is a character in the popular TV series M*A*S*H, which ran from 1972 to 1983. He always knew what the colonel needed before the colonel knew he needed it.

single-purpose gadgets we use today. They will come together over the next decade to surprise us with services previously available to only a small select group of people, all made easier by wireless connectivity and the Internet.

1.2 History and Background Leading to Cognitive Radio

The sophistication possible in a software-defined radio (SDR) has now reached the level where each radio can conceivably perform beneficial tasks that help the user, help the network, and help minimize spectral congestion. Radios are already demonstrating one or more of these capabilities in limited ways. A simple example is the adaptive digital European cordless telephone (DECT) wireless phone, which finds and uses a frequency within its allowed plan with the least noise and interference on that channel and time slot. Of these capabilities, conservation of spectrum is already a national priority in international regulatory planning. This book leads the reader through the technologies and regulatory considerations to support three major applications that raise an SDR's capabilities and make it a cognitive radio:

1. Spectrum management and optimizations.
2. Interface with a wide variety of networks and optimization of network resources.
3. Interface with a human and providing electromagnetic resources to aid the human in his or her activities.

Many technologies have come together to result in the spectrum efficiency and cognitive radio technologies that are described in this book. This chapter gives the reader the background context of the remaining chapters of this book. These technologies represent a wide swath of contributions upon which cognitive technologies may be considered as an application on top of a basic SDR platform.

To truly recognize how many technologies have come together to drive cognitive radio techniques, we begin with a few of the major contributions that have led up to today's cognitive radio developments. The development of digital signal processing (DSP) techniques arose due to the efforts of such leaders as Alan Oppenheim [1], Lawrence Rabiner [2, 3], Ronald Schaefer [3], Ben Gold, Thomas Parks [4], James McClellen [4], James Flanagan [5], fred harris [6], and James Kaiser. These pioneers[2] recognized the potential for digital filtering and DSP, and prepared the seminal textbooks, innovative papers, and breakthrough signal

---

[2] This list of contributors is only a partial representative listing of the pioneers with whom the author is personally familiar, and not an exhaustive one.

processing techniques to teach an entire industry how to convert analog signal processes to digital processes. They guided the industry in implementing new processes that were entirely impractical in analog signal processing.

Somewhat independently, Cleve Moler, Jack Little, John Markel, Augustine Gray, and others began to develop software tools that would eventually converge with the DSP industry to enable efficient representation of the DSP techniques, and would provide rapid and efficient modeling of these complex algorithms [7, 8].

Meanwhile, the semiconductor industry, continuing to follow Moore's law [9], evolved to the point where the computational performance required to implement digital signal processes used in radio modulation and demodulation were not only practical, but resulted in improved radio communication performance, reliability, flexibility, and increased value to the customer. This meant that analog functions implemented with large discrete components were replaced with digital functions implemented in silicon, and consequently were more producible, less expensive, more reliable, smaller, and of lower power [10].

During this same period, researchers all over the globe explored various techniques to achieve machine learning and related methods for improved machine behavior. Among these were analog threshold logic, which lead to fuzzy logic and neural networks, a field founded by Frank Rosenblatt [11]. Similarly, languages to express knowledge and to understand knowledge databases evolved from list processing (LISP) and Smalltalk and from massive databases with associated probability statistics. Under funding from the Defense Advanced Research Projects Agency (DARPA), many researchers worked diligently on understanding natural language and understanding spoken speech. Among the most successful speech-understanding systems were those developed by Janet and Jim Baker (who subsequently founded Dragon Systems) [12], and Kai Fu Lee et al. [13]. Both of these systems were developed under the mentoring of Raj Reddy at Carnegie Mellon. Today, we see Internet search engines reflecting the advanced state of artificial intelligence (AI).

In networking, DARPA and industrial developers at Xerox, BBN Technologies, IBM, ATT, and Cisco each developed computer-networking techniques, which evolved into the standard Ethernet and Internet we all benefit from today. The Internet Engineering Task Force (IETF), and many wireless-networking researchers continue to evolve networking technologies with a specific focus on making radio networking as ubiquitous as our wired Internet. These researchers are exploring wireless networks that range from access directly via a radio access point to more advanced techniques in which intermediate radio nodes serve as repeaters to forward data packets toward their eventual destination in an ad hoc network topology.

All of these threads come together as we arrive today at the cognitive radio era (see Figure 1.1). Cognitive radios are nearly always applications that sit on top of an SDR, which in turn is implemented largely from digital signal processors and general-purpose processors (GPPs) built in silicon. In many cases, the spectral efficiency and other intelligent support to the user arises by sophisticated networking of many radios to achieve the end behavior, which provides added capability and other benefits to the user.

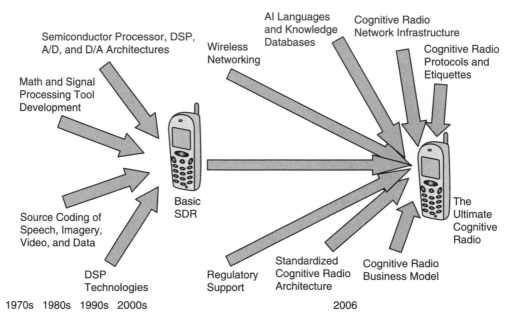

Figure 1.1: Technology timeline. Synergy among many technologies converge to enable the SDR. In turn, the SDR becomes the platform of choice for the cognitive radio.

1.3 A Brief History of SDR

An SDR is a radio in which the properties of carrier frequency, signal bandwidth, modulation, and network access are defined by software. Today's modern SDR also implements any necessary cryptography; forward error correction (FEC) coding; and source coding of voice, video, or data in software as well. As shown in the timeline of Figure 1.2, the roots of SDR design go back to 1987, when Air Force Rome Labs (AFRL) funded the development of a programmable modem as an evolutionary step beyond the architecture of the integrated communications, navigation, and identification architecture (ICNIA). ICNIA was a federated design of multiple radios, that is, a collection of several single-purpose radios in one box.

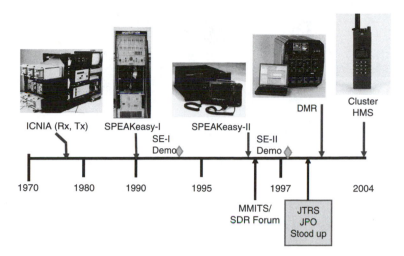

Figure 1.2: SDR timeline. Images of ICNIA, SPEAKeasy I (SE-I), SPEAKeasy II (SE-II), and Digital Modular Ratio (DMR) on their contract award timelines and corresponding demonstrations. These radios are the early evolutionary steps that lead to today's SDR.

Today's SDR, in contrast, is a general-purpose device in which the same radio tuner and processors are used to implement many waveforms at many frequencies. The advantage of this approach is that the equipment is more versatile and cost-effective. Additionally, it can be upgraded with new software for new waveforms and new applications after sale, delivery, and installation. Following the programmable modem, AFRL and DARPA joined forces to fund the SPEAKeasy I and SPEAKeasy II programs.

SPEAKeasy I was a six-foot-tall rack of equipment (not easily portable), but it did demonstrate that a completely software-programmable radio could be built, and included a software-programmable cryptography chip called Cypress, with software cryptography developed by Motorola (subsequently purchased by General Dynamics). SPEAKeasy II was a complete radio packaged in a practical radio size (the size of a stack of two pizza boxes), and was the first SDR to include programmable voice coder (vocoder), and sufficient analog and DSP resources to handle many different kinds of waveforms. It was subsequently tested in field conditions at Ft. Irwin, California, where its ability to handle many waveforms underlined its extreme usefulness, and its construction from standardized commercial off-the-shelf (COTS) components was a very important asset in defense equipment. SPEAKeasy II subsequently evolved into the US Navy's digital modular radio (DMR), becoming a four-channel full duplex SDR, with many waveforms and many modes, able to

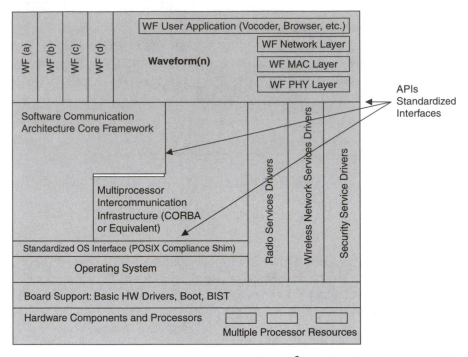

Figure 1.3: Basic software architecture of a modern SDR.[3] Standardized APIs are defined for the major interfaces to assure software portability across many very different hardware platform implementations. The software has the ability to allocate computational resources to specific waveforms. It is normal for an SDR to support many waveforms in order to interface to many networks, and thus to have a library of waveforms and protocols.

be remotely controlled over an Ethernet interface using Simple Network Management Protocol (SNMP).

These SPEAKeasy II and DMR products evolved not only to define these radio waveform features in software, but also to develop an appropriate software architecture to enable porting the software to an arbitrary hardware platform, and thus to achieve hardware independence of the waveform software specification. This critical step allows the hardware to separately evolve its architecture independently from the software, and thus frees the hardware to continue to evolve and improve after delivery of the initial product.

The basic hardware architecture of a modern SDR (Figure 1.3) provides sufficient resources to define the carrier frequency, bandwidth, modulation, any

---

[3]BIST: built-in self-test; CORBA: Common Object Request Broker Architecture; HW: hardware; MAC: medium access control; OS: operating system; PHY: physical (layer); POSIX: Portable Operating System Interface; WF: waveform.

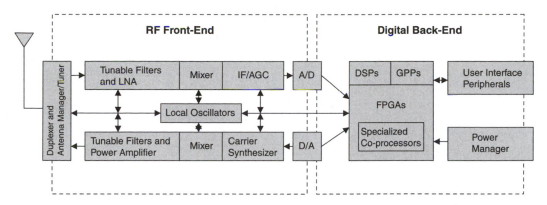

Figure 1.4: Basic hardware architecture of a modern SDR.[4] The hardware provides sufficient resources to define the carrier frequency, bandwidth, modulation, any necessary cryptography, and source coding in software. The hardware resources may include mixtures of GPPs, DSPs, FPGAs, and other computational resources, sufficient to include a wide range of modulation types.

necessary cryptography, and source coding in software. The hardware resources may include mixtures of GPPs, DSPs, field-programmable gate arrays (FPGAs), and other computational resources, sufficient to include a wide range of modulation types (see Section 1.4.1). In the basic software architecture of a modern SDR (Figure 1.4), the application programming interfaces (APIs) are defined for the major interfaces to assure software portability across many very different hardware platform implementations, as well as to assure that the basic software infrastructure supports a wide diversity of waveform applications without having to be rewritten for each waveform or application. The software has the ability to allocate computational resources to specific waveforms (see Section 1.4.2). It is normal for an SDR to support many waveforms in order to interface to many networks, and thus to have a library of waveforms and protocols.

The SDR Forum was founded in 1996 by Wayne Bonser of AFRL to develop industry standards for SDR hardware and software that could assure that the software not only ports across various hardware platforms, but also defines standardized interfaces to facilitate porting software across multiple hardware vendors and to facilitate integration of software components from multiple vendors. The SDR Forum is now a major influence in the SDR industry, dealing not only with standardization of software interfaces but many other important enabling technology

---

[4]A/D: analog to digital; AGC: automatic gain control; D/A: digital to analog; IF: intermediate frequency; LNA: low-noise amplifier; RF: radio frequency.

issues in the industry from tools, to chips, to applications, to cognitive radio and spectrum efficiency. The SDR Forum currently has a Cognitive Radio Working Group, which is preparing papers to advance both spectrum efficiency and cognitive radio applications. In addition, special interest groups within the Forum have interests in these topics.

The SDR Forum Working Group is treating cognitive radio and spectrum efficiency as applications that can be added to an SDR. This means that we can begin to assume an SDR as the basic platform upon which to build most new cognitive radio applications.

1.4 Basic SDR

In this section, we endeavor to provide the reader with background material to provide a basis for understanding subsequent chapters.

1.4.1 The Hardware Architecture of an SDR

The basic SDR must include the radio front-end, the modem, the cryptographic security function, and the application function. In addition, some radios will also include support for network devices connected to either the plain text side or the modem side of the radio, allowing the radio to provide network services and to be remotely controlled over the local Ethernet.

Some radios will also provide for control of external radio frequency (RF) analog functions such as antenna management, coax switches, power amplifiers, or special-purpose filters. The hardware and software architectures should allow RF external features to be added if or when required for a particular installation or customer requirement.

The RF front-end (RFFE) consists of the following functions to support the receive mode: antenna-matching unit, low-noise amplifier, filters, local oscillators, and analog-to-digital (A/D) converters (ADCs) to capture the desired signal and suppress undesired signals to a practical extent. This maximizes the dynamic range of the ADC available to capture the desired signal.

To support the transmit mode, the RFFE will include digital-to-analog (D/A) converters (DACs), local oscillators, filters, power amplifiers, and antenna-matching circuits. In transmit mode, the important property of these circuits is to synthesize the RF signal without introducing noise and spurious emissions at any other frequencies that might interfere with other users in the spectrum.

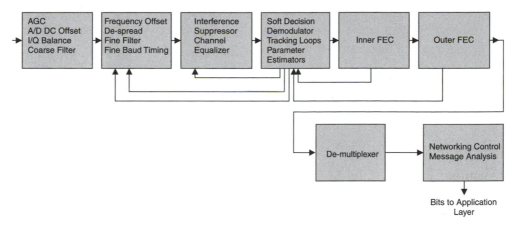

Figure 1.5: Traditional digital receiver signal processing block diagram.[5]

The modem processes the received signal or synthesizes the transmitted signal, or both for a full duplex radio. In the receive process (Figure 1.5), the modem will shift the carrier frequency of the desired signal to a specific frequency nearly equivalent to heterodyne shifting the carrier frequency to direct current (DC), as perceived by the digital signal processor, to allow it to be digitally filtered. The digital filter provides a high level of suppression of interfering signals not within the bandwidth of the desired signal. The modem then time-aligns and de-spreads the signal as required, and refilters the signal to the information bandwidth. Next the modem time-aligns the signal to the symbol or baud time so that it can optimally align the demodulated signal with expected models of the demodulated signal. The modem may include an equalizer to correct for channel multipath artifacts, and for filtering and delay distortions. It may also optionally include rake filtering to optimally cohere multipath components for demodulation. The modem will compare the received symbols with the possible received symbols and make a best possible estimate of which symbols were transmitted. Of course, if there is a weak signal or strong interference, some symbols may be received in error. If the waveform includes FEC coding, the modem will decode the received sequence of encoded symbols by using the structured redundancy introduced in the coding process to detect and correct the encoded symbols that were received in error.

---

[5] I/Q, meaning "in phase and quadrature," is the real part and the imaginary part of the complex-valued signal after being sampled by the ADC(s) in the receiver, or as synthesized by the modem and presented to the DAC in the transmitter.

The process the modem performs for transmit (Figure 1.6) is the inverse of that for receive. The modem takes bits of information to be transmitted, groups the information into packets, adds a structured redundancy to provide for error correction at the receiver, groups bits to be formed into symbols, selects a wave shape to represent each symbol, synthesizes each wave shape, and filters each wave shape to keep it within its desired bandwidth. It may spread the signal to a much wider bandwidth by multiplying the symbol by a wideband waveform which is also generated by similar methods. The final waveform is filtered to match the desired transmit signal bandwidth. If the waveform includes a time-slotted structure such as time division multiple access (TDMA) waveforms, the radio will wait for the appropriate time while placing samples that represent the waveform into an output first in, first out (FIFO) buffer ready to be applied to the DAC. The modem must also control the power amplifier and the local oscillators to produce the desired carrier frequency, and must control the antenna-matching unit to minimize voltage standing wave radio (VSWR). The modem may also control the external RF elements, including transmit versus receive mode, carrier frequency, and smart antenna control. Considerable detail on the architecture of SDR is given by Reed [14].

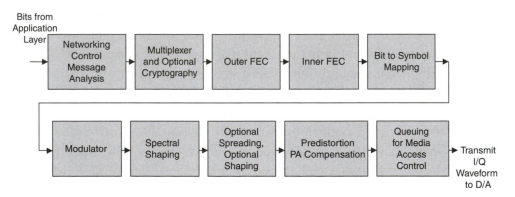

Figure 1.6: Traditional transmit signal processing block diagram.

The cryptographic security function must encrypt any information to be transmitted. Because the encryption processes are unique to each application, these cannot be generalized. The Digital Encryption Standard (DES) and the Advanced Encryption Standard (AES) from the US National Institute of Standards and Technology (NIST) provide example cryptographic processes [15, 16]. In addition to providing the user with privacy for voice communication, cryptography also plays a major role in assuring that the billing is to an authenticated user terminal.

In the future, it will also be used to authenticate transactions of delivering software and purchasing services. In future cognitive radios, the policy functions that define the radios' allowed behaviors will also be cryptographically sealed to prevent tampering with regulatory policy as well as network operator policy.

The application processor will typically implement a vocoder, a video coder, and/or a data coder, as well as selected web browser functions. In each case, the objective is to use knowledge of the properties of the digitized representation of the information to compress the data rate to an acceptable level for transmission. Voice, video, and data coding typically utilize knowledge of the redundancy in the source signal (speech or image) to compress the data rate. Compression factors typically in excess of 10:1 are achieved in voice coding, and up to 100:1 in video coding. Data coding has a variety of redundancies within the message, or between the message and common messages sent in that radio system. Data compression ranges from 10 to 50 percent, depending on how much redundancy can be identified in the original information data stream.

Typically, speech and video applications run on a DSP processor. Text and web browsing typically run on a GPP. As speech recognition technology continues to improve its accuracy, we can expect that the keyboard and display will be augmented by speech input and output functionality. On cognitive radios with adequate processors, it may be possible to run speech recognition and synthesis on the cognitive radio, but early units may find it preferable to vocode the voice, transmit the voice to the base station, and have recognition and synthesis performed at an infrastructure component. This will keep the complexity of the portable units smaller.

1.4.2 Computational Processing Resources in an SDR

The design of an SDR must anticipate the computational resources needed to implement its most complex application. The computational resources may consist of GPPs, DSPs, FPGAs, and occasionally will include other chips that extend the computational capacity. Generally, the SDR vendor will avoid inclusion of dedicated-purpose non-programmable chips because the flexibility to support waveforms and applications is limited, if not rigidly fixed, by non-programmable chips.

Currently, an example GPP selected by many SDR developers is the PowerPC. The PowerPC is available from several vendors. This class of processor is readily programmed in standard C or C++ language, supports a very wide variety of addressing modes, floating point and integer computation, and a large memory space, usually including multiple levels of on-chip and off-chip cache memory. These processors currently perform more than 1 billion mathematical operations per second (mops).[6] GPPs in this class usually pipeline the arithmetic functions and decision logic functions several levels deep in order to achieve these speeds. They also frequently execute many instructions in parallel, typically performing the effective address computations in parallel with arithmetic computation, logic evaluations, and branch decisions.

Most important to the waveform modulation and demodulation processes is the speed at which these processors can perform real or complex multiply accumulates. The waveform signal processing represents more than 90 percent of the total computational load in most waveforms, although the protocols to participate in the networks frequently represent 90 percent of the lines of code. Therefore, it is of great importance to the hardware SDR design that the SDR architecture include DSP-type hardware multiply accumulate functions, so that the signal processes can be performed at high speed, and GPP-type processors for the protocol stack processing.

DSPs are somewhat different than GPPs. The DSP internal architecture is optimized to be able to perform multiply accumulates very fast. This means they have one or more multipliers and one or more accumulators in hardware. Usually the implication of this specialization is that the device has a somewhat unusual memory architecture, usually partitioned so that it can fetch two operands simultaneously and also be able to fetch the next software instruction in parallel with the operand fetches. Currently, DSPs are available that can perform fractional mathematics (integer) multiply accumulate instructions at rates of 1 GHz, and floating point multiply accumulates at 600 MHz. DSPs are also available with many parallel multiply accumulate engines, reporting rates of more than 8 Gmops. The other major feature of the DSP is that it has far fewer and less sophisticated addressing modes. Finally, DSPs frequently utilize modifications of the C language to more efficiently express the signal processing parallelism and fractional arithmetic, and thus maximize their speed. As a result, the DSP is much more efficient at signal processing but less capable to accommodate the software associated with the network protocols.

FPGAs have recently become capable of providing tremendous amounts of multiply accumulate operations on a single chip, surpassing DSPs by more than

---

[6]The mops take into account mathematical operations required to perform an algorithm, but not the operations to calculate an effective memory address index, or offset, nor the operations to perform loop counting, overflow management, or other conditional branching.

an order of magnitude. By defining the on-chip interconnect of many gates, more than 100 multiply accumulators can be arranged to perform multiply accumulate processes at frequencies of more than 200 MHz. In addition to the DSP, FPGAs can also provide the timing logic to synthesize clocks, baud rate, chip rate, time slot, and frame timing, thus leading to a reasonably compact waveform implementation. By expressing all of the signal processing as a set of register transfer operations and multiply accumulate engines, very complex waveforms can be implemented in one chip. Similarly, complex signal processes that are not efficiently implemented on a DSP, such as Cordic operations, log magnitude operations, and difference magnitude operations, can all have the specialized hardware implementations required for a waveform when implemented in FPGAs.

The downside of using FPGA processors is that the waveform signal processing is not defined in traditional software languages such as C, but in VHDL, a language for defining hardware architecture and functionality. The radio waveform description in very high-speed integrated circuit (VHSIC) Hardware Design Language (VHDL), although portable, is not a sequence of instructions and therefore not the usual software development paradigm. At least two companies are working on new software development tools that can produce the required VHDL, somewhat hiding this language complexity from the waveform developer. In addition, FPGA implementations tend to be higher power and more costly than DSP chips.

All three of these computational resources demand significant off-chip memory. For example, a GPP may have more than 128 Mbytes of off-chip instruction memory to support a complex suite of transaction protocols for today's telephony standards.

Today's SDRs provide a reasonable mix of these computational alternatives to assure that a wide variety of desirable applications can in fact be implemented at an acceptable resource level.

In today's SDRs, dedicated-purpose application-specific integrated circuit (ASIC) chips are avoided because the signal processing resources cannot be reprogrammed to implement new waveform functionality.

1.4.3 The Software Architecture of an SDR

The objective of the software architecture in an SDR is to place waveforms and applications onto a software-based radio platform in a standardized way. These waveforms and applications are installed, used, and replaced by other applications as required to achieve the user's objectives. To make the waveform and application interfaces standardized, it is necessary to make the hardware platform present a set of highly standardized interfaces. This way, vendors can develop their waveforms independent of the knowledge of the underlying hardware. Similarly, hardware developers can develop a radio with standardized interfaces, which can subsequently be expected to run a wide variety of waveforms from standardized libraries. This way, the waveform development proceeds by assuming a standardized set of interfaces (APIs) for the radio hardware, and the radio hardware translates commands and status messages crossing those interfaces to the unique underlying hardware through a set of common drivers.

In addition, the method by which a waveform is installed into a radio, activated, deactivated, and de-installed, and the way in which radios use the standard interfaces must be standardized so that waveforms are reasonably portable to more than one hardware platform implementation.

According to Christensen et al., "The use of published interfaces and industry standards in SDR implementations will shift development paradigms away from proprietary tightly coupled hardware software solutions" [17]. To achieve this, the SDR radio is decomposed into a stack of hardware and software functions, with open standard interfaces. As was shown in Figure 1.3, the stack starts with the hardware and the one or more data buses that move information among the various processors. On top of the hardware, several standardized layers of software are installed. This includes the boot loader, the operating system (OS); the board support package (BSP, which consists of input/output drivers that know how to control each interface); and a layer called the hardware abstraction layer (HAL). The HAL provides a method for GPPs to communicate with DSPs and FPGA processors.

The US Government has defined a standardized software architecture, known as the Software Communication Architecture (SCA), which has also been adopted by defense contractors of many other countries worldwide. The SCA is a core framework to provide a standardized process for identifying the available computational resources of the radio, matching those resources to the required resources for an application. The SCA is built upon a standard set of OS features called POSIX,[7] which also has standardized APIs to perform OS functions such as file management and computational thread/task scheduling.

The SCA core framework is the inheritance structure of the open application layer interfaces and services, and provides an abstraction of underlying software and hardware layers. The SCA also specifies a Common Object Request Broker

---

[7]POSIX is the collective name of a family of related standards specified by the IEEE to define the API for software compatible with variants of the Unix OS. POSIX stand for portable operating system interface, with the **X** signifying the Unix heritage of the API [18].

Architecture (CORBA) middleware, which is used to provide a standardized method for software objects to communicate with each other, regardless of which processor they have been installed on (think of it as a software data bus). The SCA also provides a standardized method of defining the requirements for each application, performed in eXtensible Markup Language (XML). The XML is parsed and helps to determine how to distribute and install the software objects. In summary, the core framework provides a means to configure and query distributed software objects, and in the case of SDR, these will be waveforms and other applications.

These applications will have many reasons to interact with the Internet as well as many local networks; therefore, it is also common to provide a collection of standardized radio services, network services, and security services, so that each application does not need to have its own copy of Internet Protocol, and other commonly used functions.

1.4.4 Java Reflection in a Cognitive Radio

Cognitive radios need to be able to tell other cognitive radios what they are observing that may affect the performance of the radio communication channel. The receiver can measure signal properties and can even estimate what the transmitter meant to send, but it also needs to be able to tell the transmitter how to change its waveform in ways that will suppress interference. In other words, the cognitive radio receiver needs to convert this information into a transmitted message to send back to the transmitter.

Figure 1.7 presents a basic diagram for understanding cognitive radios. In this figure, the receiver (radio 2) can use Java reflection to ask questions about the internal parameters inside the receive modem, which might be useful to understand link performance. Measurements common in the design of a receiver, such as the signal-to-noise ratio (SNR), frequency offset, timing offset, or equalizer taps, can be read by the Java reflection. By examining these radio properties, the receiver can determine what change at the transmitter (radio 1) will improve the most important objective of the communication (such as saving battery life). From that Java reflection, the receiver formulates a message onto the reverse link, multiplexes it into the channel, and observes whether the transmitter making that change results in an improvement in link performance.

1.4.5 Smart Antennas in a Cognitive Radio

Current radio architectures are exploring the uses of many types of advanced antenna concepts. A smart radio needs to be able to tell what type of antenna is

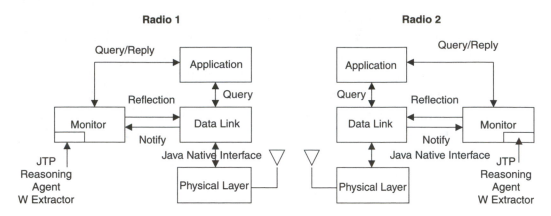

Figure 1.7: Java reflection allows the receiver to examine the state variables of the transmit and receive modem, thereby allowing the cognitive radio to understand what the communications channel is doing to the transmitted signal [19]. Copyright 2003 SDR Forum.

available, and to make full use of its capabilities. Likewise, a smart antenna should be able to tell a smart radio what its capabilities are.

Smart antennas are particularly important to cognitive radio, in that certain functionalities can provide very significant amounts of measurable performance enhancement. As detailed in Chapter 5, if we can reduce transmit power, and thereby allow transmitters to be closer together on the same frequency, we can reduce the geographic area dominated by the transmitter, and thus improve the overall spectral efficiency metric of MHz·km^2.

A smart transmit antenna can form a beam to focus transmitted energy in the direction of the intended receiver. At frequencies of current telecommunication equipment in the range of 800–1800 MHz, practical antennas can easily provide 6–9 dB of gain toward the intended receiver. This same beamforming reduces the energy transmitted in other directions, thereby improving the usability of the same frequency in those directions.

A radio receiver may also be equipped with a smart antenna for receiving. A smart receive antenna can synthesize a main lobe in the desired direction of the intended transmitter, as well as synthesize a deep null in the direction of interfering transmitters. It is not uncommon for a practical smart antenna to be able to synthesize a 20 dB null to suppress interference. This amount of interference suppression has much more impact on the MHz·km^2 metric than being able to transmit 20 dB more transmit power.

The utility of the smart antenna at allowing other radio transmitters to be located nearby is illustrated in Figure 1.8.

Figure 1.8: Utility of smart antennas. A smart antenna allows a transmitter (*T*) to focus its energy toward the intended receiver (*R*), and allows a receiver to suppress interference from nearby interfering transmitters.

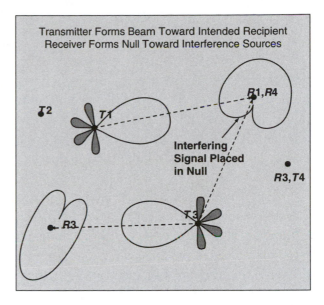

1.5 Spectrum Management

The immediate interest to regulators in fielding cognitive radios is to provide new capabilities that support new methods and mechanisms for spectrum access and utilization now under consideration by international spectrum regulatory bodies. These new methodologies recognize that fixed assignment of a frequency to one purpose across huge geographic regions (often across entire countries) is quite inefficient. Today, this type of frequency assignment results in severe underutilization of the precious and bounded spectrum resource. The Federal Communications Commission (FCC; for commercial applications) and the National Telecommunications and Information Administration (NTIA; for federal applications) in the United States, as well as corresponding regulatory bodies of many other countries, are exploring the question of whether better spectrum utilization could be achieved given some intelligence in the radio and in the network infrastructure.

This interest also has led to developing new methods to manage spectrum access in which the regulator is not required to micromanage every application, every power level, antenna height, and waveform design. Indeed, the goal of minimizing interference with other systems with other purposes may be reasonably automated by the cognitive radio. With a cognitive radio, the regulator could define policies at a higher level, and expect the equipment and the infrastructure to resolve the details within well-defined practical boundary conditions such as available frequency, power, waveform, geography, and equipment capabilities.

In addition, the radio is expected to utilize whatever etiquette or protocol defines cooperative performance for network membership.

In the United States, which has several broad classes of service, the FCC has held meetings with license holders, who have various objectives. There are license holders who retain their specific spectrum for public safety and for other such public purposes such as broadcast of AM, FM, and TV. There are license holders who purchased spectrum specifically for commercial telecommunications purposes. There are license holders for industrial applications, as well as those for special interests.

Many frequencies are allocated to more than one purpose. An example of this is a frequency allocated for remote control purposes—many garage door opener companies and automobile door lock companies have developed and deployed large quantities of products using these remote control frequencies. In addition, there are broad chunks of spectrum for which NTIA has defined frequency and waveform usage, and how the defense community will use spectrum in a process similar to that used by the FCC for commercial purposes.

Finally, there are spectrum commons and unlicensed blocks. In these frequencies, there is overlapping purpose among multiple users, waveforms, and geography. An example of spectrum commons is the 2.4 GHz band, discussed in Section 1.5.1. The following sections touch on new methods for spectrum management, and how they lead to spectrum efficiency.

1.5.1 Managing Unlicensed Spectrum

The 2.4 and 5 GHz band are popularly used for wireless computer networking. These bands, and others, are known as the industrial, scientific, medical (ISM) bands. Energy from microwave ovens falls in the 2.4 GHz band. Consequently, it is impractical to license that band for a particular purpose. However, WiFi® (802.11) and Bluetooth applications are specifically designed to coexist with a variety of interference waveforms commonly found in this band as well as with each other. Various types of equipment utilize a protocol to determine which frequencies or time slots to use and keep trying until they find a usable channel. They also acknowledge correct receipt of transmissions, retransmitting data packets when collisions cause uncorrectable bit errors.

Although radio communication equipment and applications defined in these bands may be unlicensed, they are restricted to specific guidelines about what frequencies are used and what effective isotropic radiated power (EIRP) is allowed. Furthermore, they must accept any existing interference (such as that from microwave ovens and diathermy machines), and they must not interfere with any applications outside this band.

Bluetooth and 802.11 both use waveforms and carrier frequencies that keep their emissions inside the 2.4 GHz band. Both use methods of hopping to frequencies that successfully communicate and to error correct bits or packets that are corrupted by interference. Details of Bluetooth and 802.11 waveform properties are shown in Table 1.1.

The 802.11 waveform can successfully avoid interference from microwave ovens because each packet is of sufficiently short duration that a packet can be delivered at a frequency or during a time period while the interference is minimal. Bluetooth waveforms are designed to hop to many different frequencies very rapidly, and consequently the probability of collision with a strong 802.11 or microwave is relatively small and correctable with error correcting codes.

The regulation of the 2.4 and 5 GHz bands consists of setting the spectrum boundaries, defining specific carrier frequencies that all equipment is to use, and limiting the EIRP. As shown in Table 1.1, the maximum EIRP is 1 W or less for most of the wireless network products, except for the metropolitan WiMAX service, and the FCC-type acceptance is based on the manufacturer demonstrating EIRP and frequency compliance.

It is of particular interest to note that each country sets its own spectral and EIRP rules with regard to these bands. Japan and Europe each have regulatory rules for these bands that are different from those of the United States. Consequently, manufacturers may either (a) make three models, (b) make one model with a switch to select to which country the product will be sold, (c) make a model that is commonly compliant to all regional requirements, or (d) make a model that is capable of determining its current location and implement the local applicable rules. Method (d) is an early application of cognitive techniques.

1.5.2 Noise Aggregation

Communication planners worry that the combined noise from many transmitters may add together and thereby increase the noise floor at the receiver of an important message, perhaps an emergency message. It is well understood that noise power sums together at a receiver. If a receiver antenna is able to see the emissions of many transmitters on the same desired frequency and time slot, increasing the noise floor will reduce the quality of the signal at the demodulator, in turn increasing the bit error rate, and possibly rendering the signal useless. If the interfering transmitters are all located on the ground in an urban area, the interference power from these transmitters decays approximately as the reciprocal of $r^{3.8}$

Table 1.1: Properties of 802.11, Bluetooth, ZigBee, and WiMAX waveforms.

Standard	Name/description	Carrier frequency	Modulation	Data rate	Tx Pwr; EIRP
802.11a (802.11g = both 802.11a and 802.11b)	WiFi; WLAN	5 GHz, 12 channels (8 indoor, 4 point to point)	52 carriers of OFDM, 48 data, 4 pilot, BPSK, QPSK, 16 QAM, or 64 QAM, carrier separation = 0.3125 MHz, symbol duration = 4 μs, with cyclic prefix = 0.8 μs, Viterbi $R = 1/2, 2/3, 3/4$	54, 48, 36, 27, 24, 1, 8, 12, 9, and 6 Mbps	12–30 dBm
802.11b	WiFi; WLAN	2.4 GHz, 3 channels	CCK, DBPSK, DQPSK with DSSS	11, 5.5, and 1.0 Mbps	12–30 dBm
802.15.1	Bluetooth; WPAN	2.4–2.4835 GHz, 79 channels each 1 MHz wide, adaptive frequency hopping at 1600 Hops/s	GFSK (deviation = 140–175 KHz)	57.6, 432.6, and 721 Kbps, 2.1 Mbps	0–20 dBm
802.15.4	ZigBee; WPAN	868.3 MHz, 1 channel; 902–928 MHz, 10 channels with 2 MHz spacing; 2405–2483.5 MHz, 16 channels with 5 MHz spacing	32 chip symbols for 16 ary orthogonal modulation with OQPSK spreading at 2.0 Mcps (2.4 GHz); DBPSK with BPSK spreading at 300 Kcps (868 MHz) or 600 Kcps (915 MHz)	250 Kbps at 2.4 GHz; 40 and 20 Kbps at 868 and 915 MHz	−3–10 dBm
802.16	WiMax, Wibro; WMAN	2–11 GHz (802.16a), 10–66 GHz (802.16); BW = 1.25, 5, 10, and 20 MHz	OFDM, SOFDM: 2048, 1024, 512, 256, and 128 FFT; carriers, each QPSK, 16 QAM, or 64 QAM; symbol rate = 102.9 μs	70 Mbps	40 dBm; EIRP = 57.3 dBm

(a detailed explanation of the exponents of range, r, is found in the Appendix to Chapter 5). The total noise received is the sum of the powers of all such interfering transmitters. Even transmitters whose received power level is below the noise floor also add to the noise floor. However, signals whose power level is extremely small compared to the noise floor have little impact on the noise floor. If there are 100 signals each 20 dB below the noise, then that noise power will sum equal to the noise, and raise the total noise floor by 3 dB. Similarly, if there are 1000 transmitters, each 30 dB below the noise floor, they can raise the noise floor by 30 dB (see the Appendix to Chapter 5 for an explanation). However, the additional noise is usually dominated by the one or two interfering transmitters that are closest to the receiver.

In addition, we must consider the significant effect of personal communication devices, which are becoming ubiquitous. In fact, one person may have several devices all at close range to each other. Cognitive radios will be the solution to this spectral noise and spectral crowding, and will evolve to the point of deployed science just in time to help with the aggregated noise problems of many personal devices all attempting to communicate in proximity to each other.

1.5.3 Aggregating Spectrum Demand and Use of Subleasing Methods

Many applications for wireless service operate with their own individual licensed spectra. It is rare that each service is fully consuming its available spectrum. Studies show that spectrum occupancy seems to peak at about 14 percent, except under emergency conditions, where occupancy can reach 100 percent for brief periods of time. Each of these services does not wish to separately invest in their own unique infrastructure. Consequently, it is very practical to aggregate these spectral assignments to serve a user community with a combined system. The industry refers to a collection of services of this type as a trunked radio. Trunked radio base stations have the ability to listen to many input frequencies. When a user begins to transmit, the base station assigns an input and an output frequency for the message and notifies all members of the community to listen on the repeater downlink frequency for the message. Trunking aggregates the available spectrum of multiple users and is therefore able to deliver a higher quality of service while reducing infrastructure costs to each set of users and reducing the total amount of spectrum required to serve the community.

Both public safety and public telephony services benefit from aggregating spectrum and experience fluctuating demands, so each could benefit from the ability to borrow spectrum from the other. This is a much more complex situation, however. Public safety system operators must be absolutely certain that they can get all the spectrum capacity they need if an emergency arises. Similarly, they might be able to appreciate the revenue stream from selling access to their spectrum to commercial users who have need of access during times when no emergency conditions exist.

1.5.4 Priority Access[8]

If agreements can be negotiated between spectrum license holders and spectrum users who have occasional peak capacity needs, it is possible to define protocols to request access, grant access, and withdraw access. Thus, an emergency public service can temporarily grant access to its spectrum in exchange for monetary compensation. Should an emergency arise, the emergency public service can withdraw its grant to access, thereby taking over priority service.

In a similar fashion, various classes of users can each contend for spectrum access, with higher-priority users being granted access before other users. This might be relevant, for example, if police, fire, or military users need to use the cellular infrastructure during an emergency. Their communications equipment can indicate their priority to the communications infrastructure, which may in turn grant access for these highest-priority users first.[9]

By extension, a wide variety of grades of service for commercial users may also prioritize sharing of commercially licensed spectrum. Users who are willing to pay the most may get high priority for higher data rates for their data packets. The users who pay least would get service only when no other grades of service are consuming the available bandwidth.

1.6 US Government Roles in Cognitive Radio

1.6.1 DARPA

Paul Kolodzy was a Program Manager at DARPA when he issued a Broad Area Announcement (BAA) calling for an industry day on the NeXt Generation (XG) program to explore how XG communications could not only make a significant impact on spectral efficiency of defense communications, but also significantly reduce the complexity of defining the spectrum allocation for each defense user.

---

[8]Cellular systems already support priority access; however, there is reported to be little control over the allocation of priority or the enforcement process.
[9]This technique is implemented in code division multiple access (CDMA) cellular communications.

Shortly after proposals were sent in to DARPA, Kolodzy moved to the FCC to further explore this question and Preston Marshall became the DARPA program manager. Under Marshall's XG program, several contractors demonstrated that a cognitive radio could achieve substantial spectral efficiency in a non-interfering method, and that the spectrum allocation process could be simplified. Basic principles of spectrum efficiency are discussed by Marshall in Chapter 5.

In the same time frame, Jonathan Smith worked as a DARPA project manager to develop intelligent network protocols that could learn and adapt to the properties of wireless channels to optimize performance under current conditions. Some of this work is discussed in Chapter 9.

1.6.2 FCC

On May 19, 2003, the FCC held a hearing to obtain industry comments on cognitive radio. Participants from the communications industry, radio and TV broadcasters, public safety officers, telecommunications systems operators, and public advocacy participants all discussed how this technology might interact with the existing spectrum regulatory process. Numerous public meetings were held subsequently to discuss the mechanics of such systems, and their impact on existing license holders of spectrum (see Section 1.4 and Chapter 2). The FCC has been actively engaged with industry and very interested in leveraging this technology.

1.6.3 NSF/CSTB Study

President George W. Bush has established the Spectrum Policy Task Force (SPTF) to further study the economic and political considerations and impacts of spectrum policy. In addition to the FCC's public meetings, the National Science Foundation (NSF) also has held meetings on the impact of new technologies to improve spectrum efficiency. A Committee chaired by Dale Hatfield and Paul Kolodzy heard testimony from numerous representatives, leading to the SPTF report further described in Chapter 2.

The Computer Science and Telecommunications Board (CSTB) is a specific work group of the NSF. This work group produces books and workshops on important topics in telecommunications. CSTB held numerous workshops on the topic of spectrum management since its opening meeting at the FCC in May 2003. These meetings have resulted in reports to the FCC on various cognitive radio topics. Recently, a workshop was held on the topic of "Improving Spectrum Management through Economic and Other Incentives." This activity has been guided by Dale Hatfield (formerly Chief of the Office of Science and Technology at the FCC, now Adjunct Professor at University of Colorado); William Lehr (Economist and Research Associate at Center for Technology Policy and Industrial Development at Massachusetts Institute of Technology); and Jon Peha (associate director of the Center for Wireless and Broadband Networking at Carnegie Mellon University).

1.7 How Smart Is Useful?

The cognitive radio is able to provide a wide variety of intelligent behaviors. It can monitor the spectrum and choose frequencies that minimize interference to existing communication activity. When doing so, it will follow a set of rules that define what frequencies may be considered, what waveforms may be used, what power levels may be used for transmission, and so forth. It may also be given rules about the access protocols by which spectrum access is negotiated with spectrum license holders, if any, and the etiquettes by which it must check with other users of the spectrum to ensure that no user hidden from the node wishing to transmit is already communicating.

In addition to the spectrum optimization level, the cognitive radio may have the ability to optimize a waveform to one or many criteria. For example, the radio may be able to optimize for data rate, for packet success rate, for service cost, for battery power minimization, or for some mixture of several criteria. The user does not see these levels of sophisticated channel analysis and optimization except as the recipient of excellent service.

The cognitive radio may also exhibit behaviors that are more directly apparent to the user. These behaviors may include: (a) awareness of geographic location, (b) awareness of local networks and their available services, (c) awareness of the user and the user's biometric authentication to validate financial transactions, and (d) awareness of the user and his or her prioritized objectives. This book explores each of these technologies. Many of these services will be immediately valuable to the user without the need for complex menu screens, activation sequences, or preference setup processes.

The cognitive radio developer must use caution to avoid adding cognitive functionality that reduces the efficiency of the user at his or her primary tasks. If the user thinks of the radio as a cell phone and does not wish to access other networks, the cognitive radio developer must provide a design that is friendly to the user, timely and responsive, but is not continually intruding with attempts to be helpful by connecting to networks that the user does not need or want. If the radio's owner is a power user, however, the radio may be asked to watch for multiple opportunities: access to other wireless networks for data services, notification of critical turning points to aid navigation, or timely financial information, as a few simple examples.

One of the remaining issues in sophisticated software design is a method for determining whether the cognitive services the radio might offer will be useful. Will the services be accomplished in a timely fashion? Will the attempted services be undesired and disruptive? Will the services take too long to implement and arrive too late to be usable? The cognitive radio must offer functionality that is timely and useful to its owner, and yet not disruptive. Like "Radar" O'Reilly, we want the cognitive radio to offer support of the right type at the right time, properly prioritized to the user needs given sophisticated awareness of the local situation, and not offering frequent useless or obvious recommendations. We will explore this topic in Chapter 16.

1.8 Organization of this Book

In Chapter 2, Paul Kolodzy describes the regulatory policy motivations, activities, and initiatives within US and international regulatory bodies to achieve enhanced spectral efficiency. Chapter 3, by Max Robert of Virginia Tech and Bruce Fette of General Dynamics C4 Systems, describes the details of hardware and software architecture of SDRs, and explains why an SDR is the primary choice as the basis for cognitive radios. Chapter 4, by John Polson of Bell Helicopter, is about the technologies required to implement basic services in a cognitive radio.

In Chapter 5, Preston Marshall of DARPA deals with spectrum efficiency and the DARPA programs that demonstrated the feasibility of cognitive radio principles. Chapter 6, by Robert Wellington of the University of Minnesota, introduces the cognitive policy engine. The policy engine provides an efficient mechanism to express the rules applied to the function of the cognitive radio. This includes regulatory policy, network operator policy, radio equipment capability, and real-time checking that, at the end of all the cognitive logic, the radio's planned performance is allowed within its rules.

Chapter 7, by Thomas Rondeau and Charles Bostian of Virginia Tech, provides a detailed analysis of cognitive techniques at the physical and medium access control layers. These techniques are discussed in the context of genetic algorithms that can adapt multiple waveform properties in order to optimize link performance.

In Chapter 8, John Polson and Bruce Fette describe a wide variety of methods by which a radio can determine its local position, and thereby to use time and location information to assist the network or the user. In Chapter 9, Jonathan Smith of

DARPA[10] covers cognitive techniques in network adaption. This technology allows the radio to be aware of local networks and their properties and services. Smith focuses on how networks apply intelligence to the selection of network protocols in order to optimize network performance in spite of the differences between wireless and wired systems.

Chapter 10, by Joe Campbell, Bill Campbell, Scott Lewandowski, and Cliff Weinstein of Lincoln Laboratory, is about using speech as an input/output mechanism for the user to request and access services, as well as to authenticate the user to the radio network and services. Speech analysis tools extract basic properties of speech. These properties are further analyzed in different ways to result in word recognition, language recognition, or speaker identification.

Chapter 11, by Youping Zhao, Bin Le, and Jeffrey Reed of Virginia Tech, deals with the ways in which network infrastructure can provide cognitive functionality and support services to the user, even if the subscriber's unit is of low power or small computational capability.

Chapter 12, by Vince Kovarik of Harris Corporation, provides extensive coverage of learning technologies and techniques, and how these are applied to the cognitive radio application.

In Chapter 13, Mitch Kokar, David Brady, and Kenneth Baclawski of Northeastern University give a detailed overview of how to represent the types of knowledge a radio would need to know in order to behave intelligently. This chapter deals with the storage, and analysis of this information as additional spatial, temporal, radio, network, and application data are accumulated.

In Chapter 14, Joseph Mitola III of the MITRE Corporation describes how to develop a complete radio and how to make the various radio modules work with each other as an integrated cognitive system.

In Chapter 15, Jody Neel, Jeff Reed, and Allen McKenzie of Virginia Tech provide a detailed analysis of game theory and how it is used to model the performance choices and system-level behaviors of networks consisting of a mixture of cognitive and non-cognitive radios.

Chapter 16 completes this book with an overview of the remaining key problems that must be solved to allow cognitive radio to fulfill its considerable potential and become a recognized and successful self-sustaining industry.

---

[10] Jonathan Smith is currently on the faculty of the University of Pennsylvania, PA.

References

[1]   A. Oppenheim and R. Schaefer, *Discrete Time Signal Processing*, Prentice Hall, Englewood Cliffs, NJ, 1989.

[2]   L. Rabiner and R. Schaefer, *Digital Processing of Speech Signals*, Prentice Hall, Englewood Cliffs, NJ, 1978.

[3]   L.R. Rabiner, J.H. McClellan and T.W. Parks, "FIR Digital Filter Design Techniques Using Weighted Chebyshev Approximations," *Proceedings of the IEEE*, Vol. 63, 1975, pp. 595–610.

[4]   T.W. Parks and J.J. McClellan, "Chebyshev Approximation for Nonrecursive Digital Filters with Linear Phase," *IEEE Transactions on Circuit Theory*, Vol. 19, 1972, pp. 189–194.

[5]   J. Flanagan, *Speech Synthesis and Perception*, Springer Verlag, New York, 1972.

[6]   Fred Harris, *Multirate Signal Processing for Communication Systems*, Prentice Hall, Englewood Cliffs, NJ, 2004.

[7]   http://www.mathworks.com/company/aboutus/founders/jacklittle.html

[8]   J. Markel and A. Gray, *Linear Prediction of Speech*, Springer Verlag, Berlin, Heidelburg and New York, 1976.

[9]   http://www.intel.com/pressroom/kits/bios/moore.htm

[10] http://www.dspguide.com/filters.htm

[11] F. Rosenblatt, "The Perceptron: A probabilistic Model for Information Storage and Organization in the Brain," *Cornell Aeronautical Laboratory, Psychological Review*, Vol. 65, No. 6, 1958, pp. 386–408.

[12] J. Baker, *Stochastic Modeling for Speech Recognition*, Doctoral Thesis, Department of Computer Science, Carnegie Mellon University, Pittsburgh, PA, 1976.

[13] K. Lee, R. Reddy and Hsiao-wuen, "An Overview of the SPHINX Speech Recognition System," *IEEE Transactions on Acoustics Speech and Signal Processing*, pp. 34–45, January 1990.

[14] J.H. Reed, *Software Radio: A Modern Approach to Radio Engineering*, Prentice Hall, Englewood Cliffs, NJ, 2002.

[15] http://csrc.ncsl.nist.gov/cryptval/des/des.txt

[16] http://csrc.nist.gov/CryptoToolkit/aes/

[17] E. Christensen, A. Miller and E. Wing, "Waveform Application Development Process for Software Defined Radios," *IEEE Milcom Conference*, pp. 231–235, Oct. 22, 2000.

[18] http://en.wikipedia.org/wiki/POSIX

[19] J. Wang, "The Use of Ontologies for the Self-Awareness of Communication Nodes," in *Proceedings of the SDR Forum Technical Conference*, Orlando, FL, November 2003.

CHAPTER 2

Communications Policy and Spectrum Management

Paul Kolodzy
Kolodzy Consulting
Centreville, VA, USA

2.1 Introduction

New technologies impact the worlds of commerce and policy. This is especially true of disruptive technologies that significantly alter either the realities or perceptions within these worlds. Cognitive radio technology has the potential of affecting the marketplace for radio devices and services as well as changing the means by which wireless communications policy is developed and implemented. One of the key parameters that must be addressed to enter the radio market is access to radio spectrum. Once access is obtained, the capacity to manage interference becomes a key attribute in order to increase the number of users. Throughput is critical in order to maximize benefit (for the device) or maximize revenue (for the service). Radio frequency (RF) spectrum access and interference management are thus the primary roles of spectrum management.

Cognitive radio technology has the potential of being a disruptive force within spectrum management. Spectrum management, since the dawn of radio technology, has been within the domain of management agencies, both private and government. Therefore, it has required a person-in-the-loop. The ability of a device to be aware of its environment and to adapt to enhance its performance, and the performance of the network, allows a transition from a manual, oversight process to an automated, device-oriented process. This ability has the potential to allow a much more intensive use of the spectrum by lowering of spectrum access barrier to entry for new devices and services. It also has the potential to radically change how policy should be developed in order to account for these new uses of the spectrum, and it can fundamentally change the role of the spectrum policy-maker and regulator.

In this chapter, Section 2.2 discusses the cognitive radio technology enablers. Section 2.3 addresses spectrum access and how cognitive radio needs various types of policies, depending on the density of spectral activity and the types of usages. This section also provides examples of spectral activity measurement. Section 2.4 discusses the challenges to equipment developers associated with a policy-based approach to spectrum management. Section 2.5 presents the challenges to the regulators to manage spectrum policy through radios and networks that operate based on policy. Section 2.6 discusses the global interest and activity in policy-based cognitive radio. Finally, Section 2.7 provides a summary of this chapter's major issues.

2.2 Cognitive Radio Technology Enablers

The development of wideband power amplifiers, synthesizers, and analog-to-digital converters (ADCs) is providing a new class of radios: the software-defined radio (SDR) and its software and cognitive radio cousins. Although at the early stages of development, this new class of radio ushers in new possibilities as well as potential pitfalls for technology policy. The flexibility provided by the cognitive radio class of radios allows for more dynamics within radio operations. The same flexibility poses challenges for certification and the associated liability through potential misuse.

SDRs provide software control of a variety of modulation techniques, wideband and narrowband operation, transmission security (TRANSEC) functions (such as hopping), and waveform requirements. In essence, components can be under digital control and thus defined by software. The advantage of an SDR is that a single system can operate under multiple configurations, providing interoperability, bridging, and tailoring of the waveforms to meet the localized requirements. SDR technology and systems have been developed for the military. The digital modular radio (DMR) system was one of the first SDR systems. Recently the US Defense Advanced Research Projects Agency (DARPA) developed the Small Unit Operations Situational Awareness Systems (SUO SAS), which was a man-portable SDR operating from 20 MHz to 2.5 GHz. The level of success of these programs has led to the Joint Tactical Radio System (JTRS) initiative to develop and procure SDR systems throughout the US military.

SDRs exhibit software control over a variety of modulation techniques and waveforms. Software radios (SRs) specifically implement the waveform signal processing in software. This additional caveat essentially has the radio being constructed with a RF front-end, a down-converter to an intermediate frequency (IF) or baseband, an ADC, and then a processor. The processing capacity therefore limits the complexity of the waveforms that can be accommodated.

A cognitive radio adds both a sensing and an adaptation element to the software defined and software radios. Four new capabilities embodied in cognitive radios will help enable dynamic use of the spectrum: flexibility, agility, RF sensing, and networking [1].

- *Flexibility* is the ability to change the waveform and the configuration of a device. An example is a cell tower that can operate in the cell band for telephony purposes but change its waveform to get telemetry from vending machines during low usage, or other equally useful, schedulable, off-peak activity. The same band is used for two very different roles, and the radio characteristics must reflect the different requirements, such as data rate, range, latency, and packet error rate.

- *Agility* is the ability to change the spectral band in which a device will operate. Cell phones have rudimentary agility because they can operate in two or more bands (e.g., 900 and 1900 MHz). Combining both agility and flexibility is the ultimate in "adaptive" radios because the radio can use different waveforms in different bands. Specific technology limitations exist, however, to the agility and flexibility that can be afforded by current technology. The time scale of these adaptations is a function of the state of technology both in the components for adaptation as well as the capacity to sense the state of the system. These are classically denoted as the observable/controllable requirements of control systems.

- *Sensing* is the ability to observe the state of the system, which includes the radio and, more importantly, the environment. It is the next logical component in enabling dynamics. Sensing allows a radio to be self-aware, and thus it can measure its environment and potentially measure its impact to its environment. Sensing is necessary if a device is to change in operation due to location, state, condition, or RF environment.

- *Networking* is the ability to communicate between multiple nodes and thus facilitate combining the sensing and control capacity of those nodes. Networking, specifically wireless networking, enables group-wise interactions between radios. Those interactions can be useful for sensing where the combination of many measurements can provide a better understanding of the environment. They can also be useful for adaptation where the group can determine a more optimal use of the spectrum resource over an individual radio.

- *SDR:* "A radio that includes a transmitter in which the operating parameters of frequency range, modulation type or maximum output power (either radiated or conducted), or the circumstances under which the transmitter operates can be altered by making a change in software without making any changes to hardware components that affect the RF emissions." —Derived from the US FCC's Cognitive Radio Report and Order, adopted 2005-03-10.
- *Cognitive radio:* A radio or system that senses and is aware of its operational environment and can be trained to dynamically and autonomously adjust its radio operating parameters accordingly.
  [It should be noted that "cognitive" does not necessarily imply relying on software. For example, cordless telephones (no software) have long been able to select the best authorized channel based on relative channel availability.]
- *Policy-based radio*: A radio that is governed by a predetermined set of rules for behavior. The rules define the operating limits of such a radio. These rules can be defined and implemented:
  - During manufacture
  - During configuration of a device by the user
  - During over-the-air provisioning and/or
  - By over-the-air control.
- *Software reconfigurable radio*: A SDR that: (1) incorporates software-controlled antenna filters to dynamically select receivable frequencies, and (2) is capable of downloading and installing updated software for controlling operational characteristics and antenna filters without manual intervention.
- *DFS*:
  1. "A general term used to describe mitigation techniques that allow, amongst others, detection, and avoidance of co-channel interference with other radios in the same system or with respect to other systems." —From current version of WP8A PDNR on SDR;
  2. "The ability to sense signals from other nearby transmitters in an effort to choose an optimum operating environment." —Derived from the US FCC's Cognitive Radio Report and Order, adopted 2005-03-10.

Figure 2.1: ITU GSC proposed definitions

These new technologies and radio classes, albeit in their nascent stages of development, are providing many new tools to the system developer while allowing for more intensive use of the spectrum. However, an important characteristic of each of these technologies is the ability to change configuration to meet new requirements.[1] This capacity to react to system dynamics will require the development of new spectrum policies in order to take advantage of these new characteristics.

The definitions for the various radio technologies are developed by the International Telecommunication Union (ITU) with help from many of its member organizations. Figure 2.1 provides a list of radio technology definitions

---

[1] The question of how cognitive radios would apply physical layer adaption in sensor networks that transmit and receive over very low duty cycles has not been adequately studied. Although it is assumed that they could also benefit from cognitive radio adaption, the time to reach network stability is lengthened by the low duty cycles. It seems that if the spectral dynamics exceed the time to reach stable performance, or consume more bandwidth for parametric exchange and adaption parameters than the network can effectively provide, then such sensor networks may not be able to fully benefit from cognitive radio techniques. However, if other systems operating in the same environment are cognitive radios with stable adaption strategies, then sensor networks may still benefit.

proposed by the Global Standards Collaboration (GSC) group within the ITU. Definitions for policy-based radios and dynamic frequency selection (DFS) radios are also provided. These two new radio classes are specific implementations of cognitive radios. Policy-based radios are discussed in Section 2.6 with an emphasis on how cognitive radio technology can impact the development and implementation of communications policy. DFS radios are addressed in Section 2.3.3. These advanced radio technologies are enabling a multitude of new radio concepts, as discussed in Section 2.3.2.

2.3 New Opportunities in Spectrum Access

Two general management methods allow access to the RF spectrum: spectrum access licenses and unlicensed devices. Spectrum licenses are issued by the appropriate regulatory agency within the nation. The licenses include a band, a geographic region, and the allowable operational parameters (e.g., in-band and out-of-band transmission levels). Although the licenses have a finite duration, there is an expectation of renewal. There is also a level of protection from interference from other systems accessing the spectrum. Unlicensed devices, also called licensed-free devices and licensed-by-rule devices, are provided frequency bands and transmission characteristics (albeit at much lower transmission power levels), and are not provided regulatory protection from interference.

2.3.1 Current Spectrum Access Techniques

The RF spectrum is organized by allocations and assignments. Allocations determine the type of use and the respective transmission parameters. The allocations are specified by each individual nation determined by sovereign needs and international agreements. International agreements can be bilateral or multilateral treaties or resolutions by the ITU during the periodic World Radio Communication Conferences (WRCs). Figures 2.2 and 2.3 are examples of spectrum allocations for the United States and New Zealand, respectively. Each color is an indication of a service type that is allocated to that frequency band *across the entire nation*. Many of the primary allocations (television (TV), frequency modulation (FM) radio, global positioning systems (GPS), etc.) are identical. The vertical splitting that occurs in some bands depicts the instance of multiple allocations (primary, co-primary, and secondary) within the band. The intensity of usage in the United States is much higher than in New Zealand. This is really noticeable in the figures by the large number of multiple allocations needed to provide enough capacity for the numerous wireless services for commercial and noncommercial applications, such as defense, air traffic, and scientific exploration.

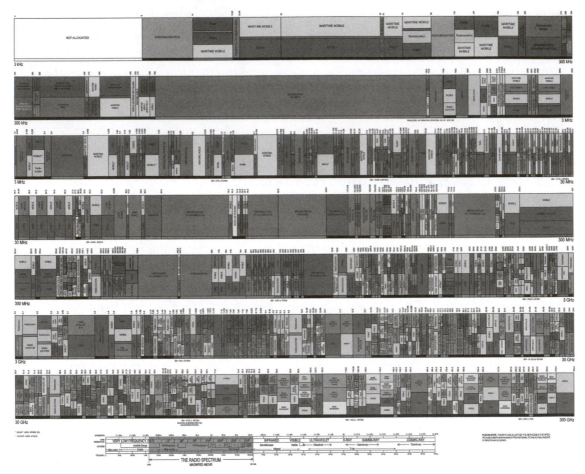

Figure 2.2: US spectrum allocations [2].

Assignments are individual licenses within an allocation. Assignments are provided by the method of choice of the sovereign state. Assignments can also be limited by geographic extent. There are many ways to obtain assignments, as depicted by Figure 2.4. Currently, three basic types of assignment methods—command and control, auctions, and protocols and etiquettes are generally employed.

- *Command and control*: Command and control assignments are provided by the regulatory agency by reviewing specific licensing applications and choosing the prospective licensee by criteria specific to the national goals. Detractors of this assignment technique call it a "beauty contest" because only a handful of regulators determines the relative value between potential services and license holders. Since the US Radio Act of 1934, the Federal Communications Commission (FCC) has had the regulatory authority to decide which firms

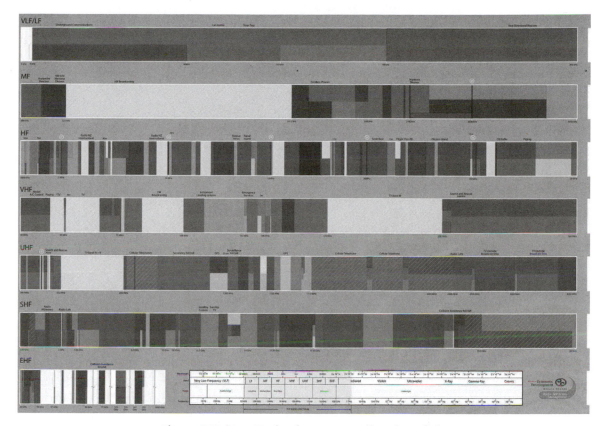

Figure 2.3: New Zealand spectrum allocations [3].

should get licenses in order to bring spectrum to its highest and best use. The FCC held public hearings to gather information to make these determinations. In the early 1980s the number of applications for licenses was growing so large that the command and control system ground to a halt. The FCC, to enable more capacity for the command and control assignment mechanism, decided to start awarding licenses by lottery. The assignment process was still in complete control of the regulators; however, there were no restrictions as to who could participate in the lottery. One year, for instance, a group of dentists won a license to run cellular phones on Cape Cod, and then promptly sold it to Southwestern Bell for $41 million. This set of events disturbed regulators because the spectrum assignment process became an investment opportunity rather than providing a service (albeit for profit). This directly challenged their authority in using spectrum for the benefit of the country (i.e., its citizens).

- *Auctions*: Spectrum assignments through auctions are fairly new. New Zealand, with the Radiocommunications Act 1989, enabled the use of market-driven

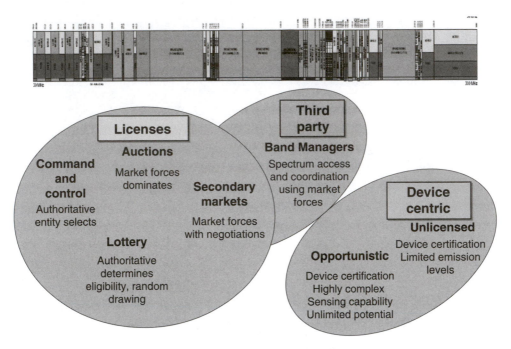

Figure 2.4: Spectrum access regimes.

allocation mechanisms for assignment of spectrum rights. Although the act did not explicitly provide the market mechanisms, by 1996 the mechanisms for selling spectrum by auction were in place. Specifically, the use of competitive bidding was authorized in granting licenses to qualified applicants.[2] In the Telecommunications Act of 1996, the United States also enabled the use of auctions for spectrum access licenses. The Wireless Telegraphy Act of 1998 enabled auctions in the United Kingdom. Auctions are now a common form for spectrum assignments throughout the industrial world. The United States, Australia, Mexico, Canada, New Zealand, and the European Union now employ auctions regularly. The current spectrum allocations in the United States and New Zealand presented in Figures 2.1 and 2.2, respectively, show the many competing applications for spectrum.

---

[2] Section 309 [47 U.S.C 309] Subsection J: "USE OF COMPETITIVE BIDDING. If mutually exclusive applications are accepted for filing for any initial license or construction permit which will involve a use of the electromagnetic spectrum described in paragraph (2), then the Commission shall have the authority, subject to paragraph (10), to grant such license or permit to a qualified applicant through the use of a system of competitive bidding that meets the requirements of this subsection."

- *Protocols and etiquettes*: Unlicensed devices and amateur licensees do not have, per se, specific frequency assignments. The allocation allows these devices to operate within a band, and the selection of a particular frequency is accomplished through protocols and etiquettes. Protocols are explicit interactions for spectrum access. Carrier Sense Multiple Access with Collision Avoidance (CSMA/CA) is a protocol for media access and control that requires information to be communicated between devices. Etiquettes are rules that are followed without explicit interaction between devices. Simple etiquettes, such as "listen before talk," DFS, and power density limits are one-sided processes. The amateur radio operators have constructed a well-tested and successful manual set of etiquettes. For example, the developers of DX tuners (for long-distance transmission and reception by ham radio operators) have etiquettes such as "If there are other users logged in, *always* ask in the chat window before you tune the receiver!" Unlicensed devices are operated on a much shorter time scale. Industry groups such as the Institute of Electrical and Electronics Engineers (IEEE), form groups in order to develop standards to promote device interoperability. The 802.3 (Ethernet) and 802.11 (wireless local area network, or WLAN) are two such protocol standards. New etiquettes, such as DFS, have been developed for the new 5–5.8 GHz 802.11a band (IEEE 802.11h).

Spectrum access for services provided by licenses is controlled directly by the service provider. In the case of public broadcast services such as FM radio and TV, the waveforms and/or protocols are defined by the rules provided by the regulators. These rules, although provided by the regulators, are developed with both the service providers and the device manufacturers. Being a broadcast service, the broadcast station controls access simply by its own use of the frequency band. In the case of providers of two-way communications or private broadcast, the waveforms and protocols are defined independently by the license holder. However, market forces tend to drive groups of license holders to compatible technology in order to increase either the interoperability with other license holders or to increase the geographic footprint of their service. A classic example of this independent development of industry standards has been the commercial mobile radio service (CMRS, or cellular service) and the development of the capacity to roam from service provider to service provider. The development of roaming has generally been considered to have been the threshold event in making radiotelephony a viable service.

The operational envelopes for unlicensed devices are defined by rules in three general groupings: unintentional, incidental, and intentional radiators. An unintentional radiator, per FCC definition, is a device that *intentionally generates RF energy for use within the device, or that sends RF signals by conduction to associated equipment via connecting wiring, but which is not intended to emit RF energy by radiation or induction.* One example would be a TV receiver with leakage from an oscillator in the receiver RF chains. Another very common example concerns a computer and display that typically exhibit harmonics up to more than 3 GHz. The first example has the emission contained in a particular band and the second example is a broadband, noise emission.

An incidental radiator, per FCC definition, is a device that *generates RF energy during the course of its operation, although the device is not intentionally designed to generate or emit RF energy.* Examples of incidental radiators are direct current (DC) motors, mechanical light switches, or even the radiation from a hair dryer.

In both the unintentional and incidental radiator cases, the regulatory agency has set limits on the emission levels. The allowable emission levels are generally very low due to the general and ubiquitous deployment of these devices. The devices may be in extremely close proximity to licensed service devices. Usually, the consumer does not know the potential impact of the device and thus cannot easily resolve any interference issues. Therefore, the aggregated interference from many such devices must remain small.

Intentional radiators are devices that are specifically constructed for a communications application. A prime example is that of baby monitors, a one-way communications system for use in the home. Other examples of intentional unlicensed radiators are the data networks using the 802.11 standard. These devices are allowed to radiate at higher levels than the unintentional radiators but still at low overall power spectral density. The low power spectral density provides some assurance that the interference potential to a primary license holder within that band is low. The regulators determine the allowable transmission parameters. There is no licensee, however, so the device manufacturer has the burden to conform to the transmission rules.

The primary difference between licensed and unlicensed device spectrum access is the afforded interference protection. The unlicensed device always operates as a secondary user within a band and is allowed to operate only in a *not-to-interfere* basis, per the transmission rules. The unlicensed device is not provided any interference assurances from either the primary use devices or other unlicensed devices. It is for this reason that device manufacturers and application designers develop standard protocols and/or etiquettes in order to reduce the potential of like-device interference. This is true with the development of the

802.11 specifications for WLANs. These specifications are widely used with the popular consumer products in the 2.4 and 5.15–5.825 GHz bands. However, unlike devices such as 2.4 GHz cordless telephones and 802.11g devices, which operate in the industrial, scientific, and medical (ISM) band, the expanded 802.11h devices operating in the 5.15–5.825 GHz band have the potential to interfere with incumbent devices. The common use of the 2.4 GHz ISM band is for microwave ovens, which are not overly susceptible to radiated energy.

2.3.2 Opportunistic Spectrum Access

The current toolbox of spectrum access techniques is limited by the capacity of both the devices and the policy that are in place. As was shown in the spectrum allocations charts depicted in Figures 2.2 and 2.3, spectrum less than 6 GHz is completely allocated for various services. However, the use of the assignments in time and space is not complete. An example of a measurement of spectrum utilization is shown in Figure 2.5. This time-frequency measurement of the 700–800 MHz band was made over an 18-hour period in Hoboken, New Jersey,

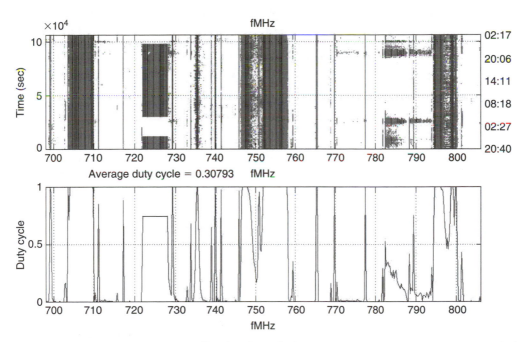

Figure 2.5: Spectrum utilization. Utilization in Hoboken, New Jersey, 700–800 MHz, made by Shared Spectrum Corporation over an 18-hour period.

which is directly across the Hudson River from Manhattan, New York. It clearly indicates that there are portions of the spectrum that are continuously accessed, portions that are never accessed, and portions that are accessed for a fraction of the time. These "white spaces" in the spectrum have created interest within the US Department of Defense (DoD) to begin research into more efficiently using spectrum. The FCC also saw the potential for more intensive use of the spectrum as well as revisiting the age-old claim of scarcity of the RF spectrum.

The onset of SDR and cognitive radio technology enables *opportunistic spectrum access* (OSA). OSA looks for "holes" in the spectrum and then adjusts the link parameters to conform to the hole. That is, the radio transmits over sections of the spectrum that are not in use. However, it has the additional complexity of listening for other transmitters in order to vacate a hole when other, nonopportunistic spectrum devices are accessing it. This technology combines flexible waveform capacity with sensing and adaptive frequency technology. The potential gain is the higher utilization of infrequently used spectrum. It has been estimated that on average, less than 5 percent, and possibly as little as 1 percent, of the spectrum less than 3 GHz as measured in frequency-space-time is used.[3] Opportunistic spectrum technology is under development by the US DoD with the goal of increasing accessible spectrum by a factor of 20. The European Union Information Society Technologies (IST) as well as the US National Science Foundation (NSF) and other research organizations, as of 2005, are actively pursuing this technology. Figure 2.6 provides a chronology of events in OSA activities.

The availability of SDR also allows high-priority users, such as in a public service context, to access the radio spectrum on an "as-needed" basis. This is called *interruptible spectrum access* because the normal user's spectrum access is interrupted in order to provide the spectral resource to a higher priority user (see Figure 2.7). For example, in a major regional disaster there will be a significant increase in public safety users due to the influx of responders from outside the immediate area. Such an influx of radio users would require additional spectrum

---

[3] This percentage is a function of where the measurement is made and what sensitivity thresholds are used. Rural areas have an extremely low utilization. In the worst-case condition of downtown New York City during the Republican Convention, the utilization rose to approximately 20 percent. The peak utilization in San Diego on one sample day was 7 percent. Because the numbers are location and emergency dependent, precise numbers cannot be stated universally, but utilization is definitely low.

- *1999–2000 (United States):* Localized set of measurements conducted by the DARPA indicated that spectrum use was not very high.
- *2002 (United States):* DARPA initiates the XG (neXt Generation) project to investigate the potential for the military to share spectrum spatially and temporally with multiple devices.
- *2002 (United States):* The FCC Spectrum Policy Task Force (SPTF) concludes that spectrum access is a more significant problem than spectrum scarcity. The SPTF recommends that new rules be developed to allow more intensive access to the spectrum, including opportunistic spectrum.
- *2003 (European Union):* IST includes DSA technologies as part of the sixth Framework Programme of R&D.
- *2003 (United States):* The NSF initiates research projects in spectrum measurements and DSA.
- *2004 (United States):* The FCC issues a Notice of Proposed Rulemaking on Facilitating Opportunities for Flexible, Efficient, and Reliable Spectrum Use Employing Cognitive Radio Technologies.
- *2004 (European Union):* End-to-End Reconfigurability (E2R) Project initiated in IST.
- *2005 (United States):* DARPA XG and NSF projects complete series of spectrum occupancy measurements indicating less than 10 percent occupancy in time-space under 3 GHz.

Figure 2.6: Chronology of opportunistic spectrum.

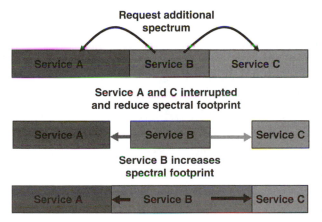

Figure 2.7: Interruptible spectrum– aperiodic increase in spectral requirements (i.e., public safety) may be addressed by interrupting part of the band of other services. This can be accomplished through central coordination or beaconing.

to cope with the increased load. Additional spectrum could be temporarily obtained from adjacent spectral bands in the immediate area of the disaster. After the need diminished, perhaps in minutes or hours, the spectrum would be released back to the primary licensee. Mechanisms may be developed to compensate primary licensees for their inconvenience or loss of revenue.

If policies can be malleable to specific conditions in order to allow greater access to the RF spectrum, then the *cost barriers for new entrants* should be reduced. Thus, new consumer products can be developed using lower-cost spectrum. The lower barriers will allow developers to use more resources in developing techniques to access the spectrum instead of investing in spectrum.

2.3.3 Dynamic Frequency Selection

OSA represents the general case for accessing the spectrum using "listen before talk" etiquette. A subset of OSA is DFS, which also is called dynamic channel selection. DFS is used to prevent a device from accessing a specific band if it is in use and to prevent co-channel interference of the primary user from a secondary user. It differs from OSA in that it is not seeking spectrum access—it is determining whether access should be allowed. DFS has been adopted in the 802.11h standard. When the ITU allowed unlicensed WLAN in 2003, 802.11h was developed in order to satisfy the primary users within the 5.25–5.725 GHz band. DFS detects other devices using the same radio channel, and it switches WLAN operation to another channel if necessary. DFS is responsible for avoiding interference with other devices, such as radar systems and other WLAN devices. Specifically, for the 5.25–5.725 GHz service, the DFS system must be able to detect a primary user above a detection threshold of −62 dBm.

2.4 Policy Challenges for Cognitive Radios

The capacity to sense, learn, and adapt to the radio environment provides new opportunities for spectrum users. However, the same sensing and adaptation also creates challenges for policy-makers. The primary concern is with the potential to have nondeterministic behaviors. Nondeterministic behaviors can be created by a variety of conditions:

- The allowance of self-learning mechanisms will create a condition in which the response to a set of inputs will be changing and thus unknown.
- The allowance of software changes will create conditions either from errors within the software or from rogue software, which can cause the device to not conform to the transmission rules.
- The allowance of frequency and waveform agility will create conditions in which devices that conform to transmission rules may cause interference due to mismatch between out-of-band receivers and the in-band transmitter waveforms.

In addition to nondeterministic behaviors, another primary concern is the impact of horizontal versus vertical service structure. Vertically integrated services, such as cellular telephony, clearly delineate responsibility for spectrum management to the service provider. The service provider has the sole responsibility for problems, interference, and all other technical and service issues. However, or horizontally integrated service, which includes device-centric systems that may be the initial focus for cognitive radio technology, there isn't a single point of responsibility for interference and other problems. One example has been the issue with secondary spectrum markets. The formal responsibility of a device creating interference is the primary licensee. The rules had to be modified to allow that responsibility to follow the usage to the secondary licensee when appropriate. The extrapolation of this approach is problematic when applied to cognitive radios, as each device is, in essence, a licensee. This is a serious problem for the policy-makers that can be addressed by rules, technology, or a combination of both.

This section addresses these concerns with the deployment of cognitive radios. In particular, the context of cognitive radio for dynamic spectrum access (DSA) and security concerns are described. The highlighted questions are areas for research and development (R&D).

2.4.1 Dynamic Spectrum Access

The development of the OSA and DFS technology is straightforward in terms of radio technology. The discussion of policy impact is much more complex. In both cases, the dynamic system must interface seamlessly, without interference, to existing spectrum licensees. Two basic questions need to be addressed:

- *Question 1: What are the rights of the license holder to prevent unauthorized use by an opportunistic device?* To answer this question, a clear definition of interference is needed. The interference definition must include some aspect of duration and persistence of the interferer. One draconian response could be that it is better to use less than 1 percent of the spectrum to ensure that a single packet is never dropped due to interference.

- *Question 2: What kind of assurance can be provided that the interference will be for only a finite and precise duration?* Answering this question is complicated by the fact that as these systems become more flexible and agile, the combination of possible configurations grows exponentially.

The ability to move the operations of a radio in frequency is quite enticing. The operational rules, however, are different in each of the separate frequency bands. Therefore, a single device must: (1) know of all of the transmission rules in the bands in which it can operate; (2) have the ability to adjust its transmission parameters accordingly; and (3) ensure that device will adhere to the transmission rules.

Defining the Rules for Dynamic Spectrum Access

The policy goal of managing interference is generally defined as to the parameters for the transmitter (e.g., center frequency, bandwidth, power spectral density, antenna gain). Interference management is determining the threshold for producing harmful interference. DSA systems can move in frequency, so fixed parameters are insufficient. The metric for interference must be defined to ensure that the DSA device does not cause harmful interference. The primary questions are:

1. What is an appropriate interference metric?

   With an explicit interference metric, the transmission rules could allow operation under a much broader set of transmitter parameters with the explicit goal of preventing interference above a particular threshold within a radius of the transmitter. The interference metric could also be contextually sensitive with respect to location, time, or condition of the RF environment. As the FCC has said, "Quantitative standards reflecting real-time spectrum use would provide users with more certainty and, at the same time, would facilitate enforcement."[4]

   The interference metric must be indicative of the impact of the transmitter on the surrounding region. In a static and well-defined geometry, a measurement at the transmitter can provide sufficient information from which to extrapolate the value of the interference metric across a wide area. However, the highest density of use of the RF spectrum is in areas with complex geometry and a large number of users, most of which are mobile. Although the technology for propagation models, inclusive of complex environments, is improving and can provide qualitative results for extrapolation, it is insufficient for quantitative interference analysis. Thus, another important question arises:

2. What kind of in situ measurement system is needed to provide the basis for interference analysis?

   Multiple measurements distributed across the operational area are needed to accurately measure the interference metric. One possible mechanism to obtain multiple measurements would be the development of monitoring

---

[4] See [3] Section VI: Interference Avoidance found at http://www.fcc.gov/sptf/reports.html

stations such as pollution monitoring devices. Many policy and technical questions arise:

3. Who would appropriate the funds to develop and deploy such a system? If it is a federally funded system, should that system be managed by a government or independent regulatory entity?

4. Who would ensure the accuracy of the measurements? Who could challenge the accuracy of the devices?

5. What are the mechanisms to disseminate the results of the measurements? Would there be an interface to all service providers and users who could obtain the data on a near real-time basis? Would the data be broadcast to all devices or should it be on a request basis?

   Liability when the policy is not followed must also be determined. Public safety communications' use of interruptible spectrum is one example in which liability would need to be explicitly addressed. The need for assured communications, free from avoidable interference, is a paramount requirement for public safety communications. So the mechanisms for obtaining and releasing spectrum need to be highly predictable. Beaconing is one such mechanism; that is, either a service would send a beacon indicating it is using the spectrum or send a beacon when the band is available for use by another entity. The question then arises as to how beaconing should be used. Propagation challenges may create shadow zones that would prevent a nonpublic safety user from hearing the beacon. So the basic technical challenge is to create the proper signaling technique. Thus, the following question arises:

6. What policy will assure the public that the system will work, determine who is liable if it does not work, and fix what type of compensation may be allowed between the public safety user and the primary licensee?

Safeguards and Incentives for Incumbent Users

Incumbent license holders that are currently using all of their rights to access the spectrum would initially see no value to having DSA [4]. The desired impact of DSA is to improve spectrum utilization and thus provide more competition and products to the market. The challenge is to determine how such impacts can be obtained while providing both assurances to current license holders and incentives for increased utilization through DSA.

7. How can assurances be made to current license holders while developing incentives for DSA?

Technology provides a potential solution to this particular challenge. The recent development of policy-enabled devices (see Section 2.5.3) can provide assurance that the device can have "fail-safe" mechanisms to prevent operation in unauthorized bands.

The problem set that has been put forward essentially requires the policy-makers to provide the infrastructure to enable a technology. Generally, this is a very difficult task because regulatory agencies are not focused on providing an infrastructure or service to the commercial world. Technologists must actively address these concerns.

2.4.2 Security

The challenges of employing cognitive radios for the policy community also include that of ensuring secure device operations. Security in this context includes enforcement of DSA rules. Enforcement for static systems is already a challenge due to the amount of resources necessary to authorize equipment, the requirement of obtaining proof that violations have occurred, and the determination of the violators' identities. As the systems become more dynamic, there is an increase in the number of potential interactions that can lead to a violation. Additionally, this leads to a decrease of the time and spatial scales of these interactions. Both of these changes will amplify the enforcement challenges.

Equipment Authorization

Initial equipment authorizations have two components that will increase significantly in complexity with the onset of dynamic policies: evaluation criteria and security certification. The capacities to modify waveforms, change operating conditions, and change transmission frequencies all contribute to an exponential growth in potentially adverse interactions between systems. Exhaustive testing becomes unrealizable due to the sheer number of combinations. Formal method techniques have been a focus to solve this problem. Formal methods provide provable atoms of code that mitigate the requirement for exhaustive testing. However, the maturity of the technology is inadequate to address the complexity of software embodied in a cognitive radio. Therefore, the challenge will be to answer a new set of questions:

1. How can realizable test plans be developed for certification that provide sufficient certainty as to the impact of the system on existing, certified equipment?
2. How can a device be certified if the software and hardware come from different manufacturers?

If self-learning mechanisms are to be employed, equipment authorization becomes even more problematic. Software and hardware certification will not provide sufficient assurances that the device conforms to the operational envelopes. New mechanisms will need to be developed, such as policy-enabled devices (see Section 2.5.3).

Software Certification

Software certification and the security of the software are also challenging areas. Cognitive radio algorithms are written in software that provides the control of dynamic systems. Software can be modified to allow policies to change on either a periodic or an aperiodic basis. The security of that software is critical to ensure that rogue behavior is not programmed into the device. If the consumer can access the device's software, then the consumer can instruct the radio to perform outside of the permitted operational parameters. Thus, basic issues arise such as:

3. How is software protected to ensure that this abhorrent behavior does not occur?

4. Security is not an absolute but a level of acceptable risk. How much protection is necessary? How is it tested to ensure that it is sufficiently secure?

5. Who is liable if there is a security failure?

Monitoring Mechanisms

The number of combinations of interactions is high and the mobility and the agility of future systems is great, so the basic issue remains:

6. How should enforcement systems be developed to observe all of these new cognitive capabilities?

Three possible mechanisms are suggested: authority-based, network-based, and infrastructure-based systems. The greatest challenges for development of such monitoring mechanisms are the equipment and analysis costs and the civil liberty concerns.

- The *authority-based* system is for a regulatory agency[5] to deploy a national monitoring system. A secondary advantage of such a national system would be

---

[5] Or an entity authorized by a regulatory agency.

additional uses. Such a system would have capability of measuring and reporting spectrum usage, interference environment, and propagation characteristics.

- The *network-based* system is to use the plethora of user devices already present to monitor the activity of the RF spectrum. The challenge would be in the methods to:
  - provide sufficient confidence in the accuracy of the measurements;
  - obtain sufficient geolocation information to make the information valuable;
  - collect and disseminate the information to enforcement organizations within the regulatory community;
  - do so with minimal overhead on network resources.

- The *infrastructure-based* system is a combination of authority-based and network-based systems. The goal would be to use preexisting infrastructure such as cell towers, Federal Aviation Agency (FAA) towers, and the like. The challenge would be to obtain the authority to equip each site and to have priority access to the network from each site.

Again, as with the policy challenges for DSA systems, the challenges for cognitive radios in the security arena are also quite great. The solution basis is also a combination of technical and policy initiatives. The technical initiatives outlined above need to be conducted and vetted through the policy communities.

2.4.3 Communications Policy before Cognitive Radio

This section addresses the role of the communications regulatory bodies on the development and deployment of cognitive radios as well as the impact of cognitive radios on regulation. Communication regulation is separate and distinct within every nation. The FCC is the regulatory agency for all wireless communications systems within the United States, except those used by the federal government (e.g., DoD, National Aeronautics and Space Administration (NASA), and FAA). Those systems are regulated by National Telecommunications and Information Administration (NTIA) within the Department of Commerce (DoC). The primary goal of the regulatory agencies is to promote technology for the public interest while preventing interference between the multitudes of systems already deployed.

2.4.4 Cognitive Radio Impact on Communications Policy

The challenges that face current telecommunication regulators are great. The rapid development of new techniques to access the spectrum and applications is quickly outpacing the ability for the policy-maker to *react*. This is apparent by how rapidly society is changing its views on spectrum use and management. The consumer has an expectation of untethered connectivity with more devices and applications. Technology is also lowering the barriers for new commercial entrants, therefore increasing both the amount and diversity of the uses for the RF spectrum. In a seemingly contradictory statement, given the consumer demands for using the RF spectrum, technology is also challenging the long-held views that spectrum is a scarce resource. Policy-makers need a clear, nonreactive vision for future management of the RF spectrum.

The mirror image is also true. The challenges that face current technologists and entrepreneurs are also great. The lifespan of an individual market has shortened from decades to years and possibly months. The regulatory approval cycle can be longer than the market lifespan, creating risks of missing a market altogether. Although the technologists can create new capabilities and products, the regulators can have a marked impact on their ability to bring those products to the marketplace. Additionally, technology has become sophisticated and the application market is intertwined with that technology. Regulators must now be technology savvy in order to comprehend all aspects of their decisions. Technologists and entrepreneurs must now be cognizant of the role regulators have on what they can develop.

2.4.5 US Telecommunications Policy, Beginning with the Titanic

It is critical to understand the underpinnings of current spectrum management. The United States provides a good example of the historical context of the development of telecommunications policy. This section briefly reviews the chronology of events that have shaped US policy.

Frequently, the impetus for new policies is a catastrophe that leads to the perception of a policy failure. The sinking of the Titanic was the pivotal point in initiating the development of wireless communications policy. In 1910, the US Congress had mandated that passenger ships carry wireless telegraphs. Investigations into the 1911 Titanic disaster indicated that amateur radio operators caused interference after initial reports of the disaster. This interference, it was thought, hampered rescue efforts.

Federal Communications Commission

The Radio Act of 1927 established the Federal Radio Commission and set forth as its intent to "maintain the control of the United States over all the channels of interstate and foreign radio transmission; and to provide for the use of such channels, but not the ownership thereof." The 1927 Act provided that the new Commission shall, "as public convenience, interest, or necessity requires" classify radio stations, prescribe the nature of the service, assign bands of frequencies or wavelengths and determine the power, time, and location of stations and regulate the kind of apparatus to be used. Licenses were to be granted by the Commission for a limited duration (3 years for broadcast licenses and 5 years for all others), but all federal government stations were to be assigned by the president.

Seven years later, the Communications Act of 1934 abolished the Federal Radio Commission and transferred the authority for spectrum management to the newly created FCC. The 1934 Act brought together the regulation of telephone, telegraph, and radio services within a single independent federal agency. The 1927 Radio Act was absorbed largely intact into Title III of the 1934 Act.

From 1934 to the early 1990s, the US Congress enacted many amendments to Title III, but there were no fundamental changes to the core provisions that can be traced back to the 1912 and 1927 acts. However, two noteworthy additions to the 1934 Act inserted in 1983 by Congress were:

- that it is the policy of the United States "to encourage the provision of new technologies and services to the public" and that anyone who opposes a new technology or service will have the burden of demonstrating that the proposal is inconsistent with the public interest; and

- "notwithstanding any licensing requirement established in this Act, [the FCC may] by rule authorize the operation of radio stations without individual licenses [in certain services]."

Therefore, new technologies will be promoted over existing services, and license by rule, or what is generally called unlicensed devices, are permissible.

National Telecommunications and Information Administration

The US DoC also has an important function in telecommunications regulation. Primarily through the NTIA, the DoC serves as the president's expert advisor on telecommunications matters and policy. NTIA is charged with reviewing policy options on behalf of the Executive Branch and communicating proposed policy decisions to the Congress. NTIA also manages and administers the portion of the RF spectrum that has been set aside for exclusive use by the federal government. NTIA is also responsible for coordinating the federal government's participation in the ITU WRCs and related national and international meetings.

State Department

The US State Department is the primary representative on foreign policy matters. Through its Economics Bureau Office of International Communications and Information Policy, the State Department represents the United States in international telecommunications forums, including bilateral and multilateral negotiations, and before international organizations, such as the ITU. The WRC delegation is led by the WRC ambassador, whose role is to help negotiate a unified US position at the conference. The US president typically confers the personal rank of ambassador in connection with this special mission for a period not exceeding 6 months.

2.4.6 US Telecommunications Policy: Keeping Pace with Technology

The regulatory methods have gone through several changes throughout the years. During the initial era of amplitude modulation (AM) radio, the numbers of systems were small, and interference avoidance was accomplished through frequency and spatial separation. Although the frequency range was limited to low-frequency (LF) through high-frequency (HF) bands, it was sufficient to accommodate the hundreds of stations. These systems included stationary transmitters and mobile receivers. Characterizing a typical receiver and using propagation models easily allowed for the computation of potential interference. These calculations worked very well because the transmitters did not move. The benefit to the system developers was that most of the regulatory requirements were placed upon the transmitters, which were few in comparison to the millions of receivers. As long as the transmitter complied with the in-band and out-of-band emission limits, interference between systems was prevented. Out-of-band emissions and antenna placement were specifically controlled to assure that the spectral and spatial distances between two potential interferers were sufficient to prevent interference.

The development of TV and FM radio saw the numbers of systems grow, as did the number of frequency assignments needed for these systems. If still limited to the HF bands, then a spectrum crisis would have precipitated. But fortunately, development of new RF component technology allowed the operational frequency bands to also grow into the very high frequency (VHF) and eventually the ultra-high-frequency (UHF) bands. Because the same topology was still being used—a single transmitter with many receivers—the same type of computations could be made to ensure a relatively interference-free operation.

Then, in the 1970s and 1980s, transistor and integrated circuit technology reduced the cost of transmitters. This provided cost-effective transmitters for the mass market in the form of Citizens' Band (CB) radios, unlicensed devices, cellular mobile radios, and numerous other devices. Protocols were developed for the devices and the users in order to provide orderly sharing of the RF channels. The channels were not all the same bandwidth. Some were wideband to accommodate high information data rate systems. Most were quite narrow. However, the main complexity of transmitter mobility had to be addressed. Therefore, the calculations for potential interference were no longer static. Out of this new complexity came ideas for minimum operational distances between potential interferers. These placed bounds on the interference. Devices that might be close in spatial proximity had to rely on spectral separation, and devices close in spectral proximity had to rely on spatial separation.

Although technology continued to march forward, the radio portion of the US communication policy has remained mostly unchanged for 70 years. Therefore, current policy is based on the technology that existed for broadcasting: one transmitter to many receivers. The evolution and revolution of radio technology over the past 15 years has significantly challenged that basis. Although the exact values can be argued, the United States had an average of more than two receivers for every person in 1980. The advent of cellular telephony, data networking, and two-way paging has now added one transceiver or more for each and every person. But the trends indicate that it will not stop there. It can be argued that a model similar to what happened to computing is taking hold in telecommunications. The evolution of technology also incurred an evolution of computer use from a corporate site, to an office group, to an individual desk, to every home, to every individual, and now to embedded-computing in many consumer household products. The largest growth occurred when computers changed from devices that required interaction with the primary user to devices that work in the background, such as a refrigerator, a TV, a phone, and the automobile. The same trend is occurring within the wireless communications environment. There are now multiple wireless devices per person within the first adopters, as more consumer products are being developed that use primarily unlicensed devices.

2.5 Telecommunications Policy and Technology Impact on Regulation

Regulations based on static broadcast geometries cannot address the spatial, numeric, and spectral dynamics of future radio technology. Technologists must begin to address not only how to construct such new technologies, but also to address how to bring dynamics into the regulatory framework.

2.5.1 Basic Geometries

Four basic geometries affect the type of technical and social/economic issues that are addressed in wireless communications policy: fixed or mobile transmitters combined with fixed or mobile receivers:

- Fixed Transmitter, Mobile Receiver(s)
- Fixed Transmitter, Fixed Receiver(s)
- Mobile Transmitter, Fixed Receiver(s)
- Mobile Transmitter, Mobile Receiver(s).

Fixed Transmitter, Mobile Receiver(s)

Fixed transmitter, mobile receiver systems include broadcasting, radio position determination, and standard time and frequency signal services. Broadcasting comprises a large fraction of the consumer devices such as radio (AM, FM, TV, etc.). Radio position determination includes radio navigation and radio beaconing services such as GPS. Standard time and frequency signal services include WWVB, the National Institute of Standards and Technology (NIST) long-wave standard time signal, which continuously broadcasts time and frequency signals at 60 kHz. The carrier frequency provides a stable frequency reference traceable to the national standard. A time code is synchronized with the 60 kHz carrier and is broadcast continuously at a rate of 1 bit per second (bps). Emission-only devices, such as those for ISM purposes, including microwave ovens, magnetic resonance equipment, and industrial heaters, are included in this category. The important feature of these systems is that there are small numbers of high power transmitters at fixed and potentially known locations. The economic challenge is to put most of the complexity (i.e., cost) in the transmitter since the ratio of receivers to transmitters is more than 1 million to 1.

The policy challenge with fixed transmitter, mobile receiver systems is to determine the allowable transmission parameters (power, location) that prevent interference at the receivers and provide for the potential of an economically viable business. The trade-offs for broadcasting services include the number of stations in a given region and the viability of the service (number of customers). The trades are exceptionally complex. The station density could be increased by reducing transmission power and greater frequency reuse. However, that would decrease the coverage area and thus the number of potential customers. The station density could be increased by using closer band spacing between stations. However, that would increase the out-of-band rejection by the receivers.

Fixed Transmitter, Fixed Receiver(s)

Fixed transmitter, fixed receiver systems include point-to-point, point-to-multipoint, and radio astronomy services. Both endpoints are in fixed locations and could either be a one-way (transmitter to receiver) or a two-way (transceiver to transceiver) configuration. A point-to-point communication system is defined as having two fixed transceivers. Private operational-fixed microwave may use an operational-fixed station, and only for two-way communications related to the licensee's commercial, industrial, or safety operations. Point-to-multipoint includes multipoint distribution systems (MDSs) and multichannel, multipoint distribution systems (MMDSs) that are generally used for one-way data broadcasting. Originally, the primary MMDS application was "wireless cable" to deliver TV programs. Advances in antenna development allowed for two-way digital subscriber link (DSL) applications to be implemented with MMDS.

Radio astronomy is the scientific study of celestial phenomena through measurement of the characteristics of radio waves emitted by physical processes occurring in space. The radio telescopes that are used for astronomical work are extremely large because the signal strength coming from the distant stellar objects is low and many of the frequencies that are observed are below 3 GHz. The challenge is in addressing the location of fixed receiver, radio astronomy systems that take decades to plan and construct. Originally located in places away from population centers to minimize the potential for interference from commercial systems, these systems now find themselves surrounded by population centers. The policy-makers are essentially facing an issue of whether to keep "radio-free zones" around the telescopes or to find other means to provide interference-free operation.

Fixed transmitter, fixed receiver systems are the most straightforward to determine the transmission parameters to prevent interference with other systems. However, the complexity occurs with mobile transmitters interfering with these systems. Because the location of the fixed transceivers is generally unknown to mobile users, mobile transmitters can potentially interfere with a receiver operating close to the noise floor due to out-of-band emissions or lack of out-of-band rejection by the receiver. However, the policy trades are quite straightforward because the RF environment and the geometry are fixed.

Mobile Transmitter, Fixed Receiver(s)

Mobile transmitter, fixed receiver systems include monostatic active as well as passive meteorological and Earth exploration systems. These systems are mobile (either airborne or space based). In the passive sensing configuration, the operational area is unknown, as it has similar characteristics to radio astronomy. In the active sensing configuration, the transmission location is unknown but the receiver is co-located with the transmitter.

Mobile transmitter, fixed receiver systems generally employ extremely sensitive receivers. Due to the mobility of the transmitter, the geographic region impacted is fixed in size and moves with the transmitter. The policy challenges are that the amount of frequency needed for these systems is large but the specified frequency use is very small. Also, the operational frequencies are specific to the physical attributes of the chemicals that are to be sensed. The sensitivity of the receivers also requires all adjacent channel systems to have an extremely low out-of-band emission. This is usually accomplished through guard bands. The challenge to the policy-maker is to determine the relative values of consumer services compared with scientific investigation and Earth exploration. Those values provide input as to whether to find mechanisms to access the unused spectrum in one location while the sensor is operating in another location.

Mobile Transmitter, Mobile Receiver(s)

Mobile transmitter, mobile receiver systems include a wide range of mobile services as well as portable unlicensed devices. These systems include radiotelephony (e.g., cellular, personal communication system (PCS), wireless communication, and specialized mobile radio (SMR)) and private land mobile radio (PLMR) services. PLMR services are for state and local governments, and for commercial and nonprofit organizations to use for mobile and ancillary fixed communications to assure the safety of life and property and to improve productivity and efficiency. Personal radio services include CB radios, Family Radio Service (FRS), and remote control. Unlicensed devices, also known as licensed-by-rule, license-free, or "Part 15" devices as denoted in the FCC rules, are included. The important feature of these systems is that there are potentially large numbers of moderate

(100 W) and extremely large numbers of low power (1 mW–1 W) transceivers that are mobile. Both the mixture of powers and the unknown geometries between receivers and transmitters make it impossible to provide absolute assurance of interference-free operation without specifying a minimum separation distance.

Mobile transmitter, mobile receiver systems involve by far the most complex geometries for policy-makers to address. Currently, all computations for interference assume a minimum separation distance between devices (and it differs from device to device). The assumption is that these distances represent the space in which the user has full control and thus can directly impact the presence of interference. As with the mobile transmitter, fixed receiver systems, the mobility of the transmitters creates the uncertainty of the geometry between the transmitters and receivers. The mobility also creates spectral regions in time, space, and frequency that are not used. The policy-making challenge is to maximize the use of the spectrum, encourage the development of new technologies and services, and to provide certainty (spectrum access, interference, etc.) for the service providers to encourage investment.

2.5.2 Introduction of Dynamic Policies

Currently, the government spectrum management rules in the United States are dynamic with respect to frequency. That is, the rules for particular spectrum-based services tend to differ based on where a device is authorized to operate in the RF spectrum. For example, licensed transmitters operating in the radio broadcasting bands from 88 to 108 MHz must conform to the FM broadcasting rules of Part 73 [5]. The frequencies of such transmitters can be changed only after lengthy regulatory review to ensure such changes will not potentially cause harmful interference. Cognitive radios can change frequency readily, in seconds or milliseconds. These devices must incorporate aspects of the governing rules or policies from each of the different spectral areas in which they might operate. New devices that incorporate wireless fidelity (WiFi®) with mobile telephones and "roam" between wide area networks (WANs) and local area networks (LANs) are examples of how multiple spectrum policies can be merged within a single device. The dynamics are quite limited, yet possible under government and industry policies. The capacity to adjust spectrum policies dynamically opens the new possibility of dynamic spectrum policies. These policies can be at the device level, by which operational envelopes can be downloaded and modified by either the regulatory agency or the primary license holder. The dynamic policies can also be at the system level, whereas the network policies that are used to optimize performance can now add the parameter of spectrum access and the associated policies for using a specific band, at a location, with the particular system load. Three operational dimensions of spectrum policy avail themselves to dynamics—specifically, time, space, and interference.

- *Time*: An example of using the time dynamic in spectrum policy was exhibited in the early days of radio. Particular AM stations would cease transmission late at night and resume early the next morning. Time-based dynamics can be extended significantly from this example. One extension is to include scheduled/ expected interactions that are quite predictable. These may include secondary market transactions, in which a secondary provider accesses the spectrum using a separate network. These also may include the flexible access of a band by the primary user for a different application, such as reusing a cell site to provide data telemetry to/from vending machines. A further extension of this concept, which would be more opportunistic in character and less predictable, yet reliable for both primary and secondary users, may include using the spectrum for a short time or within a very limited area. One example is microtransactions within the secondary market for such "spot" use. Another could be a noncooperative use of spectrum that is currently not in use. The opportunistic use would exhibit quick transactions that could be impractical for human intervention. Automated schemes will be used similar to those used in financial transactions on the New York Stock Exchange.

- *Space*: Spatial dynamics are depicted in cases where the location of a device would determine its operational characteristics. One proposal for spatial dynamics includes the allowance of higher power transmission of unlicensed devices in rural environs. Another proposal is the use of unlicensed devices in bands where the device is sufficiently far away from a UHF TV transmitter. Location sensing would be necessary for the first proposal. Signal strength sensing would be necessary for the second proposal. In either case, because the transmitters are stationary, the location information is static. Therefore, once the boundaries are determined through calculation or measurement, then these boundaries could be programmed into a device. However, extending the concept to avoid mobile transmitters creates additional complexities. The distance to mobile transmitters would be constantly changing, and thus more automated sensing and interference avoidance techniques would be required.

- *Interference*: In contrast to the spatial and temporal dynamics, interference dynamics would need to understand not only its environment but also the impact of its own transmission on the surrounding environment. The capacity to accurately measure and model the environment would be needed. A significant amount of R&D has occurred over the past decade to improve the fidelity of simulation and modeling of RF propagation. Companies such as Remcom have products that are examples of those developments. Additionally, device technology has significantly reduced the cost of RF sensing while also improving in fidelity.

There are many specific applications of dynamic spectrum policies. In the case where dynamic policies overlay current static policies, the choice for the device designer is whether to provide those new capabilities at additional cost for each device. An example of this is the case of whether to use licensed spectrum, secondary market spectrum, or unlicensed devices. Licensed spectrum has an assured quality of spectrum access and interference but is associated with higher spectrum costs. Each of the other choices has less assurance of quality but at lower spectrum costs.

It is easy to expect that with dynamic policies an explosion of new sensing devices and cooperative networks can be developed. These will be aimed at providing cost-effective solutions for both licensed and unlicensed uses. The incorporation of more processing capacity within licensed and unlicensed devices will present system developers with a large number of choices to provide new services with variable quality to the consumer.

2.5.3 Introduction of Policy-Enabled Devices

All radio devices are policy-enabled devices at the base level—the policies (or constraints) are the physical capabilities of the radio (e.g., power output, frequency range, modes of operation). Cognitive radios offer the ability to create dynamic-policy-enabled devices, or in other words, cognitive radios can realize the dynamic usage of frequency bands on an opportunistic basis by identifying and using the under-utilized spectrum. Policies that determine when spectrum is to be considered as an available opportunity and that define the possibilities of using these spectrum opportunities must be specified by the regulators but implemented into the devices. The ability to encode policy into a device, thus creating a *policy-enabled device*, can potentially have a profound impact at the policy level [6], provided the device's functionality as a radio remains trustable [7].

The technology for policy-enabled devices is derived from the development of the Semantic Web (SeW). The SeW is a machine for creating syllogisms. Thus the promise of SeW is to allow third parties to combine assertions to discover things that are true but not specified directly. The research has focused on the development of the Resource Description Framework (RDF) and the Web Ontology Language (OWL). The RDF is a structured environment for representing information about resources in the World Wide Web (WWW). It is particularly intended for meta-data about web resources, such as the title, author, and modification date of a web page, and so on. OWL can be used to explicitly represent the meaning of terms in vocabularies and the ontological relationships between those terms.[6] Challenges remain in the development of policy-enabled devices. The functionality of the policy-enabled device must be "trustable" in its own operation. There has been a great deal of work in the mathematics of trustable systems and in demonstrating the impact of a flawed (e.g., not trustable) system (see, e.g., Mitola [7]).

If the ontology for the regulations for wireless communications could be developed, then devices would be able to directly instantiate a policy into software and check that policy: (1) for self-consistency and (2) for consistency with other policies already in use.

Two profound changes for telecommunication policy-making can result from this technology: (1) policy-makers can now include policies that are distinct in both space and context and (2) policy-makers can now look at policies that change with time. The latter is much more significant. The time dimension allows for policy to expire.

These new characteristics allow for more aggressive policy concepts. Current policy-makers and regulators are extremely cautious because the rules they develop will be long lived and affect millions of users. Thus the rules are very conservative and address not only well-understood interference possibilities, but also rare and statistically remote interference possibilities. In fact, with the increased density and dynamic behavior of devices, it may be practically impossible (exponentially complex) to ensure interference-free operation. With policy-enabled devices, new policies could be tried for shorter periods of time to determine the impact of the new policies. This would be equivalent to a test of the spectrum access system.

The ability to have policies instantiated in devices that can change with respect to location and context is very desirable. Power limits that change with

---

[6]Note that the eXtensible Markup Language (XML) can also be used to define policies, but devices require the same parser for correct interpretation. OWL is better for distributing policy information among different types of cognitive radio devices with possibly different interpretation engines.

respect to signal density or proximity to high usage areas could enable more cost-effective deployments of communications to rural and underserved areas. In the same way, devices in high usage areas would have to be capable of addressing interference more robustly.

Policy-enabled devices allow for the implementation of dynamic policies. Such new capabilities for the radios afford new possibilities for the regulator.

2.5.4 Interference Avoidance

The central role of spectrum management is interference management. The design and operation of RF equipment, including communications and emitting noncommunications devices, are predicated upon preventing and/or mitigating electromagnetic interference. In today's RF environment, interference generally limits the usable range of communications signals. Interference protection has always been a core responsibility of communication regulatory bodies such as the FCC. Section 303(f) of the US Communications Act of 1934 as Amended directs the FCC to promulgate regulations it deems necessary to prevent interference between stations, as the public interest shall require [8]. This is still a critical aspect of the FCC operation. The FCC's strategic plan for the years 2003–2008 includes as a spectrum-related objective the "vigorous protection against harmful interference …" [9].

In 2002, the FCC created the Spectrum Policy Task Force (SPTF) to provide recommendations for future spectrum policy [8]. The SPTF determined that there are rising concerns that current interference management paradigms will not adequately meet future spectrum demands. Four basic challenges to spectrum management were outlined:

1. many radio communications services have grown substantially in recent years;
2. consumer demand for RF devices has exploded;
3. the technology of waveform flexibility in radios moved from a relatively small number of waveforms to widely varying signal architectures and modulation types for voice, video, data, and interactive services;
4. the use of rapidly advancing technology, such as SDR and cognitive radio, will continue to change the interference landscape.

For example, due to advances in digital signal processing and antenna technology, communications systems and devices are becoming more tolerant of interference through their ability to sense and adapt to the RF environment.

Additionally, the increased ability of new technologies to monitor their local RF environment and operate more dynamically than traditional technologies can provide new mechanisms for interference management. The predictive models used by government regulators and network operators can be updated, and perhaps eventually replaced, by techniques that take into account and assess actual, rather than predicted, interference.

2.5.5 Overarching Impact

Regulation of spectrum will undergo revolutionary changes in the near future, allowing less restricted, more flexible access to spectrum. Such flexible spectrum usage requires regulation to realize a more open spectrum. Policies that determine when spectrum is considered as opportunity and that define the possibilities of using these spectrum opportunities are to be specified. The advent of policies that change with time and space will allow greater access to the fallow spectrum as well as lower infrastructure costs. Lower infrastructure costs are obtained through the ability to have policies change when conditions are warranted. The example of rural areas being allowed to use a higher transmit power for unlicensed devices would reduce the number of access points that are needed to service an area. The optimal density of access points is related to the number of users within the footprint of the access point. Essentially a provider wants a specific number of users per access point. Therefore, current rules have a highly suboptimal number of users per access point in low usage areas and a highly suboptimal number of access points per user in high usage areas. The current set of rules allows the area covered by the access point to be reduced, but it does not allow the increase of power. Dynamic policies associated with the state of the RF environment could allow a more optimal design, and thus a lower infrastructure cost, to be used.

2.6 Global Policy Interest in Cognitive Radios

Since 2000, the interest in how technology has changed both in RF spectrum needs by the user community and RF spectrum regulatory methods has been heightened within the radio community. The push to data services within the context of wireless Internet access has been seen as a fundamental shift from the wireless services to wireless access. The spectrum needs manifested themselves with the move toward third-generation (3G) services by wireless service providers as well as the plethora of unlicensed wireless devices e.g., WiFi, cordless phones) in consumer electronics. These trends actually followed a significant increase of wireless communications development and usage within the federal users of the spectrum, such as the military. The inherent limitation of omnidirectional RF propagation limited the new uses of the spectrum to below 6 GHz.

The development of cognitive radio technology and application concepts, especially in DSA, has received significant interest worldwide. The greatest interest has been in the United States from both the DoD and the FCC.

2.6.1 Global Interest

Figure 2.8 provides a list of global regulatory activities with respect to cognitive radios and DSA. Most of the interest outside of the United States and United Kingdom has been investigatory. A great deal of activity has been ongoing within the ITU and the European Telecommunications Standards Institute (ETSI) in developing definitions, standards, and regulatory regimes for using this new technology.

ETSI is officially responsible for standardization of Information and Communication Technologies (ICT) within Europe, including telecommunications. ETSI has 688 members from 55 countries inside and outside Europe, including manufacturers, network operators, administrations, service providers, research bodies, and users. ETSI plays a major role in developing a wide range of standards and other technical documentation as Europe's contribution to worldwide ICT standardization.

2.6.2 US Reviews of Cognitive Radios for Dynamic Spectrum Access

In the United States, a number of federal institutions involved with spectrum regulation, usages, or oversight undertook a wide range of studies. These studies were broken into two categories: (1) analysis of the use of needs of the RF spectrum (Government Accountability Office (GAO), Defense Science Board (DSB), NTIA, and Toffler Associates; see under the sections *Government Accountability Office* through *Toffler Associates*) and (2) investigations of changes needed in the regulation, policy, and management of the RF spectrum (FCC, Center for Strategic and International Studies (CSIS), DoC; see under the sections *Federal Communications Commission* through *US Department of Commerce*).

Government Accountability Office

The US GAO exists to support the US legislature through evaluation and investigations of federal programs and policies. This provides additional information for the US Congress to make oversight, policy, and funding decisions. The GAO has conducted a series of studies into wireless communications policy. During the

- **United States**
  - *FCC:* Report and Order, ET Docket No. 00-47—*Authorization and Use of Software Defined Radios;* Vanu Corporation get First Approved SDR in 2004; Notice of Proposed Rulemaking, ET Docket No. 03-108—*Facilitating Opportunities for Flexible, Efficient, and Reliable Spectrum Use Employing Cognitive Radio Technologies*; Notice Proposed Rulemaking, ET Docket No. 04-186—*Unlicensed Operation in the TV Broadcast Bands and Additional Spectrum for Unlicensed Devices Below 900 MHz and in the 3 GHz Band*; Strategic Plan 2006-11—Encouraging the development of new technologies, such as cognitive radio and dynamic frequency selection; …
- **Japan**
  - *Ministry of Internal Affairs and Communications* (formally MPHPT): Promoting R&D of technologies for efficient spectrum inclusive of cognitive radio systems to search for unused spectrum.
- **New Zealand**
  - *Ministry of Economic Development*: 2005 Review of Radio Spectrum Policy in New Zealand—the role of spectrum band managers may be reduced through the use of SDRs, New Zealand's physical isolation provides for an ideal position to be a test bed for new wireless technologies, "cognitive or smart" radios will eventually have the capacity to locate and utilize any unoccupied spectrum, and none of New Zealand's current licensing types is well suited for managing SDRs and cognitive radios.
- **Australia**
  - *Australian Communications and Media Authority* (formally ACA and ABA): Vision 20/20 report on future communications—realizing the future of ubiquitous communications with wireless expected to have an increasingly central role with increases spectrum sharing and cognitive radio technologies.
- **Canada**
  - *Canadian Radio–Television and Telecommunications Commission*
  - *Industry Canada*: *Spectrum Policy Framework for Canada*—Implementation of new technologies and new spectrum management concepts using recent innovations in wireless technology such as cognitive radio and software defined radio. The department solicited for comments in determining to what extent these technologies might increase the use and access of the RF spectrum in the future.
- **ITU**
  - *GSC*: GSC-10 (Radiocommunication Items) in 2005 issued a resolution (GSC-10/6) on Global Radio Standards Collaboration on Wireless Access Systems to encourage collaboration on measurement techniques and certification requirements for cognitive capabilities including DFS; GSC-10 also called for accelerated standards development for SDR and cognitive radio; GSC-10 also resolves to study better ways to manage interference using adaptive frequency agility, listen before transmit, etc.
  - *ETSI*: Considering the impact on SDR of the Radio & Telecommunications Terminal Equipment (R&TTE) Directive with respect to electromagnetic compatibility (EMC), radio characteristics, nonconforming software, security, and integrity issues due to potential failures of the software download process.
  - *ITU-R WP8A and WP8F*: Developing definitions and application areas for SDR and cognitive radio technology. The draft reports indicate that many of the communications administrators around the globe have begun to investigate the use of these new technologies.
- **India**
  - *Telecom Regulatory Authority of India*: Consultation paper on *Issues Relating to Private Terrestrial TV Broadcasting Service* addresses alternative technologies. Comments received state that rural communications could use cognitive radio technologies to find, in situ, better bands for foliage penetration.
- **United Kingdom**
  - *Office of Communications (OFCOM)*: Technology R&D program, initiated in 2005, studies flexible, multiprotocol, multiband cognitive radio systems. Near-term investigations into band-sharing technologies.
  - *Commission for Communications Regulation (COMREG) in Ireland:* Issued the first cognitive radio/dynamic spectrum assess test license.
- **European Union**
  - *Project Team 8 (PT8)* (Postal and Telecommunications Administration, Electronic Communications Committee Working Group): Tasked with developing a report on the regulatory structure needed to enable the introduction of new radio technologies. A particular focus will be increased opportunities to share spectrum.
  - *EU Commission:* Published *a Forward-Looking Radio Spectrum Policy for the European Union—Second Annual Report* with key initiatives, including the implementation of flexible spectrum usage through the development of "smart" or cognitive radios. Such efforts will be funded under the EU Research and Technology Development (RTD) Framework Programme.

Figure 2.8: 2005 International regulatory activities on cognitive radio.

time from 2000 to 2005, the GAO conducted a series of studies on the uses and expected needs of the spectrum.

Due to the increasing demands for RF spectrum, the GAO was asked to examine whether future spectrum needs can be met given the current regulatory framework. The ensuing report [10] focused mostly on the spectrum management structure within the United States and the decision-making impediments caused by having two regulatory agencies focusing on separate constituents: federal users versus nonfederal users. The significant increase of shared bands between federal and nonfederal users requires a consensus between the two groups. However, the report indicated that SDRs and other advanced technologies can potentially alleviate many of the conflicts by making spectrum more plentiful through more efficient access.

In 2004, the GAO performed a study [11] into how agencies within the US government that access RF spectrum can use advanced technologies to improve their "spectrum efficiency." The study clearly acknowledges that technologies such as "software-defined cognitive radios can be adapted to operate in virtually any segment of spectrum and, in the future, may be able to adapt to real-time conditions and make use of under-utilized spectrum in a given location and time." The report, however, also indicated that the current allocation system does not allow these technologies to operate. The study also indicates that the lack of knowledge of the operating environment is an impediment. The report concludes that new technologies, such as cognitive radio, exhibit exceptional promise for efficient spectrum utilization but that there are few regulatory requirements or incentives to employ these technologies.

Defense Science Board

The DSB is a Federal Advisory Committee for the US DoD staffed by technology and business experts, who conduct studies pertinent to the DoD. The military is the primary source of R&D funding for advanced communications due to its changing requirements and the willingness to deploy expensive communications assets.

The DSB conducted two spectrum-related studies during the period 2000–2005 on the current and expected DoD needs for accessing the RF spectrum. The DSB reported in November 2000 [12] that the management of time, space, and modulation dimensions of the RF spectrum increases the use of the scarce RF spectrum resource. This control should share the spectrum, in real time, and thus be more efficient than fixed allocations. These real-time systems dynamically control the assigned frequencies to assure communications. The report further finds that dynamic frequency assignment by radios that can sense the spectrum could present unknown problems in spectrum management. It further questions, the technical issues concerning which technologies to develop and how to apply these new capabilities for spectrum sharing and dynamic frequency allocation and/or assignment.

In July 2003, the DSB addressed dynamic access to mobile networks in a report on wideband RF modulation [13]. In that report, the DSB viewed software radios as both a risk and an opportunity. The risk lies in the potential of developing and implementing protocols and waveforms that could not be supported by existing systems and command and control techniques. There is also an interference danger with radios that can change their configuration without knowledge of how the new transmission waveforms and frequencies would impact other deployed systems. Even with these risks, the report has a strong recommendation that the US DoD "increase and focus investment in flexible and adaptive agile wideband communications technologies to achieve necessary mission capabilities in a highly dynamic RF communications environment." That is, invest in cognitive radios.

National Telecommunications and Information Administration

Although the FCC is the official US regulatory body for RF spectrum, there are cases in which it does not have jurisdiction. By law, national security use of the spectrum is under the jurisdiction of the US president. The use of spectrum for national security has been applied broadly to include federal agencies such as the DoD, FAA, and Federal Bureau of Investigation (FBI). The jurisdiction of the federal use of the spectrum has been delegated to the Assistant Secretary of Commerce for Communications and Information, who is also the Director of NTIA. NTIA performs numerous studies to investigate the impact of new technology and the needs of the RF spectrum users within federal agencies.

Toffler Associates

In 2001, a study by Toffler Associates provided recommendations as to the next steps for spectrum policy in the United States [14]. The study focused primarily on management aspects of spectrum allocations and thus did not address specific technology impacts such as those from cognitive radio. However, the recommendations from this independent group do address overall issues such as developing a long-term strategy and how to conduct spectrum reallocations. A primary focus of the proposed strategy was to develop a framework that anticipated change and remained versatile. This includes the predication of the needs for future spectrum allocations as well as the impact of future technologies such as software defined and cognitive radios.

Federal Communications Commission

The FCC is the US regulatory body for the RF spectrum as per the US Communications Act. The FCC performs numerous studies on both RF spectrum needs and technical issues such as interference. In 2002, the FCC chartered a forward-looking study to investigate the changes occurring in technology and to recommend how the FCC should develop and implement spectrum policy; more specifically, the intent of the study was to "identify and evaluate changes in spectrum policy that will increase the public benefits derived from the use of radio spectrum. The creation of the SPTF initiated the first ever comprehensive and systematic review of spectrum policy at the FCC" [15]. A primary goal was to move from a reactive spectrum management model to one more in line with a proactive spectrum policy. The SPTF noted that technological advances, specifically in software defined and cognitive radio, are enabling both the need to change spectrum policy and the capacity to look at different paradigms to implement spectrum policy.

The SPTF concluded that technology improvements have shown that the capacity is limited by the regulatory means used to access the RF spectrum. That is, access and not technical efficiency is the limiting factor for using the spectrum. Software defined and cognitive radio technology enables accessing the spectrum in multiple dimensions, such as time, frequency, bandwidth, and power.

Traditional spectrum management techniques use models and measurements to predict the interaction between different radio transmitters and receivers. Therefore operational parameters such as transmission power, out-of-band emission, and the size of guard bands are selected to ensure interference-free interoperation of devices. The technological advances in interference rejection and digital coding will require changes to these parameters if spectrum management is to keep pace with technology. Without those changes, the operational parameters will be too conservative and thus limit the capacity to use the spectrum efficiently. Additionally, the ability to monitor the RF environment and dynamically alter radio operation makes the traditional predictive interaction model of management obsolete.

Because operational values can change with technology as well as be modified in situ, spectrum management will need to change. The current method of explicitly regulating transmission characteristics to prevent interference will need to move to a rights and responsibilities model in which interference is explicitly defined and the operational parameters must be set by the device accessing the spectrum to be within the interference limits.

The SPTF recommended 39 changes to promote more efficient spectrum policies and more intensive use of the RF spectrum. Recommendations in four broad areas have a direct impact on cognitive radios:

- Allow for maximum feasible flexibility of spectrum use by both licensed and unlicensed users.

- Adopt a more quantitative approach to interference management based on the concept of interference temperature.

- Promote access to the spectrum through secondary market policies that encourage access for "opportunistic" devices and, where appropriate, permit easements for spectral overlays and underlays.

- Promote spectrum access and flexibility in rural areas such as varying power levels, leasing mechanisms, and geographic licensing.

The overarching impact that the recommendations provided was that the FCC move from the single-use model of spectrum use to more of a flexible-use model. In that model, the FCC would clearly define the rights and responsibilities for access to the RF spectrum and then let the users optimize and interact and/or trade. It was recommended that policy-makers address the difficult challenge of defining the appropriate engineering metrics in which to enable such a policy. Such metrics (e.g., the interference metric, which is discussed in section *Defining the Rules for DSA*), could then be implemented directly within a cognitive radio.

Center for Strategic and International Studies

The CSIS is a private, bipartisan public policy research organization that provides insights and possible solutions to current and emerging issues. In 2002–2003, CSIS organized a commission to address spectrum management for the 21st century. The commission focused on four problems for US spectrum management: (1) lack of long range plans, (2) lack of mechanisms to resolve disputes between spectrum management organizations, (3) increased challenges in negotiating international spectrum agreements, and (4) the risk to security and economic growth due to "lag in the development and use of new technologies" [16]. Although the commission addressed concerns for US policy, the fourth problem is relevant to all spectrum policy organizations.

The CSIS commission noted that the development of spectrum-based technologies is increasing rapidly. In general, these technologies are initially developed within the military communications communities and then migrate to the commercial and consumer markets. Thus, technology is providing new ways to allow more intensive use of the spectrum and could alleviate the perception of a spectrum shortage. Cognitive radio technology enables the exploitation of gaps in transmission frequencies and usage times to allow such an increase of spectrum use. Additionally, such technology can be used to provide more robust behavior to interference. The commission was concerned that the license-centric spectrum management policies cannot accommodate such new capabilities.

The CSIS commission had multiple recommendations for improving spectrum management. Two recommendations addressed organizational structure through more oversight at the White House, with new National Security Council and National Economic Council positions and a spectrum advisory board. In order to address the impact of the pace of technological innovation, the commission also recommended the establishment of a research consortium to support the government and private sectors. This consortium would establish goals for research in spectrum innovations as well as provide a platform for resolving technical disputes that arise as technology changes.

US Department of Commerce

In June 2003, the US president issued a memorandum recognizing how the RF spectrum contributes to "significant innovation, job creation and economic growth." He then created a Spectrum Policy Initiative, chaired by the DoC, to develop recommendations for improving spectrum management policies and procedures [17, 18]. The subsequent report from the initiative indicates: "Given the increase in new and innovative radio communication systems seeking access to the spectrum, the most challenging issue is interference problems inherent in using the latest technologies." The report continues to address the challenges from technology by stating, "The unpredictable nature of … ingenuity is not to be solved—it is a reality to be embraced …"

Many of the recommendations put forth by the initiative are for modifications to the structure of the spectrum management community. However, there are two recommendation areas that could significantly impact the development of cognitive radios: a Spectrum Sharing Innovation Testbed and Spectrum Management Tools. The report acknowledges the need for more sharing between spectrum users is inevitable due to the increased need for spectrum and the available technology to enable such sharing. In fact, 54 percent of the spectrum allocations below 3 GHz are shared and 94 percent below 300 GHz are shared. The initiative report recommends the development of a Spectrum Sharing Innovation Testbed, which consists of spectrum that will be set aside exclusively to test and evaluate new methods for spectrum sharing.

The initiative also recognized that there will be a continued explosion of spectrum uses, and that spectrum managers need the proper tools to be effective and efficient. The employment of cognitive radios was specifically mentioned as a technology whose promise and limitations should be better understood by both the FCC and the NTIA.

2.7 Summary

By 2005, the policy community had embraced the utility of cognitive radios from the vantage points of new applications as well as new policy regimes. Due to the lack of operational prototypes, the community has been in a preparatory phase in developing definitions and potential operation envelopes in which cognitive radios may function. The greatest amount of interest is in the DSA applications because these are seen as opportunities to provide more services and new technologies without having to allocate new spectrum.

This chapter has presented a review of relevant technical and policy definitions. Key among them are the following:

- Cognitive radios consist of four new capabilities: (1) *flexibility*, to change waveform and configuration; (2) *agility*, to change the spectral band in which to operate; (3) *sensing*, to observe the state of the system; and (4) *networking*, to communicate between multiple nodes for aggregating capacity.
- The GSC within the ITU is developing a definition of cognitive radios as well as extensions to also cover policy based and DFS radios.
- Currently, there are three basic frequency assignment methods: command and control, auctions, and protocols and etiquettes. Command and control has been employed the most, but the use of auctions has been growing in popularity since its inception in 1994. Protocols and etiquettes are generally applied for unlicensed or license-free devices.
- OSA provides a new mechanism to dynamically obtain frequency assignments through sensing open spectral regions and adapting frequency selection in a

cognitive radio. This ability has been under development within the technical community since 1999 and within the regulatory community since 2002.

- Interruptible spectrum is a special subset of frequency assignment when aperiodic increase in spectral assignment is obtained by interruption of part of the band of another service.

This chapter has also addressed the policy-relevant issues that need to be addressed if cognitive radios are to be used to their fullest capacity:

- Policy challenges for cognitive radios include addressing nondeterminism of self-learning algorithms, verification/validation of software, and the impact of waveform flexibility on out-of-band receivers.
- DSA policy must address the question of the rights of the license holder to prevent unauthorized use by an opportunistic device.
- DSA uses waveform flexibility and frequency agility to optimize performance. Interference metrics to determine the impact of waveforms on receivers is needed to provide a quantitative method to limit and/or eliminate interference.
- Multiple mechanisms are employed to insure security operation of a cognitive radio: equipment authorization, software certification, and in situ monitoring of a transmitter. Equipment authorization and software certification become more impractical as the number of operations and/or software states grows exponentially. In situ monitoring is a developing technology.

Finally, this chapter provided a short tutorial as to the relevant regulatory roles with cognitive radios:

- US spectrum policy agencies are the NTIA and the FCC.
- Four basic geometries determine the type of technical and economic issues that are addressed in communications policy: (1) fixed transmitter, mobile receiver(s); (2) fixed transmitter, fixed receiver(s); (3) mobile transmitter, mobile receiver(s); and (4) mobile transmitter, fixed receiver(s).
- Spectrum management policies can be selected to optimize system and network performance given current spectrum activity and interference properties in space and time.
- Significant global interest in cognitive radio technology and policy exists, including Japan, New Zealand, Australia, Canada, the United States, the International Telecommunications Union, India, the United Kingdom, and the European Union.

References

[1] Kolodzy, P, "Dynamic Spectrum Policies: Promises and Challenges," CommLaw Conspectus, 2004.

[2] http://www.ntia.doc.gov/osmhome/allochrt.pdf

[3] http://www.med.govt.nz/rsm/img/spectrum-chart.jpg

[4] R. Lynch, "Exploring Technology Frontiers—Is There a Network in Your Future?" *Keynote address, Dyspan Conference*, Baltimore, Maryland, November 9, 2005.

[5] 47 C.F.R. § 73.200–73.600, SubPart B FM Broadcast Stations.

[6] L. Berlemann, S. Mangold and B. Walke, "Policy-Based Reasoning for Spectrum Sharing in Cognitive Radio Networks," *Proceedings of the 1st IEEE International Symposium on New Frontiers in Dynamic Spectrum Access Networks (DySPAN)*, November, 2005.

[7] J. Mitola III, "Software Radio Architecture: A Mathematical Perspective," *IEEE Journal on Selected Areas in Communications*, Vol. 17, No. 4, April 1999.

[8] "FCC Spectrum Policy Task Force Report," www.fcc.gov/sptf, 2002.

[9] *See Federal Communications Commission Strategic Plan FY 2003–FY 2008*, available at http://www.fcc.gov/omd/strategicplan2003-2008.pdf

[10] US General Accounting Office, "Comprehensive Review of U.S. Spectrum Management with Broad Stakeholder Involvement Is Needed," GAO-03-277, January, 2003.

[11] US General Accounting Office, "Better Knowledge Needed to Take Advantage of Technologies That May Improve Spectrum Efficiency," GAO-04-666, May, 2004.

[12] US Department of Defense, "Report of the Defense Science Board Task Force on DoD Frequency Spectrum Issues," Washington, DC, 2000.

[13] US Department of Defense, Report of the "Defense Science Board Task Force on Wideband Radio Frequency Modulation," Washington, DC, 2000.

[14] S. Kenney, J. O'Connor and R. Szafranski, "Creating the Future of Spectrum Allocation"; contact through tofflerassociates@toffler.com; 2001.

[15] http://www.fcc.gov/sptf/reports.html

[16] R. Galvin, J. Schlesinger, "Spectrum Management for the 21st Century," CSIS Press, Washington, DC, 2003.

[17] US Department of Commerce, "Spectrum Policy for the 21st Century—The President's Spectrum Policy Initiative: Report 1," Washington, DC, DoC, 2004.

[18] US Department of Commerce, "Spectrum Policy for the 21st Century—The President's Spectrum Policy Initiative: Report 2," Washington, DC, DoC, 2004.

CHAPTER 3

The Software Defined Radio as a Platform for Cognitive Radio

Pablo Robert

Mobile and Portable Radio Research Group, Bradley Department of Electrical and Computer Engineering, Virginia Tech, Blacksburg, VA, USA

Bruce A. Fette

Communications Networks Division, General Dynamics C4 Systems, Scottsdale, AZ, USA

3.1 Introduction

This chapter explores both the hardware and software domains of software defined radio (SDR). The span of information covered is necessarily broad; therefore, it focuses on some aspects of hardware and software that are especially relevant to SDR design. Beyond their obvious differences, hardware and software analyses have some subtle differences. In general, hardware is analyzed in terms of its capabilities. For example, a particular radio frequency (RF) front-end (RFFE) can transmit up to a certain frequency, a data converter can sample a maximum bandwidth, and a processor can provide a maximum number of million instructions per second (MIPS). Software, in contrast, is generally treated as an enabler. For example, (1) a signal processing library can support types of modulation, (2) an OS can support multithreading, or (3) a particular middleware implementation can support naming structures. Given this general form of viewing hardware and software, this chapter presents hardware choices as an upper bound on performance, and software as a minimum set of supported features and capabilities.

Cognitive radio (CR) assumes that there is an underlying system hardware and software infrastructure that is capable of supporting the flexibility demanded by the cognitive algorithms. In general, it is possible to provide significant flexibility with a series of tunable hardware components that are under the direct control of the cognitive software. In the case of a cognitive system that can support a large number of protocols and air interfaces, it is desirable to have a generic underlying hardware structure.

The addition of a series of generalized computing structures underlying the cognitive engine implies that the cognitive engine must contain hardware-specific knowledge. With this hardware-specific knowledge, the cognitive engine can then navigate the different optimization strategies that it is programmed to traverse. The problem with such knowledge is that a change in the underlying hardware would require a change in the cognitive engine's knowledge base. This problem becomes exacerbated when one considers porting the engine to other radio platforms. For example, there could be a research and development platform that is used to test a variety of cognitive algorithms. As these algorithms mature, it is desirable to begin using these algorithms in deployed systems. Ideally, one would just need to place the cognitive engine in the deployed system's management structure. However, if no abstraction were available to isolate the cognitive engine from the underlying hardware, the cognitive engine would need to be modified to support the new hardware platform. It is clear that an abstraction is desirable to isolate the cognitive engine from the underlying hardware. The abstraction of hardware capabilities for radio software architecture is a primary design issue.

SDR is more than just an abstraction of the underlying hardware from the application. SDR is a methodology for the development of applications, or *waveforms* in SDR parlance, in a consistent and modular fashion such that both software and hardware components can be readily reused from implementation to implementation. SDR also provides the management structure for the description, creation, and tear-down of waveforms. In several respects, SDR offers the same capabilities supported by OSs; SDR is actually a superset of the capabilities provided by an OS. SDR must support a variety of cores, some of which may be deployed simultaneously in the same system. This capability is like a distributed OS designed to run over a heterogenous hardware environment, where heterogenous in this context means not only general purpose processors (GPPs), but also digital signal processors (DSPs), field-programmable gate arrays (FPGAs), and custom computing machines (CCMs). Furthermore, SDR must support the RF and intermediate frequency (IF) hardware that is necessary to interface the computing hardware with radio signals. This support is largely a tuning structure coupled with a standardized interface. Finally, SDR is not a generic information technology (IT) solution in the way that database management is. SDR deals explicitly with the radio domain. This means that context is important. This context is most readily visible in the application programming interface (API), but is also apparent in the strict timing requirements inherent to radio systems, and the development and debugging complexities associated with radio design.

This chapter is organized as follows: Section 3.2 introduces the basic radio hardware architecture and the processing engines that will support the cognitive function. Section 3.3 discusses the software architecture of a SDR. Section 3.4 discusses SDR software design and development. At present, many SDRs utilize a Software Communications Architecture (SCA) as a middleware to establish a common framework for waveforms, and the SCA is covered in some detail in this section. Section 3.5 discusses applications as well as the cognitive functionality and languages that support cognitive software as an application. Section 3.6 discusses the development process for SDR software components. Section 3.7 then discusses cognitive waveform development. Finally, Section 3.8 presents a summary of the chapter.

3.2 Hardware Architecture

The underlying hardware structure for a system provides the maximum bounds for performance. The goal of this section is to explore hardware for SDR from a radio standpoint. Figure 3.1(a) shows a basic radio receiver. As one example based on the basic radio receiver architecture, Figure 3.1(b) shows a design choice made possible by digital signal processing techniques, in which the sampling process

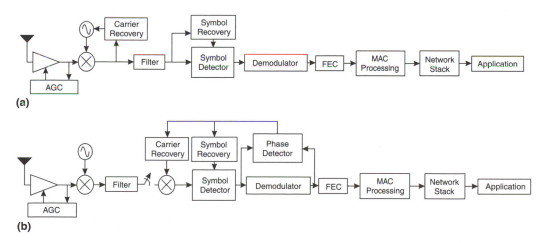

Figure 3.1: (a) Data flow and component structure of a generalized coherent radio receiver; (b) Data flow and component structure of a generalized coherent radio receiver designed for digital systems, with sampling as a discrete step (see *Design Choices* section on page 81 for an explanation).

for digital signal processing can be placed in any of several locations and still provide equivalent performance.

3.2.1 The Block Diagram

The generic architecture tour presented here traces from the antenna through the radio and up the protocol stack to the application.

RF Externals

Many radios may achieve satisfactory performance with an antenna consisting of a passive conductor of resonant length, or an array of conductors that yield a beam pattern. Such antennas range from the simple quarter-wavelength vertical to the multi-element Yagi and its wide bandwidth cousin, the log periodic antenna. Antennas used over a wide frequency range will require an antenna tuner to optimize the voltage standing wave radio (VSWR) and corresponding radiation efficiency. Each time the transceiver changes frequency, the antenna tuner will need to be informed of the new frequency. It will either have a prestored table derived from a calibration process, and then adjust passive components to match the tuning recommendations of the table, or it will sense the VSWR and adapt the tuning elements until a minimum VSWR is attained.

Some modern antennas include a number of passive components spread over the length of the radiating elements that are able to present reasonable VSWR performance without active tuning. The best such units today span a frequency range of nearly 10:1. However, for radios that are expected to span 2 MHz–2 GHz, it is reasonable to expect that the radio will need to be able to control a switch to select an appropriate antenna for the frequency band in which the transceiver is currently operating. Where beam antennas are used, it may also be necessary for the radio to be able to manage beam pointing. In some cases, this is accomplished by an antenna rotator, or by a dish gimbals. The logic of how to control the pointing depends greatly on the application. Examples include: (1) exchanging the global positioning system (GPS) position of each transceiver in the network, as well as tracking the three-dimensional (3-D) orientation of the platform on which the antenna is mounted so that the antenna pointing vector to any network member can be calculated; (2) scanning the antenna full circle to find the maximum signal strength; (3) dithering the antenna while tracking peak performance; or (4) using multiple receive feed elements and comparing their relative strength.

Another common antenna is the electronically steered antenna. Control interfaces to these antennas are quite similar in function; however, due to the ability to rapidly steer electronically, many of these antennas update their steering angle as rapidly as once every millisecond. Thus, response time of the control interfaces is critical to the expected function.

The most sophisticated antenna is the multiple input, multiple output (MIMO) antenna. In these antennas, the interface boundary between the radio and the antenna is blurred by the wide bandwidth and large number of complex interfaces between the beam-steering signal processing, the large number of parallel RF front-end receivers, and the final modem signal processing. For these complex antennas, the SDR Forum is developing interface recommendations that will be able to anticipate the wide variety of multi-antenna techniques currently used in modern transceivers.

Another common external component is the RF power amplifier (PA). Typically, the external PA needs to be told to transmit when the transceiver is in the transmit mode and to stop transmitting when the transceiver is in the receive mode. A PA will also need to be able to sense its VSWR, delivered transmit power level, and its temperature, so that the operator can be aware of any abnormal behavior or conditions.

It is also common to have a low-noise amplifier (LNA) in conjunction with an external PA. The LNA will normally have a tunable filter with it. Therefore, it is necessary to be able to provide digital interfaces to the external RF components to provide control of tuning frequency, transmit/receive mode, VSWR and transmit power-level sensing, and receive gain control.

In all of these cases, a general-purpose SDR architecture must anticipate the possibility that it may be called upon to provide one of these control strategies for an externally connected antenna, and so must provide interfaces to control external RF devices. Experience has shown Ethernet to be the preferable standard interface, so that remote control devices for RF external adapters, switches, PAs, LNAs, and tuners can readily be controlled.

RF Front-End

The RF front-end consists of the receiver and the transmitter analog functions. The receiver and transmitter generally consist of frequency up converters and down converters, filters, and amplifiers. Sophisticated radios will choose filters and frequency conversions that minimize spurious signals, images, and interference within the frequency range over which the radio must work. The front-end design will also maximize the dynamic range of signals that the receiver can process, through automatic gain control (AGC). For a tactical defense application radio, it is common to be able to continuously tune from 2 MHz to 2 GHz, and to support analog signal bandwidths ranging from 25 kHz to 30 MHz. Commercial applications need a much smaller tuning range. For example, a cell phone subscriber unit might tune only 824 MHz–894 MHz and might need only one signal bandwidth. With a simplified design, the designer can eliminate many filters, frequency conversions, and IF bandwidth filters, with practical assumptions.

The RF analog front-end amplifies and then converts the radio carrier frequency of a signal of interest down to a low IF so that the receive signal can be digitized by an analog-to-digital converter (ADC), and then processed by a DSP to perform the modem function. Similarly, the transmitter consists of the modem producing a digital representation of the signal to be transmitted, and then a digital-to-analog converter (DAC) process produces a baseband or IF representation of the signal. That signal is then frequency shifted to the intended carrier frequency, amplified to a power level appropriate to close the communication link over the intended range, and delivered to the antenna. If the radio must transmit and receive simultaneously, as in full duplex telephony, there will also be some filtering to keep the high-power transmit signal from interfering with the receiver's ability to detect and demodulate the low-power receive signal. This is accomplished by complex filters usually using bulk acoustic wave or saw filters at frequencies below 2 GHz, or yttrium-iron-garnet (YIG) circulators at frequencies above 2 GHz.

The typical software-defined RFFE begins notionally with receiving a signal and filtering the signal to reflect the range of frequency covered by the intended signals. For a spread spectrum wideband code division multiple access (WCDMA) signal, this could be up to 6 MHz of bandwidth. In order to assure that the full 6 MHz is presented to the modem without distortion, it is not unusual for the ADC to digitize 12 MHz or so of signal bandwidth. In order to capture 12 MHz of analog signal bandwidth set by the IF filters without aliasing artifacts, the ADC will probably sample the signal at rates above 24 million samples per second (Msps). After sampling the signal, the digital circuits will shift the frequency of the RF carrier to be centered as closely as possible to 0 Hz direct current (DC) so that the signal can again be filtered digitally to match the exact signal bandwidth. Usually this filtering will be done with a cascade of several finite impulse response (FIR) filters, designed to introduce no phase distortion over the exact signal bandwidth. If necessary, the signal is despread, and then refiltered to the information bandwidth, typically with an FIR filter.

Analog-to-Digital Converters

The rate of technology improvement versus time has not been as profound for A/D converters as for digital logic. The digital receiver industry is always looking for wider bandwidth and greater dynamic range. Successive approximation ADCs were replaced by flash converters in the early 1990s, and now are generally replaced with sigma-delta ADCs. Today's ADC can provide up to 105 Msps at 14-bit resolution. Special-purpose ADCs have been reported to provide sample rates over 5 giga samples per second (Gsps) at 8-bit resolution.

State-of-the-art research continues to push the boundaries of analog-to-digital (A/D) performance with a wide variety of clever techniques that shift the boundaries between DSP and ADC.

Modem

After down-conversion, filtering, and equalization, the symbols are converted to bits by a symbol detector/demodulator combination, which may include a matched filter or some other detection mechanism as well as a structure for mapping symbols to bits. A symbol is selected that most closely matches the received signal. At this stage, timing recovery is also necessary, but for symbols rather than the carrier. Then the output from the demodulator is in bits.

The bits that are represented by that symbol are then passed to the forward error correcting function to correct occasional bit errors. Finally, the received and error-corrected bits are parsed into the various fields of message, header, address, traffic, etc. The message fields are then examined by the protocol layers eventually delivering messages to an application (e.g., Web browser or voice coder (VoCoder)), thus delivering the function expected by the user.

SDRs must provide a wide variety of computational resources to be able to gracefully anticipate the wide variety of waveforms, they may ultimately be expected to demodulate. Today we would summarize that a typical SDR should be able to provide at least 266 MIPS of GPP, 32 Mbytes of random access memory (RAM), 100 MIPS of DSP, and 500 K equivalent gates of FPGA-configurable logic. More performance and resources are required for sophisticated waveforms or complex networking applications. Typically, the GPP will be called upon to perform the protocol stack and networking functions, the DSP will perform the physical layer modulation and demodulation, the FPGA will provide timing and control as well as any special-purpose hardware accelerators that are particularly unique to the waveform. It appears that SDR architectures will continue to evolve as component manufacturers bring forward new components that shift the boundaries of lowest cost, highest performance, and least power dissipation.

Forward Error Correction

In some instances, the demodulated bits are passed on to a forward error correction (FEC) stage for a reduction in the number of bit errors received. One of the interesting aspects of FEC is that it can be integrated into the demodulation process, such as in trellis-coded modulation; or it can be closely linked to demodulation, as in soft decoding for convolutional codes; or it can be an integral part of the next stage, medium access control (MAC) processing.

Medium Access Control

MAC processing generally includes framing information, with its associated frame synchronization structures, MAC addressing, error detection, link management structures, and payload encapsulation with possible fragmentation/defragmentation structures. From this stage, the output is bits, which are input to the network-processing layer. The network layer is designed for end-to-end connectivity support. The output of the network layer is passed to the application layer, which performs some sort of user functions and interface (speaker/microphone, graphical user interface, keypad, or some other sort of human-computer interface).

User Application

The user's application may range from voice telephony, to data networking, to text messaging, to graphic display, to live video. Each application has its own unique set of requirements, which, in turn, translate into different implications on the performance requirements of the SDR.

For voice telephony today, the dominant mode is to code the voice to a moderate data rate. Data rates from 4800 bps to 13,000 bps are popular in that they provide excellent voice quality and low distortion to the untrained listener. The digital modem, in turn, is generally more robust to degraded link conditions than analog voice would be under identical link conditions.

Another criterion for voice communications is low latency. Much real experience with voice communications makes it clear that if the one-way delay for each link exceeds 50 milliseconds,[1] then users have difficulty in that they expect a response from the far speaker and, hearing none, they begin to talk just as the response arrives, creating frequent speech collisions. In radio networks involving ad hoc networking, due to the delay introduced by each node as it receives and retransmits the voice signaling, it can be quite difficult to achieve uniformly low delay. Since the ad hoc network introduces jitter in packet delivery, the receiver must add a jitter buffer to accommodate a practical upper bound in latency of late

---

[1] Some systems specify a recommended maximum latency limit, such as 150 milliseconds for ITU-T G114.

packets. All of this conspires to add considerable voice latency. In response, voice networks have established packet protocols that allocate traffic time slots to the voice channels, in order to guarantee stable and minimal latency. In much the same way, video has both low error rate and fixed latency channel requirements, and thus networking protocols have been established to manage the quality requirements of video over wireless networks. Many wireless video applications are designed to accept the bit errors but maintain the fixed latency.

In contrast, for data applications, the requirement is that the data must arrive with very few or absolutely no bit errors; however, latency is tolerated in the application.

Voice coding applications are typically implemented on a DSP. The common voice coding algorithms require between 20 and 60 MIPS and about 32 Kbytes of RAM. Voice coding can also be successfully implemented on GPPs, and will typically require more than six times the instructions per second (100–600 MIPS) in order to perform both the multiply—accumulate signal processing arithmetic and the address operand fetch calculations.

Transmitting video is nearly 100 times more demanding than voice, and is rarely implemented in GPP or DSP. Rather, video encoding is usually implemented on special purpose processors due to the extensive cross-correlation required to calculate the motion vectors of the video image objects. Motion vectors substantially reduce the number of bits required to faithfully encode the images. In turn, a flexible architecture for implementing these special purpose engines is the use of FPGAs to implement the cross-correlation motion-detection engines.

Web browsing places a different type of restriction on an SDR. The typical browser needs to be able to store the images associated with each Web page in order to make the browsing process more efficient, by eliminating the redundant transmission of pages recently seen. This implies some large data cache function, normally implemented by a local hard drive disk. Recently, such memories are implemented by high-speed flash memory as a substitute for rotating electro-mechanical components.

Design Choices

Several aspects of the receiver shown in Figure 3.1(a) are of interest. One of the salient features is that the effect of all processing between the LNA and the FEC stage can be largely modeled linearly. This means that the signal processing chain does not have to be implemented in the way shown in Figure 3.1(a). The carrier recover loop does not have to be implemented at the mixer stage. It can just as easily be implemented immediately before demodulation. Another point to note is that no sampling is shown in Figure 3.1(a). It is mathematically possible to place the sampling process anywhere between the LNA and the FEC, giving the designer significant flexibility. An example of such design choice selection is shown in Figure 3.1(b), where the sampling process is shown as a discrete step.

The differences seen between Figures 3.1(a) and 3.1(b) are not at a functional level; they are implementation decisions that are likely to lead to a dramatically different hardware structure for equivalent systems. The following discussion on hardware concentrates on processing hardware selections because a discussion of RF and IF design considerations are beyond the scope of this book.

Several key concepts must be taken into consideration for the front-end of the system, and it is worthwhile to briefly mention them here. From a design standpoint, signals other than the signal of interest can inject more noise in the system than was originally planned, and the effective noise floor may be significantly larger than the noise floor due to the front-end amplifier.

One example of unpredicted noise injection is the ADC conversion process, which can inject noise into the signal through a variety of sources. The ADC quantization process injects noise into a signal. This effect becomes especially noticeable when a strong signal that is not the signal of interest is present in an adjacent channel. Even though it will be removed by digital filtering, the stronger signal sets the dynamic range of the receiver by affecting the AGC, in an effort to keep the ADC from being driven into saturation. Thus, the effective signal-to-interference and noise ratio (SINR) of the received signal is lower than might otherwise be expected from the receiver front-end. To overcome this problem and others like it, software solutions will not suffice, and flexible front-ends that are able to reject signals before sampling become necessary. Tunable RF components, such as tunable filters and amplifiers, are becoming available, and the SDR design that does not take full advantage of these flexible front-ends will handicap the final system performance.

3.2.2 Baseband Processor Engines

The dividing line between baseband processing and other types of processing, such as network stack processing, is arbitrary, but it can be constrained to be between the sampling process and the application. The application can be included in this portion of processing, such as a VoCoder for a voice system or image processing in a video system. In such instances, the level of signal processing is such that it may be suitable for specialized signal processing hardware, especially in demanding applications such as video processing.

Four basic classes of programmable processors are available today: GPPs, DSPs, FPGAs, and CCMs.

General Purpose Processors

GPPs are the target processors that probably come to mind first to anyone writing a computer program. GPPs are the processors that power desktop computers and are at the center of the computer revolution that began in the 1970s. The landscape of microprocessor design is dotted with a large number of devices from a variety of manufacturers. These different processors, while unique in their own right, do share some similarities, namely, a generic instruction set, an instruction sequencer, and a memory management unit (MMU).

There are two general types of instruction sets: (1) machines with fairly broad instruction sets, known as complex instruction set computers (CISCs); and (2) machines with a narrow instruction set, known as reduced instruction set computers (RISCs). Generally, the CISC instructions give the assembly programmer powerful instructions that address efficient implementation of certain common software functions. RISC instruction sets, while narrower, are designed to produce efficient code from compilers. The differences between the CISC and RISC are arbitrary, and both styles of processors are converging into a single type of instruction set. Regardless of whether the machine is CISC or RISC, they both share a generic nature to their instructions. These include instructions that perform multiplication, addition, or storage, but these instruction sets are not tailored to a particular type of application. In the context of CR, the application in which we are most interested is signal processing.

The other key aspect of the GPP is the use of a MMU. Because GPPs are designed for generic applications, they are usually coupled with an OS. This OS creates a level of abstraction over the hardware, allowing the development of applications with little or no knowledge of the underlying hardware. Management of memory is a tedious and error-prone process, and in a system running multiple applications, memory management includes paging memory, distributed programming and data storage throughout different blocks of memory. An MMU allows the developer to "see" a contiguous set of memory, even though the underlying memory structure may be fragmented or too difficult to control in some other fashion (especially in a multitasking system that has been running continuously for an extended period of time). Given the generic nature of the applications that run on a GPP, an MMU is critical because it allows the easy blending of different applications with no special care needed on the developer's part.

Digital Signal Processors

DSPs are specialized processors that have become a staple of modern signal processing systems. In large part, DSPs are similar to GPPs. They can be programmed with a high-level language such as C or C++ and they can run an OS. The key difference between DSPs and GPPs comes in the instruction set and memory management. The instruction set of a DSP is customized to particular applications.

For example, a common signal processing function is a filter, an example of which is the Direct Form II infinite impulse response (IIR) filter. Such a filter is seen in Figure 3.2.

Figure 3.2: Structure and data flow for Direct Form II IIR filter as a basis for an estimate on computational load.

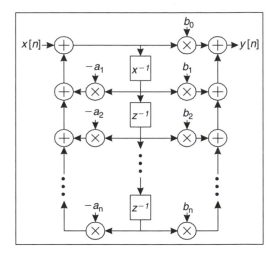

As seen in Figure 3.2, the signal path is composed of delays on the incoming signal of one sample, z^{-1}, and the multiplication of each delayed sample by a coefficient of the polynomial describing either the poles (a) or zeros (b) of the filter. If each delayed sample is considered to be a different memory location, then to quickly implement this filter, it is desirable to perform a sample shift in the circular buffer, perform a multiply and an add that multiplies each delayed sample times the corresponding polynomial coefficients, and store that result either in the output register, in this case $y[n]$, or into the register that is added to the input, $x[n]$.

The algorithm in Figure 3.2 has several characteristics. Assuming that the filter is of length N (N-order polynomials), then the total computation cost for this algorithm can be computed. To optimize radio modem performance, filters are frequently designed to be FIR filters with only the b (all zeros) polynomial. In order to implement extremely efficient DSP architectures, most DSP chips support performing many operations in parallel to make an FIR filter nearly a single instruction that is executed in a 1 clock cycle instruction loop. First, there is a loop control mechanism. This loop control has a counter that has to be initialized and then incremented in each operation, providing a set of $N+1$ operations. Within the loop, there is also an evaluation of the loop control state; this evaluation is performed N times. Within the loop, a series of memory fetch operations have to be performed. In this case, there are $N+1$ accesses to get the coefficient and store the result. There are additional $N+1$ circular accesses of the sample data plus the fetch operation of the new data point, yielding a total of $2N+2$ memory accesses. Finally, there are the arithmetic operations: the coefficient multiplied by the signal sample, and the accumulation operation (including the initial zero operation) resulting in $2N+1$ operations. Therefore, the DSP typically performs on the order of $6N$ operations per instruction clock cycle.

Assuming that a GPP performs a memory fetch, memory store, index update, comparison, multiplication, or addition in each computer clock cycle, then the GPP would require $6N+3$ clock cycles per signal sample. Using these assumptions, then an FIR filter with 32 filter taps and a signal sampled at 100 ksps would yield $(6 \times 32 + 3) \cdot 100 \times 10^3 = 19.5$ MIPS. Therefore, just for the filter mentioned, almost 20 MIPS are required for the GPP to successfully filter this narrowband signal. In reality, a GPP is likely to expend more than one cycle for each of these operations. For example, an Intel Pentium 4 floating point multiply occupies 6 clock cycles, so the given performance is a lower bound on the GPP MIPS load.

A DSP, in contrast, has the ability to reduce some of the cycles necessary to perform the given operations. For example, some DSPs have single-cycle MAC (multiple and accumulate). The Motorola 56001 is an example DSP that performs single instruction multiply and accumulate with zero overhead looping. These reductions result in a computational total of $3N+3$. Given these reductions, the computation load for the DSP is now $(3 \times 32 + 3) \cdot 100 \times 10^3 = 9.9$ MIPS.

Given its customized instruction set, a DSP can implement specialized signal processing functionality with a significantly fewer clock cycles than the more generic GPP processors. However, GPPs are attempting to erode this difference using multiple parallel execution arithmetic logic units so that they can perform effective address calculations in parallel with arithmetic operations. These are called superscalar architectural extensions. They are also attempting to raise the performance of GPP multipliers through use of many pipeline stages, so that the multiplication steps can be clocked at higher speed. This technique is called a highly pipelined architecture.

Field-Programmable Gate Arrays

FPGAs are programmable devices that are different in nature from GPPs and DSPs. An FPGA comprises some discrete set of units, sometimes referred to as logical elements (LE), logic modules (LM), slices, or some other reference to a self-contained Boolean logical operation. Each of these logical devices has at least one logic unit; this logic unit could be one or more multipliers and one or more accumulators, or a combination of such units in some FPGA chip selections. Logical devices are also likely to contain some memory, usually a few bits. The developer then has some freedom to configure each of these logical devices, where the level of reconfigurability is arbitrarily determined by the FPGA manufacturer. The logical devices are set in a logic fabric, a reconfigurable connection matrix that allows the developer to describe connections between different logical devices. The logic fabric usually also has access to some additional memory that logical devices can share. Timing of the different parts of the FPGA can be controlled by establishing clock domains. This allows the implementation of multirate systems on a single FPGA.

To program an FPGA, the developer describes the connections between logical devices as well as the configuration of each of these logical devices. The final design that the developer generates is closer to a circuit than a program in the traditional sense, even though the FPGA is ostensibly a firmware programmable device. Development for the FPGA is done by using languages such as very high-speed integrated circuit (VHSIC) Hardware Design Language (VHDL), which can also be used to describe application-specific integrated circuits (ASICs), essentially non-programmable chips. Variants of C exist, such as System-C, that allow the developer to use C-like constructs to develop FPGA code, but the resulting program still describes a logic circuit running on the FPGA.

The most appealing aspect of FPGAs is their computational power. For example, the Virtex II Pro FPGA by Xilinx has a total of 6460 slices, where each slice is composed of two look-up tables, two flip-flops, some math logic, and some memory. To be able to implement 802.11a, a communications standard that is beyond the abilities of any traditional DSP in 2005, would require approximately 3000 slices, or less than 50 percent of the FPGA's capabilities, showing the importance of a high degree of parallelism in the use of many multiply accumulators to implement many of the complex-waveform signal processes in parallel.

From a performance standpoint, the most significant drawback of an FPGA is that it consumes a significant amount of power, making it impractical for battery powered handheld subscriber solutions. For example, the Virtex II Pro FPGA mentioned above is rated at 2648 mW of power expenditure, whereas a low-power DSP such as the TMS320C55x is rated at 65–160 mW, depending on clock speed and version, and its high-performance cousin, the TMS320C64x, is rated at 250–1650 mW, depending on clock speed and version.

3.2.3 Baseband Processing Deployment

One of the problems that the CR developer will encounter when designing a system is attempting to determine what hardware should be included in the design and what baseband processing algorithm to deploy into which processors. The initial selection of hardware will impose limits on maximum waveform capabilities, while the processing deployment algorithm will present significant run-time challenges if left entirely as an optimization process for the cognitive algorithm. If the processing deployment algorithm is not designed correctly, it may lead to a sub-optimal solution that, while capable of supporting the user's required quality of service (QoS), may not be power efficient, quickly draining the system batteries or creating a heat dissipation problem.

The key problem in establishing a deployment methodology is one of scope. Once a set of devices and algorithm performance for each of these devices has been established, there is a finite set of possibilities that can be optimized. The issue with such optimization is not the optimization algorithm itself; several algorithms exist today for optimizing for specific values, such as the minimum mean square error (MMSE), maximum likelihood estimation (MLE), genetic algorithms, neural nets, or any of a large set of algorithms. Instead, the issue is in determining the sample set over which to perform the optimization.

There is no openly available methodology for establishing the set of possible combinations over which optimization can occur. However, at least one approach has been suggested by Neel et al. [1] that may lead to significant improvements in deployment optimization. The proposed methodology is partitioned into platform-specific and waveform-specific analyses. The platform-specific analysis is further partitioned into two types, DSP/GPP, and FPGA. The platform-specific analysis is as follows:

1. Create an operations audit of the target algorithms (number and type of operations).

2. For DSP:
   (a) Create a set of target devices.
   (b) Establish cycle-saving capabilities of each target device.

3. For FPGA:
   (a) Create a set of devices.
   (b) Establish mapping between different FPGA families. This mapping can be done on the basis of logical devices, available multiplies per element, or another appropriate metric.
   (c) Find logical device count for each target algorithm. FPGA manufacturers usually maintain thorough libraries with benchmarks.
   (d) Use mapping between devices to find approximate target load on devices when a benchmark is not available.

Once the platform-specific analysis is complete, the developer now has the tools necessary to map specific algorithms onto a set of devices. Note that at this stage, there is no information as to the suitability of each of these algorithms to different platforms, since a base clock rate (or data rate) is required to glean that type of information.

Given the platform-specific information assembled above, it is now possible to create performance estimates for the different waveforms that the platform is intended to support:

1. Create a block-based breakdown of the target waveform (using target algorithms from step 1 as building blocks).

2. Breakdown target waveform into clock domains.

3. Estimate time necessary to complete each algorithm.
   (a) In the case of packet-based systems, this value is fairly straightforward.[2]
   (b) In the case of stream-based systems, this value is the allowable latency.

4. Compute number of operations per second (OPS) needed for each algorithm.

---

[2] If the received signal is blocked into a block of many signal samples, and then the receiver operates on that block of signal samples through all of the receive signal processes, the process can be imagined to be a signal packet passing through a sequence of transforms. In contrast, if each new signal sample is applied to the entire receive process that is referred to here as a stream-based process.

5. Create a set of devices from the platform-specific phase that meet area and cost parameters (or whatever other parameters are appropriate for a first cut). This set of devices can be very large. At this stage, the goal is to create a set of devices or combination of devices that meet some broad criteria.

6. Cycle through the device set.
   (a) Attempt to map algorithms onto given devices in set.
      (i) For DSP:
         (1) Make sure that OPS calculated in step 4 of the waveform-specific analysis are reduced by cycle-saving capabilities outlined in step 2b of the platform-specific analysis.
         (2) The result of the algorithm map is a MIPS count for each device.
      (ii) For FPGA:
         (1) Mapping of the algorithm is a question of the number of occupied logical devices.
         (2) Make sure that clock domains needed for algorithms can be supported by the FPGA.
   (b) If a solution set of MIPS and/or LE exists for the combination of devices in the set, then save the resulting solution set for this device set; if a solution does not exist, discard the device set.

7. Apply appropriate optimization algorithm over resulting solution set/device set from step 6. Additional optimization algorithms include power budgets, and performance metrics.

The process described yields a coarse solution set for the hardware necessary to support a particular set of baseband processing solutions. From this coarse set, traditional tools can be used to establish a more accurate match of resources to functionality. Furthermore, the traditional tools can be used to find a solution set that is based on optimization criteria that are more specific to the given needs.

3.2.4 Multicore Systems and System-on-Chip

Even though several computing technologies have some promise over the horizon, such as quantum computing, it is an undeniable fact that silicon-based computing such as multicore systems and system-on-chip (SoC) will continue to be the bedrock of computing technology. Unfortunately, as technology reaches transistors under 100 nm, the key problems become the inability to continue the incremental pace of clock acceleration as well as significant problems in power dissipation. Even though the number of gates per unit area has roughly doubled every 18 months since the 1970s, the amount of power consumed per unit area has remained unchanged. Furthermore, the clocks driving processors have reached a plateau, so increases in clock speed have slowed significantly.

In order to overcome the technology problems in fabrication, a design shift has begun in the semiconductor industry. Processors are moving away from single-core solutions to multicore solutions, in which a chip is composed of more than one processing core. Several advantages are evident from such solutions. First, even though the chip area is increasing, it is now populated by multiple processors that can run at lower clock speeds. In order to understand the ramifications of such change, it is first important to recall the power consumption of an active circuit as:

$$P = \alpha \cdot C \cdot f \cdot V^2 \tag{3.1}$$

As shown in Eq. (3.1), the power dissipated, P, by an active circuit is the product of the switching activity, α, the capacitance of the circuit, C, the clock speed, f, and the operating voltage, V, squared. It is then clear from the above equation that the reduced clock speed results in a proportional reduction in power consumption. Furthermore, since a lower operating frequency means that a lower voltage is needed to operate the device, the reduction in the operating voltage produces a reduction in power consumption that follows a quadratic curve.

One of the principal bottlenecks in processor design is the input/output interface from data inside the chip to circuits outside the chip. This interface tends to be significantly slower than the data buses inside the chip. By reducing this interface capacitance and voltage swing for intercore communications, system efficiency grows. Furthermore, communication through shared memory is now possible within the chip. This capability can greatly increase the efficiency of the design.

3.3 Software Architecture

Software is subject to differences in structure that are similar to those differences seen in the hardware domain. Software designed to support baseband signal processing generally does not follow the same philosophy or architecture that is used for developing application-level software. Underlying these differences is the need to accomplish a variety of quite different goals. This section outlines some of the key development concepts and tools that are used in modern SDR design.

3.3.1 Design Philosophies and Patterns

Software design has been largely formalized into a variety of design philosophies, such as object-oriented programming (OOP), component-based programming (CBP), or aspect-oriented programming (AOP). Beyond these differences is the specific way in which the different pieces of the applications are assembled, which is generally referred to as a design pattern. This section describes design philosophies first, providing a rationale for the development of different approaches. From these philosophies, the one commonly used in SDR will be expanded into different design patterns, showing a subset of approaches that are possible for SDR design.

Design Philosophies

Four basic design philosophies are used for programming today: linear programming (LP), OOP, CBP, and AOP.

Linear Programming

LP is a methodology in which the developer follows a linear thought process for the development of the code. The process follows a logical flow, so this type of programming is dominated by conditional flow control (such as "if-then" constructs) and loops. Compartmentalized functionality is maintained in functions, where execution of a function involves swapping out the stack, essentially changing the context of operation, performing the function's work, and returning results to the calling function, which requires an additional stack swap. An analogy of LP is creating a big box for all items on your desktop, such as the phone, keyboard, mouse, screen, headphone, can of soda, and picture of your attractive spouse, with no separation between these items. Accessing any one item's functionality, such as drinking a sip of soda, requires a process to identify the soda can, isolate the soda can from the other interfering items, remove it from the box, sip it, and then place it back into the box and put the other items back where they were. C is the most popular LP language today, with assembly development reserved for a few brave souls who require truly high speed without the overhead incurred by a compiler.

Object-Oriented Programming

OOP is a striking shift from LP. Whereas LP has data structures—essentially variables that contain an arbitrary composition of native types such as float or integer—OOP extends the data structure concept to describe a whole object. An object is a collection of member variables (such as in a data structure) and functions that can operate on those member variables. From a terminology standpoint, a class is an object's type, and an object is a specific instance of a particular class. There are several rules governing the semantics of classes, but they generally allow the developer to create arbitrary levels of openness (or visibility), different scopes, different contexts, and different implementations for function calls that have the same name. OOP has several complex dimensions; additional information can be found elsewhere (e.g., Budd [2] and Weisfeld [3]).

The differences inherent in OOP have dramatic implications for the development of software. Extending the analogy from the previous example, it is now possible to break up every item on your desktop into a separate object. Each object has some properties, such as the temperature of your soda, and each object also has some functions that you can access to perform a task on that object, such as drinking some of your soda. There are several languages today that are OOP languages. The two most popular ones are Java and C++, although several other languages today are also OOP languages.

Component-Based Programming

CBP is a subtle extension of the OOP concept. In CBP, the concept of an object is constrained; instead of allowing any arbitrary structure for the object, under CBP the basic unit is now a component. This component comprises one or more classes, and is completely defined by its interfaces and its functionality. Again extending the previous example, the contents on the desktop can now be organized into components. A component could be a computer, where the computer component is defined as the collection of the keyboard, mouse, display, and the actual computer case. This particular computer component has two input interfaces, the keyboard and the mouse, and one output interface, the display. In future generations of this component, there could be additional interfaces, such as a set of headphones as an output interface, but the component's legacy interfaces are not affected by this new capability. Using CBP, the nature of the computer is irrelevant to the user as long as the interfaces and functionality remain the same. It is now possible to change individual objects within the component, such as the keyboard, or the whole component altogether, but the user is still able to use the computer component the same as always. The primary goal of CBP is to create stand-alone components that can be easily interchanged between implementations. Note that CBP is a coding style, and there are no mainstream languages that are designed explicitly for CBP.

Even though CBP relies on a well-defined set of interfaces and functionality, these aspects are insufficient to guarantee that the code is reusable or portable from platform to platform. The problem arises not from the concept, but from the implementation of the code. To see the problem, it is important now to consider writing the code describing the different aspects of the desktop components that we described before, in this case a computer. Conceptually, we have a component that contains an instance of a display, keyboard, mouse, headphone, and computer. If one were to write software emulating each of these items, not only would the interfaces and actual functional specifications need to be written, but also a wide variety of housekeeping functions, including, for example, notification of failure. If any one piece of the component fails, it needs to inform the other pieces that it failed, and the other pieces need to take appropriate action to prevent further mal-functions. Such a notification is an inherent part of the whole component, and implementing changes in the messaging structure for this notification on any one piece requires the update of all other pieces that are informed of changes in state. These types of somewhat hidden relationships create a significant problem for code reuse and portability because relationships that are sometimes complex need to be verified every time that code is changed. AOP was designed to attempt to resolve this problem.

Aspect-Oriented Programming

AOP allows for the creation of relationships between different classes. These relationships are arbitrary, but can be used to encapsulate the housekeeping code that is needed to create compatibility between two classes. In the messaging example, this class can include all the messaging information needed for updating the state of the system. Given this encapsulation, a class such as the headphone in the ongoing example can be used not only in the computer example, but also in other systems, such as a personal music player, an interface to an airplane sound system, or any other appropriate type of system. The relationship class encompasses an aspect of the class; thus, context can be provided through the use of aspects. Unlike CBP, AOP requires the creation of new language constructs that can associate an aspect to a particular class; to this end, there are several languages (AspectJ, AspectC++ , and Aspect#, among others).

Design Philosophy and SDR

The dominant philosophy in SDR design is CBP because it closely mimics the structure of a radio system, namely the use of separate components for the different functional blocks of a radio system, such as link control or the network stack. SDR is a relatively new discipline with few open implementation examples, so as the code base increases and issues in radio design with code portability and code reuse become more apparent, other design philosophies may be found to make SDR software development more efficient.

Design Patterns

Design patterns are programming methodologies that a developer uses within the bounds of the language the developer happens to be using. In general, patterns provide two principal benefits: they help in code reuse and they create a common terminology. This common terminology is of importance when working on teams because it simplifies communications between team members. As will be shown in the next section, some architectures, such as the SCA, use patterns. For example, the SCA uses the factory pattern for the creation of applications. In the context of this discussion, patterns for the development of waveforms and the deployment of cognitive engines will be shown. The reader is encouraged to explore formal patterns using available sources (e.g., Gamma et al. [4], Shalloway and Trott [5], or Kerievsky [6]).

3.4 SDR Development and Design

From the discussion above, it is clear that a software structure following a collection of patterns is needed for efficient large-scale development. In the case of SDR, the underlying philosophy coupled with a collection of patterns is called an architecture or operating environment. There are two open SDR architectures, GNURadio and SCA.

3.4.1 GNURadio

GNURadio [7] is a Python-based architecture (see section *Python*) that is designed to run on general-purpose computers running the Linux OS. GNURadio is a collection of signal processing components and supports primarily one RF interface, the universal software radio peripheral (USRP), a four channel up- and down-converter board coupled with ADC and DAC capabilities. This board also allows the use of daughter RF boards. GNURadio in general is a good starting point for entry-level SDR and should prove successful in the market, especially in the amateur radio and hobbyist market. GNURadio does suffer from some limitations—namely, (1) that it is reliant on GPP for baseband processing, thus limiting its signal processing capabilities on any one processor, and (2) it lacks distributed computing support, limiting solutions to single-processor systems, and hence limiting its ability to support high-bandwidth protocols.

3.4.2 Software Communications Architecture

The other open architecture is the SCA, sponsored by the Joint Program Office (JPO) of the US Department of Defense (DoD) under the Joint Tactical Radio System (JTRS) program. The SCA is a relatively complex architecture that is designed to provide support for secure signal processing applications running on heterogenous, distributed hardware. Furthermore, several solutions are available today that provide support for systems using this architecture, some of which are available openly, such as Virginia Tech's OSSIE [8] or Communications Research Center's SCARI [9], providing the developer with a broad and growing support base.

The SCA is a component management architecture; it provides the infrastructure to create, install, manage, and de-install waveforms as well as the ability to control and manage hardware and interact with external services through a set of consistent interfaces and structures. There are some clear limits to what the SCA provides. For example, the SCA does not provide such real-time support as maximum latency guarantees or process and thread management. Furthermore, the SCA also does not specify how particular components must be implemented, what hardware should support what type of functionality, or any other deployment strategy that the user or developer may follow. The SCA provides a basic set of rules for the management of software on a system, leaving many of the design decisions up to the developer. Such an approach provides a greater likelihood that the developer will be able to address the system's inherent needs.

The SCA is based on some underlying technology to be able to fulfill two basic goals, namely, code portability and reuse. In order to maintain a consistent interface, the SCA uses Common Object Request Broker Architecture (CORBA) as part of its middleware. CORBA is software that allows a developer to perform remote procedure calls (RPCs) on objects as if they resided in the local memory space even if they reside in some remote computer. The specifics of CORBA are beyond the scope of this book, but a comprehensive guide on the use of CORBA is available [10].

In recent years, implementations of CORBA have appeared on DSP and FPGA, but traditionally CORBA has been written for GPP. Furthermore, system calls are performed through an OS, requiring an OS on the implementation. In the case of the SCA, the OS of choice is a portable operating system interface

(POSIX) PSE-52 compliant OS, but CORBA need not limit itself to such an OS. Its GPP-centric focus leads to a flow for the SCA, as seen in Figure 3.3.

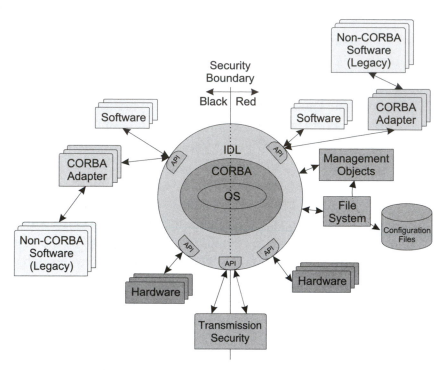

Figure 3.3: CORBA-centric structure for the SCA. The DoD implementation of the SCA requires that the RF modem (black) side of an SDR be isolated from the side where the plain text voice or data is being processed (red), and that this isolation be provided by the cryptographic device.

As seen in Figure 3.3, at the center of the SCA implementation is an OS, implying but not requiring the use of a GPP. Different pieces of the SCA are linked to this structure through CORBA and the interface definition language (IDL). IDL is the language used by CORBA to describe component interfaces, and is an integral part of CORBA. The different pieces of the system are attached together by using IDL. An aspect of the SCA that is not obvious from Figure 3.3 is that there can be more than one processor at the core, since CORBA provides location independence to the implementation. Beyond this architecture constraint are the actual pieces that make up the functioning SCA system, namely SCA and legacy software, non-CORBA processing hardware, security, management software, and an integrated file system.

The SCA is partitioned into four parts: the framework, the profiles, the API, and the waveforms. The framework is further partitioned into three parts, base components, framework control, and services. Figure 3.4 is a diagram of the different classes and their corresponding parts.

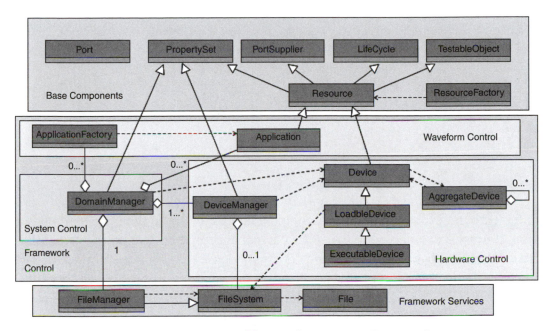

Figure 3.4: Classes making up the SCA core framework.

The SCA follows a component-based design, so the whole infrastructure revolves around the goal of creating, installing, managing, and de-installing the components making up a particular waveform.

Base Components

The base class of any SCA-compatible component is the Resource class. The Resource class is used as the parent class for any one component. Since components are described in terms of interfaces and functionality, not the individual makeup of the component, it follows that any one component comprises one or more classes that inherit from the Resource. For example, the developer may create a general Filter category for components. In this example, the filter may be implemented as an FIR or IIR filter. These filters can then be further partitioned into specific Filter implementations up to the developer's discretion, but for the purposes of this example, assume that the developer chooses to partition the filters only into finite or infinite response times (rather than some other category such as structure (e.g., Butterworth, Chebyshev) or spectral response (e.g., low-pass, high-pass, notch). If the developer chooses to partition the Filter implementation into FIR and IIR, then a possible description of the Filter family of classes can be that shown in Figure 3.5.

Figure 3.5: Sample filter component class hierarchy.

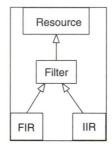

The Resource base class has only two member methods, start() and stop(), and one member attribute, identifier. A component must do more than just start and stop, it must also be possible to set the component's operational parameters, initialize the component into the framework and release the component from the framework, do some sort of diagnostic, and connect this component to other components. In order to provide this functionality, Resource inherits from the following classes: PropertySet, LifeCycle, TestableObject, and PortSupplier, as seen in Figure 3.6.

Figure 3.6: Resource class parent structure. The resource inherits a specialized framework functionality.

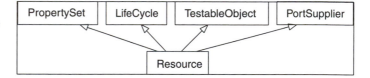

PropertySet

The PropertySet class provides a set of interfaces that allow other classes to both configure and query the values associated with this Resource. As part of the creation process of the component, the framework reads a configuration file and sets the different values associated with the component through the PropertySet interface. Later on, during the run-time behavior of the component, the Resource's properties can be queried through the query() interface and reconfigured with the configure() interface.

LifeCycle

The LifeCycle parent class provides the Resource with the ability to both initialize and release the component from the framework through the initialize() and releaseObject() interfaces. Initialization in this context is not the same as configuring the different values associated with the component. In this context, initialization sets the component to a known state. For example, in the case of a filter, the initialization call may allocate the memory necessary to perform the underlying convolution and it may populate the filter polynomial from the component's set of properties. The releaseObject performs the complimentary function to initialize. In the case of a filter, it would deallocate the memory needed by the component. One aspect of releaseObject that is common to all components is that it unbinds the component from CORBA. In short, releaseObject performs all the work necessary to prepare the object for destruction.

TestableObject

Component test and verification can take various forms, so it is beyond the scope of the SCA to outline all tests that are possible with a component. However, the SCA does provide a simple test interface through the use of the TestableObject parent class. TestableObject contains a single interface, runTest(). runTest() takes as an input parameter a number signifying the test to be run and a property structure to provide some testing parameters, allowing the inclusion of multiple tests in the component. Even though the interface provides the ability to support multiple tests, it is fundamentally a black box test structure.

PortSupplier

The final capability that parent classes provide to a Resource is the ability to connect to other components. Connections between components are performed through Ports (discussed below), not through the Resource itself. The use of PortSupplier allows the Resource to return one of any number of Ports that are defined within the component. To provide this functionality, the only interface provided by PortSupplier is getPort(). As its name implies, getPort returns the port specified in the method's arguments.

ResourceFactory

The SCA provides significant latitude concerning how an instance of a Resource can be created. In its broadest context, a Resource can be created at any time by any entity before it is needed by the framework for a specific waveform. This wide latitude is not necessarily useful because sometimes the framework needs the Resource to be explicitly created. In the cases in which it needs to be explicitly created, the specifications provide the framework with the ResourceFactory class. The ResourceFactory class has only three methods, createResource(), releaseResource(), and shutdown(). The createResource function creates an instance of the desired Resource and returns the reference to the caller, and the releaseResource function calls the releaseObject interface in the specified Resource. The shutdown function terminates the ResourceFactory. How the ResourceFactory goes about actually creating the Resource is not described in the specifications. Instead, the specifications provide the developer with the latitude necessary to create the Resource in whichever way seems best for that particular Resource.

Port

The Port class is the entry point and, if desired, the exit point of any component. As such, the only function calls that are explicitly listed in the Port definition are connectPort() and disconnectPort(). All other implementation-specific member functions are the actual connections, which, in the most general sense, are guided by the waveform's API. A component can have as many Ports as it requires. The implementation of the Port and the structure that is used to transfer information between the Resource (or the Resource's child) to the Port and back is not described in the specifications and is left up to the developer's discretion. Section 3.3.1 describes the development process and describes a few patterns that can be used to create specific components.

Framework Control

The base component classes are the basic building blocks of the waveform, which are relatively simple; the complexity in component development arrives in the implementation of the actual component functionality. When developing or acquiring a framework, the bulk of the complexity is in the framework control classes. These classes are involved in the management of system hardware, systemwide and hardware-specific storage, and deployed waveforms. From a high level, the framework control classes provide all the functionality that one would expect from an OS other than thread and process scheduling.

From Figure 3.4, the framework control classes can be partitioned into three basic pieces: hardware control, waveform control, and system control. The following sections describe each of these pieces in more detail.

Hardware Control

As discussed in Section 3.1, a radio generally comprises a variety of different pieces of hardware. As such, the framework must include the infrastructure necessary to support the variety of hardware that may be used. From a software placement point of view, not all hardware associated with a radio can run software. Thus, the classes that are described to handle hardware can run on the target hardware itself, or it can run in a proxy fashion, as seen in Figure 3.7. When used as a proxy, the hardware control software allows associated specialized hardware to function as if it were any other type of hardware.

Figure 3.7: Specialized hardware controlled by programmable interface through a proxy structure.

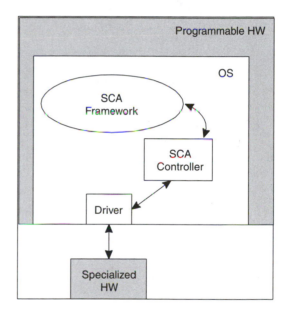

The needs of the hardware controller in the SCA are similar to the needs of components supporting a waveform. Furthermore, the component model closely fits the concept of hardware. Thus, the device controllers all inherit from the Resource base class. There are four device controllers: Device, LoadableDevice, ExecutableDevice, and AggregateDevice. Each of these classes is intended to support hardware with increasingly complex needs.

The most fundamental hardware is the hardware that performs some hardwired functionality and that may or may not be configurable. Such hardware has some inherent capacities that may be allocated for specific use. An example of such a piece of hardware is an ADC. The ADC is not programmable, but it is conceivable for an ADC to be tunable, meaning the developer may set the sampling rate or number of quantization bits. In such a case, the available capacity of this tunable hardware depends on whether it is being used for data acquisition of a particular signal or over a particular band. To this end, the only two functions that the Device class includes are allocateCapacity and deallocateCapacity.

Beyond the simple ability to allocate and deallocate capacities, a particular piece of hardware may have the ability to load and unload binary images. These images are not necessarily executable code, but they are images that configure a particular piece of hardware. For example, an FPGA, once loaded with a bit image, is configured as a circuit. The FPGA is never "run," it just acts as a circuit. The LoadableDevice class was created to deal with such hardware, where LoadableDevice inherits from Device. As would be expected, the only two functions that the LoadableDevice class contains are load() and unload().

Finally, there is the more complex hardware that not only has capacities that can be allocated and deallocated as well as memory that can be loaded and unloaded with binary images, but it also can execute a program from the loaded binary image. Such hardware is a GPP or a DSP. For these types of processors, the SCA uses the ExecutableDevice class. Much like an FPGA, a GPP or DSP has capacities that can be allocated or deallocated (like ports), and memory that can hold binary images, so the ExecutableDevice class inherits from the LoadableDevice class. As would be expected, the two functions that ExecutableDevice supports are execute() and terminate().

DeviceManager

The different hardware controllers behave as stand-alone components, so they need to be created and controlled by another entity. In the case of the SCA, this controller is the DeviceManager. The DeviceManager is the hardware booter; its job is to install all the appropriate hardware for a particular box and to maintain the file structure for that particular set of hardware. The boot-up sequence for the DeviceManager is fairly simple, as shown in Figure 3.8.

As seen in Figure 3.8, when an instance of the DeviceManager is created, it installs all the Devices described in the DeviceManager's appropriate profile and installs whatever file system is appropriate. After installing the hardware and file system, it finds the central controller, in this case the DomainManager, and installs all the devices and file system(s).

In a distributed system with multiple racks of equipment, the DeviceManager can be considered to be the system booter for each separate rack, or each separate board. As a rule, a different DeviceManager is used for each machine that has a

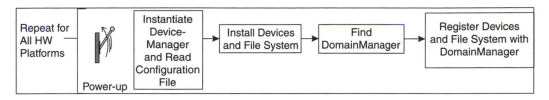

Figure 3.8: Device manager boot-up sequence.

different Internet Protocol (IP) address, but that general rule does not have to apply to every possible implementation.

Application Control

Two classes in the framework control section of the SCA provide application (waveform) control: Application and ApplicationFactory. ApplicationFactory is the entity that creates waveforms; as such, ApplicationFactory contains a single function, create().

ApplicationFactory

The create() function in the ApplicationFactory is called directly by the user, or the cognition engine in the case of a CR system, to create a specific waveform. The waveform's individual components and their relative connections are described in an eXtensible Markup Language (XML) file, and the component's specific implementation details are described in another XML file. The different XML files that are used to describe a waveform in the SCA are described in (see section *Profiles*).

Figure 3.9 shows an outline of the behavior of the ApplicationFactory. As seen in Figure 3.9, the ApplicationFactory receives the request from an external source. Upon receiving this request, it creates an instance of the Application class (discussed next), essentially a handle for the waveform. After the creation of the Application object, the ApplicationFactory allocates the hardware capacities necessary in all the relevant hardware, checks to see if the needed Resources already exist and creates them (through an entity such as the ResourceFactory) when they do not already exist, connects all the Resources, and informs the DomainManager (a master controller discussed in see section *System Control*) that the waveform was successfully created. The other steps in this creation process, such as initializing the Resources, are not included in this description for the sake of brevity. Once the ApplicationFactory has completed the creation process, a waveform comprising a variety of connected components now exists and can be used by the system.

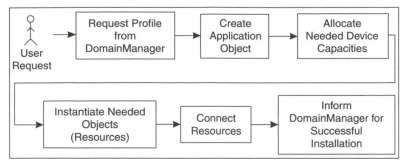

Figure 3.9: Simplified ApplicationFactory create() call for the creation of a waveform.

Application Class

The ApplicationFactory returns to the caller an instance of the Application class. The Application class is the handle that is used by the environment to keep track of the application. The Application class inherits from the Resource class, and its definition does not add any more function calls, just application identifiers and descriptors. The instance of the Application class that is returned by the ApplicationFactory is the object that the user would call start, stop, and releaseObject on to start, stop, and terminate the application, respectively.

System Control

In order for the radio system to behave as a single system, a unifying point is necessary. The specific nature of the unifying point can be as simple as a central registry, or as sophisticated as an intelligent controller. In the SCA, this unifying point is neither. The DomainManager, which is the focal point for the radio, is such an entity, and its task is as a central registry of applications, hardware, and capabilities. Beyond this registry operation, the DomainManager also serves as a mount point for all the different pieces of the hardware's system file. The DomainManager also maintains the central update channels, keeping track of changes in status for hardware and software and also informing the different system components of changes in system status.

   The point at which the DomainManager is created can be considered to be the system boot-up. Figure 3.10 shows this sequence. As seen in Figure 3.10, the DomainManager first reads its own configuration file. After determining the different operating parameters, the DomainManager creates the central file system, called a FileManager in the context of the SCA, reestablishes the previous configuration before shutdown, and waits for incoming requests. These requests can be

DeviceManagers associating with the DomainManager, or new Applications launched by the ApplicationFactory.

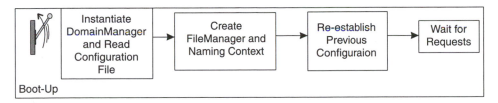

Figure 3.10: DomainManager simplified boot-up sequence.

Putting It Together

In general, the boot-up sequence of an SCA radio is partitioned into two different sets of steps, Domain boot-up, and one or more Device boot-ups. Once the platform has been installed (i.e., file system(s) and device(s)), the system is ready to accept requests for new waveforms. These requests can arrive either from the user or from some other entity, even a cognitive engine. Figure 3.11 shows a simplified boot-up sequence for the different parts of the SCA radio.

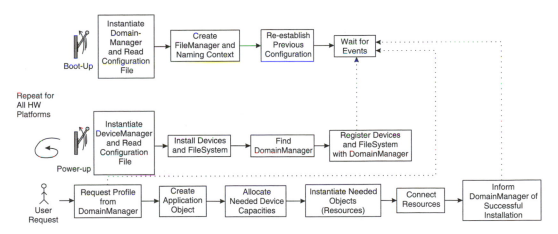

Figure 3.11: Simplified boot-up sequence and waveform creation for SCA radio.

Profiles

The SCA is composed of a variety of profiles to describe the different aspects of the system. The collection of all files describes, "the identity, capabilities, properties, inter-dependencies, and location of the hardware devices and software components that make up the system" [11] and is referred to as the Domain Profile.

The Domain Profile is in XML, a language similar to the HyperText Markup Language (HTML), which is used to associate values and other related information to tags. XML is relatively easy for a human to follow and, at the same time, easy for a machine to process.

The relationships among the different profiles is shown in Figure 3.12. As seen in this figure, there are seven file types: Device Configuration Descriptor (DCD), DomainManager Configuration Descriptor (DMD), Software Assembly Descriptor (SAD), Software Package Descriptor (SPD), Device Package Descriptor (DPD), Software Component Descriptor (SCD), and Properties Descriptor (PRF). In addition, there is Profile Descriptor, a file containing a reference to a DCD, DMD, SAD, or SPD.

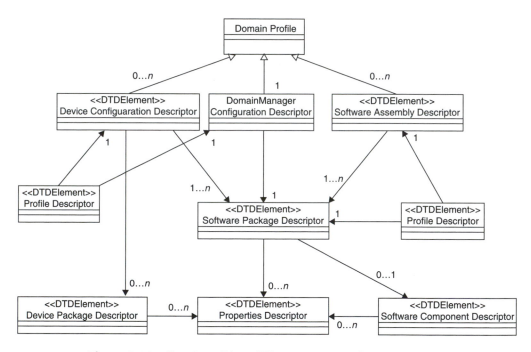

Figure 3.12: Files describing different aspects of an SCA radio.

The profiles have a hierarchical structure that is split into three main tracks, all of which follow a similar pattern. In this pattern, there is an initial file for the track. This initial file contains information about that particular track as well as the names of other files describing other aspects of the track. These other files will describe their aspect of the system and, when appropriate, will contain a reference to another file, and so on. Three files point to the beginning of a track: DCD,

DMD, and SAD. The DCD is the first file that is read by the DeviceManager on boot-up, the DMD is the first file that is read by the DomainManager on boot-up, and the SAD is the first file that is read by the ApplicationFactory when asked to create an application. Each of these files will contain not only information about that specific component, but one or more (in the case of the DMD, only one) references to an SPD. One SPD in this list of references contains information about that specific component. Any other SPD reference is for an additional component in that system: in the case of the DCD, a proxy for hardware; in the case of the SAD, another component in the waveform.

The SPD contains information about that specific component, such as implementation choices, and a link to both a PRF, which contains a list of properties for that component, and the SCD, which contains a list of the interfaces supported by that component. In the case of the DCD, an additional track exists to describe the actual hardware, the DPD. The DPD, much like the SPD, contains information about that specific component, in this case, the hardware side. The DPD, also mirroring the SPD, contains a reference to a PRF, which contains properties information concerning capacities for that particular piece of hardware.

Application Programming Interface

To increase the compatibility between components, it is important to standardize the interfaces; this means standardizing the function names, variable names, and variable types, as well as the functionality associated with each component. The SCA contains specifications for an API to achieve this compatibility goal. The structure of the SCA's API is based on building blocks, where the interfaces are designed in a sufficiently generic fashion such that they can be reused or combined to make more sophisticated interfaces.

There are two types of APIs in the SCA: real-time, shown as A in Figure 3.13, and non-real-time, shown as B in Figure 3.13. Real-time information is both data and time-sensitive control, whereas non-real-time is control information, such as setup and configuration, which does not have the sensitivity of real-time functionality. An interesting aspect of the SCA, API is that it describes interaction between different layers as would be described in the Open Systems Interconnection (OSI) protocol stack.

The API does not include interfacing information for intralayer communications. This means that in an implementation, the communications between two baseband processing blocks, such as a filter and an energy detector, would be outside the scope of the specifications. This limitation provides a limit to the level of granularity that the SCA supports, at least at the level of the API. To increase the

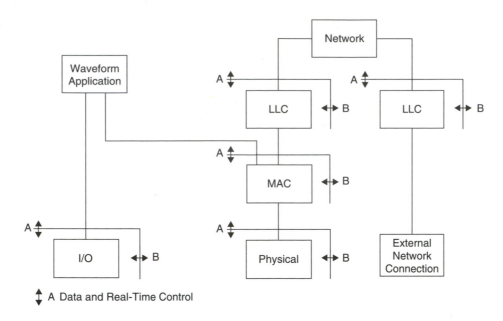

↕ A Data and Real-Time Control

↔ B Non-Real-Time Control, Setup and Initialization,
  from Applications, Other Levels, User Interface

Figure 3.13: SCA API for interlayer communications (I/O = input/output; MAC = medium access control; LLC = logical link control).

granularity to a level comparable to that shown in the example in Section 3.2, an additional API is required.

3.5 Applications

3.5.1 Application Software

The application level of the system is where the cognitive engine is likely to operate. At this level, the cognitive engine is just another application in the system. The biggest difference between a cognitive application and a traditional application such as a Web browser is that the cognitive engine has some deep connections with the underlying baseband software. Unlike baseband systems, which are generally light and mainly comprise highly optimized software, application-level design has more leeway in design overhead, allowing the use of a rich and varied set of solutions that sometimes add significant performance overhead. Beyond the typical OSs, such as products by Microsoft or products of the Linux family, and the typical toolkits, such as wxWindows for graphing, or OS-specific solutions such as Microsoft's DCOM, there exist three technologies that can have a significant impact on the application environment for CR: Java, Binary Runtime Environment for Wireless (BREW), and Python.

Java

Java is an object-oriented language developed at Sun Microsystems that allows the development of platform-independent software while also providing some significant housekeeping capabilities to the developer.

Fundamentally, Java is a compiled language that is interpreted in real-time by a virtual machine (VM). The developer creates the Java code and compiles it into bytecodes by the Java compiler, and the bytecodes are then interpreted by the Java VM (JVM) running on the host processor. The JVM has been developed in some language and compiled to run (as a regular application) on the native processor, and hence should be reasonably efficient when it runs on the target processor. The benefit from such an approach is that it provides the developer the ability to "write once, run anywhere."

Java also provides some housekeeping benefits that can best be defined as dreamlike for an embedded programmer, the most prominent of which is *garbage collection*. Garbage collection is the JVM's ability to detect when memory is no longer needed by its application and to deallocate that memory. This way, Java guarantees that the program will have no memory leaks. In C or C++ development, especially using distributed software such as CORBA, memory management is a challenge, and memory leaks are sometimes difficult to find and resolve. Beyond memory management, Java also has an extensive set of security features that provide limits beyond those created by the developer. For example, Java can limit the program's ability to access native calls in the system, reducing the likelihood that malicious code will cause system problems.

Java has several editions, each tailored to a particular type of environment:

- Java 2 Standard Edition (J2SE) is designed to run in desktop environment, servers, and embedded systems. The embedded design of J2SE relates to reduced overhead inherent to Java, such as reduced memory or reduced computing power.

- Java 2 Enterprise Edition (J2EE) is designed for large, multitier applications as well as for the development of and interactions with Web services.

- Java 2 Micro Edition (J2ME) is designed specifically for embedded and mobile applications. J2ME includes features that are useful in a wireless environment, such as built-in protocols and more robust security features.

Beyond these versions of Java, other features are available that can be useful for the radio developer. For example, under J2ME, there is a JavaPhone API. This API provides application-level support for telephony control, datagram messages (unreliable service over IP), power management, application installation, user profile, and address book and calendar. The JavaPhone API coupled with existing software components can reduce the development time associated with an application suite.

Java has been largely successful, especially in the infrastructure market. In such an environment, any overhead that may be incurred by the use of a real-time interpreter is overwhelmed by the benefit of using Java. For the embedded environment, the success of Java has been limited. There are three key problems with Java: memory, processing overhead, and predictability.

- *Memory:* Most efforts to date in Java for the embedded world revolve around the reduction of the overall memory footprint. One such example is the Connected Limited Device Configuration (CLDC). The CLDC provides a set of API and a set of VM features for such constrained environments.

- *Processing Overhead:* Common sense states that processing overhead is a problem whose relevance is likely to decrease as semiconductor manufacturing technology improves. One of the interesting aspects of processing overhead is that if the minimum supported system data rates for communications systems grows faster than processing power per milliwatt, then processing overhead will become a more pressing problem. The advent of truly flexible software-defined systems will bear out this issue and expose whether it is in fact a problem.

- *Predictability:* Predictability is a real-time concern that is key if the application software is expected to support such functionality as link control. Given that the executed code is interpreted by the JVM, asynchronous events such as memory management can mean that the execution time between arbitrary points of execution in the program can vary. A delay in execution can be acceptable because the design may be able to tolerate such changes. However, large variations in this execution time can cause such problems as dropped calls. A Real-time Extension Specification for Java exists that is meant to address this problem.

Java is a promising language that has met success in several arenas. Efforts within the Java community to bring Java into the wireless world are ongoing and are likely to provide a constantly improving product.

BREW

Qualcomm created the BREW for its CDMA phones. However, BREW is not constrained to just Qualcomm phones. Instead, BREW is an environment designed to run on any OS that simplifies development of wireless applications. BREW supports C/C++ and Java, so it is also largely language independent.

BREW provides a functionality set that allows the development of graphics and other application-level features. BREW is designed for the embedded environment, so the overall footprint and system requirements are necessarily small. Furthermore, it is designed for the management of binaries, so it can download and maintain a series of programs. This set of features is tailored for the handset environment, where a user may download games and other applications to execute on the handset but lacks the resources for more traditional storage and management structures.

Python

Python is an "interpreted, interactive, OOP language" [12]. Python is not an embedded language and it is not designed for the mobile environment, but it can play a significant role in wireless applications, as evidenced by its use by GNURadio. What makes Python powerful is that it combines the benefits of object-oriented design with the ease of an interpreted language.

With an interpreted language, it is relatively easy to create simple programs that will support some basic functionality. Python goes beyond most interpreted languages by adding the ability to interact with other system libraries. For example, by using Python one can easily write a windowed application through the use of wxWidgets. Just as Python can interact with wxWidgets, it can interact with modules written in C/C++, making it readily extendable.

Python also provides memory management structures, especially for strings, that make application development simpler. Finally, because Python is interpreted, it is OS-independent. The system requires the addition of an interpreter to function; of course, if external dependencies exist in the Python software for features such as graphics, the appropriate library also has to exist in the new system.

3.6 Development

Waveform development for sophisticated systems is a complex, multidesigner effort and is well beyond the scope of this book. However, what can be approached is the design of a simple waveform. In this case, the waveform is designed using the SCA, bringing several of the concepts described thus far into a more concrete example. Before constructing a waveform, it is important to build a component.

3.6.1 Component Development

Several structures are possible for a component because it is defined only in terms of its interfaces and functionality. At the most basic level, it is possible to create a component class that inherits from both Resource and Port. In such a case, the component would function as a single thread and would be able to respond to only one event at a time. A diagram of this simple component is seen in Figure 3.14.

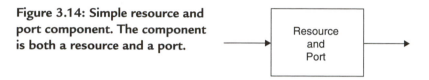

Figure 3.14: Simple resource and port component. The component is both a resource and a port.

A more sophisticated approach is to separate the output ports into separate threads, each interfacing with the primary Resource through some queue. This approach allows the creation of fan-out structures while at the same time maintaining a relatively simple request response structure. A diagram of this fan-out structure is seen in Figure 3.15.

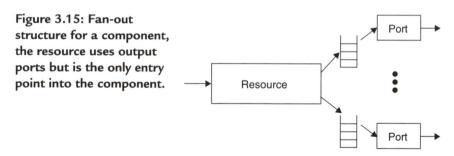

Figure 3.15: Fan-out structure for a component, the resource uses output ports but is the only entry point into the component.

One of the problems with the component structure shown in Figure 3.15 is that it does not allow for input interfaces that have the same interface name; this limitation increases the difficulty in using the API described in the specifications. To resolve this problem, a fan-in structure needs to be added to the system, even though this creates another level of complexity to the implementation. A way to implement this fan-in structure is to mimic the fan-out structure, and to place each input Port as a separate thread with a data queue separating each of these threads with the functional Resource. An example of this implementation is shown in Figure 3.16.

The approaches shown in Figures 3.14, 3.15, and 3.16 each present a different structure for the same concept. The specific implementation decision for a component is up to the developer, and can be tailored to the specific implementation. Because the components are described in terms of interfaces and functionality, it is possible to mix and match the different structures, allowing the developer even more flexibility.

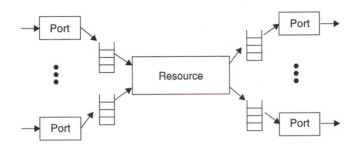

Figure 3.16: Fan-in, fan-out structure for a component. The resource uses both input and output ports.

3.6.2 Waveform Development

As an example of the waveform development, assume that a waveform is to be created that splits processing into three parts: baseband processing (assigned to a DSP), link processing (assigned to a GPP), and a user interface (assigned to the same GPP). A diagram of this waveform is shown in Figure 3.17.

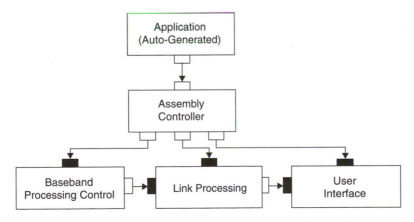

Figure 3.17: Simple SCA application. This example is made up of three functional components and one control component.

The application shown in Figure 3.17 has several pieces that should be readily recognized; the application was generated by the framework as a result of the Application Factory. Furthermore, the three proxies representing the relevant processing for the radio are also shown. Since the baseband processing is performed on a DSP, the component shown is a proxy for data transfer to and from the DSP. Link processing and the user interface, however, are actually implemented on the GPP. The assembly controller shown is part of the SCA, but not as a separate class. The assembly controller provides control information to all the different deployed components. Because the framework is generic, application-specific information, such as which components to tell to start and stop, must somehow be included. In the case of the SCA, this information is included in the assembly controller, a custom component whose sole job is to interface the application object to the rest of the waveform.

The waveform described in Figure 3.17 does not describe real-time constraints; it assumes that the real-time requirements of the system are met through some other means. This missing piece is an aspect of the SCA that is incomplete.

3.7 Cognitive Waveform Development

In the context of SDR, a cognitive engine is just another component of a waveform, so the issue in cognitive engine deployment becomes one of component complexity. In the simplest case, a developer can choose to create components that are very flexible, an example of which is seen in Figure 3.18. The flexible baseband processor seen in this figure can respond to requests from the cognitive engine and alter its functionality. A similar level of functionality is also available in the link-processing component. The system shown in Figure 3.18 is a slight modification of the system shown in Figure 3.17. The principal difference in this system is the addition of a reverse-direction set of ports and changed functionality for each component.

The principal problem with the structure shown in Figure 3.18 is that it places all the complexity of the system onto each separate component. Furthermore, the tie-in between the cognitive engine and the rest of the waveform risks the engine implementation to be limited to this specific system. An alternate structure is to create a whole waveform for which the only functionality is a cognitive engine, as seen in Figure 3.19.

The cognitive engine waveform shown in Figure 3.19 has no link to a waveform. To create this link, a link is created between the ApplicationFactory and the

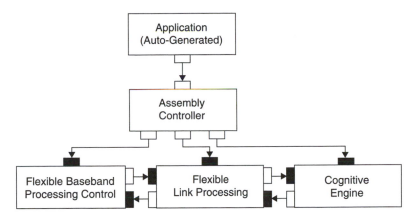

Figure 3.18: Simple cognitive waveform. The waveform performs both communications and cognitive functionality.

Figure 3.19: Cognitive engine waveform. The waveform performs only cognitive functionality. It assumes communications functionality is performed by other waveforms.

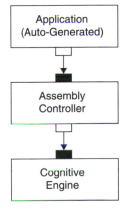

cognitive engine waveform. The cognitive engine can then request that the ApplicationFactory launch new waveforms. These new waveforms perform the actual communications work, which is evaluated by the functioning cognitive engine. If performance on the new waveform's communications link falls within some determined parameter, the cognitive engine can terminate the existing communications link waveform and request a new waveform (with different operating parameters or a different structure altogether) from the ApplicationFactory. This structure is seen in Figure 3.20.

An aspect of Figure 3.20 that is apparent is that the Port structure evident at the waveform level, as seen in Figure 3.19, scales up to interwaveform communications. Another aspect of this approach is that the cognitive waveform does not have to be collocated with the communications waveform. As long as timing

Figure 3.20: Multi-waveform cognitive support. The stand-alone cognitive waveform requests new waveforms from the SCA ApplicationFactory.

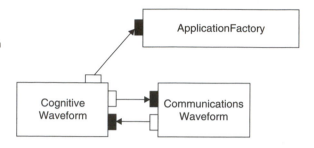

constraints are met, the cognitive waveform can be placed anywhere within the network that has access to the system. This aspect of the deployment of the waveforms allows the concept of a CR within the SCA to easily extend to a cognitive network, where a single cognitive engine can control multiple flexible radios, as seen in Figure 3.21.

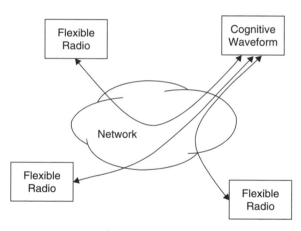

Figure 3.21: SCA-enabled cognitive network composed of multiple cognitive nodes.

With SDR and an astute selection of processing and RF hardware on the part of a developer, it is possible to create highly sophisticated systems that can operate well at a variety of scales, from simple single-chip mobile devices all the way up to multitiered cognitive networks.

3.8 Summary

Although a SDR is not a necessary building block of a CR, the use of SDR in CR can provide significant capabilities to the final system. An SDR implementation is a system decision, in which the selection of both the underlying hardware composition and the software architecture are critical design aspects.

The selection of hardware composition for an SDR implementation requires an evaluation of a variety of aspects, from the hardware's ability to support the required signals to other performance aspects, such as power consumption and silicon area. Traditional approaches can be used to estimate the needs at the RF and data acquisition levels. At the processing stage, it is possible to create an estimate of a processing platform's ability to be able to support a particular set of signal processing functions. With such an analysis, it is possible to establish the appropriate mix of general-purpose processors, DSPs, FPGAs, and CCMs for a particular set of signal processing needs.

In order to mimic the nature of a hardware-based radio, with components such as mixers and amplifiers, CBP is a natural way to consider software for SDR. In CBP, components are defined in terms of their interfaces and functionality. This definition provides the developer with significant freedom on the specific structure of that particular component.

Even though a developer may choose to use CBP for the design of an SDR system, a substantial infrastructure is still needed to support SDR implementations. This infrastructure must provide basic services, such as the creation and destruction of waveforms, as well as general system integration and maintenance. The goal of a software architecture is to provide this underlying infrastructure. The SCA is one such architecture. The SCA provides the means to create and destroy waveforms, manage hardware and distributed file systems, and manage the configuration of specific components.

Finally, beyond programming methodologies and architectures are the actual languages that one can use for development of waveforms and the specific patterns that are chosen for the developed software. The various languages have different strengths and weaknesses. C++ and Java are the dominant languages in SDR today. Python, a scripting language, has become increasingly popular in SDR applications, and is likely to be an integral part of future SDR development.

Much like the language selection, a design pattern for a particular component can have a dramatic effect on the capabilities of the final product. Design patterns that focus on flexibility can be more readily applied to cognitive designs, from the most basic node development all the way up to full cognitive networks.

References

[1] J. Neel, J.H. Reed and M. Robert, "A Formal Methodology for Estimating the Feasible Processor Solution Space for a Software Radio," in *Proceedings of the*

2005 Software Defined Radio Technical Conference and Product Exposition, November 2005, Orange Co, CA.

[2]   T. Budd, *An Introduction to Object-Oriented Programming, Third Edition*, Addison-Wesley, 2001, Boston, MA.

[3]   M. Weisfeld, *The Object-Oriented Thought Process, Second Edition*, Sams, 2003, Indianapolis, IN.

[4]   E. Gamma, R. Helm, R. Johnson and J. Vlissides, *Design Patterns: Elements of Reusable Object-Oriented Software*, Addison-Wesley, 1995, Boston, MA.

[5]   A. Shalloway and J.R. Trott, *Design Patterns Explained: A New Perspective on Object-Oriented Design, Second Edition*, Addison-Wesley, 2005, Boston, MA.

[6]   J. Kerievsky, *Refatcoring to Patterns*, Addison-Wesley, 2005, Boston, MA.

[7]   http://www.gnu.org/software/gnuradio/

[8]   http://ossie.mprg.org/

[9]   http://www.crc.ca/en/html/crc/home/research/satcom/rars/sdr/sdr

[10]   M. Henning and S. Vinoski, *Advanced CORBA Programming with C++*, Addison-Wesley, 1999, Boston, MA.

[11]   JTRS Joint Program Office, JTRS-5000, "Software Communications Architecture Specification, SCA V3.0," August 2004, San Deigo, CA.

[12]   http://www.python.org/

CHAPTER 4

Cognitive Radio: The Technologies Required

John Polson
Bell Helicopter
Textron Inc.
Fort Worth, TX, USA

4.1 Introduction

Technology is never adopted for technology's sake. For example, only hobbyists used personal computers (PCs) until a spreadsheet program, a "killer application," was developed. Then business needs and the benefits of small computers became apparent and drove PC technology into ubiquitous use. This led to the development of more applications, such as word processors, e-mail, and more recently the World Wide Web (WWW). Similar development is under way for wireless communication devices.

Reliable cellular telephony technology is now in widespread use, and new applications are driving the industry. Where these applications go next is of paramount importance for product developers. Cognitive radio (CR) is the name adopted to refer to technologies believed to enable some of the next major wireless applications. Processing resources and other critical enabling technologies for wireless killer applications are now available.

This chapter presents a CR roadmap, including a discussion of CR technologies and applications. Section 4.2 presents a taxonomy of radio maturity, and Sections 4.3 and 4.4 present more detailed discussions. Sections 4.5 and 4.6 are about enabling and required technologies for CRs. They present three classes of cognitive applications, one of which may be the next killer application for wireless devices. Conjectures regarding the development of CR are included in Section 4.7 with arguments for their validity. Highlights of this chapter are discussed in the summary in Section 4.8, which emphasizes that the technologies required for CR are presently available.

4.2 Radio Flexibility and Capability

More than 40 different types of military radios (not counting variants) are currently in operation. These radios have diverse characteristics; therefore, a large number of examples can be drawn from the pool of military radios. This section presents the continuum of radio technology leading to the software-defined radio (SDR). Section 4.3 continues the continuum through to CR.

The first radios deployed in large numbers were "single-purpose" solutions. They were capable of one type of communication (analog voice). Analog voice communication is not particularly efficient for communicating information, so data radios became desirable, and a generation of data-only radios was developed. At this point, our discussion of software and radio systems begins. The fixed-point solutions have been replaced with higher data rate and voice capable radios with varying degrees of software integration. This design change has enabled interoperability, upgradability, and portability.

It will be clear from the long description of radios that follows that there have been many additional capabilities realized in radios over their long history. SDRs and even more advanced systems have the most capabilities, and additional functions are likely to be added over time.

4.2.1 Continuum of Radio Flexibility and Capability

Basing CR on an SDR platform is not a requirement, but it is a practical approach at this time because SDR flexibility allows developers to modify existing systems with little or no new hardware development, as well as to add cognitive capabilities. The distinction of being a CR is bestowed when the level of software sophistication has risen sufficiently to warrant this more colorful description. Certain behaviors, discussed in this chapter, are needed for a radio to be considered a CR.

Historically, radios have been fixed-point designs. As upgrades were desired to increase capability, reduce life cycle costs, and so forth, software was added to the system designs for increased flexibility. In 2000, the Federal Communications Commission (FCC) adopted the following definition for software radios: "A communications device whose attributes and capabilities are developed and/or implemented in software" [1]. The culmination of this additional flexibility is an SDR system, as software capable radios transitioned into software programmable radios and finally became SDRs. The next step along this path will yield aware radios, adaptive radios, and finally CRs (see Figures 4.1 and 4.2).

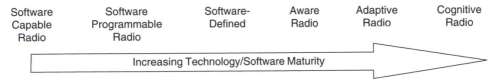

Figure 4.1: SDR technology continuum. As software sophistication increases, the radio system capabilities can evolve to accommodate a much broader range of awareness, adaptivity, and even the ability to learn.

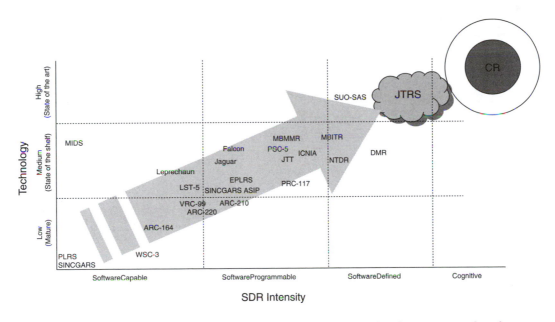

Figure 4.2: Examples of software radios. The most sophisticated software control and applications are reaching cognitive levels (see Tables 4.1–4.3 for descriptions).[1]

4.2.2 Examples of Software Capable Radios

Several examples of software capable radios are shown in Figure 4.2 and detailed in Table 4.1. The common characteristics of these radios are fixed modulation capabilities, relatively small number of frequencies, limited data and data rate capabilities, and finally the ability to handle data under software control.

---

[1] Note that the radios that are part of Figure 4.2 and Tables 4.1–4.3 are based on the best available public information and are intended only to notionally indicate a distribution of some of the well-known radios. For additional, accurate information about capabilities, contact the respective manufacturers.

Table 4.1: Selected software capable radios.

SINCGARS (Single-Channel Ground and Airborne Radio System)	A family of very high frequency—frequency modulation (VHF-FM) radios that provides a primary means of command and control. SINCGARS has frequency-hopping capability, and certain US Air Force versions operate in other bands using amplitude modulation (AM) waveforms. The SINCGARS family of radios has the capability to transmit and receive voice and tactical data, and record traffic on any of 2320 channels (25 kHz) between 30 and 88 MHz. A SINCGARS with an Internet controller is software capable [2].
PLRS (Position Location Reporting System)	A command-and-control aide that provides real-time, accurate, three-dimensional (3D) positioning, location, and reporting information for tactical commanders. The jam-resistant ultra high-frequency (UHF) radio transceiver network automatically exchanges canned messages that are used to geolocate unit positions to 15-m accuracy. Commanders use PLRS for situational awareness. PLRS employs a computer-controlled, crypto-secured master station and an alternate master station to ensure system survivability and continuity of operations. The network, under master station management, automatically uses user units as relays to achieve over-the-horizon (OTH) transmission and to overcome close-in terrain obstructions to line-of-sight (LOS) communications. When a rugged portable computer (PC) is used with the user unit, it becomes a mini-command-and-control station with a local area map along with superimposed with position and identification (ID) information. The computer interface and network control make PLRS a software capable radio system [3].

AN/WSC-3	A Demand Assigned Multiple Access (DAMA) satellite communication (SATCOM) terminal. It meets tight size and weight integration requirements. This single-waveform radio has a computer (Ethernet) interface and is software capable [4].
AN/ARC-164 HaveQuick II	A LOS UHF-AM radio used for air-to-air, air-to-ground, and ground-to-air communications. ARC-164 radios are deployed on all US Army rotary wing aircraft and provide anti-jam, secure communications links for joint task force missions in the tactical air operations band. This radio operates as a single-channel (25 kHz) or a frequency-hopping radio in the 225–399.975 MHz band. The aircraft radio transmits at 10 W output power and can receive secure voice or data. The ARC-164 data-handling capability makes it software capable. It has an embedded electronic counter-countermeasure (ECCM) anti-jam capability. The 243.000 MHz guard channel can be monitored. One model of the AN/ARC-164 is interoperable with SINCGARS [5].
AN/ARC-220	The standard high-frequency (HF) (2.0–29.999 MHz) radio for US Army aviation. It includes secure voice and data communications on any of 280,000 frequencies. The system uses software-realized DSP. Using MIL-STD-2217, software updates can be made over a MIL-STD-1553B bus. ARC-220 data processing capabilities optionally include e-mail applications. The ARC-220 is software capable. MIL-STD-148-141A automatic link establishment (ALE) protocols improve connectivity on behalf of ARC-220 users. ALE executes in microprocessors and is particularly interesting as a cognitive application because of its use of a database and its sounding of

(*Continued*)

Table 4.1: Selected software capable radios (*Continued*)

	the RF environment for sensing. GPS units may be interfaced with the radio, providing geolocation information [6].
AN/VRC-99	A secure digital network radio used for high data rate applications. This 1.2–2.0 GHz broadband direct-sequence spread-spectrum radio has National Security Agency (NSA) certified high-assurance capabilities and provides users with 31 RF channels. It supports TDMA and FDMA (frequency division multiple access). The networking capabilities are software programmable. AN/VRC-99 provides digital battlespace users with the bandwidth to support multimedia digital terminal equipment (DTE) of either an army command-and-control vehicle (C2V), army battle-command vehicle (BCV), or a marine corps advanced amphibious assault vehicle (A3V) with a single wideband secure waveform that supports voice, workstation, data, and imagery with growth for video requirements [7].
LST-5E	A software capable UHF tactical SATCOM/LOS transceiver with embedded communications security (COMSEC). The LST-5E provides the user with a single unit for high-grade half-duplex secure voice and data over both wideband (25 kHz) AM/FM and narrowband (5 kHz) 1200 bits per second (bps) binary phase shift keying (BPSK) and 2400 bps BPSK. Current applications for the LST-5E include manpack, vehicular, airborne, shipborne, remote, and fixed stations. The terminal is compatible with other AM or FM radios that operate in the 225–399.995 MHz frequency band [8].

Radio	Description
AN/PRC-6725 or Leprechaun	A handheld or wearable tactical radio. It can be programmed from a frequency fill device, laptop or PC, system base station, or it can be cloned from another radio [9].
MBITR (MultiBand Intra/Inter Team Radio)	Provides voice communications between infantry soldiers. The Land Warrior Squad Radio is a SINCGARS-compatible, eight-channel radio. MBITR is a software capable design based on the commercially available PRC-6745 Leprechaun radio [10].
CSEL (Combat Survivor/Evader Locator)	Provides UHF communications and location for the purposes of Joint Search and Rescue Center operations. CSEL uses GPS receivers for geolocation. Two-way, OTH, secure beaconing through communication satellites allows rescue forces to locate, authenticate, and communicate with survivors/evaders from anywhere in the world. The handheld receivers are one segment of the CSEL command-and-control system. The satellite-relay base station, Joint Search and Rescue Center software suite, and radio set adapter units interface with the UHF SATCOM network to provide CSEL capability. Upgrades through software loads make CSEL radios software capable [11].
MIDS (Multifunction Information Distribution System)	A direct-sequence spread-spectrum, frequency-hopping, anti-jam radio that supports the Link 16 protocol for communication between aircraft. Operates in the band around 1 GHz. Originally called Joint Tactical Information Distribution System (JTIDS), this radio waveform has been redeveloped on a new hardware platform, and is now being converted to a JTRS-compliant radio. When this conversion is completed it will be an SDR. It is currently in various stages of development and production as an interoperable waveform for Europe and the United States [12].

4.2.3 Examples of Software Programmable Radios

Several examples of software programmable radios are shown in Figure 4.2 and detailed in Table 4.2. The common characteristics of these radios are their ability to add new functionality through software changes and their advanced networking capability.

4.2.4 Examples of SDR

Only a few fully SDR systems are available, as shown in Figure 4.2 and detailed in Table 4.3. The common characteristic of SDR systems is complete adjustability through software of all radio operating parameters.

Other products related to SDR include GNURadio and the Vanu Anywave™ Base Station. GNURadio is a free software toolkit that is available on the Internet. It allows anyone to build a narrowband SDR. Using a Linux-based computer, a radio frequency (RF) front-end and an analog-to-digital converter (ADC), one can build a software-defined receiver. By adding a digital-to-analog converter (DAC) and possibly a power amplifier, one can build a software-defined transmitter [25].

The Vanu Anywave™ Base Station is a software-defined system that uses commercial off-the-shelf (COTS) hardware and proprietary software to build a wireless cellular infrastructure. The goal is simultaneous support for multiple standards, reduced operating expenses, scalability, and future proofing (cost-effective migration) [26].

4.3 Aware, Adaptive, and CRs

Radios that sense all or part of their environment are considered aware systems. Awareness may drive only a simple protocol decision or may provide network information to maintain a radio's status as aware. A radio must additionally autonomously modify its operating parameters to be considered adaptive. This may be accomplished via a protocol or programmed response. When a radio is aware, adaptive, and learns, it is a CR [27]. Only cutting-edge research and demonstration examples of aware, adaptive, or CRs are available currently.

4.3.1 Aware Radios

A voice radio inherently has sensing capabilities in both audio (microphone) and RF (receiver) frequencies. When these sensors are used to gather environmental information, it becomes an aware radio. The local RF spectrum may be sensed in pursuit of channel estimates, interference, or signals of interest. Audio inputs may

Table 4.2: Selected examples of software programmable radios.

AN/ARC-210	Provides fully digital secure communications in tactical and air traffic control (ATC) environments (30–400 MHz). Additionally, the radio provides 8.33 kHz channel spacing capability to increase the number of available ATC frequencies, and provides for growth to new very-high-frequency (VHF) data link modes. Currently provided functions are realized through software and provide integrated communications systems with adaptability for future requirements with little or no hardware changes. In other words, the ARC-210 is software programmable. The ARC-210 is integrated into aircraft and operated via MIL-STD-1553B data bus interfaces. Remote control is also available for manual operation. This radio is interoperable with SINCGARS (Single-Channel Ground and Airborne Radio System) and HaveQuick radios [13].
Racal 25	A compliant Project 25 public safety radio operating in the 136–174 MHz band. Its 5 W peak transmit power, rugged and submersible housing, and digital voice with Digital Encryption Standard (DES) make it an advanced radio. The radio uses DSP and flash memory architectures, and supports 12 and 16 kbps digital voice and data modes. Racal posts software upgrades on a protected Internet site. Software programmability enables field upgrades and is possible due to its DSP and flash memory-based architecture [14].
SINCGARS ASIP (Advanced System Improvement Program)	Interoperates with SINCGARS radios and enhance operational capability in the Tactical Internet (TI) environment. The ASIP models reduce size and weight, and provide further enhancements to operational capability in the TI environment. SINCGARS ASIP radios are software programmable and provide improved data capability, improved FEC for low-speed data modes, a GPS interface, and an Internet controller that allows them to interface with EPLRS and Battlefield Functional Area host computers. The introduction of Internet Protocol (IP) routing enables numerous software capabilities for radio systems [2].

(Continued)

Table 4.2: Selected examples of software programmable radios (Continued)

EPLRS (Enhanced Position Location Reporting System)	An AN/TSQ-158 that transmits digital information in support of tactical operations over a computer-controlled, TDMA communications network. EPLRS provides two major functions: data distribution and position location and reporting. EPLRS uses a frequency-hopping, spread-spectrum waveform in the UHF band. The network architecture is robust and self-healing. Radio firmware may be programmed from external devices. The peak data rate of 525 kbps supports application layer radio-to-radio interaction [15].
AN/PRC-117F	A software programmable, multiband, multimode radio (MBMMR). The PRC-117F operates in the 30–512 MHz band. Embedded COMSEC, SATCOM (satellite communication), SINCGARS, HaveQuick, and ECCM capabilities are standard. Various software applications, such as file transfer, Transmission Control Protocol (TCP) with IP, and digital voice, are included in this software programmable radio [7].
Jaguar PRC-116	In service in more than 30 nations, including the UK (Army) and US (Navy). It is a frequency-hopping, software programmable radio with considerable ECCM capability, resisting jamming by constantly shifting hopsets to unjammed frequencies. Security is further heightened by use of a scrambler. The Jaguar-V can also be used for data transmission at a rate of 16 kbps, and it may tolerate up to 50 radio nets, each with dozens of radios, at once; if each net is frequency hopping in a different sequence, it will still transmit to all of them [16].
JTT (Joint Tactical Terminal)	A high-performance, software programmable radio. Its modular functionality is backward- and forward compatible with the Integrated Broadcast Service (IBS). Using a software download, JTT can accept changes in format and protocol as IBS networks migrate to a common format. Subsequent intelligence terminals require total programmability of frequency, waveform, number of channels, communication security, and transmission security, making subsequent terminals SDRs [17].

NTDR (Near-Term Digital Radio) system	A mobile packet data radio network that links Tactical Operations Centers (TOCs) in a brigade area (up to 400 radios). The NTDR system provides self-organizing, self-healing network capability. Network management terminals provide radio network management. The radios interface with emerging Army Battle Command System (ABCS) automated systems and support large-scale networks in mobile operations with efficient routing software that supports multicast operations. Over-the-air programmability eliminates the need to send maintenance personnel to make frequency changes [18].
AN/PRC-138 Falcon	A manpack or vehicular-mounted HF and VHF radio set. Capabilities include frequency coverage of 1.6–60 MHz in SSB/CW/AME (single-sideband, continuous wave, AM equivalent) and FM in the VHF band and 100 preset channels. In the data mode, the AN/PRC-138 offers a variety of compatible modem waveforms that allows it to be integrated into existing system architectures. Specific features include embedded encryption, ALE, embedded high-performance HF data modems, improved power consumption management, and variable power output [19].
MBMMR (MultiBand, MultiMode Radio)	An AN/PSC-5D(C) that enhances interoperability among special operation forces units. The MBMMR supports LOS and SATCOM voice and data in six basic modes: LOS, Maritime, HaveQuick I/II, SINCGARS, SATCOM, and DAMA. Additional features include embedded TI Range Extension and Mixed Excitation Linear Prediction (MELP) voice coding. Over-the-air rekeying (OTAR), extended 30 to 420 MHz band, MIL-STD-188-181B high data rate in LOS communications and SATCOM, and MIL-STD-188-184 embedded advanced data controller are supported [20].
MBITR (MultiBand Intra/Inter Team Radio)	Designated AN/PRC-148, the MBITR provides AM/FM voice and data communications in the 30–512 MHz band. Development of the MBITR is an outgrowth of Racal's work on DSP and flash memory. MBITR is JTRS SCA 2.2 compliant and does not require a JTRS waiver. The information security (INFOSEC) capabilities are software programmable [21].

Table 4.3: Selected examples of SDRs.

DMR (Digital Modular Radio)	A full SDR capable of interoperability with tactical systems such as HF, DAMA, HaveQuick, and SINCGARS, as well as data link coverage for Link 4A and Link 11. These systems are programmable and include software-defined cryptographic functions. The US Navy is committed to migrating DMR to SCA compliance to allow the use of JTRS JPO provided waveforms. The DMR may be reconfigured completely via on-site or remote programming over a dedicated LAN or wide area network (WAN). The four full-duplex programmable RF channels with coverage from 2.0 MHz to 2.0 GHz require no change in hardware to change waveforms or security. The system is controlled, either locally or across the network, by a Windows®-based HMI [23].
JTRS (Joint Tactical Radio System)	A set of current radio procurements for fully SDRs. These radio systems are characterized by SCA compliance that specifies an operating environment that promotes waveform portability. The JTRS JPO is procuring more than 30 waveforms that will ultimately be executable on JTRS radio sets. System packaging ranges from embeddable single-channel form factors to vehicle-mounted multi-channel systems. JTRS radios are the current state-of-the-art technology and have the highest level of software sophistication ever embedded into a radio [24].
SUO-SAS (Small Unit Operations—Situational Awareness System)	Developed by DARPA to establish the operational benefits of an integrated suite of advanced communication, navigation, and situation awareness technologies. It served as a mobile communications system for small squads of soldiers operating in restrictive terrain. SUO-SAS provides a navigation function utilizing RF ranging techniques and other sensors to provide very high accuracy [22].

be used for authentication or context estimates or even natural language understanding or aural human–machine interface (HMI) interactions. Added sensors
enable an aware radio to gather other information, such as chemical surroundings,
geolocation, time of day, biometric data, or even network quality of service
(QoS) measures. The key characteristic that raises a radio to the level of aware
is the consolidation of environmental information not required to perform simple
communications. Utilization of this information is not required for the radio
to be considered aware. There is no communication performance motivation
for developing this class of aware radios, and it is expected that this set will
be sparse.

One motivation for an aware radio is providing information to the user. As an
example, an aware radio may provide a pull-down menu of restaurants within a
user-defined radius. The radio may gather this information, in the future, from
low-power advertisement transmissions sent by businesses to attract customers.
A military application may be a situational awareness body of information that
includes a pre-defined set of radios and their relative positions. As an example, the
radios exchange global positioning system (GPS) coordinates in the background,
and the aware radios gather the information for the user and provide it on request.
The radio is not utilizing the information but is aware of the situation.

One example of an aware radio is the code division multiple access (CDMA)
based cellular system proposed by Chen et al. [28]. This system is aware of QoS
metrics and makes reservations of bandwidth to improve overall QoS. Another
example of an aware radio is the orthogonal frequency division multiplexing
(OFDM) based energy aware radio link control discussed by Bougard et al. [29].

4.3.2 Adaptive Radios

Frequency, instantaneous bandwidth, modulation scheme, error correction coding,
channel mitigation strategies such as equalizers or RAKE filters, system timing
(e.g., a time division multiple access [TDMA] structure), data rate (baud timing),
transmit power, and even filtering characteristics are operating parameters that
may be adapted. A frequency-hopped spread-spectrum radio is not considered
adaptive because once programmed for a hop sequence, it is not changed. A
frequency-hopping radio that changes hop pattern to reduce collisions may be
considered adaptive. A radio that supports multiple channel bandwidths is not
adaptive, but a radio that changes instantaneous bandwidth and/or system timing
parameters in response to offered network load may be considered adaptive. If a
radio modifies intermediate frequency (IF) filter characteristics in response to
channel characteristics, it may be considered adaptive. In other words, if a radio

makes changes to its operating parameters, such as power level, modulation, frequency, and so on, it may be considered an adaptive radio.

At this time, two wireless products exhibit some degree of adaptation: the digital European cordless telephone (DECT) and 802.11a.

DECT can sense the local noise floor and interference of all the channels from which it may choose. Based on this sensing capability, it chooses to use the carrier frequencies that minimize its total interference. This feature is built into hardware, however, and not learned or software adaptive; thus, DECT is not normally considered an adaptive radio.

802.11a has the ability to sense the bit error rate (BER) of its link, and to adapt the modulation to a data rate and a corresponding forward error correction (FEC) that set the BER to an acceptably low error rate for data applications. Although this is adaptive modulation, 802.11 implementations generally are dedicated purpose-fixed application-specific integrated circuit (ASIC) chips, not software defined, and thus 802.11 is not normally considered to be an adaptive radio.

4.3.3 Cognitive Radios

A CR has the following characteristics: sensors creating awareness of the environment, actuators to interact with the environment, a model of the environment that includes state or memory of observed events, a learning capability that helps to select specific actions or adaptations to reach a performance goal, and some degree of autonomy in action.

Since this level of sophisticated behavior may be a little unpredictable in early deployments, and the consequences of "misbehavior" are high, regulators will want to constrain a CR. The most popular suggestion to date for this constraint is a regulatory policy engine that has machine-readable and interpretable policies.

Machine-readable policy-controlled radios are attractive for several reasons. One feature is the ability to "try out" a policy and assess it for impacts. The deployment may be controlled to a few radios on an experimental basis, so it is possible to assess the observation and measurement of the behaviors. If the result is undesirable, the policies may be removed quickly. This encourages rapid decisions by regulatory organizations. The policy-driven approach is also attractive because spatially variant or even temporally variant regulations may be deployed. As an example, when a radio is used in one country, it is subject to that country's regulations, and when the user carries it to a new country, the policy may be reloaded to comply in the new jurisdiction. Also, if a band is available for use during a certain period but not during another, a machine-readable policy can realize that behavior.

The language being used to describe a CR is based on the assumption of a smart agent model, with the following capabilities:[2]

- Sensors creating awareness in the environment.
- Actuators enabling interaction with the environment.
- Memory and a model of the environment.
- Learning and modeling of specific beneficial adaptations.
- Specific performance goals.
- Autonomy.
- Constraint by policy and use of inference engine to make policy-constrained decisions.

The first examples of CRs were modeled in the Defense Advanced Research Projects Agency (DARPA) NeXt Generation (XG) radio development program. These radios sense the spectrum environment, identify an unoccupied portion, rendezvous multiple radios in the unoccupied band, communicate in that band, and vacate the band if a legacy signal re-enters that band. These behaviors are modified as the radio system learns more about the environment; the radios are constrained by regulatory policies that are machine interpretable. The first demonstrations of these systems took place late in 2004 and in 2005 [30].

4.4 Comparison of Radio Capabilities and Properties

Table 4.4 summarizes the properties for the classes of advanced radios described in the previous sections. Classes of radios have "fuzzy boundaries," and the comparison shown in the table is broad. There are certain examples of radios that fall outside the suggestions in the table. A CR may demonstrate most of the properties shown, but is not required to be absolutely reparameterizable. Note that the industry consensus is that a CR is not required to be an SDR, even though it may demonstrate most of the properties of an SDR. However, there is also consensus that the most likely path for development of CRs is through enabling SDR technology.

4.5 Available Technologies for CRs

The increased availability of SDR platforms is spurring developments in CR. The necessary characteristics of an SDR required to implement a practical CR are

---

[2] A smart agent also has the ability to *not* use some or all of the listed capabilities.

Table 4.4: Properties of advanced radio classes.

Radio property	Software capable radio	Software programmable radio	Software-defined radio	Aware radio	Adaptive radio	Cognitive radio
Frequency hopping	X	X	X	X	X	X
Automatic link establishment (i.e., channel selection)	X	X	X	X	X	X
Programmable crypto	X	X	X	X	X	X
Networking capabilities		X	X	X	X	X
Multiple waveform interoperability		X	X	X	X	X
In-the-field upgradable		X	X	X	X	X
Full software control of all signal processing, crypto, and networking functionality			X	*	*	*
QoS measuring/channel state information gathering				X	X	X
Modification of radio parameters as function of sensor inputs					X	X
Learning about environment						X
Experimenting with different settings						X

* The industry standards organizations are in the process of determining the details of what properties should be expected of aware, adaptive, and CRs.

134

excess computing resources, controllability of the system operating parameters, affordability, and usable software development environments including standardized application programming interfaces (APIs). This section discusses some additional technologies that are driving CR. Even though this is not a comprehensive list of driving technologies, it includes the most important ones.

4.5.1 Geolocation

Geolocation is an important CR enabling technology due to the wide range of applications that may result from a radio being aware of its current location and possibly being aware of its planned path and destination.

The GPS is a satellite-based system that uses the time difference of arrival (TDoA) to geolocate a receiver. An overview of this system is presented in Chapter 8. The resolution of GPS is approximately 100 m. GPS receivers typically include a one-pulse-per-second signal that is Kalman filtered as it arrives at each radio from each satellite, resulting in a high-resolution estimate of propagation delay from each satellite regardless of position. By compensating each pulse for the predicted propagation delay, the GPS receivers estimate time to approximately 340 nanoseconds (ns, or 10^{-9} seconds) of jitter [31].

In the absence of GPS signals, triangulation approaches may be used to geolocate a radio from cooperative or even non-cooperative emitters. Chapter 8 discusses the classical approaches of TDoA, time of arrival (ToA), and, if the hardware supports it, angle of arrival (AOA). Multiple observations from multiple positions are required to create an accurate location estimate. The circular error probability (CEP) characterizes the estimate accuracy.

4.5.2 Spectrum Awareness/Frequency Occupancy

A radio that is aware of spectrum occupancy may exploit this information for its own purposes, such as utilization of open channels on a non-interference basis. Methods for measuring spectrum occupancy are discussed in Chapter 5.

A simple sensor resembles a spectrum analyzer. The differences are in quality and speed. The CR application must consider the quality of the sensor in setting parameters such as maximum time to vacate a channel upon use by an incumbent signal. It ingests a band of interest and processes it to detect the presence of signals above the noise floor. The threshold of energy at which occupancy is declared is a critical parameter. The detected energy is a function of the instantaneous power, instantaneous bandwidth, and duty cycle. An unpredictable duty cycle is expected. Spectrum occupancy is spatially variant, time variant, and subject to observational blockage (e.g., deep fading may yield a poor observation). Therefore, a distributed approach to spectrum sensing is recommended.

The primary problem associated with spectrum awareness is the hidden node problem. A lurking receiver (the best example is a television (TV) set) may be subjected to interference and may not be able to inform the CR that its receiver is experiencing interference. Regulators, spectrum owners, and developers of CR are working to find robust solutions to the hidden node problem. Again, a cooperative approach may help to mitigate some of the hidden node problems, but a cooperative approach will not necessarily eliminate the hidden node problem.

In addition to knowing the frequency and transmit activity properties of a radio transmitter, it may also be desirable for the radio to be able to recognize the waveform properties and determine the type of modulation, thereby allowing a radio to request entry into a local network. Many articles have been published on this topic, as well as a textbook by Azzouz and Nandi [32]. Once the modulation is recognized, then the CR can choose the proper waveform and protocol stack to use to request entry into the local network.

4.5.3 Biometrics

A CR can learn the identity of its user(s), enabled by one or more biometric sensors. This knowledge, coupled with authentication goals, can prevent unauthorized users from using the CR. Most radios have sensors (e.g., microphones) that may be used in a biometric application. Voice print correlation is an extension to an SDR that is achievable today. Requirements for quality of voice capture and signal processing capacity are, of course, levied on the radio system. The source radio can authenticate the user and add the known identity to the data stream. At the destination end, decoded voice can be analyzed for the purposes of authentication.

Other biometric sensors can be used for CR authentication and access control applications. Traditional handsets may be modified to capture the necessary inputs for redundant biometric authentication. For example, cell phones recently have been equipped with digital cameras. This sensor, coupled with facial recognition software, may be used to authenticate a user. An iris scan or retina scan is also possible. Figure 4.3 shows some of the potential sensors and their relative strengths and weakness in terms of reliability and acceptability [33].

4.5.4 Time

Included in many contracts is the phrase "time is of the essence," testament to the criticality of prompt performance in most aspects of human interaction. Even a

Biometrics in Order of Effectiveness	Biometrics in Order of Social Acceptability
1. Palm scan	1. Iris scan[a]
2. Hand geometry	2. Keyboard dynamics
3. Iris scan	3. Signature dynamics
4. Retina scan	4. Voiceprint[b]
5. Fingerprint	5. Facial scan[a]
6. Voiceprint	6. Fingerprint[c]
7. Facial scan	7. Palm scan[c]
8. Signature dynamics	8. Hand geometry[c]
9. Keyboard dynamics	9. Retina scan[a]
	[a]Requires a camera scanner [b]Utilizes a copy of the voice input (low impact) [c]Requires a sensor in the push-to-talk (PTT) hardware

Figure 4.3: Biometric sensors for CR authentication applications. Several biometric measures are low impact in terms of user resistance for authentication applications.

desktop computer has some idea about what time it is, what day it is, and even knows how to utilize this information in a useful manner (date and time stamping information). A radio that is ignorant of time has a serious handicap in terms of learning how to interact and behave. Therefore, it is important for the CR to know about time, dates, schedules, and deadlines.

Time-of-day information enables time division multiplexing on a coarse-grained basis, or even a fine-grained basis if the quality of the time is sufficiently accurate. Time-of-day information may gate policies in and out. Additionally, very fine knowledge of time may be used in geolocation applications.

GPS devices report time of day and provide a one-pulse-per-second signal. The one-pulse-per-second signal is transmitted from satellites, but does not arrive at every GPS receiver at the same time due to differences in path lengths. A properly designed receiver will assess the propagation delay from each satellite and compensate each of these delays so that the one-pulse-per-second output is synchronous at all receivers with only a 340 ns jitter. This level of accuracy is adequate for many applications, such as policy gating and change of cryptographic keys. Increased accuracy and lowered jitter may be accomplished through more sophisticated circuitry.

The local oscillator in a radio system may be used to keep track of time of day. The stability of these clocks is measured at approximately 10^{-6}. These clocks tend to drift over time, and in the course of a single day may accumulate up to 90 ns of error. Atomic clocks have much greater stability (10^{-11}), but have traditionally been large and power hungry. Chip-scale atomic clocks have been demonstrated and are expected to make precision timing practical. This will enable geolocation applications with lower CEPs.

4.5.5 Spatial Awareness or Situational Awareness

A very significant role for a CR may be viewed as a personal assistant. One of its key missions is facilitating communication over wireless links. The opposite mission is impeding communications when appropriate. As an example, most people do not want to be disturbed while in church, in an important meeting, or in a classroom. A CR could learn to classify its situation into "user interruptible" and "user non-interruptible." The radio accepting aural inputs can classify a long-running exchange in which only one person is speaking at a time as a meeting or classroom and autonomously put itself into vibration-only mode. If the radio senses its primary user is speaking continuously or even 50 percent of the time, it may autonomously turn off the vibration mode.

4.5.6 Software Technology

Software technology is a key component for CR development. This section discusses key software technologies that are enabling CR. These topics include policy engines, artificial intelligence (AI) techniques, advanced signal processing, networking protocols, and the Joint Tactical Radio System (JTRS) Software Communications Architecture (SCA).

Policy Engines

Radios are a regulated technology. A major intent of radio regulatory rules is to reduce or avoid interference among users. Currently, rules regarding transmission and reception are enumerated in spectrum policy as produced by various spectrum authorities (usually in high-level, natural language). Regulators insist that even a CR adhere to spectrum policies. To further complicate matters, a CR may be expected to operate within different geopolitical regions and under different regulatory authorities with different rules. Therefore, CRs must be able to dynamically update policy and select appropriate policy as a function of situation.

Spectrum policies relevant to a given radio may vary in several ways:

1. Policies may vary in time (e.g., time of day, date, and even regulations changing from time to time).
2. Policies may vary in space (e.g., radio and user traveling from one policy regulatory domain to another).

3. A spectrum owner/leaser may impose policies that are more stringent than those imposed by a regulatory authority.

4. The spectrum access privileges of the radio may change in response to a change in radio user.

As a result, the number of different policy sets that apply to various modes and environments grows in a combinatorial fashion. It is impractical to hard-code discrete policy sets into radios to cover every case of interest. The accreditation of each discrete policy set is a major challenge. SDRs, for example, would require the maintenance of downloadable copies of software implementations of each policy set for every radio platform of interest. This is a configuration management problem.

A scalable expression and enforcement of policy is required. The complexity of policy conformance accreditation for CRs and the desire for dynamic policy lead to the conclusion that CRs must be able to read and interpret policy. Therefore, a well-defined, accepted (meaning endorsed by an international standards body) language framework is needed to express policy. For example, if an established policy rule is constructed in the presence of other rules, the union of all policies is applicable. This enables hierarchal policies and policy by exception. As an example, suppose the emission level in band A is X dBm, except for a sub-band A' for which the emission-level constraint is Y dBm if a Z KHz guard band is allowed around legacy signals. Even this simple structure is multi-dimensional. Layers of exceptions are complex. The policy engine must be able to constrain behavior according to the intent of the machine-readable policy. An inference capability is needed to interpret multiple rules simultaneously.

In the case of spectrum subleasing, policies must be delegated from the lessor to the lessee, and a machine-readable policy may be delegated. When a CR crosses a regulatory boundary, the appropriate policy must be enabled. Policies may also be used by the system in a control function.

The policy should use accepted standard tools and languages because the policy engine must be able to access automatic interpretation to achieve the goals of CR applications. Policies may be written by regulatory agencies or by third parties and approved by regulators, but in all cases policy is a legal or contractual operating requirement and provability in the policy interpretation engine is needed for certification.

Several suggestions for policy language have emerged. The eXtensible Markup Language (XML) is not appropriate because it does not typically have inference capabilities in the interpretation engines. The Ontology Inference Layer

(OIL), Web Ontology Language (OWL), and DARPA Agent Markup Language (DAML) have all been explored as possible policy languages. DARPA's XG program cites the OWL language as an appropriate language. Tool sets are available for building policy definitions and for machine interpretation of the definitions [34].

AI Techniques

The field of AI has received a great deal of attention for decades. In 1950, Alan Turing proposed the Turing test, regarding interacting with an entity and not being able to tell if it is human or machine. The AI techniques that work are plentiful, but most are not widely applicable to a wide range of problems. The powerful techniques may even require customization to work on a particular problem.

An agent is an entity that perceives and acts. A smart agent model is appropriate for CR. Figure 4.4 explains four models of smart agents—simple reflex agents,

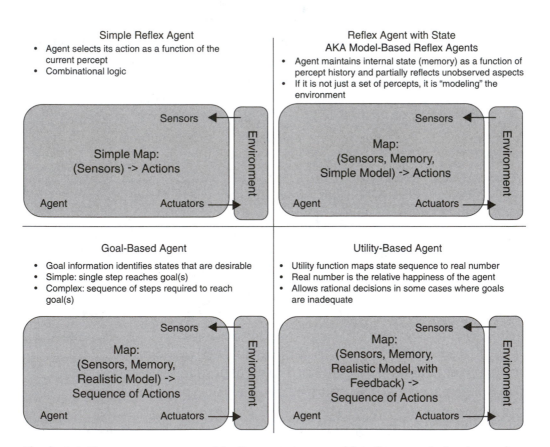

Figure 4.4: Four smart agent models. Smart agents provide a framework that is consistent with the continuum of software maturity in software radios.

model-based reflex agents, goal-based agents, and utility-based agents—defined as follows:

1. A *simple reflex agent* is a simple mapping from current sensor inputs to actuator settings. This is a stateless agent model that neither learns nor adapts to the environment.

2. A *model-based reflex agent* is still a simple mapping but now includes memory of past inputs. The actions are a function of the current sensor inputs and the recent past inputs, making it a finite memory agent model. There is still no learning. Adaptation is limited, but this is the minimum level of sophistication for an adaptable radio.

3. A *goal-based agent* adds to the memory of past inputs; it is a "realistic" model of the environment. Now a sequence of actions may be "tested" against a goal and an appropriate next action may be selected. The level of sophistication for the model of the environment is not well defined. These agents have increased capability of adapting because a prediction about the consequences of an action is available. There is no feedback, and learning is therefore limited. This is the minimal level of sophistication for a CR.

4. A *utility-based agent* maps the state sequence (memory state) to a "happiness" value and therefore includes feedback. The more sophisticated environment model may experiment with sequences of actions for selection of the best next action. This model of a CR has the ability to learn and adapt [35].

A smart agent model for CR is appropriate. The agent framework supports the continuum of radio maturity, and it allows the modular introduction of various AI techniques from fuzzy control to genetic algorithms (GAs). Agents may be tailored to the application's environment. In this sense, environment may be characterized in the following dimensions: fully observable versus partially observable, deterministic versus stochastic, episodic versus sequential, static versus dynamic, discrete versus continuous, and single agent versus multiagent.

The following is an incomplete list of AI techniques likely to find applicability to CR, and further described in subsequent chapters:

- State space models and searching
- Ontological engineering
- Neural networks in CR
- Fuzzy control in CR

- GAs in CR
- Game theory in CR
- Knowledge-based reasoning in CR

Signal Processing

Digital signal processing (DSP) technology enables rapid advances in CRs. Intellectual property resources are widely available for signal processing functions. In GPP (general-purpose processor) or DSP resources, libraries of routines realize functions in efficient assembly language. In FPGA (field programmable gate array) or ASIC resources, licensable cores of signal processing engines are available in very-high-speed integrated circuit (VHSIC) Hardware Design Language (VHDL) or Verilog. Signal processing routines are available for communication signal processing (modulation/demodulation, FEC, equalization, filtering, and others); audio signal processing (voice coding, voice generation, natural language processing); and sensor signal processing (video, seismic, chemical, biometric, and others).

Synthesizing signal processing functions together to form a system is a complex task. A process for algorithm development, test case generation, realization, verification, and validation eases the process of building a waveform or a cognitive system. Integrated tools for system development are available. Many of the tool sets include automatic generation of high-level language source code or hardware definition language code. A bit-level accurate simulation environment is used to develop the system algorithms and to generate test cases for post-integration verification and validation. This environment may be used for CRs to synthesize communications or multimission waveforms that enable a CR to achieve specific goals.

Networking Protocols

Cooperative groups (a multiagent model) have the potential to increase capabilities in a variety of ways. For example, a lone CR is limited in its ability to access spectrum, but a pair of CRs can sense the spectrum, identify an unused band, rendezvous there, and communicate. A network of CRs enables other significant increases in capabilities. Software for Mobile Ad hoc Networking (MANET), although maturing slowly, is a key enabling technology.

The medium access control (MAC) layer is critical in CR networks. If the CR is employing advanced spectrum access techniques, a robust MAC that mitigates the hidden node problem is needed. In a "static spectrum access" environment, a more traditional MAC is possible. A carrier sense collision detection (802.11 MAC) mode is not possible because a radio cannot receive in the exact same band in which it is transmitting, so a carrier sense collision avoidance approach is frequently used. Request-to-send (RTS) and clear-to-send (CTS) messaging are popular wireless MACs amenable to MANETs. Other approaches include TDMA or CDMA MACs.

The architecture for routing packets is important for performance in MANETs. The approaches are generally divided into proactive and reactive algorithms. In a proactive-routing environment, routing data, often in the form of a routing table, are maintained so that a node has a good idea of where to send a packet to advance it toward its final destination, and a node may know with great confidence how to route a packet even if one is not ready to go. Maintaining this knowledge across a MANET requires resources. If the connection links are very dynamic or the mobility of the nodes causes rapid handoff from one "network" to another, then the overhead to maintain the routing state may be high. In contrast, reactive-routing approaches broadcast a short search packet that locates one or more routes to the destination and returns that path to the source node. Then the information packet is sent to the destination on that discovered route. This causes overhead in the form of search packets. Proactive and reactive routing both have pros and cons associated with their performance measures, such as reliability, latency, overhead required, and so on. A hybrid approach is often best to provide scalability with offered network load.

An interesting application of CR is the ability to learn how to network with other CRs and adapt behavior to achieve some QoS goal such as data rate below some BER bound, bounded latency, limited jitter, and so forth. Various cognitive-level control algorithms may be employed to achieve these results. As an example, a fixed-length control word may be used to parameterize a communications waveform with frequency, FEC, modulation, and other measures. The deployment of a parameterized waveform may be controlled and adapted by using a generic algorithm and various QoS measures to retain or discard a generated waveform.

Software Communications Architecture

The primary motivations for SDR technology are lower life cycle costs and increased interoperability. The basic hardware for SDR is more expensive than for a single-point radio system, but a single piece of hardware performs as many radios. The single piece of hardware requires only one logistics tail for service, training, replacement parts, and so on. One of the driving costs in SDR development is that of software development. The JTRS acquisitions are controlling these costs by ensuring software reuse. The approach for reuse is based on the Software Communications Architecture (SCA), which is a set of standards that describes the software environment. It is currently in release 2.2.1. Software written to be SCA compliant is more easily ported from one JTRS radio to another. The waveforms are maintained in a JTRS Joint Program Office (JPO) library.

A CR can be implemented under the SCA standards. Applications that raise the radio to the level of a CR can be integrated in a standard way. It is expected that DARPA's XG program will provide a CR application for policy-driven, dynamic spectrum access on a non-interference basis that executes on JTRS radios. XG is the front-runner in the race to provide the first military CR.

4.5.7 Spectrum Awareness and Potential for Sublease or Borrow

The Spectrum Policy Task Force (SPTF) recommends that license holders in exclusive management policy bands be allowed to sublease their spectrum. Figure 4.5 shows a sequence diagram for spectrum subleasing from a public safety spectrum owner. During the initial contact between the service provider and the public safety spectrum manager, authentication is required. This ensures that spectrum use will be accomplished according to acceptable behaviors and that the bill will be paid [36].

For a subleasing capability to exist in a public safety band a shut-down-on-command function must be supported with a bounded response time. There are three approaches to this: continuous spectrum granting beacon, time-based granting of spectrum, and RTS–CTS–inhibit send. Figure 4.5 shows the time-based granting of spectrum, described as a periodic handshake confirming sublease.

Even though public safety organizations may not sublease spectrum, other organizations may choose to do so. Subleasing has the benefit of producing an income stream from only managing the resource. Given proper behavior by lessees and lessors, the system may become popular and open up the spectrum for greater utilization.

4.6 Funding and Research in CRs

DARPA is funding a number of cognitive science applications, including: the XG Program, Adaptive Cognition-Enhanced Radio Teams (ACERT), Disruption Tolerant Networking (DTN), Architectures for Cognitive Information Processing (ACIP), Real World Reasoning (REAL), and DAML. DARPA research dollars

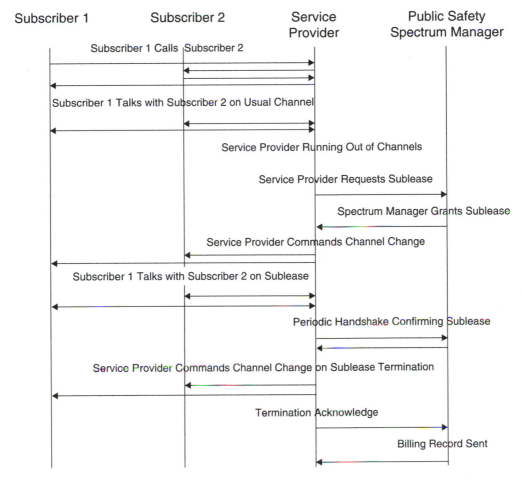

Figure 4.5: Spectrum subleasing sequence. One motivation for CR is the potential income stream derived from subleasing idle spectrum on a non-interference basis.

under contract are easily in the tens of millions. Good results have been achieved in many of these efforts.

The National Science Foundation (NSF) is also funding cognitive science applications including grants to the Virginia Polytechnic Institute and State University (Virginia Tech or VT). Additionally, NSF has sponsored Information Theory and Computer Science Interface workshops that communicate CR research results.

The SDR Forum has a CR Working Group that is investigating various CR technologies such as spectrum access techniques, and a Cognitive Applications Special Interest Group that is working with regulatory bodies, spectrum owners, and users to communicate the potential of CR technologies. Numerous organizations participate at SDR Forum meetings. It is expected that the SDR Forum will solicit inputs from industry through a Request for Information (RFI) process in the near future.

Both the Federal Communications Commission (FCC) and the National Telecommunications and Information Administration (NTIA) have interest in CR. The FCC has solicited various comments on rule changes and an SPTF report. NTIA has been involved in the discussions as they relate to government use of spectrum.

4.6.1 Cognitive Geolocation Applications

If a CR knows where it is in the world, myriad applications become possible. The following is a short set of examples. Figure 4.6 shows a use case level context diagram for a CR with geolocation knowledge.

Figure 4.6: CR use case for geolocation. Several interactions between the CR and the world are enabled or required by geolocation information.

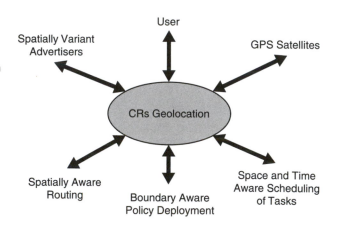

Establishing the location of a CR enables many new functions. A cognitive engine for learning and adapting can utilize some of the new functions. There are multiple methods for determining location. For example, GPS receiver technology is small and inexpensive. Given an appropriate API, numerous applications including network localization (discussed next) and boundary awareness are available to the cognitive engine in the CR. Other methods for determining location are discussed in Chapter 8.

Network localization is a term to describe position aware networking enhancement. For example, if a radio knows it is in a car, that it is morning rush hour, and that the path being taken is the same as that of the last 200 days of commuting, it can predict being in the vicinity of the owner's office wireless local area network (LAN) within a certain time period. The radio may wait until it is in the office to download e-mail and upload the pictures of the accident taken a few minutes ago. This is an example of the highest level of management of radio functions (assuming the radio is used for e-mail, photos, etc.).

Spatial awareness may be used for energy savings. In a multiple short hop-routing algorithm with power management just closing the wireless link with multiple short hops is usually more energy efficient than one long hop. Knowing the position of each node in a potential route allows the CR to take energy consumption into consideration when routing a packet.

Spatially variant searching is a powerful concept for increasing a user's operating efficiency. If it is time for supper, the CR may begin a search for the types of restaurants the user frequents and sort them by distance and popularity. Other spatially variant searches are possible.

A radio aware of boundaries may be able to invoke policy as a function of geopolitical region. When passing from one regulatory jurisdiction to another, the rules change. The radio can adopt a conservative operation mode near the boundaries and change when they are crossed. The radio must have the ability to distinguish one region from another. This may require a database (which must be kept up-to-date) or some network connectivity with a boundary server. Figure 4.7 shows a simplified sequence diagram in which a CR accesses spectrum as a function of a spatially variant regulatory policy.

A sequence of position estimates may be used to estimate velocity. Take a scenario in which teenagers' cell phones would tattle to their parents when their speed exceeds a certain speed. For example, suppose tattle mode is set. The radio is moving at 45 mph at 7:30 a.m. The radio calls the parent and asks a couple of questions such as: "Is 45 mph okay at this time? Is time relevant?" The questions will be in the same order each time and the parent won't have to wait for the whole question, just the velocity being reported. Then the parent keys in a response. After a few reports, the radio will develop a threshold as a function of time. For example, during the time the radio is heading for school, 45 mph is okay. During the lunch break (assuming a closed campus), 15 mph might be the threshold. An initial profile may be programmed, or the profile may be learned through tattling and feedback to the reports. Vehicle position and velocity might also be useful after curfew.

The CR application uses special hardware or customized waveforms that return geolocation information. This information is used to access databases of policies or resources to make better decisions. Dynamic exchange of information

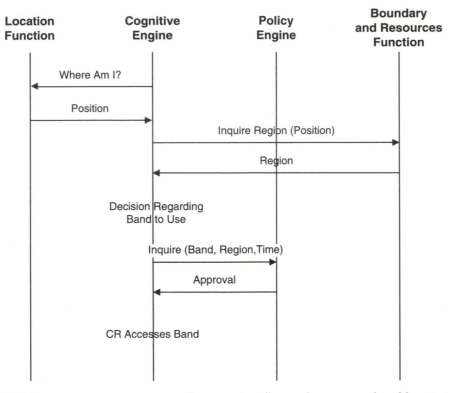

Figure 4.7: Spectrum access sequence diagram. A policy engine uses regional inputs to select spatially variant policies to approve or disapprove a requested spectrum access.

may be used for other networking actions. The set of CR applications that are enabled by geolocation capability is large and has many attractive benefits.

4.6.2 Dynamic Spectrum Access and Spectrum Awareness

One of the most common capabilities of CRs is the ability to intelligently utilize available spectrum based on awareness of actual activity. Current conservative spectrum management methods (static spectrum assignments) are limited because they reduce spatial reuse, preclude opportunistic utilization, and delay wireless communication network deployment. Without the need to statically allocate spectrum for each use, however, networks can be deployed more rapidly. A CR with spectrum sensing capability and cooperative opportunistic frequency selection is an enabling technology for faster deployment and increased spatial reuse.

Spectrum access is primarily limited by regulatory constraints. Recent measurements show that spectrum occupancy is low when examined as a function of frequency, time, and space [36]. CRs may sense the local spectrum utilization either through a dedicated sensor or by using a configured SDR receiver channel. Uses of this information may create increased spectrum access opportunities.

One of the primary considerations for such a cognitive application is non-interference with other spectral uses. Figure 4.8 shows local spectrum awareness and utilization. If the regulatory body is allowing CRs to utilize the unoccupied "white space," increased spectral access can be achieved. The CR can examine the signals and may extract detailed information regarding use. By estimating the other uses and monitoring for interference, two CRs may rendezvous at an unoccupied "channel" and communicate.

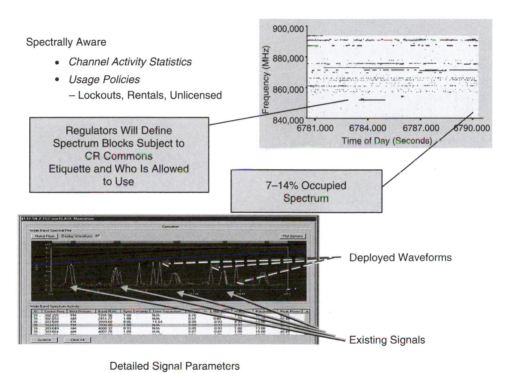

Figure 4.8: Spectrum awareness. A CR, or a set of CRs, may be aware of the spectrum and may exploit unoccupied spectrum for its own purposes.

Sophisticated waveforms that have the ability to periodically stop transmitting and listen for legacy users are called for in this application, as well as waveforms that can adapt their spectral shape over time. Dynamic selection of channels to utilize or vacate is important. Simulations of these cooperating CRs already exist, and additional field demonstrations are expected in the near future. Another advantageous waveform characteristic is discontinuous spectrum occupancy. This allows a wideband communication system to aggregate available spectrum between other existing signals. Careful analysis is needed to ensure that sufficient guard bands are utilized.

Figure 4.9 shows five suggested alternatives for utilizing spectrum in and around legacy signals. The characteristics of the legacy signals may be provided to the cooperating CRs by federated sensors. An alternative method for characterizing the legacy signals is time division sharing of a channel as a sensor and providing a look-through capability by duty-cycling transmit and monitor functions. The five methods shown in Figure 4.9 avoid the legacy signal in various ways.

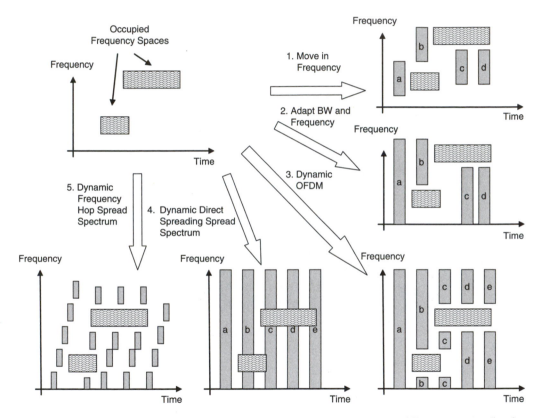

Figure 4.9: Non-interference methods for dynamic spectrum access. Different strategies for deploying non-interfering waveforms have been proposed.

The first method assumes a fixed bandwidth waveform: the center frequency may be adapted. The second method assumes a variable bandwidth signal, such as a direct-sequence spread-spectrum waveform, where the chip rate is adapted and the center frequency is adapted. The third method uses a water-filling method to populate a subset of the carriers in an OFDM waveform. These three methods impact legacy signals very little if appropriate guard bands are observed. The fourth method is a direct-sequence spread-spectrum waveform that underlies the legacy signals. The interference from this underlay must be very small so that legacy systems do not experience noise; thus the processing gain of the spread spectrum underlay must be very high. The last method shown also avoids the legacy signals by frequency hopping into only unoccupied channels.

The spread-spectrum method deserves some elaboration at this point. Legacy receivers will perceive the spread-spectrum signal as an increase in the noise floor, which may result in a decrease in the link margin of the legacy signal. The spread-spectrum receiver will perceive the legacy waveform(s) as a narrowband interference. The de-spreading of the desired signal will cause the narrowband signals to spread. This spreading will cause the narrowband signals to appear at a signal level reduced in power by the spreading gain, and thus appear as noise to the CR, resulting in reduced link margin. The ability of each of these communication systems to tolerate this reduced link margin is link specific and therefore a subject of great concern to the legacy system operators.

An OFDM waveform, the third method in Figure 4.9, has several benefits including flat fading subchannels, elimination of equalizer due to long symbol time and guard intervals, the ability to occupy a variety of bandwidths that "fit to the available opportunity," and the ability to null subcarriers to mitigate interference. Variable bit loading enables pre-nulling and dynamic nulling. Table 4.5 compares several methods for dynamic bit loading an OFDM waveform. Not loading subcarriers occupied by legacy signals with guard bands around them minimizes interference between CR and non-CR systems [37–39].

Because spectrum utilization is a spatially and temporally variant phenomenon, it requires repeated monitoring and needs cooperative, distributed coordination. The familiar hidden node and exposed node problems have to be considered. Figure 4.10 shows a context diagram in which external sensor reports are made available to the CR and may be considered when selecting unoccupied bands.

Figure 4.11 shows a sequence diagram in which a set of CRs is exchanging sensor reports and is learning about local spectrum occupancy. At some time, a pair of CRs wishes to communicate and rendezvous at a band for that purpose. When a legacy signal is detected, the pair of CRs must vacate that band and relocate to another.

The sensor technology utilized for spectrum awareness should be of high quality to mitigate the hidden node problem. For example, if a CR wants to use a TV band, its sensor should be significantly more sensitive than a TV set so that if it detects a TV signal, it will not interfere with local TV set reception, and if it does

Table 4.5: Comparison of variable bit loading algorithms.

Method	*Characteristic*	*Complexity*
Water filling	Original approach, optimal, frequently used for comparison	$O(N^2)$
Hughes-Hartogs [37]	Optimal, loads bits serially based on subcarrier energy level, slow to converge, repeated sorts	$O(SN^2)$
Chow [38]	Suboptimal, rounds to integer rates using signal-to-noise gap approximations, some sorting required	$O(N \log N + N \log S)$
Lagrange (unconstrained) Krongold [39]	Optimal, computationally efficient, efficient table lookup, Lagrange multiplier with bisection search, integer bit loading, power allocation, fast convergence	$O(N \log N)$, revised $O(N)$

Note: N: number of subcarriers; S: number of bits per subcarrier.

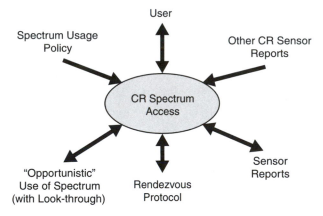

Figure 4.10: Spectrum access context diagram. A spatially diverse sensing protocol is required to mitigate such problems as hidden node or deeply fading RF channels for CR accessto spectrum on a non-interference basis.

not detect a signal, there is a high probability that no TV set is near enough to demodulate a TV signal on that channel.

Dynamic spectrum access as a result of learning about the spectrum occupancy is a strong candidate for a CR application. Numerous discussions in regulatory organizations have involved whether to allow this behavior. If implemented

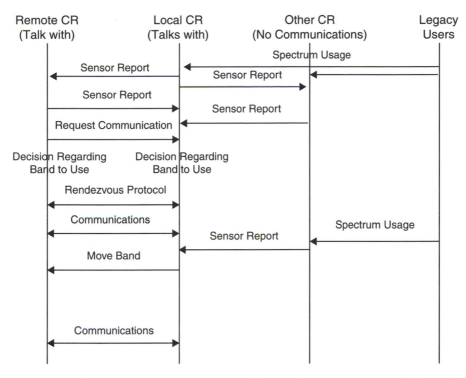

Figure 4.11: Dynamic spectrum access sequence diagram (assumes a control channel). A pair of CR communicating on an unoccupied channel must vacate when interference potential is detected.

correctly, it is a win–win situation in which more spectrum is utilized and very little additional interference is suffered.

4.6.3 The Rendezvous Problem

The difficulty for the CR network is the radios locating each other and getting the network started. Before any transmission can occur, all radios must survey the spectrum to determine where the available "holes" (frequency reuse opportunities) are located. However, each receiver–sensor will perceive the spectrum slightly differently because each will see different objects shadowing different transmitters, and will either see each transmitter with a different signal strength or will not see some transmitters that other nodes are able to see. So while node A may see an open frequency, node B may consider that frequency to be in use by a distant transmitter node or network L. To get the network started, the CRs must agree on a protocol to find each other.

There are several possible methods by which to do this. These methods depend strongly on whether there is an infrastructure in place to help start up a CR network, or whether no infrastructure can be assumed.

Infrastructure-Aided Rendezvous

If we assume that there is an infrastructure component, we must also assume it behaves just like the CR and does not interfere with legacy systems. We assume, however, that it periodically transmits a beacon signal, and we assume that this beacon includes reference time, next frequency-hop(s), and a description of frequencies in use within the local region. Furthermore, we assume that the infrastructure beacon is followed by an interval in which CRs request an available time-frequency slot and associate a net name to that slot, as well as to their location and transmit power, followed by a response from the infrastructure recommending which frequency to use and when to check back. Subsequent requests by other net members can be directed to the proper time-frequency slot, and the geographic distribution of net members can be tracked, allowing the infrastructure to assess interference profiles.

Unaided Rendezvous

Defense systems are rarely able to assume support infrastructure. Similarly, early deployments of commercial CR equipment will not be able to assume infrastructure. Consequently, it is important to have an unaided method for rendezvous. Several methods exist for systems to find each other.

The problem is somewhat like two men searching for each other when they are a mile apart on a moonless dark night out in the desert. Each has a flashlight the other may see, but only when it is pointed in the right direction. They can finally signal each other, but only when each has noticed the other's flashlight, so each must look in the proper direction at the proper time.

In the case of CRs trying to find each other, one must transmit and the other must receive in the proper frequency "hole" at the proper time. Several methods are feasible. All involve one node transmitting "probe" signals (a uniquely distinguishing waveform that can be readily correlated) until receiving a response from a net member. In all cases, the problem is that the frequencies the transmitter sees as being "usable holes" are different from those the receiver sees. Thus, neither can be stationary and just sit on one frequency hoping the other will find it:

Case 1: Node A transmits probes in randomly selected frequency holes, while
node B listens to random frequencies considered to be unoccupied. Preferably,

node B listens about three times as long as a probe pulse. The search time can be dramatically reduced if node B is able to listen to many simultaneous frequencies (as it might with an OFDM receiver). Node B responds with a probe response when node A is detected.

Case 2: Node A selects the five largest unoccupied frequency blocks, and rotates probe pulses in each block. Similarly, node B scans all of the unoccupied frequency blocks that it perceives, prioritized by the size of the unoccupied frequency block, and with a longer dwell time.

Both of these cases are similar, but minor differences in first acquisition and robustness may be significant.

After node B hears a probe and responds with a probe acknowledge, the nodes will exchange each node's perception of locally available frequencies. Each node will "logical OR" active frequencies, leaving a frequency list that neither node believes is in use. Then node A will propose which frequency block to use for traffic for the next time slot, at what data rate, and in what waveform.

From this point forward, the nodes will have sufficient connectivity to track changes in spectral activity and to stay synchronized for which frequencies to use next.

4.6.4 CR Authentication Applications

A CR can learn the identity of its user(s). Authentication applications can prevent unauthorized people from using the CR or the network functions available to the CR. This enhanced security may be exploited by the military for classified communications or by commercial vendors for fraud prevention.

Because many radios are usually used for voice communications, a microphone often exists in the system. The captured signal is encoded with a VoCoder (voice coder) and transmitted. The source radio can authenticate the user (from a copy of the data) and add the known identity to the data stream. At the destination end, decoded voice can be analyzed for the purposes of authentication, and the result may be correlated with the sent identity.

Other sensors may be added to a CR for the purposes of user authentication. A fingerprint scanner in a push-to-talk (PTT) button is not intrusive. Automatic fingerprint correlation software techniques are available and scalable in terms of reliability versus processing load required. Additionally, cell phones have been equipped with digital cameras. This fingerprint sensor coupled with facial recognition software may be used to authenticate a user. Again, the reliability is scalable with processor demands.

Figure 4.12 shows a context diagram for a CR's authentication application. The detailed learning associated with adding a user to an access list through a "third-party" introduction is not shown. The certificate authority could be the third party, or another authorized user could add a new user with some set of authority. The CR will learn and adapt to the new set of users and to the changing biometric measures of a user. For example, if a user gets a cold, his voice may change, but the correlation between the "new voice" and the fingerprint scanner is still strong, and the CR may choose to temporarily update the voice print template.

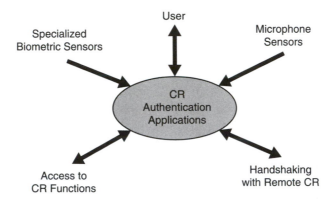

Figure 4.12: Authentication context diagram. A CR application that learns the identity of its user(s) and has an Access List Security Protocol can minimize fraud and maximize secure communications.

4.7 Timeline for CRs

CR development will be a spiral effort. The when, where, how, and who of this development are discussed below:

When: Currently, several CR initiatives are under way. Progress is evident every day, and more and more sophisticated demonstrations are imminent. Some of these demonstrations will include better policy conformance characteristics.

Where: The FCC and NTIA are currently discussing a test band for CR. They are suggesting a 10 MHz experimental chunk of spectrum to allow developers to experiment. This promising development must be exploited. As policy is written for this band, it can be deployed and policy violations may be assessed.

Who: Vendors, regulators, service providers, and users are highly interested in CR systems. A great deal of discussion on exactly what that means has already taken place and continues. Academic researchers using COTS SDR demonstrations and government-sponsored demonstrations using custom-developed SDRs will reduce CR to practice, for at least the first systems. Where the technology will progress is difficult to predict. As an example, the following organizations are working with CR: General Dynamics; Shared Spectrum; Raytheon; Lockheed Martin; Bolt, Beranek, & Newman; Rockwell Collins; Harris; Virginia Tech; and Northeastern University (and the list has doubled during the preparation of this book).

How: The easiest experiments utilize COTS SDR hardware and execute cognitive applications as demonstrations. Custom-developed hardware is more expensive, but is better tailored to show the benefits of CR. Spectrum awareness using a spectrum sensor has the best information to exploit unoccupied spectrum. This custom capability also has computational resources sized to execute a policy constraint engine.

4.7.1 Decisions, Directions, and Standards

Numerous organizations and standards bodies are working in the area of CR. The SDR Forum, the Institute of Electrical and Electronics Engineers (IEEE), the Federal Communications Commission (FCC), the National Telecommunications and Information Administration (NTIA), and the International Telecommunication Union (ITU) all have interests in this area.

4.7.2 Manufacture of New Products

Many products have new sensors and actuators in them. Cellular telephone handsets are a high-volume, highly competitive product area, from which innovation is driven by these characteristics. In fact, a large fraction of new cell phones have an integrated digital camera. These are manually operated today, but CR applications may take advantage of this sensor for more "cognitive" operation modes.

Chemical sensors have been integrated to some cell phone models. The purpose of the chemical sensors is to report on important "problems" such as "bad breath" and "blood alcohol level." This is a manually operated device, but future applications may be more autonomous.

Among the new applications in cellular telephones is a Bluetooth-like waveform to introduce single people. "Flirting radios" may subsequently need to have AI technology added to filter out the "undesirable" introductions.

Much to the dismay of teenagers and employees everywhere, phone tracking applications are now available. Although these capabilities are used "manually" now, learning algorithms can be applied to the interface to create a new "filtered" report stream. There is serious interest in tracking other sets of people, such as first responders, delivery people, service people, or doctors.

4.8 Summary and Conclusions

Radio evolution has taken the path toward more digital realizations and more software capabilities. The original introduction of software made possible software capable radios that communicate and process signals digitally. In the pursuit of flexibility, software programmable radios, and the even more flexible SDR, have become the standard in the military arena and are starting to gain favor in the commercial world, as explained in Section 4.2. We are now seeing the emergence of aware radios, adaptive radios, and finally CRs, and we have traced the continuum among these various degrees of capability, as well as providing a few examples of each. Sections 4.3 and 4.4 explored the properties and capabilities of each of these classes of radios.

Section 4.5 outlined the enabling technologies for CRs. Numerous technologies have matured to the point where CR applications are possible and even attractive. The ability to geolocate a system, sense the spectrum, know the time and date, sense biometric characteristics of people, access new software capabilities, and determine new regulatory environments are all working together to enable CR.

Geolocation through the use of GPS or other methods is now available. Chapter 8 discusses how a radio can know where it is located. This enabling technology allows a radio to make spatially variant decisions, which may be applied to the selection of policy or networking functions.

Sensing of the local RF environment is available. This information may be used to mitigate a deeply fading channel or may be used to access locally unoccupied spectrum. Non-interference is particularly important, and protocols for sensing, deciding, and accessing spectrum are being designed, developed, and demonstrated today.

Increased robustness in biometric sensor technology provides a whole new dimension to CR applications. The most likely initial use of this technology is in user authentication applications, such as the purchasing of services and products.

Knowledge of time has been available in many forms, but integration into a broader range of full-function capabilities will enable all new applications. Stable time knowledge enables a CR to plan and execute with more precision. Using this capability for non-infrastructure-based geolocation, dynamic spectrum access, or AI planning is envisioned for near-term CR functions.

A smart agent model of CRs is attractive. An agent is an entity that perceives and acts on behalf of another. This is where CRs are going. Smart agent models of the world enable a radio to provide services for its user. Improved performance or new capabilities may be provided. As the CR's smart agent model of the world becomes more sophisticated and realistic, situational awareness will increase. As the models improve, the ability of the CR to act effectively over a broader range of user services will improve.

Maybe the most important software technology is a policy engine that enables a CR to know how to behave where it is right now, given the priorities of current circumstances. AI applications at a very high level, networking services in the middle levels, and signal processing primitives at a very low level are all available for a CR developer to utilize in creating new capabilities. Finally, middleware technology enables greater software reuse, which makes CR development economical.

Modern regulatory philosophy is starting to allow CR to deploy new services and capabilities. As the trend continues, there are economic motivations for deploying CR systems.

Section 4.6 covered research in CR technologies, and presented three significant classes of CR applications. Geolocation-enabled applications and authentication applications were discussed in some detail. The most promising CR application is dynamic spectrum access. Suggestions for using OFDM waveforms along with dynamic bit loading are included in this chapter. Solutions to the rendezvous problem are suggested, and the hidden node problem is described.

Section 4.7 covered the timeline in which these technologies will roll out and be integrated into radio equipment and products. Many of the technologies required to provide some of the useful and economically important CR functions already exist, so some of these features should begin to appear within the timeline of the next development cycle.

The bottom line is that the enabling technology for CR applications is available. There is interest in integrating the technologies to build cognitive applications. Finally, the emergence of CRs and their cognitive applications is imminent.

References

[1] Federal Communications Commission, Notice of Proposed Rule Making, August 12, 2000.
[2] http://www.fas.org/man/dod-101/sys/land/sincgars.htm
[3] http://www.fas.org/man/dod-101/sys/land/.htm

[4]   http://www.fas.org/spp/military/program/com/an-wsc-3.htm

[5]   http://www.fas.org/man/dod-101/sys/ac/equip/an-arc-164.htm

[6]   http://www.fas.org/man/dod-101/sys/ac/equip/an-arc-220.htm

[7]   http://www.harris.com

[8]   http://www.columbiaelectronics.com/motorola_LST_5b___LST_bc.htm

[9]   http://army-technology.com

[10] http://www2.thalescomminc.com

[11] http://www.fas.org/man/dod-101/sys/ac/equip/csel.htm

[12] www.jcs.mil/j6/cceb/jtidsmidswgnotebookjune2005.pdf

[13] http://www.fas.org/man/dod-101/sys/ac/equip/an-arc-210.htm

[14] http://www.afcea.org/signal

[15] http://www.fas.org/man/dod-101/sys/land/eplrs.htm

[16] http://www.nj7p.org/history/portable/html

[17] http://raytheon.com/products/jtt_cibs/

[18] http://www.acd.itt.com/case4.htm

[19] http://jtrs.army.mil/

[20] http://jtrs.army.mil/

[21] http://www2.thalescomminc.com

[22] www.comsoc.org/tech_focus/pdfs/rf/06.pdf

[23] http://www.gdc4s.com/

[24] http://jtrs.army.mil/

[25] http://www.gnu.org/software/gnuradio

[26] http://vanu.com/technology/softwareradio.html

[27] J. Polson, "Cognitive Radio Applications in Software Defined Radio," in *Software Defined Radio Forum Technical Conference and Product Exposition*, Phoenix, AZ, November 15–18, 2004.

[28] H. Chen, S. Kumar and C.-C. Jay Kuo, "QoS-Aware Radio Resource Management Scheme for CDMA Cellular Networks Based on Dynamic Interference Guard Margin (IGM)," *Computer Networks*, Vol. 46, 2004, pp. 867–879.

[29] B. Bougard, S. Pollin, G. Lenoir, L. Van der Perre, F. Catthoor and W. Dehaene, "Energy-Aware Radio Link Control for OFDM-Based WLAN." Available at http://www.homes.esat.kuleuven.be/~bbougard/Papers/sips04-1.pdf.

[30] http://www.darpa.mil

[31] P.H. Dana, *The Geographer's Craft Project*, Department of Geography, University of Colorado, Boulder, 1996.

[32] E. Azzouz and A. Nandi, *Automatic Modulation Recognition of Communication Signals*, Springer, New York, 1996.

[33] S. Harris, *CISSP All-in-One Exam Guide*, 2nd Edition, McGraw-Hill, New York, 2002.

[34] http://www.bbn.com

[35] S.J. Russell and P. Norvig, *Artificial Intelligence: A Modern Approach*, 2nd Edition, Pearson Education, London, 2003.

[36] Federal Communications Commission. *Spectrum Policy Task Force Report*. ET Docket No. 02-135, November 2002.

[37] Hughes-Hartogs, "Ensemble Modem Structure for Imperfect Transmission Media," United States Patent 4,679,227, July 7, 1987.

[38] P. Chow, J. Cioffi and J. Bingham, "A Practical Discrete Multitone Transceiver Loading Algorithm for Data Transmission Over Spectrally Shaped Channels," *Transactions on Communications*, Vol. 43, February/March/April (2/3/4), 1995, 773–775.

[39] B. Krongold, K. Ramchandran and D. Jones, *Computationally Efficient Optimal Power Allocation Algorithms for Multicarrier Communications Systems*, University of Illinois, Urbana–Champaign.

CHAPTER 5

Spectrum Awareness

Preston Marshall
Defense Advanced Research Projects Agency
US Department of Defense Arlington, VA, USA

Spectrum is the "lifeblood" of RF communications.

5.1 Introduction

The wireless designer's adaptation of the classic New England weather observation could be "Everyone complains about spectrum availability (or at least the lack of it), but no one does anything about it!" Cognitive radio, however, offers the opportunity to do something about it. Spectrum aware radios offer the opportunity to fundamentally change how we manage interference, and thus transit the allocation and utilization of spectrum from a command and control structure—dominated by decade-long planning cycles, assumptions of exclusive use, conservative worst-case analysis, and litigious regulatory proceedings—to one that is embedded within the radios, each of which individually and collectively, implicitly or explicitly, cooperates to optimize the ability of the spectrum to meet the needs of all the using devices. As this chapter looks at this opportunity, it investigates solutions that range from local brokers that "deal out" spectrum, to totally autonomous systems that operate completely independently of any other structures. In its ultimate incarnation, it is possible to actually use spectrum awareness and adaptation to relax the hardware physical (PHY) layer performance requirements by avoiding particularly stressing spectrum situations. As such, a cognitive radio could ultimately be of lower cost than a less intelligent, and more performance stressed, conventional one.

Note: Approved for public release, distribution unlimited.

5.2 The Interference Avoidance Problem

Before discussing a cognitive spectrum process, let us consider the classical spectrum management and assignment case. Once radios moved beyond spark gap techniques (the original impulsive ultra-wideband (UWB) radio), use of the spectrum has been deconflicted to avoid interference. Spectrum and frequency managers assign discrete frequencies to individual radios or networks and attempt to ensure that the emissions from one do not adversely impact others. A not insignificant legal (and seemingly smaller technical) community has grown up around this simple principle. Such planners are inherently disadvantaged by a number of factors. For one, they have to assume that:

- Interfering signals will propagate to the maximum possible range.
- Desired signals will be received without unacceptable link margin degradation.

In practice, this means that interference analysis is often driven by two unlikely conditions: maximal propagation of interfering signals and minimal propagation of the desired signal. Simplistically, if propagation was a simple R^2 condition (from a tower to a close-by remote), and the receiver needed 12 dB signal-to-noise, then we could consider that the interference would extend for a distance of four times the maximum range of the link, at which point the received signal power would be equal to the noise power.[1] Thus, the interference would be an area 16 times the usable coverage area, as shown in case 1 in Figure 5.1. Operation close to the ground would have an $R^{3.8}$ propagation, and that ratio

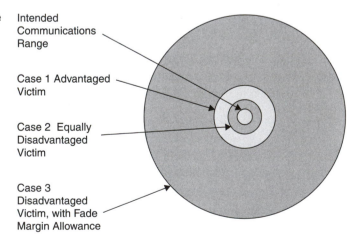

Figure 5.1: Practical interference margins. Several cases of conservative assumptions about propagation to intended and unintended receivers are illustrated.

Intended Communications Range

Case 1 Advantaged Victim

Case 2 Equally Disadvantaged Victim

Case 3 Disadvantaged Victim, with Fade Margin Allowance

---

[1] For an explanation of propagation energy loss, refer to the Appendix at the end of this chapter.

would be reduced to about four, as shown in case 2, which is slightly less stressing. However, when we consider that one of the receivers or transmitters may be at an advantaged position and one at a disadvantaged position (and thus has losses approaching $R^{3.8}$), and when we must include a multipath fading allowance on the desired link, the relative gap opens up immensely. This is shown in case 3. If we assume a worst-case fade margin (a possibility of 20 dB loss to the disadvantaged intended receiver operating in an $R^{3.8}$ regime, while having other unintended receivers operating at R^2 in the same region), it increases the area of interference to more than 150 times the area of reliable communications. Figure 5.1 illustrates the impact of ever-increasing conservative assumptions regarding signal propagation, and imputed spectral usage for conservative spectrum planning situations.

When we add mobility to this analysis, the situation gets even more restrictive. In this case, the conventional frequency manager must deconflict the entire range of motion of the emitters and receivers. All this greatly consumes effective spectrum, increasing the already low (only several percent average utilization) values of the pessimistic and static case, to add the considerations of pessimistic assumptions about possible mobility.

It is important to recognize that this pessimism is not inherent in the operation of the radio links; statistically, it is important because it represents a set of cases wherein many radio systems would have no ability to operate, because without adaptation, even if the radio recognizes that an interference condition exists, it cannot implement and coordinate a strategy for migrating to a clear channel. In this chapter, we consider spectrum strategies that use awareness to locate spectrum holes, often the result of the essential conservative nature of the planning process. However, an equally important rationale for the inclusion of awareness in real systems is the ability to locally resolve interference by using the same behaviors to locate new, and unblocked, spectrum. This feature of interference adaptive radios offers all users of the spectrum the ability to migrate from the currently conservative assumptions that underlie spectrum planning.

5.3 Cognitive Radio Role

We postulate that cognitive radio offers the ability to manage this situation more effectively by utilizing the ability to sense the actual propagation conditions that occur, and to adjust the radio dynamically to best fit these conditions. To do this, we distinguish between two radio operating objectives. In the first, the radio attempts to minimize its own spectral "footprint," consistent with the environment and needs of the networks it supports. In the second, it adapts itself to fit within whatever spectrum is available, based on local spectral analysis. When we put these two together, we can conceive of a radio that can find holes, and then morph its emissions to fit within one or more of these holes. Such radios could offer radio services without any explicit assignment of spectrum, and still be capable of providing high-confidence services.

5.4 Spectral Footprint Minimization

There are two sides to the cognitive radio problem. The first is fitting into the spectral footprint of other radios. The second, and more subtle, is to minimize the radio's own footprint, as discussed here.

For years, modulation designers have defined spectrum efficiency as the number of bits per hertz (Hz) of bandwidth, and have often used this metric as a scalar measure of the best modulation. The assumption has been that the design that utilized the least spectrum was intrinsically less consumptive of shared spectrum resources. This proposition is worth examining. The Shannon bound argues that essentially infinite bits per Hz can be achieved, but it can be accomplished only by increasing the energy per bit (Eb) exponentially, and thus the radiated spectral energy increases at the third power of the spectral information density.[2] We broaden our view of spectrum impact to include not only the amount of spectrum used, but also the area over which it propagates.[3]

One problem involves how to define spectral efficiency metrics. Classically, digital radio engineers have attempted to minimize the spectrum used by signals through maximization of bits per Hz. This is a simple and readily measured metric, but is it right to apply this to a new generation of radios? The Shannon bound shows a basic relationship between energy and the maximum possible bits per Hz, as shown in Figure 5.2.

An immediate observation is that although the bits per Hz increase linearly, the energy required per bit goes up exponentially, as energy increases an average of 2 dB per bit per Hz over the entire range of spectrum efficiencies. We can see that the proportionate cost in energy of going from one bit to two bits per Hz is essentially the same as that required to go from six to seven. Essentially arbitrary bits per Hz are possible if the channel is sufficiently stable and power is available. But this is tough on battery-powered devices!

---

[2] The Shannon bound (or limit) is the best that can be done at this time.

[3] It is worth noting that volume would be a more generalized measure, but most spectrum usage is on the surface of Earth, so we will limit our consideration to two, rather than three, dimensions.

Figure 5.2: Shannon EB for various spectral efficiency values.

Figure 5.3: Impact of bits per Hz on spectrum footprint.

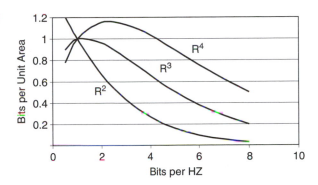

It is equally tough on the other users of the spectrum, in the form of suppressing adjacent channel interference, as well as large dynamic range in the analog functions of the receiver. In order to increase the bits per Hz, the radio must now transmit both slightly more bits, and vastly more Eb, in order to meet the Eb/No requirements of the receiver, where No is noise power.

Using Shannon's limit, we can compute the increase in spectral energy required to increase spectrum efficiency from one-half to eight bits per Hz. If simple spectrum were the measure, such a radio strategy would be effective. However, for most frequencies, spectrum is a valuable asset over a given area. Therefore, bits per unit area is an appropriate measure for how effectively we use spectrum. Extending this bound to consider propagation is important if we want to understand how this denies spectrum usage to the radios. Ignoring multipath and absorption, propagation between terrestrial antennas can be simply modeled as R^{α}, with α varying from 2 to 4, depending on antenna height and frequency. Applying Shannon's limit to this simplistic propagation yields the graph in Figure 5.3. In this case, we compute the change in effective interference region for each point in the Shannon curve, and for cardinal values of the propagation constant. Clearly, simplistic spectrum strategies that only look at bits per Hz are not effective, and in fact are quite counterproductive for all the spectrum users as a whole. The better the propagation, the more ineffective increasing energy is as a strategy for spectrum efficiency.

For most propagation conditions, it is clear that increasing the spectral efficiency of one radio will disproportionately reduce the ability of the spectrum to support other users. The value of α generally increases with frequency (for propagation in very high frequencies (VHF) and above), which implies that more sophisticated strategies are needed. Increasing modulation constellation depth is a poor solution for the radio because it greatly increases the energy it needs, as well as a poor solution for the other radios sharing the spectrum because it essentially raises the noise floor throughout a greatly increased region, or precludes operation at rapidly increasing radii from the transmitter.

This can lead us to a measure that recognizes this trade-off. In this case, bits per Hz per area reflects that the critical issue in spectrum optimization is not spectrum use; rather, it is *spectrum reuse*, and we want to measure and optimize not only how the radios themselves perform, but how the radios allow other radios to share the spectrum in a close to globally optimal manner.

The preceding discussion assumes that the power is perfectly matched to the channel; in practice, such an alignment is impossible to achieve or maintain over any degree of time-varying channel, so the real result will fall below these values. In the extreme case of no power management, the results are insensitive to many of these considerations.

5.5 Creating Spectrum Awareness

One of the key considerations for whether a radio is cognitive is its ability to create awareness of its environment. Certainly, the key environment of a radio, or network, is the spectrum in which it operates.

5.5.1 Spectrum Usage Reporting

The simplest implementation of a cognitive spectrum architecture could be based on principles already established for the upper layers. Each radio would coordinate the use of spectrum via spectrum usage (spectral power density, directionality, location) reporting and distribution. Reporting to local neighbors or to a local spectrum management infrastructure would have to be provided not only by the transmitters, but by the receivers as well. Receiver reporting would enable the sharing scheme to be more aggressive because the system would have to protect only locations in which a possible interference could occur. Without receiver reporting, the interference analysis would have to assume receivers throughout the entire radiated pattern. A collection of all such reports (and their associated locations) would, in theory, provide each member of the network with an exact understanding of the interference it would face and the interference it would cause if it radiated any specific emission. Such architectures could resemble the command and control model of spectrum, if the coordination was provided to a single authority, or "band manager," that would control the dynamic assignment of spectrum.

Alternatively, the control could be more peer to peer, with individual radios computing the effect on other radios as well as on themselves. This approach differs from the current mechanisms not so much by its intrinsic character, but by its automation, time cycle, and "localness." It meets the definition of cognitive in that it makes decisions based on a level of awareness provided by other users of the spectrum. In this model, individual radios are blind. Such a scheme has the advantages cited above if (and only if) the structure can meet the following conditions:

- Fellow radios are assumed to be trustworthy.
- The general characteristics of the propagation environment are well understood, or the bounds on the best possible performance are well understood.
- Reporting requirements are imposed on 100 percent of the emitters (and optimally, the receivers).

The process is certainly appropriate for application in bands in which the usage is homogeneous and controlled by a single entity (or possibly a set of cooperating entities), and in which the antennas in at least one end of the link are generally advantaged, in order to have the bounding equations somewhat correspond to likely propagation.

5.5.2 Spectrum Sensing

There are two current technologies for spectrum sensing: (1) Conventionally, spectrum analyzers have been rapidly tuned through a band, with the scan frequency adjusted to provide adequate "dwell" on each frequency bin, but scanning through all of the channels rapidly enough to be "real time" in terms of the signals of interest. (2) More recently, analog-to-digital conversion (ADC) has made the use of the fast Fourier transform (FFT) practical. This sensing technology has the advantage that it provides instantaneous analysis of all the frequencies within the FFT window. Both technologies have advantages, depending on the application, cost, energy available, and type of resident signal. Table 5.1 illustrates some of the qualitative advantages of each.

Table 5.1: Comparison of spectrum sensing approaches.

Technology	*Spectrum analyzer*	*Wideband FFT*
Complexity	Lower complexity if it can share receiver components	Adds requirement for wideband ADC and extensive DSP
Short signal detection	Low dwell time on each frequency can fail to see short, pulsed signals. Typical dwell is only microseconds to milliseconds, and below 1%	Higher probability of detecting short signals, based on duty cycle of the sensing, which could reach 50%
Bandwidth	Can scan large ranges of signal	Limitations in ADC constrain instantaneous bandwidth
Speed	Slower, based on dwell time on each channel. May be difficult to interleave "listen through" without penalty to the node performance	With appropriate filters, sample sub-Nyquist to minimize time delay and interleaving time. These short intervals can be compatible with MAC layer timing
Power	Mostly classical analog components, potentially shared with the mission receiver	Digital processing adds to the inherent analog energy usage

MAC: medium access control; DSP: digital signal processing.

5.5.3 Potential Interference Analysis

We can consider interference in two categories. The first category is direct interference with ongoing communications of the primary user, resulting in degraded communication functionality. The second category is interference to the channel when it is not in use, causing the primary user to think there is something wrong with the channel or with the equipment.

The first category involves direct interference with the quality of the signal at one or more layers of the receiver, as shown in Table 5.2. In conventional

Table 5.2: Interference effects on digital processing layers.

Layer	*Impact*	*Mitigation*
PHY	Creation of higher uncorrected BER	Allow higher layers to resolve a short interval of interference Adjust coding dynamically
MAC	Complete and uncorrectable blockage of a packet	MAC layer acknowledgment retransmits. Blocked RTS are automatically retried
NETWORK	Complete and uncorrectable blockage of a packet	Protocol operates effectively with missing data. For example, VoIP can suffer some loss as long as it is not correlated
TRANSPORT	Network fails to route packet	Transport layer recognizes missing sequence and requests retransmission

BER: bit error rate; VoIP: Voice over Internet Protocol.

spectrum practice, interference has generally been very quantitatively defined as the ratio of signal energy to the signal plus interference. Although this definition is scientific, the regulatory community has extensive filings from spectrum users demonstrating either extensive additional energy, or minimal energy, often in the same situation. Unprotected receivers have a cascading effect. Very small errors in the lower layers can have disproportionate impacts on the upper layers. For example, a few short bursts of noise may not impact a voice channel, but if it introduced a 10^{-3} error rate in a digital packet communications link, it would essentially deny functionality. Similarly, a packet loss rate of 10^{-1} might have minimal impact on the network layer, but could, if timed badly, essentially shut down a transport layer, such as Transmission Control Protocol (TCP). In contrast, if cognitive radios are the norm, these same layers can assist by reducing the unavoidable temporary interference that may occur at lower layers. Table 5.2 shows examples of how the various layers can magnify interference events or can mitigate the effects. The eventual adoption of interference-tolerant systems creates an environment in which less conservative practices can be applied, because in the rare cases when interference is caused, the victim radio can address this locally without losing its effectiveness.

A second potential source of impact is more subtle. This is the effect of the use of given spectrum on other (and assumed more primary) users. In this case, we contemplate impacts that are short of specific denial of communications, and that may not be measurable at the product of the communications system. For example, if a push-to-talk (PTT) frequency is not in use by any transmitter, and we "borrow" it and thereby cause all of the voice receivers to emit a horrible shrieking sound as the squelch is broken, we have caused interference to another communications network. This problem is unique in that it requires us to consider that the absence of any signal constitutes communications in and of itself, and that any use of the frequency is therefore interfering with the communications. This implicit interference is one that cannot be dealt with in the same way as direct interference.

One way to accommodate the non-quantitative nature of interference is to consider the evolution of bands shared by cognitive radios. In the initial deployment of such systems, existing legacy users will be the dominant users of the bands. These radios were not designed to mitigate the very unnatural effects of cognitive radio sharing, so cognitive radios may require care and extensive measures to ensure minimal impact. (An exception may be military radios, which are designed to have extensive interference tolerance to avoid electronic attack, such as jamming.) As cognitive radios become the majority, a principle such as "5 mile an hour bumpers" could be adopted. In this principle, radios would have some assumed ability to mitigate the effects of interference by using adaptation at all layers so that they could be tolerant of the less than perfect environment that extensive use of cognitive radios may create.

This posits the following cascading set of questions:

- If (using 802.11 as an example) I interfere with a single request-to-send (RTS) message, but you get the next one through, have I interfered with you?
- If I added Gaussian noise, but it was within your power management capability, have I interfered with you?
- If I added some Gaussian noise to your channel, but it was well within your margin and was within the range of your error correction, have I impacted you?

Spectrum owners that paid billions for spectrum would certainly say yes to all of these because it impacted the capability that they could achieve with their investment, but a community of band sharers may consider it just a cost of being in a shared band. Measuring spectrum is fundamentally a technical process, but assessing the spectrum is a much more complex perceptual, legal, and relativistic process.

One of the best educations that can be achieved in the analysis of interference is provided by reading the voluminous filings before national regulatory agencies. Well-qualified, and (we should assume) honest engineers can take the same set of facts, apply well-known engineering principles, and can reach diametrically different conclusions on the simple question of whether a certain emission will cause interference to another!

5.5.4 Link Rendezvous

When an adaptive radio is first turned on, it faces a unique burden: how to find the other radios with which it is intended to communicate. Conventional radios generally have some prior knowledge of frequency, or the frequency is provided by the operator. In the case of a cognitive radio, no one knows what the best frequency is until the environment is sampled. This topic is further discussed in Chapter 8.

5.5.5 Distributed Sensing and Operation

Distributed operations occur when the spectrum awareness problem leaves the domain of signal processing and enters the realm of cognitive processes. There are inherent limitations in the ability of any radio to fully understand its environment. One of the fundamental issues with the idea of radios creating spectral awareness is that individual radios, typically close to the ground, are subject to a wide range of propagation effects, all of which appear to be conspiring to maximally disrupt the radio's ability to create an understanding of the spectral environment. The transition from direct to diffracted communications is a major effect driving uncertainty. Direct propagation follows something like the PHY layers R^2 rule, whereas diffracted (non-direct) propagation attenuates at the rate of approximately the fourth power of distance. The inability of the radio to know which of these propagation conditions is present affects its ability to sense, determine interference, and assess the consequences of its own communications. The propagation of signals involves five parties, including a victim transmitter, victim receiver(s) (of unknown location or even existence), a cognitive transmitter, a cognitive receiver, and a spectrum sensor (often assumed to be, but not necessarily located at the Defense Advanced Research Projects Agency (DARPA) NeXt Generation (XG) transmitter location). A matrix of possible propagation conditions between these nodes can be developed, with the possible values of the matrix characterizing the nature of the propagation between each (R^2, R^4, R^4 plus 20 dB multipath fade, etc.).

At the onset, the cognitive radio has no awareness of how to fill in this matrix, and so must assume worst-case conditions on each link, which might be:

- R^2 propagation from the cognitive transmitter to the victim receiver;
- R^4 plus 20 dB multipath fade from the victim transmitter to the spectrum sensor;
- R^4 plus 20 dB multipath fade from the cognitive transmitter to the cognitive receiver.

Removal of uncertainty can greatly reduce the power that the cognitive transmitter can emit with certainty of non-interference to the victim receiver. In this case, the assumption of multipath cannot be eliminated, but multiple sensors measuring the same emission can greatly reduce the potential range of the assumptions that the interference analysis must address. For that reason, collective sensing, particularly in a mixed direct and diffracted multipath region, may be essential in allowing cognitive radios to emit appropriate energy levels. This implies that additional algorithms will be needed to fuse this sensor data and establish acceptable confidence bounds for the inter-radio propagation assumptions.

5.6 Channel Awareness and Multiple Signals in Space

The implicit assumption of most spectrum-sharing approaches is that a given frequency can be used by only one spectrally efficient signal (narrow bandwidth) at a time. The concept of multiple input, multiple output (MIMO) signaling, first described as BLAST, has been subsequently generalized and is beginning the process of commercial exploitation. In MIMO, multiple signal paths create separable orthogonalized channels between multiple transmit and multiple receive antennas. Although this process at first appears to violate Shannon's bound, in fact it is only sidestepping Shannon because in theory, each of the reflected paths is an independent channel. Essentially, this technique turns multipath into multiple channels [6].

Receivers nearly always benefit from knowledge of the channel. Active equalization, RAKE filters and other techniques exploit this awareness to improve performance at the link layer of the radio. These techniques are invisible to the upper layers. MIMO provides an important distinction from the traditional modem design. In the case of MIMO, we use the channel awareness much more architecturally, in that we design the links, and potentially the topology of the network, to be dependent on this multipath, rather than just tolerant of it. A very large body of published literature describe MIMO more fully [1–6], and therefore this discussion is not elaborated here.

Certainly MIMO can be implemented without cognitive features; however, these implementations are incapable of providing an assured level of performance because the repairability of the channel cannot be known in advance. No amount of processing can create 10 independent channels in environments that have no reflective component. However, when we combine MIMO with other spectrally adaptive techniques, it is now possible to create assured services by balancing the techniques of spectrally adaptive and spatially adaptive techniques.

Consider the reasoning at the radio and network levels. The radio operates best (least energy investment per bit) in a clear channel, without the complexity and induced noise of MIMO techniques. A "greedy" radio would never choose to be a MIMO radio. It is only when the needs of individual radios are balanced with the spectrum usage of other devices that the radio benefits from MIMO. Chapter 11 further discusses this from a network perspective, but even from a radio perspective there are advantages to MIMO when the spectrum available to the radio is capped. As an example, MIMO is attractive in unlicensed bands, such as used by 802.11, because growth in capability must be achieved within a fixed spectral footprint.

Spectrum awareness therefore has two components. If adaptive spectrum techniques are permitted, then the spectrum awareness function provides an assessment of the relative scarcity of spectrum resources. This relative availability measure provides an implicit or explicit metric regarding the cost that the radio should be willing to pay to achieve the spectrum usage reductions from MIMO. The other aspect of spectrum awareness is the measure of path differentiation available from the apertures of the radio. This measure is more specific to the MIMO algorithm, but in general can yield a metric reflecting the costs and benefits of various degrees of MIMO usage.

Cognitive technology is critical if the radio is to apply techniques more complex than simply utilizing all of its resources to the maximum extent in MIMO processing. In theory, MIMO provides a linear growth in throughput as the number of antennas are increased. In practice, the channels between the elements may be less than ideally separated. Figure 5.4 illustrates the assessment of throughput versus channel usage for a spectrum- and power-constrained radio. The upper curve represents the theoretical bound of "perfect MIMO," and the lower curve represents a situational assessment that a radio might make, given its ability to learn the degree of separation. Note that the marginal benefit of channels

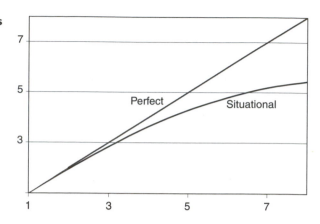

Figure 5.4: MIMO capacity versus transmit/receive resources.

decreases while the resource expenditure increases linearly. A cognitive radio should use its awareness of channel characteristics and spectrum availability to assess that it has a number of options beyond "brute force" to employ its resources.

Taking the data from Figure 5.4, we can create a number of different strategies that a cognitive radio could consider. Each strategy trades some resources (e.g., spectrum, throughput, etc.), but all yield more effective use of resources. (Chapter 11 investigates network layer cognitive approaches that could further mitigate these choices by use of the network selection of alternative topologies.) Table 5.3 shows some obvious choices between MIMO and spectrum usage. An even more complex, and potentially beneficial, extension of this approach would be to consider varying the Eb in order to also adapt the channel rate. Even this simple example shows that an adaptive mix of MIMO and spectrum usage is vastly superior to each approach applied separately.

The same data can be plotted to demonstrate the flexibility available to a radio that can adapt its performance across these two exemplar dimensions, as shown in Figure 5.5.

5.7 Spectrally Aware Networking

The prior discussion treats the PHY layer as a given that must be constructed in essentially a fixed fashion. Certainly, this is the current practice, as we direct the link layer to form certain topologies with explicit, and generally fixed, capacity between each of the modes. In the ignorance of spectrum conditions, this is as good an approach as any. However, when we introduce the capability to assess the costs of each link (e.g., amplifier mode, Eb, etc.), we find that these simplistic notions of the radio's role fail to fully exploit the options that we can now create.

Table 5.3: Alternative approaches for mixed spectral and MIMO adaptation.

	Channels used	*Resources*	*Throughput*	*Throughput/ resource*	*Throughput/ channel*
Pure 8 × 8 MIMO	1	8	5.04	0.63	5.04
6 × 6 MIMO	1	6	4.49	0.75	4.49
2 sets of 3 × 3 MIMO	2	6	5.68	0.95	2.84
2 sets of 4 × 4 MIMO	2	8	7.18	0.90	3.59
4 sets of 2 × 2 MIMO	4	8	7.80	0.98	1.95
8 independent channels	8	8	8.00	1.00	1.00

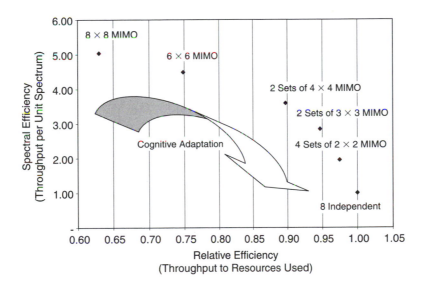

Figure 5.5: Adaptive trades from MIMO and spectrum usage.

These opportunities are somewhat hidden if we approach networking from the perspective of the technologies that have dominated wired networking because in a wired structure the capabilities of each link are intrinsic in the topology. Only in wireless networks can the network dynamically redefine the capability of each link based on the needs of the network and the availability of resources.

It is anticipated that early development of adaptive radios will generally result in "greedy algorithms" that obtain sufficient spectrum for their own needs and are not cooperative with other spectrum users. It is likely that policy-makers would force some spectrum churn in order that the spectrum not be filled with first-in devices that leave no room for new entrants. However, as adaptive devices grow in use, the opportunity for cooperation is increased, and the ability of the devices to cooperatively determine networking strategies that are more globally optimized is presented to the community.

As an example, imagine that we desire to set up a fully adaptive, spectrally aware cognitive network around the Washington, DC, area. Nodes are distributed on the ground in density proportional to the population of the region. If we form the network based on closest neighbor, it would probably tend to route traffic in and out of the densest areas: into and out of the heart of the city. Although topologically convenient, this network may be unsupportable because it concentrates both nodes and routed traffic in the same region. A spectrally aware network might force a quite different organization. This network formation would not route traffic toward the city; it would route out of the city, around the beltway (perimeter or ring road for Washington) and back in. This would reduce the peak spectrum needs in the city and make use of the lower density in the outer regions for high bandwidth "backbone" communications. This architecture is dependent on the development of spectrum sensing, automated organization of individual links, and then the imposition of a global set of objectives for the network that can be implemented by the distributed nodes.

An unanswered question for research is how similar in technology the network adaptation is to the link-level adaptation. Because of the policy implications of spectrum, the link technology has often been driven by the need to comply with policy, thus favoring declarative technologies, such as reasoning engines. Network optimization has been a subject of a greater range of technology. The literature reports a very wide range of techniques, from genetic algorithms, to learning engines, to declarative solvers as candidate technologies [7–9].

5.8 Overlay and Underlay Techniques

The preceding discussion assumes that communication is organized into discrete channels that are orthogonal by frequency or time; that is, only a single signal can occupy a given piece of spectrum at one time. In fact, a number of techniques allow us to violate this assumption. In the simplest case, the use of one of the forms of direct sequence spread spectrum (DSSS) offers the ability to transmit multiple signals on the same frequency. Although a number of signals can share the same spectrum, the coding added to each signal results in, at best, a zero-sum benefit. There are significant reasons to use this approach, but they lie in the enhanced access control, not intrinsic spectrum usage sharing. Furthermore, the coordination required to meet the stressing requirements of signal power management (to avoid near/far interference) clearly puts many of these technologies in the medium access control (MAC) layer as much as in the link layer. This chapter will not further address this technology, although the topic is well developed in the literature [10–14].

Of more interest to cognitive radio are the techniques that offer the ability to have signals coexist in space, time, and frequency, without the introduction of coding, access resolution or mediation, or other mandated control protocols. The US Federal Communications Commission (FCC) Spectrum Policy Task Force (SPTF) report [15] describes a general approach, referred to generally as interference temperature.[4]

The concept of interference temperature created an assurance to band users that total noise introduced into the band would not exceed a certain spectral noise temperature, measured as an equivalent radiation temperature. Devices would have to measure the existing noise temperature and then determine if they could add more noise (heat in the temperature metaphor) without exceeding the maximum temperature for the band. This is an important and fundamental change in the concept of spectrum coexistence in that it has the devices determine their own effect on the other users of the band and act accordingly, rather than perform operations to a set limit that was predetermined to have no impact on other users. Although the algorithms are not necessarily complex, it certainly crosses the cognitive threshold because not only is the radio aware of its environment, it's behavior is guided by a set of constraints on how its operation is permitted to impact the environment. Such a regime would inherently resolve the issue of aggregation of devices that is so much a concern over spectrum sharing.

UWB is the logical extension of direct sequence beyond the point where the spectral energy density is considered to be a factor in the operation of other UWB or narrower band receivers. Currently, two general approaches to this technology are being developed, one using an impulsive signal that is wideband by the nature of the step function modulation, and one using very wide spreading codes. For narrowband receivers (seeing only a fraction of the signal energy and for durations significantly longer than the pulse width), both appear as Gaussian noise.

---

[4]Noise temperature is directly related to spectral noise density by the equivalent black body radiation.

Shannon's limit promises great benefits from using such wideband signaling, and in the absence of other users, such approaches offer the potential to remove any need for spectrum management at all. The inherent constraints of UWB limit its use to very local signaling, or specialized applications. As such, it appears to be an extension of networking, and thus exists on the edge of the cognitive networks that are the focus of this chapter.

5.9 Adaptive Spectrum Implications for Cognitive Radio Hardware

Another aspect of cognitive radio operation will be the minimization of channel interference due to spurious artifacts inside the radio itself. If all components of the radio are operating in a linear region, then interference avoidance will be accomplished by the inherent operation of the radio's interference avoidance algorithms because channels that contain interference generally are not used, thus avoiding interference to other users. Although little work has been done on mitigation of nonlinear effects, it is a critical aspect of performance for wideband, handheld, energy-constrained devices that must operate in the presence of high-power neighbors. In this case, the cognitive process can also be utilized to select frequencies that have the least intermodulation or internal spur artifacts. Such an approach offers the ability to maintain connectivity performance even with reduced analog performance, as long as suitable preselector technology is available. This application of cognitive radio has had relatively little research, and will not be explicitly addressed further here. A note of interest is that techniques that were developed to avoid interference to other devices may also have the effect of providing self-management of internally generated interference as well.

Most of the implications of supporting cognitive radio technology within a radio have focused on the additional capability required in the radio, and thus the strong implication that a cognitive radio is inherently more expensive, and thus limited in application, compared to less capable devices. All other performance held constant, this implication is certainly true. It is worth investigating the performance drivers for wireless hardware. If we consider a radio, we can partition it into the analog and digital components. Digital logic is subject to Moore's Law, and increasing capability, lower cost, and lower energy are generally available, just by waiting. For the analog portion, the technology is much more mature, and it is unlikely that technology by itself will resolve shortfalls with currently available components. Starting with the radio frequency (RF) front-end, several driving performance characteristics generally drive energy consumption and cost. A very cursory summary is presented in Table 5.4.

Table 5.4: A summary of critical analog performance characteristics.

Metric	*Meaning*	*Impact*	*Constraints*
LNA Intercept Points (IIP2, IIP3)	Measures the linearity of the receiver's amplifier	Poor front-end will cause the receiver to overload, generating mixing products	Increasing IP has severe penalty in energy use, thermal load, and cost
SFDR	The maximum dynamic range with no self-generated signals	Spurs appear as (generally) tonal signals that can look like occupied signals, and can add noise to received signals	Reduction in SFDR requires significant investments in receiver organization, packaging, and components
Noise figure	Measure of self-generated noise in the receiver	Limits sensitivity to detect other spectrum users	Lower sensitivity leads to reduction in the power that can be transmitted for a given confidence of non-interference

LNA: low-noise amplifier; SFDR: spur-free dynamic range; IP: Internet Protocol.

Although minimal research is currently being performed to validate this assumption, many of these limitations should be amenable to spectrum awareness and cognitive reasoning. If successful, this approach could reverse the assumption that a cognitive radio is inherently more expensive than a less "intelligent" one. If it can be shown that these adaptation technologies and techniques are effective, then it is possible that a cognitive radio could enable relaxation of analog performance requirements, and thus make a cognitive radio actually less costly than a less adaptive one.

5.10 Summary: The Cognitive Radio Toolkit

What sets a cognitive radio apart from a conventional radio may be that, in the end, it fundamentally changes the way we design radios. Ideally, in fact, we would not design a cognitive radio at all. We would load it with the range of tools discussed in this and other chapters, and let it design and redesign itself continuously as both the environment and needs changed. Today, digital communications engineers attempt to system engineer radio operating points in order to draw the best compromise among competing demands. Because most networks operate within a relatively static range of characteristics, such operating points are compromises that are (hopefully) reasonable across the range of operating conditions, but it would be unreasonable to believe that these points would be optimal in any but a rare number of situations.

It is highly likely that future networking will continue to move to the adoption of more and more cognitive technologies. However, if this adaptation is driven by the needs of fixed networks, it is unlikely to address the unique aspects of wireless communications. In the wired world, each network link is independent and an abstraction. In the wireless world, they are an interconnected ecosystem, where each action by one has some impact on the others. The issues of spectrum deconfliction, nonlinear receiver responses, sensing, and the other topics covered in this chapter have no analogs in the wired world and thus will go unaddressed unless the wireless community becomes more involved in extending its practice into the more general networking problem space.

References

[1] G.J. Foschini, "Layered Space Time Architecture for Wireless Communications in a Fading Environment Using Multiple Antennas," *Bell Labs Technical Journal*, Vol. 1, No. 2, Autumn 1996, pp. 41–59.

[2] G.J. Foschini and M.J. Gans, "On limits of Wireless Communications in a Fading Environment When Using Multiple Antennas," *Wireless Personal Communications*, Vol. 6, No.3, March 1998, p. 311.

[3] http://www.nari.ee.ethz.ch/commth/pubs/p/proc03

[4] http://www.pimrc2005.de/Conferences_en/PIMRC+2005/Tutorials/T4.htm

[5] http://userver.ftw.at/~zemen/MIMO.html

[6] A. Gershman, *Space-Time Processing for MIMO Communications*, Wiley, May 2005.

[7] S. Sharma and A. Nix, "Dynamic W-CDMA Network Planning Using Mobile Location," *Vehicular Technology Conference, 2002 Proceedings*, VTC 2002, Fall 2002, September 24, 2002, pp. 182–1186.

[8] C. Tschudin, "Fraglets—A Metabolistic Execution Model for Communication Protocols," in *Proceedings of the 2nd Annual Symposium on Autonomous Inelligent Networks and Systems*, Menlo Park, 2003.

[9] C. Tschudin and L. Tamamoto, "Self-Healing Protocol Implementations," *Dagstuhl Seminar Proceedings*, 24 March 2005.

[10] http://www.cs.ucsb.edu/~htzheng/cognitive/

[11] Q. Wang and H. Zheng, "Route and Spectrum Selection in Dynamic Spectrum Networks," *Dyspan Conference*, Baltimore, November 2005.

[12] P. Kyasanur and N. Vaidya, "Protocol Design Challenges for Multi-hop Dynamic Spectrum Access Networks," *Dyspan Conference*, Baltimore, November 2005.

[13] V. Naware and L. Tong, "Cross Layer Design for Multiaccess Communications over Rayleigh Fading Channels," *IEEE Transactions on Information Theory and Networking*, submitted March 2005.

[14] V. Naware and L. Tong, "Smart Antennas, Dumb Scheduling for MAC," in *Proceedings of the 2003 Conference on Information Sciences and Systems*, Baltimore, March 2003.

[15] Federal Communications Commission, *Spectrum Policy Task Force Report*, 15 November 2002.

Appendix: Propagation Energy Loss

Radio waves propagating in free space lose energy over the distance that the signal travels. This loss of energy is called propagation loss. It is usually measured in decibels (dB) of loss on a logarithmic scale, in which 3 dB means half as much power is received as was transmitted, 6 dB means one-fourth as much power, 10 dB means one-tenth the power, 20 dB means one-hundredth the power, 30 dB means one-thousandth as much power, and so forth. Propagation loss may commonly range up to 100 dB (one ten-billionth the power received as transmitted) for practical communication ranges.

$$dB = 10 \, Log_{10}(\text{power})$$

Free space power loss can be easily understood with the example of a light bulb. Consider the light falling on the inner surface of a sphere surrounding the light bulb at a distance 1 m from the light bulb. With that radius, the surface area of the sphere is $A = 4\pi R^2 = 12.566 \, m^2$. So the percentage of the light falling on $1 \, m^2$ is $(1/12.566) = 7.96$ percent of the energy transmitted. At twice the distance (or radius), the surface area of the sphere is four times larger, so the amount of energy falling on $1 \, m^2$ is one-fourth as large; in the case of radio wave propagation, that would be called a 6 dB loss. The antenna of a radio receives the transmitted radio energy much as the light on an area on the inner surface of a sphere. Therefore, in free space the power received falls another 6 dB at each doubling of transmitted distance. Because propagation loss in this condition is proportional to the propagation range squared, this is usually referred to as R^2 conditions.

Radio waves propagating across the surface of Earth, however, lose energy more rapidly than in free space. This occurs in part because the signals bounce off Earth's surface and back up to the receive antenna, but have traveled a different distance. At certain distances where the reflected signal has traveled exactly a half wavelength farther than the signal arriving without a reflection, the signals add together but at opposite phase and cancel each other, resulting in a very high propagation loss. In addition, trees and buildings absorb and reflect the signal propagating along the surface of Earth as well. As a result, propagation losses along the surface of Earth increase not as the square of distance, but as the third or even fourth power of distance, resulting in a 9 dB or even a 12 dB increase of propagation loss per doubling of range. This is commonly called R^3 and R^4 conditions, respectively. To account for the average propagation loss in suburban conditions, the industry often chooses to average the loss per doubling of range, and refers to this as $R^{3.8}$ propagation conditions.

Cognitive Policy Engines

Robert J. Wellington

Department of Physics, University of Minnesota
Bloomington, MN, USA

6.1 The Promise of Policy Management for Radios

In familiar usage, policies are procedural statements expressing administrative conventions that are adopted by various organizational entities. The concept of automatic policy management of resources has its commercial roots in the administration of information systems and networks. Policy management refers to a particular approach for automating network management activities by specifying organizational objectives that can be interpreted and enforced by the network itself. The automatic application of management policies provides flexibility to change the configuration of network devices at run-time to satisfy administrative goals and constraints regarding security, resource allocation, application priorities, or quality of service (QoS). A "policy engine" is a program or process that is able to ingest machine-readable policies and apply them to a particular problem domain to constrain the behavior of network resources.

This chapter concerns the application of policy management to cognitive radio technology in general, and to spectrum management for frequency-agile radios in particular. It focuses on what lessons can be learned from prior applications of policy management to network resource problems. It reviews and leverages previous standards, research, and commercial implementations for policy engines and applies them to the architecture and design of policy engines for cognitive radios.

6.2 Background and Definitions

The policy engine is the main inference component that triggers responses to events that require changing the resource configuration. Often the output of the policy engine amounts to configuration commands or authorizations that are tailored to specific kinds of network devices. In this sense, the policy engine bridges the gap between domain-specific objectives and device-specific capabilities. Despite the intrinsic need for interfacing with particular vendor devices, a popular research trend has been to postulate the policy engine as a general-purpose tool capable of deductive reasoning based on rules. Seen in this manner, policy engines are descendents of the rule-based programming frameworks that were popular in the 1980s. Bemmel et al. [1] describe an expert system as simulating human reasoning using heuristic deduction rules, where knowledge is stored as facts and new facts are derived by using a set of deduction rules.

Much of the research in the area of policy-based networking has focused on the specification of formal languages for expressing complex policies for various domains and network management problems. Policies are expressed as sets of rules about how to change the behavior of the network. Chadha et al. [2] define a policy to be "a persistent specification of an objective to be achieved or a set of actions to be performed in the future or as an on-going regular activity." Carey et al. [3] explain that "policies are expressed in terms of an event that triggers the evaluation of a policy rule, a set of conditions that must be met prior to changing the behavior, and a set of actions that are performed to change the behavior." Two trends are evident in the literature: (1) the deconstruction of policies into sets of conditional rules of varying degrees of complexity, and (2) the use of object-oriented representations to support machine readability.

Figure 6.1 calls out commonly recognized functions and relations in a conceptual architecture for a network policy management system. The interpretation and application of these functions for cognitive radio networks requires revisions to this conceptual model, which are explored in this chapter. Evidently, the network resource is the cognitive radio, and the policy decision point (PDP) and policy enforcement point (PEP) represent new functions that enable policy management of cognitive radio networks. The purpose of the PDP and PEP functions are to interpret policies to control the behavior of the network devices to satisfy both the users and administrators of network resources.

To implement the policy management architecture shown in Figure 6.1, the system must monitor real-time network events and trigger the policy engine to decide how current device states (policy conditions) should be mapped into desired policy actions that can be quickly enforced by controlling specific device operations. Carey et al. [3] note that policies allow less centralized and more flexible management architectures by enabling administrative decisions to be made

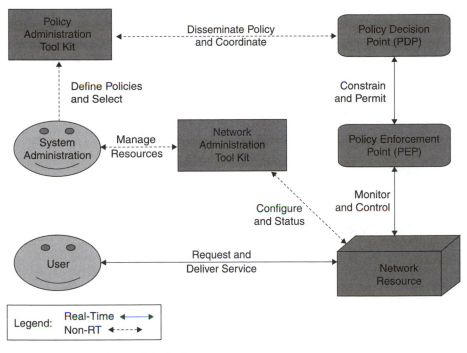

Figure 6.1: Policy management system concept.

closer to where the event and conditions are actually detected. We must examine how this approach applies to the particular problem of spectrum management for cognitive radios.

6.3 Spectrum Policy

The usable frequency range of radio spectrum is divided into frequency bands called allocations for particular types of use. The US frequency allocations are available online [4]. Within a particular allocation, an allotment is a frequency channel designated for a particular user group or service in some country. An assignment is a license that grants authority to a specific party to operate a transmitter on a specific channel under specific conditions. The allotments and assignments are associated with a particular geographic area.

Historically, the allotments for broadcast services have been deconflicted to ensure that the signal strengths in one area will not create interference for signals in another area. However, it is not easy to define interference unless certain assumptions are made about the capability of radio receivers to reject interference in adjacent bands. Other technical considerations will include uncertain propagation characteristics, locations of nearby receivers or transmitters, and limitations of waveforms.

6.3.1 Management of Spectrum Policy

The International Telecommunication Union (ITU) Radiocommunication sector [5] plays a global role in the management of the radio frequency (RF) spectrum and satellite orbits. Around the world, RF spectrum is considered a finite natural resource that is increasingly in demand from a large number of services, such as fixed, mobile, broadcasting, amateur, space research, meteorology, global positioning systems (GPS), environmental monitoring, and last but not least, those communication services that ensure safety of life at sea and in the skies. World Radiocommunication Conferences (WRCs) are held every 2-3 years to review, and, if necessary, revise the Radio Regulations, the international treaty governing the use of the RF spectrum.

Within the United States, the Federal Communications Commission (FCC) decides spectrum policy for commercial radio communications, and the National Telecommunications and Information Administration (NTIA) plays a complementary role for the federal sector. Due to the perceived spectrum shortage resulting from prior policies and the burgeoning demand, there has been great interest in rethinking the leasing and allocation of spectrum. The FCC's Spectrum Policy Task Force (SPTF) [6] reported in 2002 that "spectrum policy is not keeping pace with the relentless spectrum demands of the market." Two recurring recommendations have been to "migrate from the current command and control model to [a] market-oriented exclusive rights model and unlicensed device/commons model" and to "implement a new paradigm for interference protection." The FCC intends to facilitate cognitive radio technologies to this end. In particular, the discussion has often focused on "spectrum enhancing technologies," including software-defined radios, leasing certain spectrum bands or "white space," and sensory or adaptive devices that could find unused spectrum. For example, to identify unused RF channels, a cognitive radio might be able to measure an "interference temperature" for the ambient spectral environment.

Deciding which policies apply to a particular cognitive radio will require understanding the role the radio is playing for the user of some service at a particular location. The policy language must be rich enough to express the semantics of the possible spectrum management policies, and the cognitive policy engine must be able to automatically interpret and enforce the applicable policies. The FCC and NTIA are contemplating pilot projects to explore options for encoding selected spectrum policies in a machine-readable language. Section 6.4 looks deeper into these languages.

6.3.2 System Requirements for Spectrum Policy Management

A comprehensive policy management system that could autonomously control interference created by frequency-agile radios would have to include an extensive syntactic capability to specify policies and policy engines that can interpret policy semantics expressing technical concepts such as authorization, frequency bands, channels, propagation conditions, signals and noise, waveforms, geopolitical boundaries, geographic locations, dates, times, types of services, and possible roles for the various radios in the environment.

The policy engine must be able to automatically modify the run-time configuration of the cognitive radio to sense the spectrum environment; to detect other local radio networks; and to control its own transmissions by adjusting frequency, power, modulation, signal timing, data rate, coding rates, and antenna.[1] In summary, the spectrum policy management system for cognitive radios shall:

1. Be distributed across multiple radio platforms with autonomy at the radio level.
2. Permit network administrators to determine applicable spectrum policies.
3. Represent spectrum policy rules in a machine-readable format.
4. Resolve conflicts and inconsistencies in sets of policy rules.
5. Designate specific roles and services for each radio.
6. Require authorization for policy changes or updates for radio.
7. Monitor spectrum utilization and detect RF interference.
8. Control run-time configuration of cognitive radios to satisfy spectrum policies.
9. Support heterogeneity and diversity of legacy vendor radios.
10. Support continual growth and increasing complexity of cognitive radios.

6.4 Antecedents for Cognitive Policy Management

The last decade has seen significant research and applications for policy management technology in the area of network management. Tacit assumptions about the

---

[1] Multiple policy engines—one for the physical layer, one for the network, and one for the user— are possible; this chapter mainly covers the physical layer policy engine. In addition, there may be engines for equipment-specific implementation, behavior-specific policies, or other policies.

design of the Internet and related information technology (IT) infrastructures have greatly influenced the development of policy management approaches. It is not surprising that much of the research has focused on the limitations of descriptive ontologies for web-enabled applications and packet network resources. However, Kagal et al. [7] rightly caution that reasoning about policies generally requires application-specific information, forcing researchers to create policy languages that are bound to the domains for which they were developed.

Chadha et al. [2] have surveyed policy-based network and distributed systems management approaches that have been the subject of extensive research over the last decade (see also [8]). The Internet Engineering Task Force (IETF) [9] has sponsored standardization efforts for object-oriented models for representing policies as well as a framework and protocols for managing Internet Protocol (IP) networks.

This section examines how well the existing state of the art can be adapted to support the performance characteristics of radio networks and the specialized requirements for spectrum resource management. This section draws from government projects, commercial applications, academic research, and standardization efforts that provide a context for designing a cognitive policy engine specialized for spectrum management.

6.4.1 Defense Advanced Research Projects Agency Policy Management Projects

The Defense Advanced Research Projects Agency (DARPA) has funded research and development (R&D) efforts that push forward the technology frontiers in the area of network policy management in general and even the particular application to spectrum management. This section reviews some of that work that is in the public domain.

Funding for contractors involved in the DARPA NeXt Generation (XG) radio communications program [10] has been a very significant, if not the principal, driving force behind the development of policy management techniques for radios. The FCC now has complementary projects to define policies for frequency-agile radios. The XG program followed on the heels of the DARPA Policy-Based Survivable (PolySurv) communications program, which demonstrated that increased military survivability and real communication performance gains could be brought about by downloading dynamic mission policies to automatically manage radio networks. XG is now focused on system concepts and enabling technology to dynamically redistribute allocated spectrum in operating radio networks in order to address rapidly growing requirements for communications bandwidth. The program goals are to enable radios to automatically select spectrum and operating modes in a manner that increases the survivability of communication networks and minimizes disruption to existing users.

The DARPA Dynamic Coalitions program [11] has strongly influenced the technical approach for XG by offering policy representations and policy engines that support other network management techniques. For example, Phillips et al. [12] describe constraint-based models and the application of role-based access control (RBAC) for implementing security policies in the context of Dynamic Coalitions. Uszok et al. [13] describe a Semantic Web (SeW) language [14] called KAoS that has proliferated with support from DARPA. In fact, KAoS was based on the DARPA Agent Markup Language (DAML) [15], and KAoS has capabilities that support both the expression and enforcement of policies in a software agent context. The policy language has always been based on the eXtensible Markup Language (XML) to support common Web services, but due to shortcomings of the DAML description logic, KAoS now relies on the Web Ontology Language (OWL) [16] to represent knowledge about domains and rule-based policies. In the KAoS environment [17], domain managers act as PDPs and are responsible for administering policy for entire domains.

DARPA has also been active in funding more traditional network policy management approaches for the Next Generation Internet (NGI), and Stone et al. [18] provide background information that is relevant for policy management of cognitive radio networks.

6.4.2 Academic Research in Policy Management

This section looks at what a spectrum management implementation might leverage from the research community concerning existing policy languages and frameworks for network policy management. Carey et al. [3] include an overview of the state of the art in policy languages that addresses access control and resource management. Rules governing spectrum access can be enforced by a radio that subjects itself to access control policies. In computer network management systems, only users associated with accounts included in an access control list (ACL) can access the resource. The next enhancement is association of users with groups and roles, leading to RBAC systems. Spectrum resources are already assigned to particular radio services, so the cognitive policy engine can process attributes associated with roles for the radio to provide a context for evaluating spectrum access control rules. Ideally, we would want users to authenticate themselves to the radio and associate different types of authentication with different roles for the user, the radio, the network, and network resources.

Strauss [19] describes the requirements and architecture of a policy management system based on the IETF script management information base (MIB) infrastructure. An MIB is a device-specific database for remotely managing a network resource using the Simple Network Management Protocol (SNMP) based on IP communications. Most IP-capable devices support an MIB that provides a standard interface to monitor and configure the device. For the cognitive radio, we could envision a "spectrum MIB" with a standard interface supported by an underlying device-specific mechanism to actually configure the network elements. Strauss [19] used this approach to implement network QoS control with the Jasmin Script MIB agent [20]. In this case, the policy engine is just the Java run-time engine executing policy scripts supported by policy-class libraries for different device capabilities.

Ponder is a different policy management framework that includes a relatively mature policy language with a suite of tools and source code that has been freely available to download from the Internet [21]. Ponder tools support administration of domain hierarchies, positive and negative authorization policies, delegation policies, and event-triggered condition-action rules. Dulay et al. [22] describe how to use the Ponder framework to encode, disseminate, and process security and management policies for distributed applications. In fact, Ponder would be an initial starting point for developing an administrative toolkit that could be used to specify, compile, maintain, and disseminate spectrum policies for cognitive radios. It integrates with a domain server and supports role abstractions that could be used to manage spectrum policies for multiple communications services that might eventually be supported by cognitive radios.

Ponder has been well tested in various applications [23], and "back-ends" (i.e., application-specific PDP and PEP functions) have been implemented to generate firewall rules, Windows access control templates, Java security policies, and obligation policies for a policy agent. The Ponder language for representing policies has been described as a declarative, object-oriented language that can express both "obligation" and "authorization" policies [24]. To be specific, Damianou et al. [23] explain that "policies define choices in behavior in terms of the conditions under which predefined operations or actions can be invoked rather than changing the functionality of the actual operation themselves." Obligation policies are defined as "event-triggered condition-action rules that can be used to define adaptable management actions," and authorization policies are "used to define what services or resources a subject (user or role) can access."

The policy research community in general is particularly concerned about the difficult task of analyzing the meaning of groups of policies to determine the implications for particular agents and resolving possible conflicts between policies. Even if spectrum management objectives are clearly stated in a policy, the implications for device configurations or required actions are not always obvious in practice. For example, consider a long-standing policy that authorizes a particular frequency band for some type of messaging service, and specifies service-specific protocols for users to share airtime. Suppose a newer policy for cognitive radios specifically authorizes a class of cognitive radio users to share an overlapping band of spectrum subject to different limitations on availability of channels for legacy users. Is it clear what the airtime restrictions are for a particular cognitive device that performs a similar type of messaging using the legacy channels and protocols? Which rule takes priority, or must both usage restrictions be observed by the cognitive radio? Do permissions take precedence over prohibitions? Policy refinement is the process of deriving lower-level, more specific policies that the device can enforce in order to completely meet the requirements of a group of management policies.

Damianou et al. [24] describe other problems with policy refinement, and Ponder provides tools for policy analysis and refinement to assist administrators in detecting and resolving policy conflicts. In particular, Ponder supports the introduction of priorities and preferences. A simple method to resolve policy conflicts for a device is to assign explicit priorities to every policy so it is clear which policies overrule others. Locally, rules can be prioritized for the device in order to reflect local management priorities, such as ensuring that efficiency is more important than reliability, or vice versa. For a specific device, sets of rules are also ordered by update times, particularly if partial updates of the rule base are accepted practice. DAML relies on update times as well as numeric priorities to determine priority [24].

Stone et al. [18] suggest the idea of differentiating policies "by their granularity, such as the application level, user level, class level, or service level," and letting spectrum managers designate certain mission-applications for priority. Hierarchical policy management and domain groupings provide another degree of flexibility, permitting a PDP to branch beyond the linear ordering of priorities. The device may give priorities to policies originating within a more local domain, given an inheritance hierarchy. Similar to the manner in which federal policies overrule state policies, Uszok et al. [17] anticipate a "policy harmonization" process that invalidates portions of the lower-priority policy to resolve the conflict. Decisions about how a PDP should handle inheritance must be made at the time that the policy hierarchy is established. In addition, these decisions should belong to the human realm of policy administration. Experience tells us that it takes a human judge to decide how to invalidate portions of state policies to eliminate conflicts with federal rules.

Another way that a PDP can handle nonlinear priorities and resolve conflicts and ambiguities between overlapping rules is to permit the specification of "meta-policies" constraining the interpretation of groups of policies. For example, Kagal et al. [7] describe another policy language, Rei, that supports meta-policies for conflict resolution. Rei was designed for general application and permits domain-specific information to be added without modification. Tonti et al. [25] provide a comparison of the capabilities and shortcomings of KAoS, Ponder, and Rei, rendering a valuable perspective of the various approaches.

Clearly, the techniques for resolving policy conflicts could be a fruitful area of research for a long time to come. As far as spectrum management is concerned, it appears that the research community has already done enough work to get started expressing FCC policies in a machine-readable format. A number of abstract policy languages already exist, and the task ahead is to introduce terminology that is directly applicable to spectrum management and the cognitive radio domain. The challenge will be to come up with a reasonably useful and self-consistent rule base for cognitive radios that does not present an opportunity for litigation about the meaning, implications, or applications of the rules.

6.4.3 Commercial Applications of Policy Management

This section examines what can be learned from commercial products that could be useful for designing cognitive policy engines. No one will be surprised to discover that such major network vendors such as Cisco, Nortel, and Lucent Technologies have already developed policy management products to support administration of local and wide area networks (LAN and WAN).

Damianou et al. [23] indicate that Lucent has used a Policy Definition Language (PDL), similar to Ponder, to program Lucent switching products. PDL "uses the event-condition-action (ECA) rule paradigm of active databases to define a policy as a function that maps a series of events into a set of actions." This approach is interesting because in addition to policy rules, there are policy-defined event propositions that allow groups of simple events to trigger more complex events.

Nortel [26] advertises its Optivity Policy Services with system components consisting of a policy server, management console, directory server, and policy-enabled network components. This distributed IP network architecture interoperates using Lightweight Directory Access Protocol (LDAP), Common Open Policy Service (COPS), and Command Line Interface (CLI). The administrator is required to select from a number of predefined policies (i.e., "if condition, then action" rules) that relate to the roles of various network devices identified in the directory server. The policy server then issues COPS or CLI commands to configure devices such as routers, IP telephone gateways, and firewalls.

Damianou et al. [23] observe that a common component of commercial tools is a graphical user interface (GUI), which typically allows the administrator to visually select a network device or other managed element from a hierarchically arranged tree-view of policy targets, and specify the policies in the form of "if conditions, then action rules for the selected targets."

What comes across in all the commercial examples is that the network devices must be designed to support policy rules that can configure their behavior by some mechanism. The sophistication of the devices determines the nature of the interface with a policy server that either directly issues device configuration commands, supports a COPS dialog to make policy decisions for the device, or disseminates defined policies that are recognized by the device. Ultimately, the nuances of the policy language seem to be relatively unimportant compared to the sophistication of the network devices.

6.4.4 Standardization Efforts for Policy Management

The proliferation of vendor architectures for policy management of telecommunications networks has motivated the IETF to address standards for interoperability. Using the Policy Core Information Model (PCIM), Moore et al. [27] begin the process of standardizing policy management terminology and representations of network management policies to provide an accepted framework for vendor-specific implementations.

To what extent can these standardization efforts be applied to policy engines for spectrum resource management for cognitive radio operations? Initial cognitive radio implementations will necessarily focus on radio-specific features, and as the technology proves successful, the focus will extend to larger radio networks rather than individual radios. The IETF policy domain is already oriented toward management of large-scale telecommunications networks, with the goal to assist human network administrators to define machine-readable policies and architectures for the automatic control of network resources. The cognitive radio is analogous to a particular network device.

Snir et al. [28] envision a physical network architecture in which the PDP translates abstract policy constructs into configuration commands for multiple devices (e.g., a router, switch, or hub) where policy decisions are actually enforced (i.e., PEP). Although there may be compelling arguments for the architectural assumption that one PDP services multiple PEPs in the case of high-speed, high-reliability network architectures, this is not so clear for the cognitive radio application. Stone et al. [18] point out that an underlying assumption of PCIM is that policies are stored in a centralized repository, and the PDP is the entity in the network where policy decisions are made using information retrieved from policy repositories. When the PEP requires a policy decision about a new flow of traffic or authentication, for example, "the PEP will send a request to a PDP that may reside on a remote server."

Moore et al. [27] indicate the PCIM standard fits into an overall framework for representing, deploying, and managing policies that is being developed by the IETF Policy Framework Working Group. In Figure 6.1, the link between the PDP and PEP has two characteristics: (1) it needs to operate in near real-time for timely enforcement decisions and (2) it is conceived to be a query–response dialog. For the cognitive radio application, due to concerns about link reliability and bandwidth, the first assumption is tenable only if the PDP and PEP functions are colocated on the radio platform. Furthermore, the query–response design is also very natural, given the prevalence of client–server and three-tier database transaction architectures in the Internet. In fact, two important Internet applications for policy management involve access control (i.e., security) and admission control (i.e., QoS), and both involve permission. However, there is no reason to assume this Internet design is optimal for a cognitive radio application in which the PDP and PEP functions are colocated.

In PCIM, the policy-controlled network is modeled as a state machine, in which policy rules control which device states are allowed at any given time. Each policy rule consists of a set of conditions and a set of actions. Policy conditions are constructs that can select states according to complex Boolean expressions. Policy actions are device behaviors, such as selecting or prohibiting certain frequency bands, bandwidths, protocols, coding, or data rates. When events lead to certain conditions, then certain actions can or must be performed by the device (depending on whether the actions are obligated or simply permitted for the device).

Bemmel et al. [1] describe these policy rules as examples of ECA rules, by which changes within a system or the environment trigger adaptation of the system's behavior. Such systems are called event-driven, and many programming languages and environments support compatible software development techniques. For example, in the Microsoft Windows® architecture, events are basically messages that are routed between processes called event handlers. Unfortunately, if the policy rules were coded in this familiar manner, the component's behavior would basically be hard-coded when the program is compiled. We need a more flexible way, however, to bind actions to the event handlers at run-time for example, Java Remote Method Invocation (Java RMI) or Common Object Request Broker Architecture (CORBA).

The PCIM defines class representations and abstract attributes for policies, but not the algorithms or design of the policy engine. Figure 6.2 depicts the ontology for the object-oriented design of policy classes in the PCIM. Policy rules are aggregated into policy group classes. These groups may be nested to represent a hierarchy of policies. Although retaining all the same attributes of the policy classes is not particularly important, a similar class diagram will be suitable for representing policies for the cognitive radios.

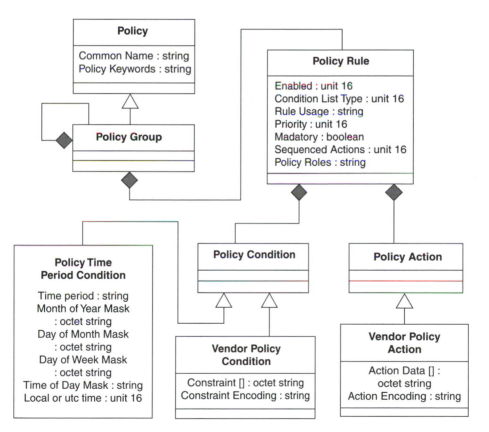

Figure 6.2: Unified Modeling Language (UML) diagram for PCIM policy classes.

One important attribute of the policy rule is the ability to associate a "role" for the radio. The policies (e.g., assigned frequency bands) for land mobile radios (LMRs) are different from those for air traffic control (ATC). Moore et al. [27]

stress that rather than configuring—and then later having to update the configuration of—hundreds or thousands (or more) of resources in a network, a policy administrator assigns each resource to one or more roles, and then specifies the policies for each of these roles. The policy framework is then responsible for configuring each of the resources associated with a role in such a way that it behaves according to the policies specified for that role. When network behavior must be changed, the policy administrator can perform a single update to the policy for a role, and the policy framework will ensure that the necessary configuration updates are performed on all the resources playing that role.

6.5 Policy Engine Architectures for Radio

6.5.1 Concept for Policy Engine Operations

This section begins to synthesize a concept for how the cognitive policy engine will operate. Section 6.4 shows how network policy management systems operate in a distributed fashion with administrators using policy languages to express network objectives, policy servers that act as PDPs deciding how to configure devices to achieve these goals, and policy-enabled devices that are configured to enforce the policy (i.e., PEP). The spectrum management objectives for the cognitive radio are slightly different than network access and resource control, for which the primary goals are security and network application performance. The primary interest here is ensuring that the frequency-agile radio is utilizing available spectrum resources in accordance with regulatory licenses and prohibitions while acting as a good neighbor to avoid interference with other users. So there is a need to distinguish between a PDP that can optimize global network performance, and a PEP that configures the cognitive radio to obey locally enforceable policy constraints. In other words, the cognitive policy engine is responsible for device configuration.

The relationship between the PDP and PEP functions is also different for the cognitive radio. Carey et al. [3] describe a typical network policy management dialog in which events detected by the device cause the PEP to formulate a request for a policy decision and send it to the PDP (usually a policy server), such as when Resource ReSerVation Protocol (RSVP) is employed by a user application to advertise its QoS requirements. COPS may be used as a policy transaction protocol between the PDP and PEP for transporting the policy requests and decisions. The PDP returns the policy decision and the PEP then enforces the policy decision and responds to the RSVP from the user-application accordingly.

In our concept, the cognitive policy engine will perform both PDP and PEP functions for the radio using local platform resources. The operation will generally accord with Figure 6.1, and involve the steps shown in Table 6.1. There is no necessity for radio device functions to request a policy decision from the policy engine because the policy engine is already responsible for enforcing the policy on the platform. As shown in Figure 6.1, the policy functions (PDP and PEP) act to monitor and control the radio platform to satisfy any policy constraints or obligations.

Table 6.1: Concept of operation for the cognitive policy engine.

Step 1	The policy engine receives a download of policies to be observed by the radio.
Step 2	The policy engine determines what information to monitor on the radio platform (e.g., location, time)
Step 3	The cognitive radio provides the requested situation data (e.g., location, time) to the policy engine as required
Step 4	The arrival of new information is an event triggering some processing of policy rules
Step 5	The policy engine formulates applicable constraints on spectrum usage for the radio
Step 6	The policy engine configures the radio to observe spectrum constraints and perform obligatory activities (e.g., communication protocols)
Step 7	The cognitive radio operates within defined constraints and performs obligatory activities
Step 8	The policy engine performs any obligatory hierarchical reporting activities for network management

If we assume the cognitive radio is administered by a central policy authority in a domain that recognizes the capabilities of the particular cognitive radio and disseminates all enforceable policies directly to the cognitive radio, then the complexity of the policy engine can be reduced. For example, the policy engine does not have to determine whether the policies can be enforced on the platform. Any ECA rule sent to the platform could (but not should!) even be compiled and executed by a Java engine on the radio.

In the case of radio networks, such as an Internet community, no centralized authority will own all of the resources. It will not even be possible to guarantee efficient connectivity among the worldwide users of cognitive radio technology. Feeney et al. [29] argue against a centralized policy administrative authority in cases of great organizational diversity, and propose a concept of operations with hierarchies of policy authorities. This approach reflects a real-world community of users, but it leads to the possibility of policy conflicts that must be resolved among organizations. These conflicts must be resolved at the network level, and the cognitive radio should not be forced to handle this difficult problem.

6.5.2 Technical Approaches for Policy Management

This section proposes a technical approach for designing the cognitive policy engine to satisfy the concept of operations presented so far in this chapter. We look first at what must be done to design the policy language and then examine alternatives for the technology behind the policy engine.

Multiple approaches for policy specification have been proposed that range from formal policy languages that can be processed and interpreted easily and directly by a computer, to rule-based policy notation using an if-then-else format, to the representation of policies as entries in a table consisting of multiple attributes [30]. This chapter has already looked at alternatives, including the compiled policies of Ponder, Java scripts supported by interpreted classes, KAoS with its OWL semantics, RBAC, Rei, and PDL.

Whatever language is selected, the syntax will have to be extended to support the technical jargon of spectrum policy with such attributes as frequency bands, channels, propagation conditions, signals and noise, waveforms, geopolitical boundaries, geographic locations, dates, times, and types of services. The language should support authorization and obligation policies, and roles for the various radios in the environment. In addition, the language should make a clear distinction between management policies and the resources and activities being managed [31]. The plan from the beginning should consider eventual "use of domains as a means for grouping resources with dependencies reflecting both hierarchical interactions (e.g., control of resources, authority delegation to subordinate managers) as well as peer-to-peer interactions (e.g., negotiations between peer managers to prevent/resolve management conflicts)" [31].

Looking forward, the ontology should support standardization efforts for multiple vendors of cognitive radios, even though it will likely be redesigned in the future. For this reason, we should eschew compiled and interpreted languages, and start with the flexibility provided by SeW languages such as OWL. Then the ontology will be built up as capabilities and understanding increase over time.

The choice to start with a SeW language does not exclude the use of compilers and interpreters; it just focuses this technology on the implementation of actions and behaviors with processes named in the policies. For example, useful radio protocols should be named in the ontology, and policies relating to standard protocols should be defined. As more device-specific actions are represented with new terminology, the generality of the policy architecture will decrease because the language must support increasingly complex and specialized actions for proprietary device behaviors [32]. Another shortcoming of this approach is that the policy engine and ontology have difficulty reasoning about complicated, composite behaviors. Selecting the right balance between generality and specificity in referencing cognitive radio behavior is an area for further research that need not obstruct initial efforts to create a spectrum policy language for the policy engine.

For policy engine development, there are also several approaches to consider. Note that the engine itself need not internally use the policy language as its input, because the policies can be interpreted when they are downloaded to the cognitive radio. Named behaviors can be bound in the radio to specific procedures and algorithms. For example, a compiler or just-in-time interpreter can create "tokens" from the input and link together whatever processes or logical structures are used internally to the cognitive radio.

Turning to the technical approach for the cognitive policy engine, it is important to recognize that the policy engine will be performing both PDP and PEP functions on the same platform. The concept of operation proposed here is that the policy engine will be responsible for both interpreting and enforcing spectrum policies, as well as monitoring platform events to trigger changes in the configuration of the radio functions. Again, according to Figure 6.1, the functional interface between the policy engine and the radio platform is defined as a relationship of monitoring and controlling the radio. This is the situation shown in Figure 6.3(a).

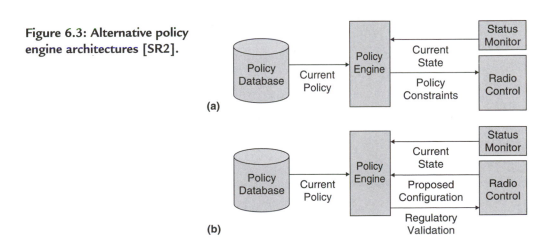

Figure 6.3: Alternative policy engine architectures [SR2].

An alternative approach characterizes the purpose of the interface as validating that the operation of the radio complies with all relevant policies. This situation is shown in Figure 6.3(b). Uszok et al. [17] argue that the interface between a PEP and a native environment can be standardized to answer the question "is a given action authorized or not?" This approach posits a query–response dialog between the policy engine and the native device. Thus the policy engine explicitly polices actions by the radio, requiring the radio to create objects that describe proposed actions so that the policy engine can pass judgment on whether the action is permissible.

Some involved parties (but not members of the FCC) have publicly voiced the opinion that validation is the only technical approach that will support certification of a frequency-agile radio by the FCC. Were this opinion correct, the policy engine architecture would have to combine both functions in Figure 6.3. Function (a) performs vendor-specific configuration and control of a proprietary device and function (b) certifies policy compliance based on names for generic behaviors. In the latter case, the policy engine can have only a partial understanding of the device's proprietary behavior. This leads to requirements for default authorization policies in which actions that are not defined in the policy language should generally be denied, or are assumed to have been deemed permissible when the radio platform was certified by the FCC.

The technical approach presented here for the cognitive policy engine starts with configuration rather than validation. The emphasis is on supporting obligation policies and ECA rules in addition to authorization policies typical of security platforms. As Stone et al. [18] state, "a policy specifies what action(s) must be taken when a set of associated conditions are met." Bemmel et al. [1] consider this policy-based approach with ECA rules to be "useful in cases when a controlled part has many choices/options and a controller intervenes, enforcing a choice in accordance with a particular policy goal." This seems to be the likely situation, as cognitive radio technology will increase in agility over time.

An important part of this approach at both the device level and the network level is status monitoring. According to Lee et al. [33], "Feedback is a critical part of the process where monitoring activities by the management system lead to changes in the low-level policies." This permits the policy management system to adjust the radio according to changes in demand for bandwidth or interference conditions.

6.5.3 Enabling Technologies

This section completes the examination of what particular technologies will be required to implement this technical approach. Carey et al. [3] summarize two challenges for cognitive policy management: "Currently policy-based management suffers from fragmentation of approach and there is no commonly accepted policy language and no common approach to the engineering of policy based systems."

Regarding the technology behind the policy engine, Khurana and Gligor [34] approach the problem by modeling the network as a state machine and using policy to place limitations on the state transitions that are allowable at any given time. Any model for the cognitive radio that is internal to the policy engine will tend to become more complex and specialized over time. The same issue arose with the increasing specialization of the policy language over time. More R&D will determine whether this requires a custom design for the policy engine that is matched to the radio's capabilities.

The state transitions in the model can occur only when the constraints are satisfied, and this may require logical deductions at some point. Damianou et al. [23] note that logic-based policy languages have a well-understood formalism that is amenable to analysis. For example, a policy engine supporting RBAC can be implemented in the Prolog language because the policy constraints are equivalent to restricted first-order predicate logic (RFOPL) statements [35]. Kagal et al. [7] say that the Rei policy engine was developed in Java and used Prolog as a reasoning engine. Stone et al. [18] summarize the situation: "One clear method is to use formal logic to represent network policies. Although this method would make conflict detection much easier with the use of existing theorem-provers, most network policy implementers are not as comfortable with this representation."

The policy engine behind KAoS is an online theorem-prover that permits logical reasoning about domains and policies [17]. The enforcement mechanisms are Java classes that are specific to a resource platform, but capable of interpreting and enforcing policies from the PDPs. The Java Theorem Prover (JTP) [36] supports queries consisting of properties defined in a given namespace and selection of all possible values. The JTP will also provide a response regarding whether a given assertion is true. For conditional policies, JTP can store a state consisting of certain assertions that are true. These assertions can be removed when the condition is determined to be false by the radio or event monitor.

Another area for R&D is motivated by Carey et al. [3]: "Problems with policy decomposition and hierarchical translation still present difficulties." Damianou et al. [23] characterize the situation: "Despite significant efforts in developing different policy specification techniques, the ability to refine such goals into concrete policy specification would be useful. Policies can be used to support adaptability at multiple levels in a network." "Research is needed on defining interfaces for the exchange of policies between these levels. It is not easy to map the semantics of the policies between the different levels." The basic problem is that the higher-level management applications "will not be aware of the network components and cannot specify policies to be interpreted by them" [23].

At this time, the need for extensive logical reasoning capabilities in the policy engine has been oversold, particularly at the level of the device. Whenever a problem requires complex reasoning, it is generally solved with a special-purpose algorithm or application. Two techniques are now available for use to help resolve ambiguities in the rule base. First, policy rules can be associated with a priority value to resolve conflicts between rules [24]. Second, "meta-policies" are policies about policies, and may be used to detect semantic conflicts between policies [3]. Kagal et al. [7] describe this capability in Rei.

6.6 Integration of Policy Engines into Cognitive Radio

In the end, the integration of the policy engine into the cognitive radio is primarily an issue of software system engineering. The process will involve deriving a complete set of software requirements from the system concepts; defining a software architecture that is compatible with the cognitive radio platform software architecture; defining the software interfaces; and designing, coding, and testing the software components. Some assumptions must be made about the software architecture and application-programming interface (API) provided by the services on the host platform. This section, for example, makes the assumption that the first cognitive radio implementation will likely evolve from the efforts of the Software Defined Radio Forum (SDR Forum [37]). It begins with platform integration issues and then moves on to issues related to integration into a policy-managed network.

6.6.1 Software Communications Architecture Integration

The Joint Tactical Radio System (JTRS) JPO defines the Software Communications Architecture (SCA) based on CORBA middleware [38]. The SCA is a layered architecture with well-defined APIs for integration with new applications (called components). The applications are supported by a core framework that provides interfaces for exchanging information, controlling software processes, configuring the platform, and accessing files. The policy engine will be another component in this architecture supported by the core framework with interfaces to other components. One of the most important architectural considerations is the definition of the interface between the policy engine and software environment of the radio.

    The policy engine can be divided into three main policy management functions: policy service, policy decisions, and policy enforcement:

- The *policy service* maintains the policies that are downloaded to the radio frame by a policy authority, and it requires an external network interface communication. The policy service also performs any necessary policy life cycle management operations, such as parsing the policy language, keeping track of whether policies are enable or disabled, and removing policies if they become obsolete.

- The *policy decision* function fulfills the requirement that the device have a local PDP. To make these decisions, it needs information about the communication conditions and spectrum events in the radio environment. Thus, it must interface to some kind of status monitor that can report conditions and real-time events that require a policy decision.

- The *policy enforcement* function acts as the PEP for the device, and it needs a control interface to constrain the radio's behavior and perform any obligated actions demanded by policy.

Figure 6.4 shows how the policy engine functions are integrated into the SCA environment by creating the necessary functional interfaces. The coupling between the policy engine and the radio will be rather complex because there is likely to be a big gap between abstract policy syntax and names, and because of the detailed design of the radio software functions and interfaces. At least one of the interfaces should be relatively simple to accomplish. The radio is already capable of network communications, presumably through an API for data

Figure 6.4: Software design for policy management.

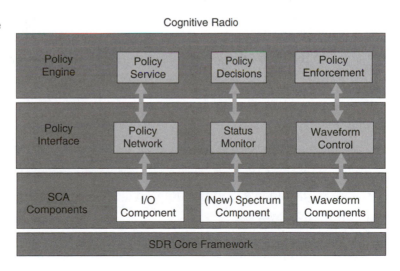

205

input/output (I/O), and a policy network interface can be implemented by software that will interact with the SCA to securely download policies to the policy server.

The policy enforcement interface will be more complicated because the policy actions will not generally correspond to particular controls for the attributes of the waveforms that are being transmitted by the radio. The waveform control interface must interact with the SCA waveform components to bridge this gap between behavior specifications and control parameters. Depending on the type of policies that will be enforced, it is likely that the waveform component may have to be adapted to be "policy-enabled." This adaptation would likely involve exposing new API methods that were designed to support the actions required by the policies. For example, the waveform control interface will have to communicate any limitations on transmission frequencies and power levels, limitations on protocols, or times when transmission is permitted.

What is entirely new is the spectrum component, which is the focus of the status monitor. The spectrum component will require signal-processing capabilities to determine channel occupancy and any other information about radio conditions and events that will be necessary to evaluate the ECA policy rules. The status monitor interface must be able to "observe" the radio operation and its spectrum environment and maintain information about the state of the device. It will have to notify the policy decision function about any relevant changes to the state that may trigger policy rules. This functional interface must aggregate digital signal processing results into higher-level event objects and provide event notifications to the policy engine. Because the events and conditions of concern depend on the policy specification, the policy engine must inform the status monitor what radio and spectrum attributes and conditions to observe, and what kind of notifications to return for consideration by the policy engine. Examples of the types of data that may flow across this interface are current frequency band, current geographic location, current time or data, and the communication roles of the radio in the network.

6.6.2 Policy Engine Design

Having discussed the external software interfaces for the policy engine, this section now looks deeper into the internal functional design of the policy engine by looking at what others have done, proposing a top-level design, trying to define software requirements, and building up a detailed design introducing new functionality as necessary.

Montanari et al. [39] describe a useful programming environment called POEMA (Policy Enabled Mobile Applications), which gives one example of designing policy-based "middleware." Boutaba and Znaty [31] insist that a policy engine design must have three logical components: execution engine, situation matcher, and observer. The observer is responsible for defining the situation in terms of "combinations of the values of observables (an attribute of some object which has a value which can be measured) and times of observations." The situation matcher makes policy decisions by "matching the observed state of the system with stored situation specifications" that correspond to the policy conditions and policy events discussed previously.[2] "The language used to express situations must be able to represent observables and provide operators for obtaining the value associated with an observation, and the time an observation was made" [31]. The execution engine operates on an "algorithmic block" (i.e., policy action) and produces a "sequence of instructions" that the policy enforcement function discussed previously will use to control the radio.

Figure 6.5 offers a tentative functional design for the cognitive policy engine. Due to the importance of the status monitor interface for the policy engine, it is included in the figure because Boutaba and Znaty [31] consider it to be an integral part of the policy engine. Note that two types of policy actions are produced by the policy decisions function. Stone et al. [18] remark that "Policies can be triggered in two ways, either statically or dynamically." "Static policies apply a fixed set of actions in a predetermined way according to a set of predefined parameters

Figure 6.5: Functional design for policy engine.

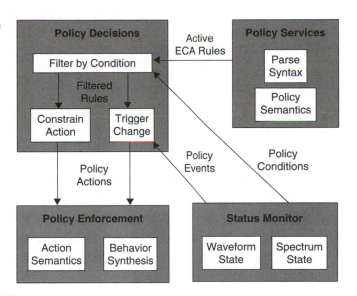

---

[2] The situation matcher may be implemented with a variety of technologies, ranging from complex Boolean specification, to fuzzy logic, to neural nets, or even a Markov model.

that determine how the policy is used." In other words, these are ECA rules with no policy events to trigger a change in policy. "Dynamic policies are only enforced when needed, and are based on changing conditions … actions can be triggered when an event causes a policy condition to be met" [18].

Some lessons about software design that have been learned from prior efforts to prototype a policy engine are informative. Strauss [19] lists some useful design requirements for a policy management engine, which are adapted to the cognitive radio application and listed here without further comment.[3]

1. A policy condition must allow read access to variable attributes in a way that the policy action can reference those attributes that matched the condition and the attributes of the event that triggered the rule. Thus the run-time system must bind variables to instances when passed to the condition and action.

2. There must be a construct to specify the value space in which the free variables of conditions are evaluated that span all instances within a certain table or all instances of a class.

3. A class that models a certain object must support accessory methods that allow a policy to retrieve and manipulate an element and support computations. This is not a requirement for the policy engine architecture itself but for the design of interface.

4. Three types of time events may be required: Periodic events that trigger continuously for a given period, calendar events that trigger periodically at points in time specified by calendar-type attributes, and one-shot events that trigger exactly at a point in time specified by calendar-type attributes.

5. Another type of event is based on the reception of external notification, like such as state reports, and these must be mapped to specific events. Details of the initiating notifications should be accessible through accessory methods of the events.

6. The policy run-time engine must support a mechanism to report errors and optional tracing/debugging information so that users can monitor the policy engine and the authors of policies can test and debug their policy codes.

7. The access to variables in conditions and actions may fail, and the policy run-time engine must be able to handle these situations in a way that policy code can catch the error conditions and bring the variable to a determined state.

---

[3] Any errors in adaptation are the responsibility of this author.

8. It must be possible to store and execute multiple policies independently. Their codes must not share any name space.

9. A security mechanism is required to differentiate which users have access to which roles on which policies (build on the existing SCA security component).

10. The notation of policies should remain declarative as much as possible, but the programmatic policy system has to implement complicated actions with code instead of just the goal.

11. It should be possible to avoid redundancy in a way that policies or policy groups sharing rules and rules-sharing conditions or actions can be built by referring common code instead of copying code fragments to increase reusability and avoid some errors.

12. The policy engine should be able to pass arguments to policies when they are activated by conditions based on messages or shared memory to reduce redundancy.

6.6.3 Integration of the Radio into a Network Policy Management Architecture

Once PDP and PEP functionalities are accommodated in the policy engine, the first question concerns what can be gained by embedding the cognitive radio in networked management architecture. The discussion of the policy service functionality makes it clear that, at a minimum, the network policy architecture needs to be concerned with policy dissemination. As Stone et al. [18] put it, "Communication is needed to and from the policy repository.... In many proposals the policy repository is a directory, and therefore the appropriate access protocol would be the Lightweight Directory Access Protocol."

Of course, the actual "minimum" must do a little bit more. There must be some kind of secure authorization to download policies to the radio. One way to handle this is to architect the cognitive radio in such a manner that the only way to update the policy database is to physically wire a connection to a special interface, thus relying on physical security to ensure the integrity of the policy database. This makes it very tedious and inconvenient to update the policy database, however, particularly if a large number of radios are involved.

Since the purpose of the radio is to communicate, it makes sense to rely on network communications for policy updates. Sheridan-Smith [40] points out that peer-to-peer management communications also eliminates problems with synchronization of policy repositories because the radios can be informed of changes. Whenever radios enter a new network, part of the session initiation can include setting up a policy management interface to synchronize policy databases as required.

Naturally, no one would permit an unknown radio to remotely update the policy database without some kind of exchange of credentials to resolve trust issues. Li et al. [41] introduce a family of role-based trust management languages for representing policies and credentials. They define a "decentralized collaborative system" being formed by several autonomous organizations desiring to cooperate and share resources for their mutual benefit. Sheridan-Smith [40] also studies synchronization techniques for distributing the responsibility for policy-based network management (PBNM) across a set of independent autonomous PDPs: "The use of the pull model is more efficient than a push model. Alternatively, the network management system is required to poll each device in turn to ensure that they are operating correctly." The push model is particularly problematic for the cognitive radio because it will often be turned off and restarted or reinitialized.

Damianou et al. [23] introduce the next level of complexity: "It should be possible to dynamically update the policy rules interpreted by distributed entities to modify their behavior." When the concept of roles is introduced into the policy language, the following premise is tacitly accepted: "It is not practical to specify policies relating to individual entities—instead it must be possible to specify policies relating to groups of entities and also to nested groups such as sections within departments, sites within organizations and within different countries" [23].

A simple view of policy in regard to cognitive radio networks is that network policies constrain network communications. Specifically, network policy defines the relationship between clients (users, applications or services) using spectrum resources and the radios that provide access to those resources. (See Chapter 9 for additional details on how the policy engine may manage network protocols.) Lee et al. [33] take the argument to the next step: "Ideally, the management system will use feedback from monitoring and network measurement to influence its decisions about how to control the network configurations to improve efficiency, manage service quality, or to deal with changes in the network environment."

Thus a network policy management should incorporate a policy engine operating on network policies, receiving status updates from the policy service on individual cognitive radios, and downloading policy instructions to the members of the network. Now the policy management architecture is self-similar, and, as Boutaba and Znaty [31] state: "The whole network policy management system is then logically constructed in a hierarchical domain structure where low level domains provide their services to those of the upper layers. Domains are managed according to a set of harmonized policies."

At the network level, the policy manager must be focused on managing services by creating policies for individual devices based on the roles they will play in the network architecture. This is the root of the distinction between network-level policies and device-level policies. Sheridan-Smith [40] comments: "In general the network policies can apply to more entities and are easier to read, write and comprehend, precisely because they are not specific about how the goal should be achieved. Device-level policies can be specific about what needs to be done, but they are difficult to read and complex to write and can apply only to a subset of entities in the system."

This, then, is the state of the art in automated policy management of complex networks. The difficulty involves how to automatically refine higher-level policies into more specific management policies for domain members. This process involves planning for network operations and deriving specific policies for all the network elements to satisfy network goals. Stone et al. [18] define the research challenge as: "refining the goals, partitioning the targets the policies affect or delegating responsibility to another manager who can perform this derivation. The main motivation for understanding hierarchical relationships between policies is to determine what is required for the satisfaction of policies. If a high-level policy is defined or changed, it should be possible to decide which lower-level policies must be created or changed."

6.7 The Future of Cognitive Policy Management

Having examined functionality and designs for cognitive policy management, this section returns to the fundamental question "Why bother?" Damianou et al. [23] reply that the "motivation for policy-based services is to support dynamic adaptability of behavior by changing policy without recoding or stopping the system." This chapter has already indicated how this is accomplished. This section considers promising military and commercial applications of policy management technology and examines the challenges.

6.7.1 Military Opportunities for Cognitive Policy Management

Military network communications are managed by a hierarchy of network operations centers that fit naturally into a recursive management model. In fact, the military is unique in that it can hierarchically manage both the network resources and the demand by users for resources, unlike commercial market applications in which the financial goals always involve increasing user demand for services.

Figure 6.6 depicts the policy management hierarchy for communications resources. This hierarchy also handles application performance, but a parallel command structure exists for military personnel (users). Higher-level network

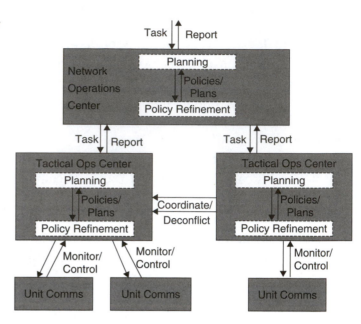

Figure 6.6: Military policy management hierarchy.

domains delegate communication management tasks to their subordinates to be planned and executed on lower-level network resources. In this type of architecture, coordination is necessary, as Boutaba and Znaty [31] note: "Peer-to-peer interactions take place in case of overlapping between domains to optimize plans and deconflict policies."

To handle the complexities of policy refinement that occur in this hierarchical architecture, "the concept of management policy is introduced as an intermediate step between goals and plans" [31]. In this case, "plans" refer to complex coordinated actions, so the refinement process becomes: (1) lower-level policies are derived from goals and missions passed down the hierarchy as tasks, (2) officers define action plans consistent with the policies, and (3) subordinates are tasked to the plans.

The fact that DARPA has invested significant funds to support the development of various policy management applications for the military suggests that this technology will filter down to the operational level. As network communications systems such as the tactical Warfighter's Internet (WIN-T) and the JTRS cluster procurements come online, policy management should play an increasing role.

6.7.2 Commercial Opportunities for Spectrum Management

There is wide recognition that the FCC's spectrum licensing and leasing practices have become highly inefficient and have led to artificial shortages due to the dedication of frequency bands for underutilized services. Due to the significant financial investments in this area, progress will probably depend primarily on the perception of financial incentives for the current spectrum lessees.

In particular, there has been some discussion of the possibility of spectrum "micro-leases" that would permit occasional opportunistic use of certain spectral bands in specific locations. Chiang et al. [42] describe protocol primitives to support a "Vickery auction" as a rational allocation of resources in a distributed system [43]. The concept requires creation of a "broker space" and an "auction space," by which buyers and sellers could automatically purchase portions of the time/space/ frequency spectrum. In this protocol, the necessary steps are:

1. A selling agent informs the market of its intent to sell spectrum by issuing an *offer* message.
2. Brokers link buyers to market opportunities.
3. Buyers submit *bid* messages to the seller.
4. The selling agent chooses its preferred buyer.
5. The transaction is *settled* (with a message) and the lease and moneys are conveyed between buyer and seller.

In some manner, it would be necessary to tie the policy management system into this market to enable and disable the use of spectrum resources.

6.7.3 Obstacles to Adoption of Policy Management Architectures

This section summarizes a few of the challenges to widespread adoption of policy management techniques for cognitive radios. Concerns about potential difficulty in getting FCC certification for a radio managed by a cognitive policy engine have already been addressed and will not be further discussed here.

It is one thing to design a policy management architecture and another thing to demonstrate that its functionality is worth the effort to implement it. Lee et al. [33] propose the design of experiments to test hypotheses and evaluate whether the actions specified actually improve or worsen system performance: "the evaluation should proceed by examining a particular approach and making quantitative measurements according to a well-defined methodology." The DARPA PolySurv program made some efforts in this direction, but the published literature consists only of marketing brochures and advertisements from vendors.

Verma [44] recalls the need for policy-transformation logic that translates high-level network administration policies to device-specific policies that can be automatically disseminated. The approach also ensures that high-level policies are mutually consistent, correct, and feasible. "The heart of policy management lies in the policy translation logic as to how the policies will be represented and how they will be managed" [44].

Boutaba and Znaty [31] see that "Today's challenge is to provide automated support not only for detecting and responding to trivial network and system events, but also for the process of planning and policy making to handle more complex situations." They enlarge the scope of the automation problem: "Peer negotiations for conflict avoidance should be handled as part of the proactive management process, whereas negotiations for conflict resolution should be handled as part of the reactive management process. Both approaches should involve all derivation of goals, policies, plans, and actions."

Sheridan-Smith [40] is also concerned: "If multiple autonomous management entities are making independent desiccations then each participant in the network might behave in a manner that is not aligned with other parts of the network, or other functions of the network which are independently managed. Coordination will ensure that the behavior that is stipulated in the policies is correctly met by all of the individual management entities."

The perspective put forth in this chapter is that the adoption of a policy-based spectrum management approach for cognitive radios will require standardization of policy languages and dissemination techniques designed specifically for cognitive radios and driven by market concerns. It seems likely that a cognitive policy engine can be designed to function usefully in the context of cognitive radios. The real question is probably not the technical feasibility of automatic policy management architectures. Instead, it is whether market forces will be favorable to the adoption of this technology and lead to integration across multiple platforms and vendors.

6.8 Summary

The focus of this chapter has been on the design of a practical cognitive policy engine that would enable the cognitive radio to operate reasonably within a suitable policy space. The approach has been to explore the engineering requirements for the policy engine in the context of a radio policy management domain that might reasonably reflect the operational environment for a cognitive radio. The scholarly literature shows what enabling technology is available and necessary to realize this objective in the near term.

The basic syntactic and semantic mechanisms for representing spectrum and network management policies were found to already be in existence in several forms. Although there are limits on the semantic capabilities of extant policy engines and languages, it is difficult to argue convincingly that there exists a significant technology gap in this area that would prevent cognitive radios from being deployed in a fruitful manner.

If anything, the opposite conclusion is much more compelling. There is no demonstrated requirement that calls for the representation of highly complex logical constraints on the operation of the cognitive radio. In fact, there is every reason to believe that implementations for policy engines in the first few generations of cognitive radios will be more than capable of enforcing restrictions on policy to constrain and optimize the behavior of the radio. It seems highly unlikely that communication regulations will call upon the radio to dynamically resolve complex and sometimes inconsistent policies.

This chapter has reviewed the application of policy engines in commercial networking environments and has found that the most difficult challenge lies not in processing or enforcing the policies, but rather in coherently, unambiguously specifying what behavior is desired. For this reason, existing policy-enabled network systems are generally focused on well-defined access control or network configuration and performance optimization problems for which the desired behaviors can be defined and agreed upon by users of the system.

The critical engineering problems that remain to be resolved for implementing cognitive policy engines are primarily software design questions. Almost all of these questions will be answered in terms that are specific to the particular kinds of real-time software platforms and policy instantiations that have already been addressed in this chapter. The interesting academic questions addressed primarily relate to theoretical limitations of policy languages and abstract processing architectures that are much more complex than what will be initially embodied in the cognitive radio. They are worthy of note, but should not be construed as obstacles for implementing a cognitive policy engine. The real challenge lies in unambiguously specifying what behaviors are required of the cognitive radio, and achieving consensus that such behaviors are sufficient for licensing the operation of the radio.

References

[1]   J. Bemmel, P. Costa and I. Widya, "Paradigm: Event-Driven Computing," *Freeband Awareness Deliverable D2.7a*, October 2004; http://awareness.freeband.nl

[2]   R. Chadha, G. Lapiotis and S. Wright, "Policy-Based Networking," *IEEE Network*, Vol. 16, No. 2, March 2002, pp. 8–9.

[3]   K. Carey, K. Feeney and D. Lewis, "State of the Art: Policy Techniques for Adaptive Management of Smart Spaces," *State of the Art Surveys*, Release 2, Trinity College Dublin, May 2003, pp. 58–66.

[4] www.ntia.doc.gov/osmhome/allochrt.html

[5] http://www.itu.int/ITU-R/

[6] http://www.fcc.gov/sptf/reports.html

[7] L. Kagal, T. Finin and A. Joshi, "A Policy Language for a Pervasive Computing Environment," in *Proceedings of the IEEE 4th International Workshop Policies for Distributed Systems and Networks*, June 2003, pp. 63–77.

[8] http://www-dse.doc.ic.ac.uk/research/policies/

[9] http://www.ietf.org/html.charters/policy-charter.html

[10] http://www.darpa.mil/ato/programs/XG/index.htm

[11] http://www.darpa.mil/ato/programs/dynamiccoal.htm

[12] C. Phillips, S. Demurjian and T. Tang, "Towards Information Assurance for Dynamic Coalitions," in *Proceedings of the 2002 IEEE Workshop Information Assurance*, June 2002.

[13] A. Uszok, et al., "KAoS Policy Management for Semantic Web Services," *IEEE Intelligent Systems*, Vol. 19, No. 4, July/August 2004, pp. 32–41.

[14] http://www.swsi.org

[15] http://www.daml.org

[16] http://www.w3.org

[17] A. Uszok, et al., "KAoS Policy and Domain Services: Toward a Description-Logic Approach to Policy Representation, Deconfliction, and Enforcement," in *Proceedings of the IEEE 4th International Workshop Policies for Distributed Systems and Networks*, June 2003, p. 93.

[18] G. Stone, B. Lundy and G. Xie, "Network Policy Languages: A Survey and a New Approach," *IEEE Network*, Vol. 15, No. 1, January 2001, pp. 10–21.

[19] F. Strauss, *Java Policy Management System: Design and Implementation Report*, Computer Science Department, Technical University Braunshweig, September 2001.

[20] http://www.ibr.cs.tu-bs.de/projects/jasmin/

[21] http://www-dse.doc.ic.ac.uk /research/policies

[22] N. Dulay, et al., "A Policy Deployment Model for the Ponder Language," in *Proceedings of the IEEE/IFIP International Symposium on Integrated Network Management (IM'2001)*, May 2001.

[23] N. Damianou, et al., *A Survey of Policy Specification Approaches*, Imperial College of Science Technology and Medicine, April 2002 (submitted for publication).

[24] N. Damianou, et al., "The Ponder Policy Specification Language," in *Proceedings of the Policy 2001: Workshop Policies for Distributed Systems and Networks*, Springer-Verlag, LNCS, Vol. 1995, 2001, p. 18.

[25] G. Tonti, et al., "Sematic Web Languages for Policy Representation and Reasoning: A Comparison of KAoS, Rei, and Ponder," in *Proceedings of the 2nd International Semantic Web Conference* (ISWC 2003), Springer-Verlag, October 2003.

[26] www.nortel.com

[27] B. Moore, et al., "Policy Core Information Model—Version 1 Specification," *IETF RFC 3060*, February 2001; http://www.rfc-editor.org/rfc/rfc3060.txt

[28] Y. Snir, et al., "Policy Quality of Service (QoS) Information Model," *IETF RFC 3644*, November 2003; http://www.rfc-editor.org/rfc/rfc3644.txt

[29] K. Feeney, D. Lewis and V.P. Wade, "Policy Based Management for Internet Communities," in *Proceedings of the 5th IEEE International Workshop Policies for Distributed Systems and Networks (POLICY'04)*, June 2004, p. 23.

[30] A. Uszok, et al., "Policy and Contract Management for Semantic Web Services," in *Proceedings of the AAAI Spring Symposium on Semantic Web Services*, March 2004.

[31] R. Boutaba and S. Znaty, "An Architectural Approach for Integrated Network and Systems Management," *ACM-SIGCOM Computer Communication Review*, Vol. 25, No. 5, October 1995, pp. 13–39.

[32] A. Uszok, et al., "Applying KAoS Services to Ensure Policy Compliance for Semantic Web Services Workflow Composition and Enactment," in *Proceedings of the 3rd International Semantic Web Conference (ISWC 2004)*, Springer-Verlag, LNCS, Vol. 3298, November 2004, pp. 425–440.

[33] S. Lee, et al., "Managing the Enriched Experience Network—Learning-Outcome Approach to the Experimental Design Life-Cycle," in *Proceedings of Australian Telecommunications, Networks and Applications Conference (ATNAC'03)*, December 2003.

[34] H. Khurana and V.D. Gligor, "A Model for Access Negotiations in Dynamic Coalitions," in *Proceedings of the Enterprise Security Workshop, 13th IEEE International Workshop Enabling Technologies: Infrastructures for Collaborative Enterprises (WETICE)*, June 2004.

[35] J.W. Lloyd, *Foundations of Logic Programming*, 2nd Edition, Springer-Verlag, 1987.

[36] http://www.ksl.stanford.edu/software/JTP

[37] www.sdrforum.org

[38] *Software Communications Architecture*, Prepared by Joint Tactical Radio System Joint Program Office, version 2.2.1, April 2004.

[39] R. Montanari, G. Tonti and C. Stefanelli, "Policy-Based Separation of Concerns for Dynamic Code Mobility Management," in *Proceedings of the 27th Annual International Computer Software and Applications Conference (COMPSAC'03)*, IEEE Computer Society Press, November 2003.

[40] N. Sheridan-Smith, *A Distributed Policy-Based Network Management (PBNM) System for Enriched Experience Networks (EENs)*, Assessment of Proposed Doctoral Research, University of Technology, Sydney, November 2003.

[41] N. Li, J.C. Mitchell and W.H. Winsborough, "Design of a Role-Based Trust Management Framework," in *Proceedings of the 2002 IEEE Symposium on Security and Privacy*, IEEE Computer Society Press, May 2002.

[42] F. Chiang, et al., "Autonomic Service Configuration for Heterogeneous Tele-
communication MASs with Extended Role-Based GAIA and JADEx," in
*Proceedings of the 2005 International Conference on Service Systems and Service
Management (ICSSSM 2005)*, June 2005.

[43] http://teleholonics.eng.uts.edu.au

[44] D. Verma, "Simplifying Network Administration using Policy Based Management,"
IEEE Network, Vol. 16, No. 2, March 2002, pp. 20–26.

CHAPTER 7

Cognitive Techniques: Physical and Link Layers

Thomas W. Rondeau and Charles W. Bostian
Bradley Department of Electrical and Computer Engineering
Virginia Tech, Blacksburg, VA, USA

7.1 Introduction

This chapter discusses the expectation of a fully functional cognitive radio, including the cognitive decision-making process using case-based theory and genetic algorithms (GAs), to solve the multi-objective optimization problem posed by such a radio. The presentation has as its basis the cognitive engine developed at the Virginia Polytechnic Institute and State University (Virginia Tech)—Center for Wireless Telecommunications (VT-CWT).

This chapter focuses on the intelligent cross-layer optimization of physical (PHY) and link (or medium access control, MAC) layers. The reader is encouraged to think beyond these two layers and consider how other layers, particularly network and transport, can be adapted and optimized by using the same techniques.

Section 7.2 defines optimization for a cognitive radio. A discussion of the cognitive radio as a mix of artificial intelligence (AI) and wireless communications follows in Section 7.3. Section 7.4 addresses the PHY and MAC layers, and considers which measurable radio settings and specifications ("knobs") and which radio and channel performance measures ("meters") fall into which layer. Section 7.5 introduces multi-objective decision-making (MODM) theory to analyze the radio's performance, and presents the analogy of GA to represent the methodology. In Section 7.6, the tiered algorithm structure of the cognition loop, based on modeling, action, feedback, and knowledge representation, is explored in detail.

Tailoring GA techniques for a cognitive radio, based on the simple GA discussed in these earlier sections, is addressed in Section 7.7 by looking at case-based reasoning (CBR) and case-based decision theory (CBDT). The need for a higher level of intelligence is the topic of Section 7.8, and Section 7.9 shows how the collection of techniques addressed in this chapter creates a cognitive engine capable of controlling and adapting a cognitive radio. Section 7.10 then summarizes the processes and ideas brought forth in this chapter.

7.2 Optimizing PHY and Link Layers for Multiple-Objectives Under Current Channel Conditions

The goal of a cognitive radio is to optimize its own performance and support its user's needs. But what does "optimize" mean? It is not a purely selfish adaptation where the radio seeks to maximize its own consumption of resources.

Consider the "tragedy of the commons" metaphor as it is often applied to wireless communications [1].[1] If two pairs of nodes are communicating on different networks using transmissions that overlap in time and frequency, they will interfere. The nodes observe the interference as a low signal-to-interference and noise ratio (SINR), and the classic response is then to increase transmitter power to obtain a corresponding increase in SINR. As the transmitter on one link increases its power, the other link will experience a lower SINR and respond by increasing its power. Each radio will respond in turn to maximize its SINR at its intended receiver by increasing its own transmit power. Each transmitter will ultimately increase its power to the limitations of the hardware. At this point, either both links will have low SINR and therefore poor performance, or the link with the more powerful transmitter will completely drown out the other. This is obviously a poor solution. Even in the latter scenario, a second glance shows this to be bad for all concerned because now each radio is transmitting much more power than is required, raising power consumption, reducing battery life, and increasing potential interference to other users. This scenario is regularly reenacted in the 2.4 GHz industrial, scientific, and medical (ISM) band in which Bluetooth and IEEE 802.11 devices are constantly creating cross-interference. Here 802.11 has a higher transmit power, but Bluetooth has a protocol to continually repeat packet transmissions until a successful transmission occurs.

---

[1] Hazlett's article [1] does not give a highly technical overview of this concept, but rather a regulatory analysis of the spectrum issues in general.

In contrast, a radio capable of understanding its environment and making intelligent adaptations will recognize the problem it encounters with a competing link trying to use the same band. While observing the other link, the cognitive radio will not be limited by a simplistic understanding that "low SINR means I should increase my transmitter power." Instead, it will try other solutions, such as altering the modulation or channel coding in ways that will improve frame error rate (FER) performance in the channel. Or it will seek a channel free of interference and change its center frequency, thus relieving both radios of the burden of fighting for the spectrum.

In a heavily congested spectral environment, changing frequency might not be an acceptable solution. This is why it is important to look at all the possible adjustments to the PHY and MAC layers to improve performance. A situation might arise when all possible frequencies are in use, or the bandwidth required is not available free from interference. At such times, a mix of cooperative techniques could be used, perhaps involving quadrature amplitude (and phase) modulation (QAM), spread spectrum techniques, orthogonal frequency division multiplexing (OFDM), clever timing mechanisms, smart antenna beam forming or null forming, or other operations that will allow sharing. This chapter addresses methods of how to find a local or global optimum for the current channel environment. A generalized solution is not usually the answer; for example, spreading techniques might better share spectrum than narrowband techniques, but a spread spectrum system has its limits and might unnecessarily waste system resources. A dynamic combination of techniques is really required to best adapt the radio in real time for the specific local problems at hand. A well-designed cognitive radio understands situations and analyzes how to best adapt all available radio communications parameters to present conditions for optimum performance.

First, a definition of optimization is in order. Within this chapter, a radio is optimized when it achieves a level of performance that satisfies its user's needs while minimizing its consumption of resources such as occupied bandwidth and battery power. To apply this, we need to understand what needs the user has and how the radio performance can be adapted to meet these needs. The bottom line, in general, is that the radio not overoptimize its external performance (its performance as observed by other radios) because this will have a negative impact on its internal performance (computational power, complexity, and available internal resources).[2] The next section clearly defines the cognitive radio. The succeeding

---

[2] Other, more sophisticated considerations are beyond the scope of this chapter.

sections then discuss what parameters can be altered and how a cognitive radio can use them to optimize its performance.

7.3 Defining the Cognitive Radio

Cognitive radios merge AI and wireless communications. The field is highly multidisciplinary, mixing traditional communications and radio work from electrical engineering while applying concepts from computer science. Interestingly, Claude Shannon, one of the early giants of communication theory, spent some of his time thinking about and discussing intelligent machines, specifically, chess-playing machines. In an article published in 1950, he discusses how computers could be made to intelligently play chess, but he also lays out reasons why this has implications in other areas, such as designing filters, equalizers, relays, and switching circuits, routing, translating languages, organizing military operations, and making logical deductions [2].

The cognitive radio architecture envisioned in the discussion in this chapter is shown in Figure 7.1. Here, the intelligent core of the cognitive radio exists in the cognitive engine. The cognitive engine performs the modeling, learning, and optimization processes necessary to reconfigure the communication system, which appears as the simplified open systems interconnection (OSI) stack [3]. The cognitive engine takes in information from the user domain, the radio domain, the policy domain, and the radio itself. The user domain passes information relevant

Figure 7.1: Generic cognitive radio architecture. This architecture has a cognitive engine to observe behaviors of the OSI protocol stack and propose optimizations based on the current environment. The policy engine determines whether the hardware can support those optimizations as well as whether it is allowed to by regulatory and network control.

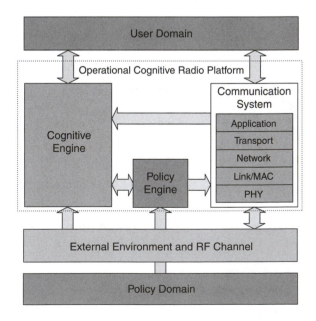

to the user's application and networking needs to help direct the cognitive engine's optimization. The radio domain information consists of radio frequency (RF) and environmental data that could affect system performance such as propagation or interference sources. The policy engine receives policy-related information from the policy domain. This information helps the cognitive radio decide on allowable (and legal) solutions and blocks any solutions that break local regulations.

Most of the topics shown in Figure 7.1 are covered in more detail in this chapter as their relationship to the cognitive engine are developed. The policy engine and policy domain will be left to experts of that field, but are included here for completeness.

7.4 Developing Radio Controls (Knobs) and Performance Measures (Meters)

The first problem in dealing with cognition in a system is to understand (1) what information the intelligent core must have and (2) how it can adapt. In radio, we can think of the classical transmitters and receivers as having adjustable control parameters (knobs) that control the radio's operating parameters. Think of a frequency modulation (FM) broadcast receiver with a tuning knob to select which station you are listening to as well as equalizer knobs to adjust the sound quality to your liking. Radio performance metrics are referred to as meters. What follows is an analysis of the knobs and meters important to a cognitive radio on the PHY and link layers.

Huseyin Arslan has developed a useful layered classification of knobs and meters (*observable parameters* and *writable parameters* in his notation), which is summarized and expanded in Table 7.1.[3]

7.4.1 PHY- and Link-Layer Parameters

Knobs

The knobs of a radio are any of the parameters that affect link performance and radio operation. Some of these are normally assumed to be design parameters, and others are usually assumed to be under real-time control of either the operator or the radio's real-time control processes. Figure 7.2 shows a simple system diagram of the PHY- and link-layer portions of a transmitter. In the PHY layer, center

---

[3] Huseyin Arslan, personal communication.

Table 7.1: Example tabulation of knobs and meters by layer.

Layer	Meters* (observable parameters)	Knobs (writable parameters)
NET	Packet delay Packet jitter	Packet size Packet rate
MAC	CRC check ARQ FER Data rate	Source coding Channel coding rate and type Frame size and type Interleaving details Channel/slot/code allocation Duplexing Multiple access Encryption
PHY	BER SNR SINR RSSI Path loss Fading statistics Doppler spread Delay spread Multipath profile AOA Noise power Interference power Peak-to-average power ratio Error vector magnitude Spectral efficiency	Transmitter power Spreading type Spreading code Modulation type Modulation index Bandwidth Pulse shaping Symbol rate Carrier frequency Dynamic range Equalization Antenna beamshape

*AOA: angle of arrival; ARQ: automatic repeat request; CRC: cyclic redundancy check; MAC: medium access control; NET: network layer; PHY: physical layer; RSSI: received signal strength indicator; SINR: signal-to-interference and noise ratio; SNR: signal-to-noise ratio.

frequency, symbol rate, transmit power, modulation type and order, pulse-shape filter (PSF) type and order, spread spectrum type, and spreading factor can all be adjusted. On the link layer are variables that will improve network performance, including the type and rate of the channel coding and interleaving, as well as access control methods such as flow control, frame size, and the multiple access technique.

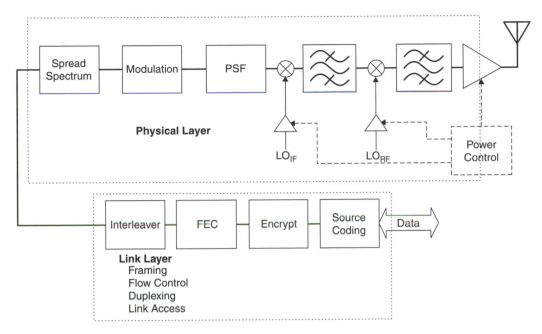

Figure 7.2: Generic transmitter PHY and MAC layers. Many of the radio control parameters (knobs) apply to these elements of the block diagram, resulting in profound impact on radio performance (meters). (*Note*: FEC: forward error correction.)

Meters

Once we understand what knobs are available to optimize the radio system performance, we must understand how these changes affect the radio channel and system performance to allow an autonomous, intelligent decision-maker to adapt the radios.

Performance is a measure of the system's operation based on the meter readings. In optimization theory, the meters represent utility and cost functions that must be maximized or minimized for optimum radio operation. All of these performance analysis functions constitute objective functions. In an ideal case, we can find a single-objective function whose maximization or minimization corresponds to the best settings. However, communication systems have complex requirements that cannot be subsumed into a single-objective function, especially if the user or network requirements change. Metrics of performance are as different for voice communications as they are for data, e-mail, web browsing, or video conferencing.

The types of meters represent performance on different levels. On the PHY layer, important performance measurements deal with bit fidelity. The most obvious meters are the signal-to-noise ratio, or a more complex SINR. The SINR has a direct consequence on the bit error rate (BER), which has different meanings for different modulations and coding techniques, usually nominally determined by the SINR ratio, Eb/(N0 + I0), where Eb is energy per bit, N0 is noise power per bit, and I0 is interference power per bit. On the link layer, the packet fidelity is an important metric, specifically the packet error rate (PER).

There are more external metrics to consider as well, such as the occupied bandwidth and spectrum efficiency (number of bits per hertz) and data rate. The growth of complexity to optimize multiple metrics quickly becomes apparent. Each metric has unique relationships with the other metrics, and different knobs alter different metrics in different ways. For example, altering the modulation type to a higher order will increase the data rate but worsen the BER.

Internal metrics also are involved in decision-making. To decrease the FER, we could use a stronger code, but this increases the computational complexity of the system, increasing both latency as well as the power required to perform the more complex forward error correction (FEC) operation. Decreasing the symbol rate or modulation order will decrease the FER as well without increasing the demands of the system, but at the expense of the data rate. Figure 7.3 begins to

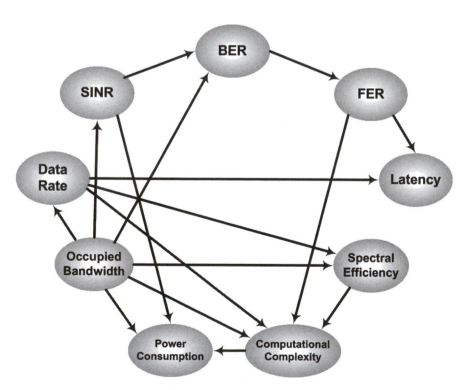

Figure 7.3: Directed graph indicating how one objective (source) affects another objective (target).

expand upon these relationships, where the direction of the arrow indicates that optimization in the source objective affects the target objective. Ongoing work fully defining these relationships should lead to more knowledge for the adaptation and learning system to use.

7.4.2 Modeling Outcome as a Primary Objective

The basic process followed by a cognitive radio is that it adjusts its knobs to achieve some desired (optimum) combination of meter readings. Rather than randomly trying all possible combinations of knob settings and observing what happens, it makes intelligent decisions about which settings to try and observes the results of these trials. Based on what it has learned from experience and on its own internal models of channel behavior, it analyzes possible knob settings, predicts some optimum combination for trial, conducts the trial, observes the results, and compares the observed results with its predictions, as summarized in the adaptation loop of Figure 7.4. If results match predictions, the radio understands the situation correctly. If results do not match predictions, the radio learns from its experience and tries something else.

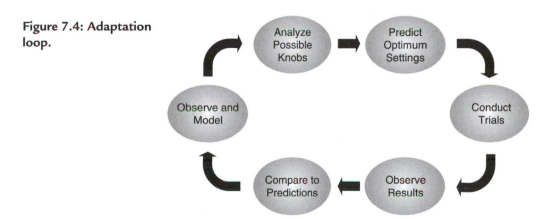

Figure 7.4: Adaptation loop.

This operational concept employed for the cognitive radio resembles closely some of the current thinking about how the human brain works [4]. The argument holds that human intelligence is derived from predictive abilities of future actions based on the currently observed environment. In other words, the brain first models the current situation as perceived from the sensor inputs, and it then makes a prediction of the next possible states that it should observe. When the predictions do not match reality, the brain does further processing to learn the deviation and incorporate it with its future modeling techniques. Although knowledge of how the human brain actually works is still uncertain, this predictive model is a good one to work from because it brings together the necessary behavior required from the cognitive radio.

As an example of the mathematics involved in this process, consider observations of BER and SINR. BER formulas are generally represented by the complementary error function or the Q function (Eq. (7.1)) as a function of the SINR:

$$\text{erfc}(x) = \frac{2}{\sqrt{\pi}} \int_x^\infty e^{-t^2} dt$$

$$Q(x) = \frac{1}{\sqrt{2\pi}} \int_x^\infty e^{-t^2} \qquad x \geq 0 \qquad\qquad (7.1)$$

$$Q(x) = \frac{1}{2} \text{erfc}\left(\frac{x}{\sqrt{2}}\right)$$

where x is the SINR. A computationally efficient calculation for BER formulas uses an approximation to the complimentary error function. Eq. (7.2) shows two approximations, one for small values of x (<3) and one for large values of x (>3) [5, 6].

Figure 7.5 compares the results of the formulas to the actual analytical function:

$$\text{erfc}(x) = \begin{cases} \left| 1 - \frac{1}{\sqrt{\pi}}\left(x - \frac{x^3}{3} + \frac{x^5}{10} + \cdots \right) \right| & \text{for } x < 3 \text{ with 40 items in the series} \\ \left| \frac{e^{-x^2}}{x\sqrt{\pi}}\left(1 - \frac{1}{2x^2} + \frac{1.3}{2^2 x^4} - \frac{1 \cdot 3 \cdot 5}{2^3 x^6} + \cdots \right) \right| & \text{for } x \geq 3 \text{ with 10 items in the series} \end{cases} \qquad (7.2)$$

Figure 7.5: erfc approximation Eq. (7.2)—compared to analytical formula—Eq. (7.3).

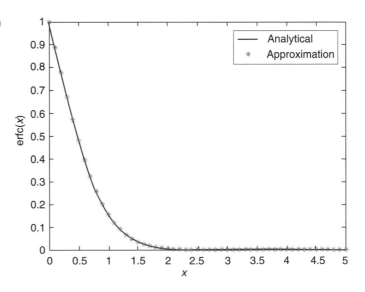

These formulas are useful because the normal approximation for the erfc function is valid only for large x ($x > 3$), and the Q function is too computationally intensive to calculate. Because a cognitive radio needs to perform a lot of these calculations, we need efficient equations; these equations trade accuracy for computational time based on the number of terms included in the series expansion. Similar lines of thought must go into developing each objective calculation.

The exact representation of the BER formula depends on the channel conditions and modulation being used. A standard BER formula for binary phase shift keying (BPSK) signals in an additive white Gaussian noise (AWGN) channel is:

$$P_e = \frac{1}{2} \operatorname{erfc}\left(\sqrt{T_0 B \frac{C}{N}} \right) \tag{7.3}$$

where T_0 is the symbol period, B is the bandwidth, C is the signal carrier energy, and N is the noise power.

In a fading channel with a probability density function (PDF) of $p(x)$, the BER of a signal is defined as:

$$P_e = \int_0^\infty P_{\mathrm{AWGN}}(x)p(x)\, dx \tag{7.4}$$

where $P_{\mathrm{AWGN}}(x)$ is the BER formula in an AWGN channel.

The radio observes the BER and SINR value. If these are consistent according to the above formulas, the radio can assume that the channel is behaving predictably. It can then turn knobs that directly affect SINR, for example starting with the easiest, transmitter power.[4] If the transmitter power is already at the allowable limit, the radio may lower the data rate to change the occupied bandwidth and therefore increase the average energy per bit. If the BER and SINR are not consistent with the known formulas, the radio might assume, for example, that

---

[4]The difference between predicted link performance and actual link performance includes both errors in the estimate of propagation channel losses, and nonlinear effects arising from interference and multipath. Performance on an AWGN channel in the absence of multipath is predictable. When this performance difference is significantly large, it may become clear that transmit power alone is inadequate to achieve the necessary performance. Thus, the performance difference is a good indicator of the need to invoke these cognitive radio techniques to optimize performance in the presence of unusual channel behavior.

the channel is dispersive and opt to change the carrier frequency rather than the transmitter power.

This analysis has dealt with only a single objective. The radio can, in fact, read a number of meters, and each of these can be some objective function we may wish to optimize. Standard communications theory can lead us to the methods of mathematically modeling each objective [7]. The communications analysis tools are fairly standard, and another aspect is the consideration for how to efficiently realize each objective function. For each, we must carefully choose the proper analytical expression that is not too computationally complex.

7.5 MODM Theory and Its Application to Cognitive Radio

The wireless optimization concept has already been described through an analysis of the many objective functions (dimensions) of inputs (knobs) and outputs (meters). In this scenario, the interdependence of the objectives to each other and to various knobs makes it difficult to analyze the system in terms of any one single objective. Furthermore, the needs of the user and of the network cannot all be met simultaneously, and these needs can change dramatically with time or between applications. For different users and applications, radio performance and optimum service have different meanings. As a simple example, e-mail has a much different performance requirement than voice communications, and a single-objective function would not adequately represent these differing needs.

Without a single-objective function measurement, we cannot look to classic optimization theory for a method to adapt the radio knobs. Instead, we can analyze the performance using MODM criteria. MODM theory allows us to optimize in as many dimensions as we have objective functions to model.

MODM work originated about 40 years ago and has application in numerous decision problems from public policy to everyday decisions (e.g., people often decide where to eat based on criteria of cost, time, value, customer experience, and quality). An excellent introduction to MODM theory is given in a lecture from a workshop held on the subject in 1984 [8]. Schaffer then applied MODM theory to create a GA capable of multi-objective analysis in his doctoral dissertation [9]. Since then, GAs have been widely used for MODM problem-solving. GAs are addressed in detail in Section 7.5.5.

7.5.1 Definition of MODM and Its Basic Formulation

At their core, MODMs are a mathematical method for choosing the set of parameters that best optimizes the set of objective functions. Eq. (7.5) is a basic representation of a MODM method [10]:

$$\min/\max\{y\} = f(\overline{x}) = [f_1(\overline{x}), f_2(\overline{x}), \ldots, f_n(\overline{x})]$$
$$\text{subject to: } \overline{x} = (x_1, x_2, \ldots, x_m) \in X \qquad (7.5)$$
$$\overline{y} = (y_1, y_2, \ldots, y_n) \in Y$$

Here all objective functions are defined to either minimize or maximize y, depending on the application. The x values (i.e., x_1, x_2, etc.) represent inputs and the y values represent outputs. The equation provides the basic formulation without prescribing any method for optimizing the system. Some set of objective functions combined in some way will produce the optimized output. There are many ways of performing the optimization. Section 7.6 discusses one of the more complex but useful methods of solving MODM problems for cognitive radios.

7.5.2 Constraint Modeling

An added benefit of MODM theory implicit in its definition is the concept of constraints. The inputs, x, are constrained to belong to the allowed set of input conditions X, and all output must belong to the allowed set Y. This is important for building in limitations for hardware as well as setting regulatory bounds.

7.5.3 The Pareto-Optimal Front: Finding the Nondominated Solutions

In an MODM problem space, a set of solutions optimizes the overall system, if there is no one solution that exhibits a best performance in all dimensions. This set, the set of nondominated solutions, lies on the Pareto-optimal front (hereafter called the *Pareto front*). All other solutions not on the Pareto front are considered dominated, suboptimal, or locally optimal. Solutions are nondominated when improvement in any objective comes only at the expense of at least one other objective [11].

The most important concept in understanding the Pareto front is that almost all solutions will be compromises. There are few real multi-objective problems for which a solution can fully optimize all objectives at the same time. This concept has been referred to as the utopian point [12]; this point is not considered further here because in radio modeling problems only very rarely do situations have a utopian point. One only has to consider the most basic radio optimization problem to see this: simultaneously minimize BER and power. Figure 7.6 shows the ideal BER curve of a BPSK signal. Here, point A is a nondominated point that minimizes power at the expense of the BER, and moving down the curve to point B, which is also nondominated, optimizes for BER at the expense of greater power consumption. Point C is a dominated point that represents a suboptimal solution of using differential phase shift keying (DPSK) to minimize the complexity of carrier phase tracking in high multipath mobile applications. The MODM problem then reduces to a trade-off decision between low BER, low power consumption, and complexity due to other system constraints.

Figure 7.6: Pareto front of a BPSK BER curve compared to a dominated solution of DPSK. Condition A is least power, condition B is lowest BER, and condition C is least complexity.

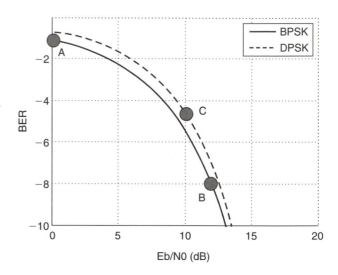

7.5.4 Why the Radio Environment Is a MODM Problem

The primary objectives developed thus far have different meanings and importance, depending on the user's needs and channel conditions. To optimize the radio behavior for suitable communications, we must optimize over many or all of the possible radio objectives. Take, as an example, the BER curve. In the two-dimensional plot, the result of the using a differential receiver was suboptimal because we were concerned with BER and power. But if we add complexity as an objective, or the need for a solution that does not require phase synchronization, such as might be necessary in highly complex and dynamic multipath, then using a differential receiver for lower complexity becomes an important decision along a third dimension. The resulting search space is N-dimensional and, due to the complex interactions between objectives and knobs, the space is difficult to define and certainly not linear, or even simply convex as is desirable in optimization theory [13]. These interactions are often difficult to characterize and predict, and so we must analyze each objective independently and use MODM theory to find an optimal aggregate set of parameters.

What further enhances the complexity of the search space is that it will change depending on the user and the application. For certain users or applications, different objectives will mean different levels of quality. As the overall optimization is to provide the best quality of service (QoS) to the user, there is no single search space that can account for all the variations in needs and wants from a given radio.

From this analysis emerge a few important points about how to analyze the multiple-objectives used in optimizing a radio:

- Many objectives exist, creating a large *N*-dimensional search space.
- Different objectives may be relevant for only certain applications/needs.
- The needs and subjective performances for users and applications vary.
- The external environmental conditions determine what objectives are valid and how they are analyzed.
- We may search for regions where multiple performance metrics meet acceptable performance, rather than searching for optimal performance.

This leads to a need for an MODM algorithm capable of robust, flexible, and online adaptation and analysis of the radio behavior. The clearest method of realizing all the needs of the problem statement is the GA, which is widely considered the best approach to MODM problem-solving [9, 10, 14–16]. Section 7.5.5 discusses the approach to GAs and Section 7.9.1 shows how to add user/application flexibility into the algorithm.

7.5.5 GA Approach to the MODM

Analyzing the radio by using a GA is inspired by evolutionary biological techniques. If we treat the radio like a biological system, we can define it by using an analogy to a chromosome, in which each gene of the chromosome corresponds to some trait (knob) of the radio. We can then perform evolutionary-type techniques to create populations of possible radio designs (waveform, protocols, and even hardware designs) that produce offspring that are genetic combinations of the parents. In this analogy, we evolve the radio parameters much like biological evolution to improve the radio "species" through successive generations, with selection based on performance guiding the evolution. The traits represented in the chromosome's genes are the radio knobs, and evolution leads toward improvements in the radio meters' readings.

GAs are a class of search algorithms that rely on both directed searches (exploitation) and random searches (exploration). The algorithms exploit the current generation of chromosomes by preserving good sets of genes through the combination of parent chromosomes, so there is a similarity between the current search space and the previous search space. If the genetic combination is from two highly fit parents, it is likely that the offspring is also highly fit. The algorithms also allow exploration of the search space by mutating certain members of the population that will form random chromosomes, giving them the ability to break the boundaries of the parents' traits and discover new methods and solutions. While providing the iterative solution through genetic combination, the randomness helps the population escape a possible local optimum or find new solutions never before seen or tried, even by a human operator. In effect, this last quality provides the algorithm with creativity.

Introduction to GAs

GAs are often useful in large search spaces, which can enable their use in many situations. A GA is a search technique inspired by biological and evolutionary behavior. The GAs use a population of chromosomes that represent the search space and determine their fitness by a certain criterion (fitness function). In each generation (iteration of the algorithm), the most fit parents are chosen to create offspring, which are created by crossing over portions of the parent chromosomes and then possibly adding mutation to the offspring. The crossover of two parent chromosomes tries to exploit the best practices of the previous generation to create a better offspring. The mutation allows the search algorithm to be "creative": that is, it can prevent the GA from getting stuck in a local maximum by randomly introducing a mutation that may result in improved performance metrics possibly closer to the global maximum, according to the optimization criteria.

To realize the GA, we follow the practices described by Goldberg [17]:

1. Initialize the population of chromosomes (radio/modem design choices)
2. Repeat until the stopping criterion
    (a) *Choose* parent chromosomes
    (b) *Crossover* parent chromosomes to create offspring
    (c) *Mutate* offspring chromosomes
    (d) *Evaluate* the fitness of the parent chromosomes
    (e) *Replace* less fit parent chromosomes
3. Choose the best chromosome from the final generation

This process is illustrated in detail in the next section.

The Knapsack Example

To explain the operation of a simple GA, we examine the knapsack problem [18], which is a classic nondeterministic polynomial time (NP) complete[5] problem [19], also called the subset-sum problem (SSP). The knapsack problem is defined by the task of taking a set of items, each with a weight, and fitting as many of these items into the knapsack while coming as close to, but not exceeding, the maximum weight the knapsack can hold. Mathematically, the knapsack problem is shown by Eq. (7.6), where K is the maximum weight the knapsack can hold, and N_s is the number of items in the set, S. The problem is represented by a weight vector w and a vector x that is a vector of 1's and 0's that indicates whether an item is present in the knapsack:

$$\max \sum_{i=1}^{N_s} x_i w_i$$

$$\text{subject to: } \sum_{i=1}^{N_s} x_i w_i \leq K$$

(7.6)

Following the practices just enumerated, use the following steps.

Step 1. Initialize Chromosomes

We introduce radio chromosome selection as a problem similar to knapsack selection. In this case, the problem consists of choosing the right set of items to place in the knapsack, so the chromosome will represent the vector x and consist of 1's and 0's as shown in Figure 7.7. Each gene is 1-bit wide and so very compact and easy to manipulate mathematically.

Figure 7.7: Chromosome representation of knapsack item vector.

x_1	x_2	x_3		$\cdots$	x_{N_s}

Step 2a. Choose

The choice of the parent chromosomes determines how random the population will remain and how much memory the biological system retains of

---

[5] NP is a class of decision problems whose positive solutions can be verified in polynomial time given the right information. It is a description of the difficulty of a computational problem, where NP-complete problems are the most difficult.

its fitness for its environment. Like biological evolution, the most fit parents are more likely to be chosen to produce offspring; however, it is still possible to choose an unfit parent. The more random the selection process, the more random the population will be in each generation. Conversely, the more the decision is weighted toward the most fit parents, the faster the convergence to local optima will occur. This trade-off is referred to as selection pressure.

As in all GA properties, the choice of how parents are chosen comes down to how random the population will remain and how fast convergence will occur. Although convergence time is highly important, fast convergence may not always be beneficial, depending on the shape of the fitness landscape, that is what the solution space looks like. If the fitness landscape has many local maxima and the GA is searching for the global maximum, fast convergence is more likely to lock on to a local maximum, and take many generations and mutations to get beyond the local maximum to find the global maximum. A more diverse population will generate many more solutions that are more likely to find the global maximum. Five different methods of selection were analyzed by De Jong in his doctoral dissertation [20].

The method used in this GA is called tournament selection. During reproduction, this selection scheme chooses as many parents as there are members in the population with replacement (i.e., the same parent may be chosen multiple times). In the selection, two parents are randomly chosen and the more fit parent wins the tournament and is selected for reproduction.

Step 2b. Crossover

Crossover is performed on two parents to form two new offspring. The GA has a crossover probability that determines if crossover will happen. A randomly generated floating-point value is compared to the crossover probability, and if it is less than the probability, crossover is performed; otherwise, the offspring are identical to the parents. If crossover is to occur, one or more crossover points are generated, which determines the position in the chromosomes where parents exchange genes. In this GA, once two parents are selected, two crossover points are randomly generated.

Figure 7.8 illustrates the crossover operation, but, for the simplicity of the figure, the genes represent a knapsack problem with only eight items. The genes after the first crossover point and before the second crossover point are interchanged in the parents to form the new offspring.

Figure 7.8: Parent chromosomes crossover. Crossover occurs at points 2 and 6 to create offspring chromosome sets.

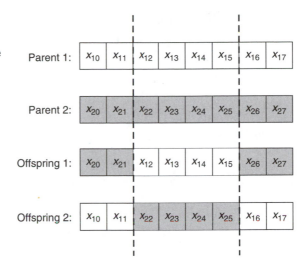

Step 2c. Mutate

After the offspring are generated from the selection and crossover, the offspring chromosomes may be mutated. Like crossover, there is a mutation probability. If a randomly selected floating-point value is less than the mutation probability, mutation is performed on the offspring; otherwise, no mutation occurs. Mutation is performed by randomly selecting a gene in the offspring's chromosome and generating a new value based on some PDF (often a uniform or Gaussian distribution). In the knapsack problem, a gene may be reset to either a 1 or 0 at random (other techniques invert the gene).

Steps 2d and 2e. Evaluate and Replace

Evaluation (step 2d) is done on the multi-objective performance of the offspring from the most recent genetic cycle to determine whether to continue the genetic process (step 2e). Evaluation is probably the most important piece of the GA aside from the initial chromosome definition. Choice of the fitness evaluation is vital to convergence and proper application.[6] The best set of offspring from among all those evaluated are ordered by performance and added to the rank-ordered list of chromosomes—say, for example, the top 10 or the top 100. Evaluation determines whether the gene pool is improving with the previous cycles—say, the last five

---

[6]For the knapsack problem, we try to find the chromosome with fitness defined by Eq. (7.6). If the constraint condition is not met, the fitness is set to 0.

cycles. If the performance is sufficiently above the goal behavior, the iterative process may terminate and the procedure goes to step 3. If, however, the performance over the previous cycles continues to improve but the goal performance is not yet met, the algorithm iterates back to step 2a. If the performance is unable to improve in previous cycles, the mutation rate may be adapted to a higher mutation rate. The evaluation function is very problem-specific, and is one of the main sources of research in the multi-objective GA to optimize cognitive radios.

Step 3. Results: Choose Best Chromosomes

To analyze the GA performance, we often graphically represent the performance over many generations, such as the knapsack example shown in Figure 7.9. This figure shows the general trends of a GA as a maximization problem with steady increase in the fitness over generations. In this example, we set the crossover probability at 0.95 and mutation probability at 0.05. There are 20 members of the population, and in each generation 17 members of the population are replaced by offspring. The problem contains 200 items, and each item could take on any floating-point weight value between 0 and 20. The maximum weight the knapsack can hold is a randomly generated integer of 810. The trends shown in Figure 7.9 show steady improvement toward the goal of 810, coming very close by the 155th generation with a value of 809.87.

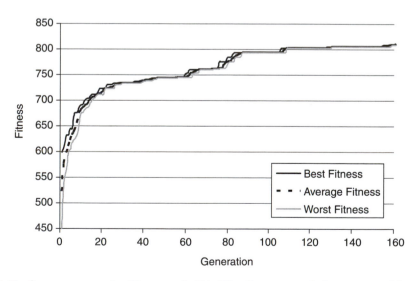

Figure 7.9: Performance graph of knapsack GA. The fitness metric improves with each generation, nearly reaching the maximum value after 155 generations.

As shown in Figure 7.9, a GA performs by making great initial progress, but then slows down as it approaches the optimal value, and even then, it does not perfectly achieve the optimum. Section 7.5 explores the use of GAs in radio optimization problems. Radio optimization must be real time, but it does not have to completely optimize under real-time constraints. Instead, we want to provide *better* responses as opposed to a *best* response for the immediate future. GAs quickly improve their performances in the first few generations; by the time the knapsack problem reaches its first plateau around generation 30, we have a useful solution. Looking at what we *need* out of the performance of a radio instead of just what we *want*, the GA can give us very usable solutions very quickly. As time and resources permit, the final iterations of the GA can find increasingly better solutions; other techniques, introduced in Section 7.5, such as CBDT, adaptive GAs, and distributed computing, can improve the solutions even faster.

Another point to consider involves the temporal features of optimum solutions. In the future, situations envisioned for the cognitive radio, the environments, both propagation and interference, as well as the user's and application's needs, will change, and with them the optimum solution. We therefore want a system that will continue to change and adapt with the needs of the situation for continuous improvement in performance.

7.6 The Multi-objective GA for Cognitive Radios

7.6.1 Cognition Loop

The primary goal of the cognitive engine is to optimize the radio, and the secondary functions are to observe and learn in order to provide the knowledge required to perform the adaptation. A cognitive radio becomes a learning machine through a tiered algorithm structure based on modeling, action, feedback, and knowledge representation, as shown in the cognition loop of Figure 7.10.

This section presents the high-level structure of the Virginia Tech cognitive radio solution as well as the background of the basic theories and procedures that were followed in its development. The following sections discuss the system in detail.

Modeling

Modeling means that the machine must have some representation of the outside world to which it can respond. Models represent external influences, such as the

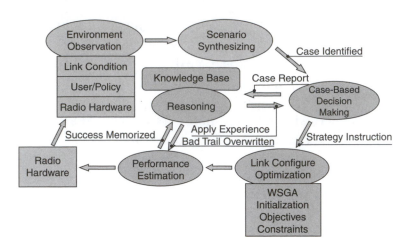

Figure 7.10: Cognition loop. Note the outer loop to observe and adapt a radio and the inner loop for learning.

user's network needs and actions, the radio spectrum and propagation environment, and the governing regulatory policy. As each of these influences changes, the radio must adjust itself to satisfy the new operating environment.

A primary function of the modeling system is to provide environmental context to the cognitive radio, such as propagation effects and the presence of other radios, which might be either cooperative or potential interferers.

Another goal of the modeling system is to monitor the data sent by the user and use it to determine the QoS parameters the cognitive radio must provide. This is a learning domain concept that a neural network could be employed to help solve, and the research group at VT is currently investigating the possibilities. The regulatory policy modeling and interpretation will come from other efforts, such as the Defense Advanced Research Projects Agency (DARPA) NeXt Generation (XG) project [21].

Figure 7.11 provides background about how Virginia Tech's work fits into the literature of machine learning and AI. It shows the VT-CWT cognitive engine in more detail, emphasizing its three major subsystems.

The cognitive system module (CSM) is responsible for learning and the wireless system genetic algorithm (WSGA) handles the behavioral adaptation of the radio, based on what it is told to do by the CSM. The modeling system observes the environment from many different angles to develop a complete picture. The CSM holds two main learning blocks: the evolver and the decision maker, which takes feedback from the radio that allows the evolver to properly update the knowledge base to respond to and direct system behavior.

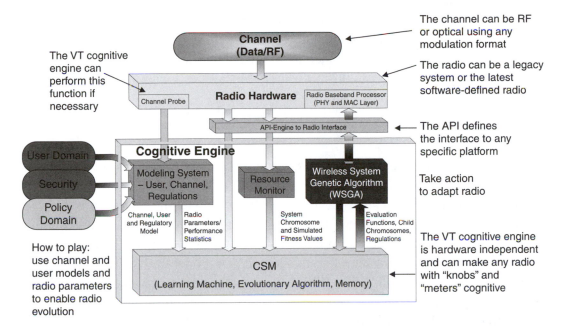

Figure 7.11: The VT cognitive engine. This hardware-independent modeling, learning, and adaptation system interfaces to radio hardware via a radio-specific application programmable interface.

Action

Actions are taken by the WSGA in creating a new radio configuration. This system is instructed by the CSM when a new model is observed and change is required in the radio operation. The WSGA creates a set of parameters that best optimize the radio for the new settings.

The WSGA treats the radios like biological creatures, where genes represent such radio traits as power, frequency, modulation, FEC coding, filtering, spreading, and so on. Groups of genes constitute chromosomes. The WSGA applies a multi-objective GA to a population of chromosomes and evaluates the results by using the objective functions and weights provided by the knowledge base. The functions determine which outputs are important, such as the BER, PER, power consumption, interference potential, and so forth, and the weights determine the relative importance of each function. After running the WSGA through some number of generations (terminating after convergence has been observed or following a predetermined number of generations, whichever occurs first), the chromosome that performs best in the largest number of objectives is chosen as the new set of radio parameters. Throughout this process, the chromosomes will be analyzed continually with respect to validation routines to ensure no combination is chosen for the final radio setting that would violate regulatory authority rules or the network operator's operational procedures. Detailed information about the WSGA and experimental results have been published [22], and a summary is provided in an example in Section 7.6.5.

Feedback

Unsupervised learning is accomplished through feedback and a series of rewards and punishments. Alan Turing proposed this idea in his article on computational intelligence [23], and it is a major theme in expert system and neural network learning algorithms [24]. In the VT learning machine, feedback information containing the radio's actual observations of its operation is provided. These actual observations of the performance parameters are then compared to simulated parameters from the WSGA. The WSGA uses the performance parameters within its GA to hypothesize the best new radio settings. When the simulated parameters are compared to the actual meters, the evolver in the CSM can determine how well the simulated performance matched the actual performance. This comparison helps the cognitive engine to understand how well its action analysis and selection block are working. Deviations between the actual and simulated performance measurements are penalized. The feedback and comparison mechanism allows the evolver to update the knowledge base to better represent the environment for improved future performance. This process requires an analysis of the model, action, and feedback system to know what is not modeled accurately, and then it specifically corrects this part of the knowledge base—probably by taking a new action from another part of the knowledge base that has performed well in these areas. This type of adaptation has been studied and used previously in GA work through techniques such as chromosome tagging and templates or in knowledge-based genetic techniques [17]. The evolver is a highly directed adaptation mechanism to correct for specific mistakes in judgment.

Knowledge Representation

Knowledge is represented as a database of past models and actions to take, along with past actions taken, and any estimates of the level of success associated with these actions. The knowledge base is a set of these models and actions that allow the WSGA to perform better as improved knowledge is gathered or generated by the evolver. As models are observed, actions are taken that best respond to the model's requirements for the radio. As these correlations are learned, the cognitive engine can more easily respond to situations as they arise. Likewise, past actions are kept in short-term memory to provide the WSGA with a set of recent actions; when a change is required, it may not have deviated too much from the immediate past, and the recently used actions may already work to guide the WSGA. The knowledge base is designed to provide the WSGA with as much relevant information as possible to reduce the computational complexity of the GA. If the population is primed with good solutions, the algorithm can terminate early and adapt the radio faster.

Case-Based Decision Theory

The core of the decision-making process is rooted in CBDT [25], a technique that grew out of economic decision-making research and is similar to the CBR used in some AI implementations. CBR recognizes similarity between the current event and past events in its memory, and actions taken by the machine for that past event are used to determine the actions the machine will take in the new event (these actions may not be the same). CBDT, in contrast, not only calculates a similarity of events in memory, but also assesses the utility of the past actions. The added knowledge of CBDT allows the decision-maker to act autonomously, more intelligently, and with better results than a CBR system or other memory-based approaches. CBDT is discussed in more detail in Section 7.7.2.

The key responsibility of the CSM is to maintain the knowledge base of model–action pairs, which ties the abstract models developed by the modeling system with the set of actions taken by the implementation system, which is the WSGA of Figure 7.11. The actions consist of genetic parameters such as crossover and mutation rates; previously used chromosomes that are related to both the specific channel and network model (exploiting benefits from long-term memory) and the most recently used chromosomes (exploiting benefits from short-term memory); and a set of objective (or fitness) functions and function weights. The actions provided by the CSM define the fitness landscape [17] and prime the GA.

Learning

Learning is the aggregation of the above processes. This learning machine mechanism is similar to processes known to occur in human learning: sensing, acting, reasoning, feedback, and accumulating knowledge and experience. It is a loop that continuously improves the cognitive radio's ability to perform. An even more powerful technique arises when the knowledge and learning capabilities are shared across the network. As any one radio gains knowledge about the models and actions to take, all radios benefit by the distribution of this knowledge. Likewise, the computational processes involved in algorithms such as the WSGA can be shared among many radios in the network, reducing the computations any one radio must perform as well as increasing the speed of the algorithm. Another benefit to distributing the cognitive engine is to share radio capabilities. For instance, if one radio in the network has a sounder such as that described by Rondeau et al. [26], then the modeling capabilities could be shared among all other radios on the network without such equipment.

7.6.2 Representing Radio Parameters as Genes in a Chromosome

The previous sections of this chapter introduced the possible knobs of the PHY and link layers that a cognitive radio can adapt, and argued that we can treat the radios as biological systems with chromosomes that use genes to represent the different knobs. Here, both of these thoughts are expanded to show the representation.

The greatest difficulty in the representation is the relatively large difference in the range of possible values (settings) for each knob. With a frequency-agile radio spanning frequencies from, say, 1 MHz to 6 GHz and a step size of 1 Hz, we must represent approximately 6×10^9 discrete frequencies in the gene. In contrast, the number of different modulation types is limited to a few dozen. Within each of these are only a handful of modulation indices or orders. Thus, modulation offers far fewer genetic possibilities than does operating frequency.

For the frequency representation, we could use clever programming techniques to realize different length genes for different traits; however, this quickly becomes problematic when a 37-bit gene represents the frequency (37 bits allows 1 Hz resolution from direct current (DC) to 100 GHz)—a gene of this size is very difficult to explore in a GA because a random search through this space is unrealistic (2^{37} possibilities). One practical approach is to segment these into multiple genes, where one gene provides a certain bandwidth of spectrum in which to search and the other provides the resolution within that bandwidth. Because both of these are genes, the search responsibility is distributed between the genes, and a usable band and center frequency can be found faster.

Knobs with smaller search spaces can be realized with smaller genes, and some traits can be combined. Modulation is segmented into the type of modulation—such as amplitude shift keying (ASK), frequency shift keying (FSK), phase shift keying (PSK), QAM, and so forth—and the order (2-, 4-, 8-, 16-point constellations, etc.). A 16-bit gene representation (65,536 possibilities) is far too large for either type or order, but the gene could be split into two 8-bit pieces. A system could then adapt these parameters together (choosing BPSK or 16-QAM) or as two separate genes (PSK with order 2 or QAM with order 16).

The challenge lies in scaling the GA to the full-scale factors of real-world applications, dealing with the difference between generality and domain-specific knowledge. Segmenting the genes into different sizes depending on traits leaves us with the problem of deciding how to best do this. If we overshoot our domain range, we can end up with large genes and overly complex search spaces. For example, if we decide to have a 37-bit gene for frequency, what do we do if we are applying the algorithm to a radio that can only realize 1–2 GHz center frequencies? We could easily represent the frequency range with a 30-bit gene (for just over 1×10^9 possibilities), but the remaining search space would be detrimental to the algorithm. However, setting a static field of 30, or the easier-to-realize 32-bit values, we could lose resolution for more complex or flexible radios.

A way around this is to use dynamic programming techniques and meta-level intelligence. With dynamic programming, we can instruct the GA to alter its gene size in real time for more appropriate adaptation to the current radio system. Understanding how to adjust the gene size can come from a higher-layer intelligence that understands the radio capabilities and environmental situation to adjust the GA behavior appropriately. Part of the responsibilities of the modeling system and the cognitive system previously discussed in section *Modeling* is to realize the meta-level intelligence. The various techniques to do this can change from platform to platform and between programming languages.

7.6.3 Multi-dimensional Analysis of the Chromosomes

Section 7.5 showed how different applications and users will have different performance criteria from the radio. This section discusses ways to capture these effects in the GA-based cognitive engine. When evaluating each chromosome in all of the objectives, we must find some way to analyze the performance of all radio chromosomes in such a way as to reflect the user needs and radio performance under current channel conditions.

Again, by using dynamic programming techniques, the CSM of the cognitive engine can learn the user and application needs and instruct the WSGA about which objectives matter and how much they matter. This information is passed from the CSM to the WSGA as a set of objective functions to use along with relative weighting values for each objective. For a video link, BER will receive a smaller weight than data rate and latency, and a standard data application, such as File Transfer Protocol (FTP) or HyperText Transfer Protocol (HTTP) will have a higher weighting for BER and PER. It is the job of the modeling and learning machines to make the necessary distinctions and learn the appropriate responses.

When evaluating the chromosomes based on the relevant objectives, the most obvious way to compare chromosomes is based on a simple summation of the weighted objectives. The next section discusses and shows how, once again, the trade-off between generalization and domain-specific knowledge causes us to rethink this standard method of evaluating MODM problems.

Linear Combination of the Objective Function Results

When trying to compare different solutions, it is useful to fall back on a single metric of evaluation. Because a MODM problem develops multiple metrics for each objective, summing over all dimensions to present a single solution presents obvious advantages—it is quick and simple.

Although the summation method is simple, however, it is not always wise, especially when comparing functions of different dimensions. Cost and efficiency are difficult to compare in a summation if efficiency has a domain that spans 0 to 1, whereas cost can span dimensions of hundreds to millions of dollars. In these situations, a multiplicative or a ratio combination may be useful. With cost and efficiency multiplied together, we get an efficiency-weighted cost value useful in comparing different solutions.

As the solution space increases to more than two dimensions, however, it becomes less straightforward to combine the solutions. Summation will not work due to the varying dimensions of each objective, and multiplication at this level of complexity brings its own problems. Two main issues arise here. The first, and easiest to handle, is that multiplying many dimensions of large quantity may result in exceedingly large values that overrun the standard data structures of the computer. This problem is quite simple to avoid (section *Normalizing Objectives* on normalizing the values to the value range diminishes this issue). The second, and more complicated issue of a multiplicative combination, is that different combinations of values in each dimension can lead to the same solution.

Take, for example, a situation in which we are trying to compare two Gaussian PDFs by trying to find the PDF that maximizes both the mean and the variance. Which is a better solution: a mean of 5 and variance of 2, or a mean of 10 and variance of 1? Without putting any constraints or values on the different dimensions, both are potential solutions, and comparison is difficult because the multiplicative combination of the objectives is 10 for both solutions. With equivalent solutions, we have a problem in deciding which one to choose; yet in some cases, maximizing the mean might be more important than maximizing the variance, or the variance's range of values might be less and so a difference of 1 between the two variances is a lot more than a difference of 5 in the mean.

With an additive combination of objectives, we can still achieve the same result with different values of objectives. As a trivial example, consider comparing Gaussian PDFs by which a mean of 5 and variance of 1 has the same result as a mean of 4 and variance of 2; that is, both have linearly additive objectives of 6. In this case, however, we can start to put some weighted value on each dimension. If maximizing the mean were more important to the solution, we could weight the mean value by 2 and the variance value by 1, giving our example here overall solutions of 11 and 10 as opposed to 6 for both. Although we can still see problems with similar solutions, we have flexibility in how we treat each dimension, and we have spread out the solution domain to make comparison easier. This same solution cannot be achieved with the multiplicative approach, however, because multiplying one dimension just adds another term into the multiplication and effectively multiplies all dimensions. The concept of valuing objectives unequally was covered in Section 7.5.1.

Another problem with these methods is that we have been assuming we are optimizing all objectives in the same direction. That is, we are either maximizing or minimizing in each dimension. Furthermore, the range of values possible for the different objectives may be to an extreme that a simple multiplication does not properly connote the differences in the solutions. To handle these problems, the next section presents an analysis of how normalizing the objectives can help to solve this problem.

Normalizing Objectives

We would like to try and fit our solutions into some comparable and workable solution space. As we have seen, comparing different objectives with varying domains is problematic. Also problematic would be trying to maximize one dimension while minimizing another. Both of these issues are important in the radio optimization problem. First, power consumption is a minimization problem valued in dBm or possibly mW, and BER is a minimization in frequency of errors per unit of received bits. The two values have large variation in their domains, with power having values of 10^{-3}, whereas BER for a data system would have values on the order of 10^{-6} and lower. If we then add data rate, which is a maximization problem on the order of 10^3 (Kbps) to 10^6 (Mbps), the problem is compounded. In order to compare all of these possible dimensions, we need some methods other than the linear combination techniques discussed in section *Linear Combination of the Objective Function Results*.

For these problems, it is possible that we can normalize all the objective's domains to comparable ranges. For the objectives we wish to minimize, we can take the reciprocal values and then try to maximize them. If we assume that a typical BER of 10^{-6}, a power of 10^{-3} W, and a data rate of 1 Mbps are appropriate values for our radio solution, we can divide the values by these typical values and then invert the solutions for the power and BER objectives. We now have three comparable values for the maximization problem.

This concept, unfortunately, works only when we have enough domain knowledge to set the typical values to normalize the solutions [10]. In communications, a typical data link should have around 10^{-6} BER, but an audio link is sustainable up to and around a BER of 10^{-3}. In a cognitive radio, we have to account for all of these applications and their different performance demands on the radio to properly normalize them. This requires a lot more information than we wish to store because the domain knowledge required is great and flexible. We need some other way of comparing the solutions, preferably on a dimension-by-dimension basis.

Relative Tournament Evaluation

We use a relative tournament selection method similar to that proposed by Lu and Yen [15], except that the fitness of the winner from a single comparison is scaled by the weight associated with that objective. After all of the single comparisons in all dimensions, the winning member, and the one that becomes a parent to the next generation, is the one with the highest fitness. Although this method does not guarantee that all winners are the best, or nondominated, members of the population (only better relative to their ancestors), it maintains species diversity within the population while still pushing toward a Pareto front (see Section 7.5.3). Diversity in the population allows different solutions to be tried and helps to prevent the algorithm from getting stuck in a local optimum.

Unlike the linear combinations of objectives, this method compares like objectives only. There are no issues, then, of improperly comparing dimensions and ranges. If we are trying to minimize BER, 10^{-6} is obviously better than 10^{-4} without any obfuscation by additively or multiplicatively combining these values with data rates of 10^4 and 10^6. We are now comparing objectives of similar dimensions and ranges. There will be no need to normalize the BER values to compare against transmit power values of 10 dBm or 1 dBW.

If we define $F(C_i)$ as the overall fitness of chromosome i, and w_k as the weight associated with an objective, $k \in K$, then we compare the chromosomes in each dimension k with the algorithm shown in Eq. (7.7). The value I is 1 if the fitness of a chromosome of interest is greater than another chromosome, else it is 0 (i.e., the chromosome is awarded points for winning the tournament).

$$F(C_i) = \sum_k I_i \cdot w_k \qquad I_i = \begin{cases} 1, & f_{i,k} > f_{j,k} \\ 0, & f_{i,k} > f_{j,k} \end{cases}$$

$$\text{where} \tag{7.7}$$

$$F(C_j) = \sum_k I_j \cdot w_k \qquad I_j = \begin{cases} 1, & f_{j,k} > f_{i,k} \\ 0, & f_{j,k} > f_{i,k} \end{cases}$$

So the winning chromosome has its weight adjusted by the weight associated with the objective. The weightings then help the decision-maker adjust its choices by putting more or less emphasis on objectives as their importance to the solution changes.

7.6.4 Relative Pooling Tournament Evaluation

In our tournament selection process, we randomly choose between two chromosomes to select one as a parent. Another method is to randomly choose two parents and a random pool of chromosomes from the remaining population [27]. Each parent will compete independently with each member of the population pool. Whichever parent wins the most number of tournaments wins the selection and ability to reproduce. This method leads to parents who are more fit for the user's needs and the communication channel within the population as a whole, and puts a lot of selection pressure on the algorithm.

Here, we did not apply Horn et al.'s [27] method of fighting two individuals against a subset of the population because their method, although they claim it produces better results, calls for a larger population and more fitness comparisons. This becomes computationally intensive and we have yet to test to see if the improvement is worth the computational cost.

7.6.5 Example of the WSGA

This section presents the results described by Rondeau et al. [22]. The first test conducted employed a real hardware test bed. However, these radios were Proxim Tsunami radios, which are hardware-based platforms with limited knobs and tuning range; all adjustable PHY-layer properties are listed in Table 7.2.

Even with this limited radio platform, the GA optimization technique was found to be viable. The scenario design consisted of a point-to-point radio link

Table 7.2: Proxim Tsunami adaptable parameters.

Parameter	Range
Frequency	5730–5820 MHz
Power	6–17 dBm
Modulation	QPSK, QAM8, QAM16
Coding	Rate 1/2, 2/3, 3/4
TDD	29.2–91% (base station unit to subscriber unit)

QPSK: quadrature phase shift keying; TDD: time division duplex.

controlled by the WSGA and a third radio of the same type that acted as an interferer. The scenario is shown in Figure 7.12 and results are shown in Figure 7.13.

The test consisted of setting up an initial video-streaming link with high-quality throughput (Figure 7.13(a)). When the interference started, the signal quality quickly

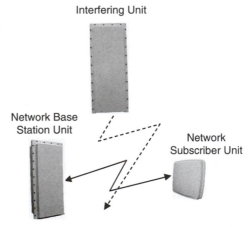

Figure 7.12: Initial test of WSGA hardware platform on Proxim Tsunami radios. The test used two radios as friendly net members and another radio as an interference source.

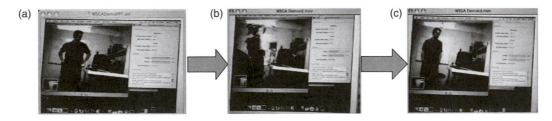

Figure 7.13: Results of the initial test of the WSGA hardware platform. The radio experienced degraded performance with onset of interference and then found acceptable waveform modifications to avoid interference.

dropped and became barely distinguishable (Figure 7.13(b)). The WSGA was then run with the objectives set to minimize BER, minimize transmit power, and maximize data rate, yet the radio was prevented from switching frequencies because that was the easy solution and the test was for how the radio would handle its other parameters. The result was that it automatically reestablished the video link (Figure 7.13(c)).

Although this test showed good performance in a live test, a more flexible platform was desired, so a PHY-layer simulation of a software-defined radio (SDR) was developed with more knobs to tune and larger tuning ranges, as listed in Table 7.3. The resulting radio simulation's objective functions were BER, bandwidth occupancy, spectral efficiency, power, data rate, and amount of interference.

This test had three goals: minimize spectrum occupancy, maximize throughput, and avoid interference. The weighting values for the available dimensions are shown in Table 7.4.

Table 7.3: Simulation adaptable parameters.

Parameter	Range
Power	0–30 dBm
Frequency	2400–2480 MHz
Modulation	M-PSK, M-QAM
Modulation, M	2–64 M
PSF roll-off factor	0.01–1
PSF order	5–50
Symbol rate	1–20 Msps

Table 7.4: Simulation test conditions.

| Functions | Weights | | |
	Minimize spectral occupancy	Maximize throughput	Interference avoidance
BER	255	100	200
Bandwidth (BW)	255	10	255
Spectral efficiency	100	200	200
Power	225	10	200
Data rate	100	255	100
Interference	0	0	255

The results are graphically illustrated in Figure 7.14, which shows that each objective was met, and each test resulted in a BER of 0. The first test (Figure 7.14(a)) minimized the spectrum occupancy to 1 MHz; the second test (Figure 7.14(b)) maximized the throughput to 72 Mbps; and the third test (Figure 7.14(c)) found a solution that fit into the white space left by the interferers.

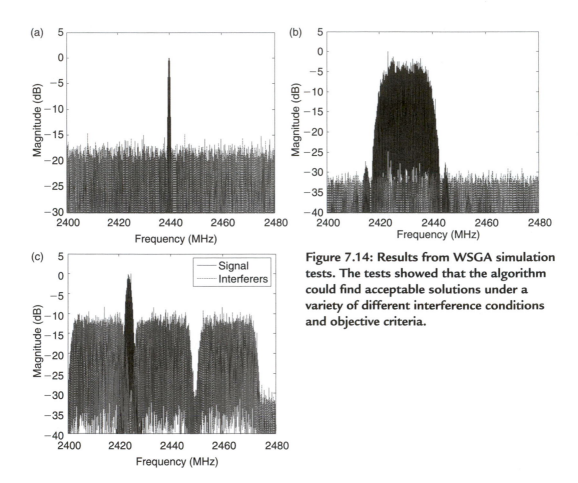

Figure 7.14: Results from WSGA simulation tests. The tests showed that the algorithm could find acceptable solutions under a variety of different interference conditions and objective criteria.

7.7 Advanced GA Techniques

The previous sections have introduced the GA mostly through the example of a simple GA. This GA is only the starting point for real GAs, and adjustments have already been made in the simple GA to make the multi-objective GA. However, far more techniques are available to improve the performance or tailor the GA performance to a specific problem. Rather than presenting here an entire treatise on the subject of GA techniques, this section is restricted to some of the more popular and useful techniques. The reader is referred to the literature for details; two excellent general sources of GA techniques are Chambers [28] and Goldberg [17].

Niching is a popular technique to ensure population diversity throughout the GA. In a population, we wish to spread our chromosomes around the solution space, especially if the space is multimodal (has many local optima). A niche is a particular range within the entire solution space, and we want the GA to maintain some presence in all niches to create many localized search spaces in hopes of finding the global optimum. Niching has a long history in GAs, going back to De Jong's crowding [20] in 1975. A multiniche-crowding algorithm is described in detail with many useful examples in Vemuri and Cedeno [29]. Fonseca and Fleming [30], and Horn et al. [27] use niching in multi-objective GA searches. Niching is one way of maintaining diversity along the Pareto front.

Dividing a total population into some smaller subset of populations is a technique used to reduce the computational complexity of each algorithm. These techniques have been tested to run different groups of solutions simultaneously to find the same optimum. These methods have been done both on the same platform or in a parallel method on multiple parallel platforms. Not only do these populations allow more searches to occur, they also have migrating populations between the "islands" to share highly fit members and maintain diversity. Many researchers have studied this issue [17, 31, 32].

GAs are sensitive to parameters such as crossover and mutation rate, population size, and replacement size, but nothing stops a GA from altering its parameters throughout its lifetime [33, 34]. One method is to monitor the performance improvement over the generations. As the improvement slows down, we might have converged to a local optimum or a global optimum. Not being sure, it would be wise to keep looking, for a while at least, but an increased mutation rate would enhance the exploration features of a GA to let it search in new areas of the search space. If the GA was caught in a local optimum, a high mutation rate will help break out of that cycle to find the path toward a new optimum, hopefully the global optimum this time.

7.7.1 Population Initialization

Although these different methods have been proposed to enhance the performance of GAs, overall only a rather small subset focuses on the population's initialization strategy. A few good representatives of the initialization methods include biasing the initial population using domain knowledge [35], creating unbiased or diverse populations [36], and using case-based initialization techniques [37]. The case-based systems, in particular, have been shown to lend themselves to successful operation in changing situations, or "anytime learning," as Ramsey and Grefenstett [37] define it. These systems are very similar to the online learning we wish to accomplish. The technique in this chapter is similar in that it presents an advanced understanding of the operation and benefits that case-based, or memory-feedback, GAs offer.

7.7.2 Priming the GA with Previously Observed Solutions

The method of priming the GA with previously observed solutions falls under the category of case-based intelligence in the AI literature. Here, we can look into the two methods discussed in section *Case-Based Decision Theory*: CBR [38] and CBDT [25]. The application of CBDT to the initialization problem, as presented here, adds to other case-based methods [37]. Those systems are usually inspired by the field of CBR [38], which uses case look-up based on similarity, but the application in this chapter is a decision-making theory based in both similarity and utility [22]. The added dimension aids the decision-making process such that cases in memory are not only similar to the current situation, but they have also performed very well in past situations. This section presents the basics behind the CBDT application to GAs and its improvement to online GA optimization systems. In doing so, it provides a more complete analysis of how the memory system works.

CBDT was introduced by Itzhak Gilboa and David Schmeidler in the mid-1990s [25]. The theory is a method of accessing past knowledge to help make decisions for present situations. The theory defines the set of cases to include all possible problems, actions, and results in which a decision-maker contains a memory M, which is a subset of the total number of possible cases, C. Current decisions are based on the knowledge of past cases by defining the amount of similarity of the current situation to the past cases and the utility of each case in memory. The case that maximizes the desirability based on the utility and similarity is chosen as the case that best applies to the current situation.

Formally, CBDT is defined to have $q \in P$ set of problems, $a \in A$ set of actions, and $r \in R$ set of outcomes. A case, c, is a tuple of a problem, an action, and a result such that $c \in C$, where $C \in P \times A \times R$. Furthermore, memory, M, is formally defined as a set of cases c currently known such that $M \subset C$.

Similarity is defined in Eq. (7.8) as how similar two decision problems are to each other:

$$s : P \times P \to [0,1] \tag{7.8}$$

Utility is defined in Eq. (7.9) as the level of desirability the outcome represents:

$$u{:}R \to \Re \tag{7.9}$$

We concentrate on the simplest formulation, where the most desirable act is chosen from the set of known acts in memory by using the product of similarity and utility [25], as defined in Eq. (7.10):

$$U(a) = s(p,q)u(r) \quad \text{where } (q,a,r) \in M \tag{7.10}$$

For a given problem, p, this equation determines the desirability of action a. The action that maximizes this function is the chosen action to respond to problem p. Section 7.7.3 introduces how the GA is aided by CBDT.

7.7.3 CBDT Initialization of GAs

The Theory

If the cognitive radio system builds a memory database that represents the outcomes of continually running the GA for real-time optimization, then as the system learns and optimizes, each new optimization would take significantly less time to run, or, conversely, would find a more optimal solution in the same amount of time. For the GA, we represent a new input as the problem p. We have previously observed problems, q, in a database associated with a set of chromosomes, a, used to model the problem along with their fitness values, r.

When the system receives this new problem, it finds the action within its memory that maximizes the product of the utility and similarity functions. The action is a set of chromosomes that partially initializes the GA population while the rest of the population is randomly generated.

Chou and Chen [36] discuss the use of uniformly generating a population to have an equal representation of all points in space, which should lead to at least one chromosome close to the global optimum. As such, by initializing with both the case-based solutions and randomly generated solutions, we maintain some diversity for proper searching while focusing the search in a region of presumably high fitness, the concept illustrated in Figure 7.15.

Figure 7.15: Search space. The square is the new target optimum, the triangle is the previous target optimum, and the stars are the initialized population. The initialized solutions from the CBDT system (dark gray) concentrate around the previous optimum while other initial solutions (white) are spread out randomly to cover the whole search space.

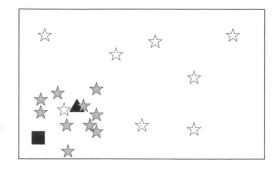

The initial population is of size N_p, where M_a members are initialized from the case-base and $N_p - M_a$ members are randomly initialized. In order to maintain the diversified population, we ensure that M_a is always less than N_p.

We define the action associated with each case in memory to consist of a number of chromosomes, M, and the results are therefore the fitness of each chromosome as fed back from the GA along with the respective chromosome. The utility of the case k, shown in Eq. (7.11), is then defined as one minus the sum of all of the solutions' fitnesses stored in each case:

$$u(r_k) = 1 - \sum_{i=1}^{M} f_{k_i} \qquad (7.11)$$

The winning case is the case that maximizes the product of the similarity and utility functions as defined by Eq. (7.10).

From here, we define two initialization methods. First, some number of solutions, $M > 1$, are associated with each case, and all M solutions are used to initialize the GA. The other method sets $M = 1$ and the solution of the winning case is used to initialize the GA, along with some number of surrounding cases. Other ways to mix these two techniques exist, but are not discussed here.

The Implementation

The design of the information flow is shown in Figure 7.16. Here, the input is compared to the cases in memory through both similarity and utility calculations. The chromosomes (action space) of the winning case are sent to the GA as well as the input. The GA then runs to optimize the result within a set period of time. The GA outputs a solution and its fitness value, which are both fed back to memory.

The initialization can happen in two ways. Either the case-base stores multiple solutions to one case in memory and sends all these solutions to the GA, or the

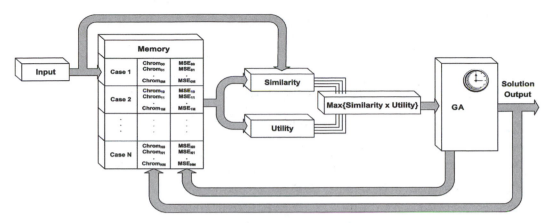

Figure 7.16: System diagram of CBDT initialization of GA.

case-base stores one solution per case and sends solutions from multiple cases to the GA.

The number of chromosomes associated with any case in memory is limited to some number, M, and when a new chromosome is inserted because it is locally optimal for current conditions, the oldest chromosome (or least suitable) is replaced. When inserting a new case to the finite memory space of size N, the oldest case in the memory is forgotten and replaced by the most recently observed case. The new memory case is then associated with both its most recent solution and the solutions of the case originally selected from memory if $M > 1$. Through this update procedure, the memory is kept current and positive actions used in the past are preserved.

Other possible methods for replacing cases and solutions include replacing the case with the worst overall utility or a case in memory that has been used less frequently than the others. Both long-term and short-term memories with different forgetting methods to exploit both time (short-term) and utility (long-term) benefits have been considered.

Two important issues to consider are the size of the memory and the initialization method. The size of the memory has a direct effect on the computational complexity of the decision-maker. The choice of the memory size should reflect the trade-off between a large memory that will add to the time required to both learn and process the input signal versus the increased efficiency to find solutions similar to current conditions. If this added time is too great, it begins to break down the benefits achieved through the system. The initialization method is the choice between associating multiple solutions with a single case and sending only those solutions to the GA, or associating only a single solution per case and sending multiple cases to the GA.

7.8 Need for a Higher-Layer Intelligence

7.8.1 Adjusting Parameters Autonomously to Achieve Goals

An important goal of the cognitive engine's higher-layer intelligence is to autonomously set the weights for the GA in order to both reduce limitations in design and remove operator interaction with the radio. The preference is that the cognitive engine has the ability to set weights on its own. By observing the user and the needs of the system, the weights should be able to adjust themselves accordingly, with purpose and intelligence. The system designer could establish a preset table of weights for different applications, but these presets are limited by the imagination of the designer and current understanding of the uses of the cognitive radio. If new situations and needs arise in the field, creative solutions should be developed as they are needed.

The cognitive engine should reflect the user's needs without burdening the user with developing the weights themselves or constant communication between the radio and the user. Purpose, intent, and lower-layer needs should be inferred from the higher layers and not from the user answering questions from the radio such as, "Would you like to increase your data rate?"

The last requirement, to infer user and application needs, is possibly the most challenging concept introduced so far and is a rich subject for research. The current design uses a simplified model, as discussed in Section 7.8.2.

7.8.2 Rewards for Good Behavior and Punishments for Poor Performance

In his landmark article, "Computing Machinery and Intelligence," Alan Turing [23] recalls his British public school days, where discipline was taught through physical punishment, which he himself both gave and received, but we can adapt this and think of it as positive and negative reinforcement. Provided the input from the user and the radio, and having produced the WSGA output, the engine now requires a means to tie the information together and learn from its past behavior, both its mistakes and successes.

A reinforcement loop must exist between the WSGA solution based on its simulated meters when solving for the objectives and the actual meters observed by the radio while operating under the parameters set by the WSGA. If the WSGA's optimization was incorrect, the system should be adjusted (punished) for improved performance next time. If the WSGA's optimization was correct, the system should be rewarded to make sure it does the same in future situations. Two ways of distributing these rewards and punishments are discussed next.

Rewards and Punishments Can Be Radio Algorithm-Inflicted

The cognitive radio must have some sense of its performance and adjust itself accordingly to improve its performance. The cognitive radio must know its current internal performance and capabilities in its approach to intelligent optimization, an approach that has been called "self-aware" in current cognitive radio circles. This does not mean that the radio is self-aware in any sense that humans are self-aware; instead, it refers to the radio's ability to keep track of its performance and available resources (battery life, geography, computational power, use-case scenario) and alter its behavior to account for needs and changes required by this understanding. On this level, the radio will use the knowledge of itself in the optimization problem, where battery life and computational power of waveform generation can play major roles in the radio performance.

On another level, the radio must be conscious of its external performance measures. When the GA adjusts the radio, the adjustments are expected to produce some known result, such as a target BER or data rate. The cognitive radio then monitors the actual observed performance and will either reward or punish itself for differences between the actual and simulated performances.

For this to work, the radio must be able to read the same meters used in the GA optimization process. The farther a meter deviates from the simulated performance, the more the cognitive radio process is punished. Again, punishment results in changes to the radio decision-making process to help it change itself for improved performance later.

Because the radio will be able to measure some meters itself, such as SINR and PER, the amount of deviation between these observed meters and the WSGA's simulated meters should tell the cognitive engine how best to adapt. It would do this by asking questions such as, "Is this PER right for this SINR?" If the answer is "yes" then the PER can possibly be improved by increasing transmitter power. If the answer is "no" (the PER is too high for the given SINR), then the problem is interference or multipath, and the PER can possibly be improved by changing frequency or modulation parameters. This method suggests the use of an expert system AI technique [24]. By programming just a few basic rules of radio operation, we can infer a lot and adjust the system appropriately. Expert systems are difficult to apply, especially to a general problem setting, because they use very specific, preprogrammed knowledge. Mixing expert systems with fuzzy logic [39, 40] allows much more flexible reasoning. Also, an implementation of an expert system in this case should be very general, using only the most basic wireless communication rules to make inferences. Other techniques, such as evolutionary strategies [24], could then be tiered with the expert system to make more creative assertions about the problem and how to fix it.

Rewards Can Be User-Inflicted

Although we advocate minimizing interaction between the user and the radio, this can be helpful in teaching the cognitive radio. The self-reflexive method just described is unsupervised learning, but sometimes a supervised learning process is necessary, especially when trying to tailor behavior to a particular user. Using the same punishment metric of credit deduction as used above, credits will be altered based on user satisfaction. This is another area of great interest and research in the cognitive radio community. Rather than promoting a particular method here, Section 7.9 looks to the field human–computer interface (HCI) for how to best facilitate the interaction and collect the data required to perform the credit adjustments.

7.9 How the Intelligent Computer Operates

This chapter has discussed many aspects we expect to see from a fully functional cognitive radio capable of altering the PHY and MAC layers with respect to the user and external environment. It has discussed the cognitive decision-making process using case-based theory and GAs to solve the multi-objective optimization problem. It has also referred to concepts of user and channel modeling, which require some advanced computer techniques, and mentioned the need for self-awareness in the radio. In the end, though, there is no single method of AI capable of performing all possible tasks required by the cognitive radio. Therefore, we look to a tiered approach of computational intelligence and machine learning algorithms to perform their tasks to create a fully cognitive radio. This section revisits these concepts to show how the collection of techniques creates a cognitive engine capable of controlling and adapting a cognitive radio.

Ultimately, all cognitive radio work really comes down to the multi-objective optimization problem, which is that the MODM search space:

1. Has no *utopian point*; that is, it has no point that fully optimizes all objectives.
2. Is nonlinear and nonconvex; instead, it is a very complex plane to model with complex relationships between all inputs and outputs.

All other aspects of the cognitive radio are inputs to the MODM problem in order to either set up the optimization problem or speed up convergence. (Refer back to Figure 7.1 for the graphical representation of the items discussed here.)

7.9.1 Sensing and Environmental Awareness

The radio must first understand what is required of it and what it must be capable of, thereby defining the optimization space. This requires sensing or awareness on four levels: (1) recognizing the needs of the user, (2) understanding the limitations imposed on the radio operation by the channel and external environment, (3) realizing its own limitations in flexibility and power, and (4) conforming to local regulations and policy. Although this chapter discusses a few implementation methods, this is not the central focus, and they are only briefly covered here.

User Awareness

The user domain provides subjective information to the cognitive radio, by which the cognitive radio is given instructions on how to behave and what to optimize. User awareness could be either passive or active. Passive awareness requires learning user needs and behavior by inferring concepts from the applications being used and data being sent through the radio. Active learning is done through direct communication with the user and HCI techniques to collect information.

Inference can be accomplished by sampling the packet headers as well as the packet and bit rate required by the application sending the data. Such learning could be accomplished using case-based techniques or even online, unsupervised neural network learning through bidirectional associative memories [41]. These techniques could learn that a header containing the protocol values for File Transfer Protocol (FTP), Transmission Control Protocol (TCP), Internet Protocol (IP), and so forth is a file transfer application requiring full packet fidelity, retransmission for failed packets, and high data rate with acceptable tolerances in latency. A similar packet that contains a protocol value for Research ReSerVation Protocol (RSVP) instead of FTP would be remembered as dealing with real-time operation, which would require different services. Some of these protocols and applications can be preprogrammed into the user modeling system, but a true unsupervised learning machine for new and unknown protocols required by the user is preferred.

Environmental Awareness

The external, RF, and geographical domains provide the environmental context. Sensing and modeling the channel can be done in a number of ways. Sending packets of training data through a channel and observing the bit errors and burst error rates is one method. Another uses channel sounding to capture a snapshot of the channel's propagation information [26]. Other methods exist for calculating noise floor and interference power to understand and recognize the presence of other users and radios [42].

Neural networks are well known for their ability to classify information and for pattern recognition. Neural networks have been used to classify modulation types in signals with great accuracy. Understanding not only how many other radios are nearby, but also the types of communication the radios employ can be of great use in determining the optimization. If a frequency is in use by another radio network, finding an orthogonal modulation, antenna polarization, and so on could lead to highly efficient spectrum sharing [43–45].

External radio and propagation information is also useful when looking at the current internal radio resources. If the radio is operating in a clean channel with few interferers, then a high-order, narrowband modulation technique could be chosen over advanced spread spectrum signals to reduce computational complexity (and therefore prolong the battery life) while providing high data rates.

7.9.2 Decision-Making and Optimization

Using the knowledge gathered in the sensing mechanisms, the cognitive radio must then make a decision as to how to optimize radio performance. The optimization problem is multi-objective and solved by using the well-known GA. GAs are not, of course, the only method to do this, but they are well understood and easy to implement as well as flexible and robust.

7.9.3 Case-Based Learning

We now have a modeling and adaptation system, but we now need some way to tie them together. A knowledge-based system to store information about the modeled environment can be tied to a set of objectives to be solved by the GA. Another entry to the knowledge base is memory of past actions taken by the GA. When similar environmental situations are observed by the modeling systems, similar actions should produce comparable results. If these past solutions gave positive results, we can aid the GA optimization process by starting with this knowledge. GAs are notorious for their length of time to find an optimal solution, but this chapter has shown how this knowledge-based approach reduces the problem quite nicely to a localized optimization search.

7.9.4 Weight Values and Objective Functions

The cognitive radio bases its decisions on a set of metrics to best represent the requirements of the whole radio system, the network, and the user. To do this properly, the decision-maker must value objectives differently. In the optimization process, different needs of the radio should be handled by using different objectives. The objectives used for a given optimization problem can be learned through successes and failures, and the weights associated with each objective can likewise be altered to best represent the situation.

7.9.5 Distributed Learning

Learning situations arise through modeling of the environment, inferring needs of the user, or simply asking the user for input. However, another way of learning that could increase the power of the cognitive radio greatly is peer learning. Children learn through their own play [46] and through teacher instruction; they also learn from each other [47]. Similarly, what one radio has already experienced and learned, it can share with newer, inexperienced radios, which removes their need for trial and error, possibly successive mistakes, and greatly accelerates optimal network performance behaviors.

7.10 Summary

The goal of a cognitive radio is to optimize its own performance and support its user's needs. A radio is optimized when it achieves a level of performance that satisfies its user's needs while minimizing its consumption of resources such as occupied bandwidth and battery power. The intelligent core of a cognitive radio exists in the cognitive engine, which performs the modeling, learning, and optimization processes necessary to reconfigure the communication system in which the radio operates.

The first problem in dealing with cognition in a system is to understand (1) what information the intelligent core must have and (2) how it can adapt. In radio, we can think of the classical transmitters and receivers as having adjustable control parameters (knobs) that control the radio's operating parameters, and observable metrics (meters) that measure its performance. The knobs are any of the parameters that affect link performance and radio operation. Meters are indicators of performance on a particular level; thus, at the link layer, PER is an important metric.

The basic process followed by a cognitive radio is this: it adjusts its knobs to achieve some desired (optimum) combination of meter readings. Rather than randomly trying all possible combinations of knob settings and observing what happens, it makes intelligent decisions about which settings to try and observes the results of these trials. Based on what it has learned from experience and on its own internal models of channel behavior, it analyzes possible knob settings, predicts some optimum combination for trial, conducts the trial, observes the results, and compares the observed results with its predictions as summarized in an adaptation loop.

Without a single-objective function measurement, we cannot look to classic optimization theory for a method to adapt the radio knobs. Instead, we can analyze the performance using MODM criteria. MODM theory allows us to optimize in as many dimensions as we have objective functions to model. Cognitive radio operation requires an MODM algorithm capable of robust, flexible, and online adaptation and analysis of the radio behavior. The clearest method of realizing all of these needs is the GA, which is widely considered the best approach to MODM problem-solving. In it, we represent the radio parameters as genes in a chromosome and select the fittest chromosomes through a process called relative tournament evaluation.

The primary goal of the cognitive engine is to optimize the radio, and the secondary functions are to observe and learn in order to provide the knowledge required to perform the adaptation. A cognitive radio becomes a learning machine through a tiered algorithm structure based on modeling, action, feedback, and knowledge representation, as shown in the cognition loop of Figure 7.10. In the Virginia Tech cognitive engine, these functions are realized through the algorithmic structure of Figure 7.11. Its main parts are the CSM, responsible for learning, and the WSGA, which handles the behavioral adaptation of the radio based on what it is told to do by the CSM. The modeling system observes the environment from many different angles to develop a complete picture. The CSM holds two main learning blocks: the evolver and the decision-maker, which takes feedback from the radio that allows the evolver to properly update the knowledge base to respond to and direct system behavior.

The algorithm structure discussed in this chapter provides the groundwork for solving the problems of the PHY- and MAC-layer adaptation. Many areas are left to explore in the machine learning for these layers, and many potential improvements and research directions have been outlined and discussed.

Bringing all of these ideas together is one of the great challenges of cognitive radio. By its nature, it is a multidisciplinary problem, just as the subjects of the other chapters in this book suggest. We need solutions that can mix these disciplines into a coherent system solution that learns and adapts to the user to satisfy all of the problems encountered when attempting to optimize the point-to-point link, as the PHY- and MAC-layer algorithms presented here do, as well as the end-to-end solution of a cognitive radio network—all the while both enhancing and protecting the regulatory structure of spectrum allocation. This chapter has provided an analysis of part of this solution and presented one possible approach to it.

Acknowledgments

The authors of this chapter wish to thank all of the researchers, colleagues, and friends that have contributed to our work. Specifically, we are pleased to recognize the members of the Virginia Tech Research Group, including Ph.D. students Bin Le, David Maldonado, and Adam Ferguson; master's students David Scaperoth and Akilah Hugine; and faculty members Allen MacKenzie and Michael Hsiao. Finally, a very big thank you goes to three former colleagues who helped start this research: Christian Rieser, Tim Gallagher, and Walling Cyre.

References

[1] T.W. Hazlett, "Spectrum Tragedies," *Yale Journal on Regulation*, Vol. 22, 2005, pp. 2–5.

[2] C.E. Shannon, "Programming a Computer for Playing Chess," *Philosophical Magazine*, Vol. 41, 1950, pp. 256–275.

[3] "Open Systems Interconnection—Reference Model (OSI-RM). FED-STD-1037C: Telecommunications: Glossary of Telecommunication Terms," National Communications Systems (NCS), 1996.

[4] J. Hawkins, *On Intelligence*, Times Books, New York, 2004.

[5] D.L. Decker, "Computer Evaluation of the Complementary Error Function," *American Journal of Physics*, Vol. 43, 1975, pp. 833–834.

[6] M. Abramowitz and I.A. Stegun, *Handbook of Mathematical Functions with Formulas, Graphs, and Mathematical Tables*, Dover, New York, 1972.

[7] J.G. Proakis, *Digital Communications*, 4th Edition, McGraw-Hill, New York, 2000.

[8] S. Zionts, "Multiple Criteria Mathematical Programming: An Overview and Several Approaches," in *Mathematics of Multi-objective Optimization*, P. Serafini (Ed.), Springer-Verlag, Wien, 1985, pp. 227–273.

[9] J.D. Schaffer, "Multiple Objective Optimization with Vector Evaluated Genetic Algorithms," *Proceedings of the International Conference on Genetic Algorithms*, Vol. 1, 1985, pp. 93–100.

[10] E. Zitzler and L. Thiele, "Multiobjective Evolutionary Algorithms—A Comparative Case Study and the Strength Pareto Approach," *IEEE Transactions on Evolutionary Computation*, Vol. 3, 1999, pp. 257–271.

[11] P. Fleming, "Designing Control Systems with Multiple Objectives," *IEE Colloquium Advances in Control Technology*, 1999, pp. 4/1–4/4.

[12] Y. Lim, P. Floquet and X. Joulia, "Multiobjective Optimization Considering Economics and Environmental Impact," *ECCE2*, Montpellier, 1999.

[13] P. Pedregal, *Introduction to Optimization*, Springer, New York, 2004.

[14] C.M. Fonseca and P.J. Fleming, "Genetic Algorithms for Multiobjective Optimization: Formulation, Discussion, and Generalization," *Proceedings of the International Conference on Genetic Algorithms*, Vol. 5, 1993, pp. 416–423.

[15] H. Lu and G.G. Yen, "Multiobjective Optimization Design via Genetic Algorithm," *IEEE Proceedings of the International Conference on Control Applications*, 2001, pp. 1190–1195.

[16] T. Hiroyasu, M. Miki and S. Watanabe, "Distributed Genetic Algorithms with a New Sharing Approach in Multiobjective Optimization Problems," *IEEE Proceedings of the Congress on Evolutionary Computation*, Vol. 1, 1999, pp. 69–76.

[17] D.E. Goldberg, *Genetic Algorithms in Search, Optimization, and Machine Learning*, Addison-Wesley, Reading, MA, 1989.

[18] R. Spillman, "Solving Large Knapsack Problems with a Genetic Algorithm," *IEEE Proceedings of the Systems, Man and Cybernetics*, Vol. 1, 1995, pp. 632–637.

[19] M.R. Garey and D.S. Johnson, *Computers and Intractability: A Guide to the Theory of NP-Completeness*, W.H. Freeman & Company, New York, 1979.

[20] K.A. De Jong, *An Analysis of the Behavior of a Class of Genetic Adaptive Systems*, Ph.D. Dissertation, University of Michigan, Ann Arbor, MI, 1975.

[21] http://www.darpa.mil/ato/programs/xg

[22] T.W. Rondeau, B. Le, C.J. Rieser and C.W. Bostian, "Cognitive Radios with Genetic Algorithms: Intelligent Control of Software Defined Radios," *Software Defined Radio Forum Technical Conference*, Phoenix, AZ, 2004, pp. C-3–C-8.

[23] A.M. Turing, "Computing Machinery and Intelligence," *Mind*, Vol. 59, 1950, pp. 433–460.

[24] M. Negnavitsky, *Artificial Intelligence: A Guide to Intelligent Systems*, Addison-Wesley, Harlow, England, 2002.

[25] I. Gilboa and D. Schmeidler, *A Theory of Case-Based Decisions*, Cambridge University Press, Cambridge, 2001.

[26] T.W. Rondeau, C.J. Rieser, T.M. Gallagher and C.W. Bostian, "Online Modeling of Wireless Channels with Hidden Markov Models and Channel Impulse Responses for Cognitive Radios," *International Microwave Symposium*, Fort Worth, TX, 2004.

[27] J. Horn, N. Nafpliotis and D.E. Goldberg, "A Niched Pareto Genetic Algorithm for Multiobjective Optimization," *IEEE Proceedings of the World Congress on Computational Intelligence*, Vol. 1, 1994, pp. 82–87.

[28] L. Chambers, *Practical Handbook of Genetic Algorithms: New Frontiers*, CRC Press, Boca, Raton, FL, 1995.

[29] V.R. Vemuri and W. Cedeno, "Multi-niche Crowding for Multi-modal Search," in *Practical Handbook of Genetic Algorithms: New Frontiers*, Vol. 2, L. Chambers (Ed.), CRC Press, Boca, Raton, FL, 1995, pp. 5–29.

[30] C.M. Fonseca and P.J. Fleming, "Multiobjective Optimization and Multiple Constraint Handling with Evolutionary Algorithms. I. A Unified Formulation," *IEEE Transactions on Systems, Man and Cybernetics*, Vol. 28, 1998, pp. 26–37.

[31] J.P. Cohoon, W.N. Martin and D.S. Richards, "Punctuated Equilibria: A Parallel Genetic Algorithm," *Proceedings of the Second International Conference on Genetic Algorithms*, Vol. 1, 1987, pp. 148–154.

[32] W.-Y. Lin, T.-P. Hong and S.-M. Liu, "On Adapting Migration Parameters for Multi-population Genetic Algorithms," *IEEE Proceedings of the Systems, Man and Cybernetics*, Vol. 6, 2004, pp. 5731–5735.

[33] M. Srinivas and L.M. Patnaik, "Adaptive Probabilities of Crossover and Mutation in Genetic Algorithms," *IEEE Transactions on Systems, Man and Cybernetics*, Vol. 24, 1994, pp. 656–666.

[34] Y.J. Cao and Q.H. Wu, "Convergence Analysis of Adaptive Genetic Algorithms," *IEE Proceedings of the Genetic Algorithms in Engineering Systems: Innovations and Applications*, Vol. 2, 1997, pp. 85–89.

[35] J. Arabas and S. Kozdrowski, "Population Initialization in the Context of a Biased Problem—Specific Mutation," *IEEE Proceedings of the Evolutionary Computation World Congress on Computational Intelligence*, 1998, pp. 769–774.

[36] C. Chou and J. Chen, "Genetic Algorithms: Initialization Schemes and Genes Extraction," *IEEE Proceedings of the Fuzzy Systems*, 2000, pp. 965–968.

[37] C.L. Ramsey and J.J. Grefenstette, "Case-Based Initialization of Genetic Algorithms," *Proceedings of the Fifth International Conference on Genetic Algorithms*, Vol. 5, 1993, pp. 84–91.

[38] J. Kolodner, *Case-Based Reasoning*, Morgan Kaufmann Publishers, San Mateo, CA, 1993.

[39] M. Black, "Vagueness: An Exercise in Logical Analysis," *Philosophy of Science*, Vol. 4, 1937, pp. 427–455.

[40] L.A. Zadeh, "Fuzzy Sets," *Information and Control*, Vol. 8, 1965, pp. 338–353.

[41] B. Kosko, "Bidirectional Associative Memories," *IEEE Transactions on Systems, Man and Cybernetics*, Vol. 18, 1988, pp. 49–60.

[42] T. Mo and C.W. Bostian, "A Throughput Optimization and Transmitter Power Saving Algorithm for IEEE 802.11b Links," *IEEE Proceedings of the WCNC*, Vol. 1, 2005, pp. 57–62.

[43] B. Le, "Modulation Identification Using Neural Network for Cognitive Radios," *Software Defined Radio Forum Technical Conference*, Anaheim, CA, 2005.

[44] A.K. Nandi and E.E. Azzouz, "Algorithms for Automatic Modulation Recognition of Communication Signals," *IEEE Transactions on Communications*, Vol. 46, 1998, pp. 431–436.

[45] V. Cheung, K. Cannons, W. Kinsner and J. Pear, "Signal Classification Through Multifractal Analysis and Complex Domain Neural Networks," *IEEE Proceedings of the CCECE*, Vol. 3, 2003, pp. 2067–2070.

[46] J. Piaget, *Biology and Knowledge*, University of Chicago Press, Chicago, 1971.

[47] L.S. Vygotsky, *Mind in Society: The Development of Higher Psychological Processes*, Harvard University Press, Cambridge, MA (original works published 1930, 1933, 1935), 1978.

Cognitive Techniques: Position Awareness

John Polson

Bell Helicopter, Textron Inc., Fort Worth, TX

Bruce A. Fette

*Communications Networks Division, General Dynamics,
C4 Systems, Scottdale, AZ, USA*

8.1 Introduction

For cognitive radio (CR) to reach its full potential as an efficient member of a network or as an aid in users' daily tasks, and even to conserve the precious spectrum resource, a radio must primarily know its position and what time it is. From position and time, a radio can: (1) calculate the antenna pointing angle that best connects to another member of the network; (2) place a transmit packet on the air so that it arrives at the receiver of another network member at precisely the proper time slot to minimize interference with other users; or (3) guide its user in his or her daily tasks to help achieve the user's objectives, whether it be to get travel directions, accomplish tasks on schedule, or any of a myriad of other purposes. Position and time are essential elements to a smart radio. Furthermore, from position and time, velocity and acceleration can be inferred, giving the radio some idea about its environment.

Geolocation applications are also a key enabling technology for such applications as spatially variant advertisement, spatially aware routing, boundary-aware policy deployment, and space- and time-aware scheduling of tasks. These capabilities enable a CR to assist its user to conveniently acquire goods and services as well as to communicate with other systems using minimal energy (short hops) and low latency (efficient directional propagation of packets through a network). Geolocation applications in a CR enable the radio to be carried throughout the world and used without any manual adjustment or modification to maintain compliance with local regulations. Finally, space- and time-aware scheduling of tasks improves the efficiency of CR operations by managing vital resources and accomplishing goals "at the right place" and "just in time."

Section 8.2 covers various techniques for a CR to geolocate itself and other systems, and Section 8.3 addresses network-aided position awareness. Section 8.4 presents the mathematics of time of arrival (ToA), time difference of arrival (TDoA), and frequency difference of arrival types of systems. Section 8.5 discusses the method of converting global positioning system (GPS) *x*-, *y*-, *z*-coordinate locations into latitude and longitude, or geopolitical region localization. Section 8.6 provides an example of boundary decisions, and an example for 911 geolocation for first responders is given in Section 8.7. Section 8.8 discusses interfaces with other cognitive subsystems. Finally, Section 8.9 summarizes this chapter.

8.2 Radio Geolocation and Time Services

Several radio transmission services are located throughout the world to aid in geolocation and accurate time tracking. The best-known system is GPS[1] [1], based on a constellation of satellites constantly broadcasting time and position (see the next section for details on GPS). The National Institute of Standards and Technology (NIST) radio station, call sign WWV, which continuously broadcasts time with high accuracy, is somewhat well known in the Western Hemisphere. (WWV consists of stations call signs WWVB and WWVH [2–4]). However, without knowing position, it is difficult to determine how long the transmission took to propagate to the receiver, and thus it renders only a coarse time.

Less well known are the very high frequency (VHF) omnidirectional ranging (VOR) transmitters used by aircraft to locate current position [5], LOng RAnge Navigation (LORAN) used by ships at sea to calculate position, and the geopositional services of cellular telephone systems [6]. Also quite widely deployed is geolocation by wireless local area network (WLAN) Internet Protocol (IP) address [6], in which the IP address of a WiFi® access point is directly translated into a geolocation [7]. The most recently published technique is the use of television (TV) broadcasts for time, frequency, and position [8]. The attempt to address these location techniques in this section will be presented only to the degree that it is clear that a software-defined radio (SDR) can participate in these time and location processes, and can thereby know geolocation and time sufficiently to aid its network, to aid its use of spectrum, and to aid its user.

---

[1] Similar systems have also been launched by Europe (Galileo) and by Russia (Glonass).

8.2.1 GPS

GPS is without a doubt the best-known location system in the world. GPS is a satellite navigation system funded and controlled by the US Department of Defense (DoD) and the Department of Commerce (DoC). The system comprises a constellation of satellites, ground control stations, and GPS receivers. At most points on Earth (other than in the deep urban canyons between skyscrapers), there is a high probability of line-of-sight (LOS) contact with multiple GPS satellites. Given LOS with four or more satellites, three-dimensional (3-D) position and time can be measured.

Satellite System Architecture

The GPS system is readily divided into three segments: space, control, and user.

Space Segment

Not counting orbiting spares, there are normally 24 active GPS satellites. The orbital period is nominally 12 hours. The 24 satellites are distributed evenly in six orbital planes with 60-degree separation between each of the four satellites in each plane. The inclination is about 55 degrees off the equator. This geometric distribution provides between five and eight satellites in view from any point on Earth. A line-of-sight (LOS—meaning unobstructed) view of four or more satellites is needed to process the signals and calculate location.

Control Segment

Ground tracking stations are positioned worldwide to monitor and operate the constellation of GPS satellites. The master control station is located at Schriever Air Force Base in Colorado. The stations monitor the satellites' signals, incorporate them into orbital models, and calculate ephemeris data that are transmitted back to the space vehicles. Ephemeris data are, in turn, transmitted to GPS receivers.

User Segment

GPS receivers and their operators form the user segment. The receivers process the signals from four or more satellites into 3-D position and time. In the differential mode, a reference GPS receiver communicating with another GPS receiver, where the position of one node is known to high accuracy, can then improve upon the inherent accuracy of another stand-alone GPS receiver. GPS receivers also produce a precise one-pulse-per-second signal.

The one-pulse-per-second timing pulse as seen by the receiver experiences a propagation time delay from each satellite. Because each satellite is at a different distance from the satellite, these time pulses are seen at different times for different satellites. Once the receiver solves for its location and the corresponding propagation time delay, it can then attribute an additional delay to all of these pulses that will exactly align all of them to when the next pulse should occur. Thus, a sophisticated receiver can reproduce one-pulse-per-second time-aligned pulses on all ground units over a large regional area by adjusting the pulses to compensate for propagation delay, and then identify the time that corresponds to those pulses, just as if the atomic clock in the satellite were connected by a wire to the receiver. As a result, all receiver units will be able to have precision time, regardless of their position. Given the internal logic that performs this alignment, the precision of the output pulse is reduced to around 340 nanoseconds (ns, equal to 10^{-9} seconds).

The operators of GPS receivers include military users as well as civilian users. The military operators, including those of the United States and its allies, use a different signal processing architecture that includes cryptographic decoding and increased accuracy.

Accuracy Obtained and Coordinate System

The two classes of GPS geolocation capabilities are the precise positioning service (PPS) and the standard positioning service (SPS). The PPS capability requires cryptographic technology and achieves 22 m horizontal accuracy, 27.7 m vertical accuracy, and 200 ns time accuracy. The SPS capability is available to any user and achieves 100 m horizontal accuracy, 156-m vertical accuracy, and 340 ns time accuracy. These specifications are defined by the 1999 Federal Radionavigation Plan and are 95 percent accuracy figures (two standard deviations of radial error).

GPS Satellite Signals

GPS satellites transmit two spread spectrum signals, one at 1575.42 MHz (L1 for SPS) and the other at 1227.60 MHz (L2 for PPS). There is a unique 1-MHz-wide, 1023-chip-long coarse acquisition pseudorandom (Gold code) spreading code for each satellite [9]. A Gold code is a spreading code synthesized by exclusive ORing of the output of two linear feedback shift register (LFSR) pseudorandom (PN) generators together. Each LFSR uses a carefully selected tap and initialization to assure that the autocorrelations of the Gold code are small at all delays except perfect time alignment, and are not confused with cross correlation from other spreading codes.

The coarse acquisition code on L1 is used for civilian GPS. There is also a precision code (P-code), 10 MHz wide and repeating every 7 days, which is superimposed on L1 and L2. The P-code is used in precise mode. Encryption converts the P-code into the Y-code. The P- and Y-codes are used for military GPS. Additionally, there is a 50 Hz signal on the L1 coarse acquisition code that transmits such ancillary data as orbit parameters and clock data.

GPS Navigation Message

A data frame (1500 bits) is transmitted every 30 seconds and consists of five 300-bit subframes.

- Subframe 1 is Telemetry Word | Handover Word | Space Vehicle Clock Correction Data.
- Subframe 2 is Telemetry Word | Handover Word | Space Vehicle Ephemeris Data part 1.
- Subframe 3 is Telemetry Word | Handover Word | Space Vehicle Ephemeris Data part 2.
- Subframe 4 is Telemetry Word | Handover Word | Other Data.
- Subframe 5 is Telemetry Word | Handover Word | Almanac Data for All Space Vehicles.

Other data and almanac data are spread over 25 data frames and take 12.5 minutes to transmit. The 12.5-minute period is the complete navigation message.

Signal Processing of GPS Signals

The GPS receiver correlates known coarse acquisition spreading codes (with a 1-millisecond period of 1023 chips) from each of the GPS satellites with the processed signal from the GPS satellites. The known spreading codes are very short and may be generated or stored in memory. Because each satellite uses a different Gold-word spreading code, when the receiver has a peak correlation it knows which satellite sent the signal. This despreading produces a full-power signal. This signal is tracked using a phase locked loop (PLL), and the 50 Hz navigation message is demodulated from each satellite. ToA information is extracted when a correlation peak is measured.

Given the ToA information measured from the correlation peak and the GPS time embedded in the signal, the GPS receiver can measure range to each satellite in view. An intersection of multiple range spheres determines where the GPS receiver is located. Four satellites must be in view to estimate *x*, *y*, and *z* coordinates along with a time estimate. A precise estimate of the position of each space vehicle in view is determined from the broadcast ephemeris data.

Reference Axes

The *x*, *y*, and *z* estimates are computed in Earth-centered fixed (ECF) coordinates. ECF is a right-hand orthogonal Cartesian coordinate system with the origin at the center of Earth, the *z*-axis increasing through the rotational North Pole of Earth, the *x*-axis increasing through the prime meridian (Greenwich, England) at latitude zero and longitude zero, and the *y*-axis increasing through 90 degrees longitude and zero degrees latitude.

Differential GPS

Position accuracy may be improved through the use of differential GPS processing. Correcting bias errors using a known location accomplishes this. A known location receiver measures its position and calculates a correction for each satellite that is passed to other GPS receivers in the local area. This is a sophisticated solution that requires more capability at both the reference and mobile GPS receiver and a data link between the reference receiver and the mobile receiver.

Another form of differential GPS is the measurement of carrier phase. This capability is used in surveying and can generate subfoot accuracies over short distances. Again, at least two receivers, a reference and a mobile, are required. Due to ionospheric effects, the set of receivers doing carrier phase geolocation must be relatively close together (approximately a 30-km limit).

Potential GPS position error sources are summarized in Table 8.1.

Table 8.1: GPS errors. The intentional degradation from selective availability is a significant source of position error in GPS systems.

Error source	Approximate error (m)
Noise (PN-code and receiver)	2
Selective availability*	100
Uncorrected clock errors (in space vehicle clocks)	1
Ephemeris data errors	1
Tropospheric delays	1
Unmodeled ionosphere delays	10
Multipath	0.5

*Selective availability is now off.

8.2.2 Coordinate System Transformations

Satellite positions and GPS receiver position are reported in ECF coordinates (x, y, z). Navigators, however, are frequently interested in latitude, longitude, and height. Eq. (8.1) is the conversion from ECF to latitude, longitude, and height (φ, λ, and h, respectively):

$$\text{Latitude} \quad \phi = \tan^{-1}\left(\frac{Z + e^2 b \sin^3 \theta}{P - e^2 a \cos^3 \theta}\right)$$

$$N(\phi) = \frac{a}{\sqrt{1 - e^2 \sin^2 \phi}}$$

$$\text{Longitude} \quad \lambda = \tan^{-1}\left(\frac{y}{x}\right) \tag{8.1}$$

$$\text{Height} \quad h = \frac{P}{\cos \phi} - N(\phi)$$

where

$$P = \sqrt{x^2 + y^2}$$
$a = $ semimajor Earth axis (ellipsoid equatorial radius)
$b = $ semiminor Earth axis (ellipsoid polar radius)
$$\theta = \tan^{-1}\left(\frac{za}{Pb}\right)$$
$$e^2 = \frac{a^2 - b^2}{b^2}.$$

8.2.3 GPS Geolocation Summary

A GPS receiver in a CR is one way to let a CR know where it is. Adding this information to an inter-radio data stream enables other CRs to know where a particular radio is located. Even though GPS is not tremendously precise without differential mode or precision mode modules, it is good enough for policy enabling and relatively long-range (similar to cell phone to base station) communication optimization.

8.3 Network Localization

8.3.1 Spatially Variant Network Service Availability

Geolocation of the radio (subscriber or user) units enables a number of useful services. There may, for example, be services that the user wishes to find in his immediate area. Consider the international traveler who has just landed after a long flight from a foreign country and who is unfamiliar with the airport. He may have several immediate needs including changing his currency, finding a restroom, finding food, finding a power plug adapter, and finding a train to the final destination. If he is not fluent in the local language, his radio's geolocation may be able to give him the relative coordinates to find such services nearby, and perhaps even help him negotiate the correct elevators.

Such processes require the ability to geolocate the user and then the requested services. The requested services may come with caveats such as "the closest," "the least expensive," or "near the route or path to be followed to some other service or objective." Having located the user, the system needs to be able to access local networks and perform inquiries about available services, their respective locations, and then sort to the user's preference criteria for presentation.

The above examples do not involve unsolicited advertising. However, there may be a way for a CR to accept unsolicited advertising, filter it against a list of topics of interest to the user, and present only those advertisements that pass the interest filter. Those advertised services that do pass the interest filter test can be offered to the user, prioritized with the user's other objectives and any route planning so as to make them convenient.

Mechanization of service offerings becomes the major issue to enable this degree of cognitive service based on geolocation. Assuming the CR has geolocated itself with one or more of the techniques described above, it needs to find a gateway to local networks. If the CR has several personal, local, regional, and cellular network access points available to it, it may then query among these networks to find those that offer cognitive support to the types of queries that support the user.

Assume, for the moment, that the CR has found an access point on a local network that offers such query services. Then the two radios may refine the location of the user to sufficient accuracy to provide appropriate directions to find the requested services. Those directions should be scaled by a scale factor appropriate to the distance as well as the local knowledge and experience of the user. Once the radio finds a service access point with the proper query service, the user may in turn be directed to a different access point where that specific type of knowledge is served. If necessary, it may include a change to whichever wireless access point is used to access the required service, as directed by the access process defined from the first server.

Finally, notice that this database may be built in part by the success of other users who have found useful services and captured the location of those services into the database. Thus, the query engine may be able to build a significant library of services without a significant system initialization effort.

Example 8.1 _____

A student is connected through his WLAN to a campus-wide server that answers a query about a restaurant with a special musical offering this evening, and the time of the date. When seeking a new restaurant in an otherwise familiar city, the student need only know to go to a nearby street location, and then be given the final guidance of where that restaurant is to be found (e.g., "The stairs to the second floor restaurant are around the back of the clothing store on the first floor, and you must go around the left side of the store to get to them."). However, when the student is off campus, but near the restaurant, the CR switches to regional service, by which the restaurant has a local query responder that can provide an applet to help find the restaurant. The CR requires 3-D positional knowledge, as well as 3-D navigation driving or walking, and, of course, the ability to understand the positional uncertainty of the user, the user's current orientation and velocity, and the relative position between the user and the objective. It is likely that local navigation may consist of applets offered by the endpoint server, in order to provide for an endless supply of unique specialized access requirements (e.g., "take the last elevator because it's the only one that leads to the top floor").

_____

Example 8.2 _____

Another example of geolocation-based services is emergency alert messaging. Suppose that all users in a geographic region must be alerted to imminent danger. Radios that know their position can be alerted, and the priority of all tasks in that region can be completely changed. It may be necessary to provide such messages over multiple networks, so there is a high probability that all CR radio users will receive the message. It is also feasible to determine how many users remain in an area and what their locations are, to expedite any remaining evacuation.

_____

277

8.3.2 Geolocation-Enabled Routing

In addition to the user's geolocation support, the radio network functionality may benefit from geolocation knowledge. Chapter 11 provides a detailed analysis of the radio environment map (REM), which is an infrastructure to enable cognitive network functionality.

However, in addition to the REM, the routing functionality of an ad hoc network, and the cellular handover of cellular networks, may be improved by explicit geolocation knowledge, velocity vector knowledge, and planned route path knowledge.

One can envision that an ad hoc network could use destination location addressing rather than medium access control (MAC) and IP addressing. In such a network, messages propagate only to nodes that know they are on a path to that location. It is expected that by reducing the number of nodes that are off course for a destination, the end power drain from all nodes and the radio interference level can be reduced. Cellular networks can improve their handover process to anticipate handovers and reduce dropouts and dead zones by being aware of geolocation, velocity vector, and planned path.

8.3.3 Miscellaneous Functions

A number of niche applications have been proposed that would be enabled by CRs with geolocation knowledge, including radios that allow users to:

1. flirt or reject flirtation;
2. recognize when someone nearby is either desirable to meet or should be avoided;
3. recognize individuals with common or special interests who are nearby;
4. identify individuals with limited time budgets who cannot support any extraneous interactions; and
5. determine the location of criminals who have been released after serving time, but who may still pose a threat under certain conditions.

This list is likely to grow as the economics of CR applications begin to benefit users.

8.4 Additional Geolocation Approaches

Terrestrial radio geolocation is accomplished through one or more of the following techniques: ToA, TDoA, angle of arrival (AOA), or received signal strength (RSS). Normally, a strong LOS signal is needed for accurate measurements.

8.4.1 Time-Based Approaches

Time-based approaches for geolocation may be divided into ToA and TDoA approaches. Both approaches require a high-resolution system clock. ToA and TDoA approaches to geolocation are based on the propagation speed of light, which is defined to be $3 \times 10^8 \, \text{m s}^{-1}$, and "straight line" LOS propagation paths so that the signal time delay relates directly to the LOS distance. Propagation is analyzed by multiple receivers or by multiple antennas converting time delay to phase difference, or time difference, or frequency shift. These are then translated into equations of range, range ratio, AOA, and/or other parameters. Most systems also track not only the estimated value but the error bounds, so the location can also be represented as a circular error probability (CEP) ellipse. At the propagation speed of light, 1 ns of time measurement error creates 0.3 m of ranging error.

ToA Approach

The ToA approach is centered on the ability to time-tag a transmitted signal and measure the exact ToA of that signal at a receiver. The propagation time, at the speed of light assuming LOS propagation, is a direct measure of the propagation distance. This provides a receiver with an iso-range sphere for a given transmitted and received signal. If multiple receivers at known locations receive the same signal, generally at different times, the multiple iso-range spheres intersect at the transmitter's location. It requires four receivers to geolocate one transmitter in three dimensions. A reverse problem may be constructed in which four transmitters provide their location in a time-tagged transmitted signal and the receiver can geolocate itself. More complex situations for which only relative positions can be determined can be constructed. This problem is useful in sensor fields and also for large numbers of CRs. A two-dimensional (2-D) depiction of this process is shown in Figure 8.1.

Round-Trip Timing and Distance-Measuring Equipment

A valuable capability is being able to locate a transmitter, for example, to rescue a sinking ship or a wrecked aircraft. Some aircraft are equipped with distance-measuring equipment (DME) transponders. These systems respond to a transmission by transmitting a response pulse whose timing is precise relative to detecting an arriving signal. By measuring the round-trip time of the interrogation pulse and its arriving response, and then subtracting the response time of the transceiver's receiver–detector–transmitter, the total distance between the two radios can be estimated. If the search transponder also uses three or more antennas, it can

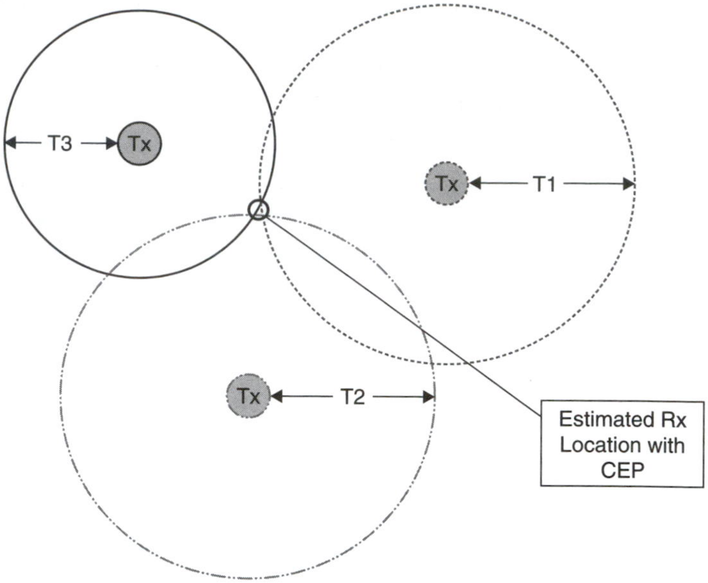

Figure 8.1: 2-D ToA: The intersection of iso-range spheres (3-D) or iso-range circles (2-D) may not be a point, introducing a CEP (Rx: receiver; Tx: transmitter).

estimate both the AOA and the distance to the distressed ship or aircraft relative to its current location and can immediately fly directly to it or otherwise dispatch aid directly to it. This has resulted in such products as the General Dynamics PRC-112 transceiver and its interrogator.

Certain cellular radio systems have developed a similar DME approach to subscriber localization. In these systems, the cellular subscriber responds to transmissions from each of several cellular base stations of known position. Each cellular base station can then estimate the range to that subscriber. By combining the range estimates from each base station, the subscriber can be accurately located. This method eliminates the requirement to install a separate GPS receiver function in each subscriber unit and shifts the location complexity to the base station, thus minimizing subscriber unit cost.

TDoA Approach[2]

The TDoA approach is centered on the ability to measure the time difference between the reception of a signal at one location and the time of reception of the same signal at another location. Again, propagation time, at the speed of light,

---

[2] This section on *Time Difference of Arrival Approach* was contributed by Nicholas W. Fette, July 2005, personal communication.

is assumed to provide a direct measure of the LOS propagation distance. The constant difference in arrival time produces a hyperboloid surface with the foci at the two receivers. If multiple pairs of receivers at known locations receive the same signal, the multiple iso-range hyperboloids intersect at the transmitter's location. It requires three surfaces to geolocate one transmitter in a 2-D plane, and it requires four surfaces to geolocate one transmitter in three dimensions.

Fitting a TDoA Curve with Two Receivers

Suppose two radio transmitter–receivers (named 1 and 2), separated by a known distance $D_{1,2}$ in meters, each record the ToA of a signal from a source (named S) in an unknown location, indicated by time stamps t_1 and t_2 in nanoseconds. Assuming the signal propagated uniformly at the speed of light, $c = 0.300\,\text{m}\,\text{ns}^{-1}$, then the speed of propagation multiplied by the difference between the values of these time stamps is the difference of the distances from the source to the two radio transmitter–receivers:

$$c(t_1 - t_2) = c\Delta t_{1,2} = D_{1,S} - D_{2,S} \tag{8.2}$$

This is one form of an equation for a hyperbola, the graph of which shows the possible locations of S based on the two radio receivers. If both receivers and the source are taken to be in the same plane, the hyperbola is roughly V-shaped (Figure 8.2), with an axis of symmetry through the two receivers.

The hyperbola can be written in terms of a coordinate system (x', y') with radio 1 (R_1) at $(0, -\tfrac{1}{2}D_{1,2})$ and radio 2 (R_2) at $(0, \tfrac{1}{2}D_{1,2})$. A familiar form for a hyperbola with these foci is:

$$y'^2 = a^2 x'^2 + k^2 \tag{8.3}$$

with the constraints that y', a, and k carry the same sign as $\Delta t_{1,2}$. The constants a and k can be determined by inspection and are dependent only on $D_{1,2}$ and $c\Delta t_{1,2}$.

Consider the end behavior of the hyperbola. The signal from a hypothetical distant transmitter (named HT) placed on the hyperbola at $(x'_{\text{HT}}, y'_{\text{HT}})$ for some infinitely high x'_{HT} value travels in parallel paths through radios 1 and 2 and the origin. A right triangle can then be formed (Figure 8.3) with the hypotenuse as the segment from radio 1 (R_1) to radio 2 (R_2), one leg as a segment of length $c\Delta t_{1,2}$ along the path to R_1, and the second leg closing the right triangle at R_2. Let θ be

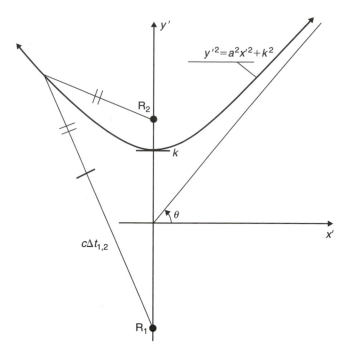

Figure 8.2: Hyperbola with two receivers and source in the same plane.

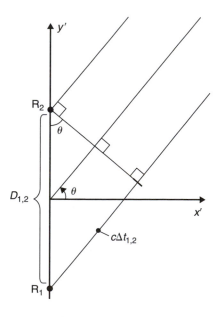

Figure 8.3: Right triangle in the coordinate system (x', y').

the angle between the hypotenuse and the second leg. By trigonometry, $\sin \theta = C\Delta t_{1,2}/D_{1,2}$, and defining $F = C\Delta t_{1,2}/D_{1,2}$, the value of θ is:

$$\theta = \arcsin f \tag{8.4}$$

By geometry, θ is also the angle from the x'-axis to the infinite-distance transmitter, so $\tan \theta = y'_{HT}/x'_{HT}$. Thus, $\lim_{x \to \infty} y'/x' = a$, a constant such that:

$$a = \tan \theta = \tan(\arcsin f) = f\Big/\sqrt{1 - f^2} \tag{8.5}$$

whereas constant a is evaluated by considering infinite x', k is evaluated by moving the hypothetical transmitter to $(0, k)$. Since HT is in line with radios 1 and 2, $D_{1,2} = D_{1,HT} + D_{2,HT}$. Thus, $D_{1,HT} = 1/2 D_{1,2} + k$ and $D_{2,HT} = 1/2 D_{1,2} - k$. By Eq. (8.2):

$$c\Delta t_{1,2} = D_{1,S} - D_{2,S} = 2k$$

or

$$k = \tfrac{1}{2} c\Delta t_{1,2} \tag{8.6}$$

In summary, the hyperbola in this system is:

$$y'^2 = \frac{c^2 \Delta t_{1,2}^2}{1 - c^2 \Delta t_{1,2}^2} x'^2 + \tfrac{1}{4} c^2 \Delta t_{1,2}^2 \tag{8.7}$$

where y' carries the sign of $\Delta t_{1,2}$.

Transforming to a Common Coordinate System

The usefulness of the coordinate system used so far in Section 8.4 is limited to communication between two radios only. To locate the signal source, at least three radios must be employed. Therefore, all pair-wise hyperbolas must be transformed into a common coordinate system to solve for the source location. That system may be chosen arbitrarily such that it enables conversion to and from GPS coordinates recognized by each radio set; however, the latitude–longitude system is warped and adds difficulty to the algebra in solving for the source location. For example, the common coordinate system (x, y) could be oriented with $+y$ directed north, $+x$ directed east, and the origin $(0, 0)$ at the location of R_1, and it would form a valid system if the curvature of Earth is negligible, given fairly even terrain. To transform Eq. (8.7) to this form, a rotation followed by a translation would suffice. The counterclockwise angle of rotation ψ is determined by the angle formed by the ray from R_1 through R_2 and the ray from R_1 pointing northward (Figure 8.4).

Figure 8.4: Geometry for determining counterclockwise angle of rotation ψ. This process would be used to convert the hyperbolic equations in radio-centered coordinate system to a north centric coordinate system.

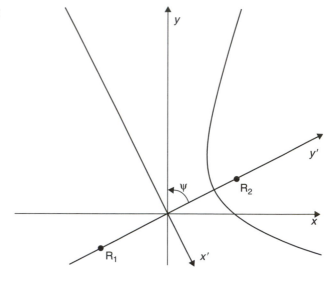

In this example, $\tan\psi = \Delta x_{1,2}/\Delta y_{1,2}$, and the old temporary axes convert by:

$$x' = x\cos\psi - y\sin\psi$$
$$y' = y\cos\psi + x\sin\psi$$

(8.8)

where these substitutions are made into Eq. (8.3), in the manner:

$$(-a^2\sin^2\psi + \cos^2\psi)y^2 + (a^2\sin 2\psi + \sin 2\psi)xy +$$
$$(\sin^2\psi - a^2\cos^2\psi)x^2 - k^2 = 0$$

(8.9)

The solution of a system of at least two distinct equations in the form of Eq. (8.9) can determine the possible position(s) (x, y) of S, the source of the transmission.

Solving for Position of Source Transmitter

The number of solutions that can result from systems of equations such as Eq. (8.9) depends on the arrangement of the receivers and which pairs of receivers were fitted with hyperbolas. As a rule, if all the receivers are in a line, there will be two solutions, one on either side of the line. If all the receivers are on a single plane, then there may be two height solutions, one of which may be underground.

LORAN

LORAN systems transmit a known burst signal from multiple transmitters with a known and published periodicity. Furthermore, the exact location of each transmitter is known. Three such transmitters cooperate to enable TDoA measurements. Ships at sea receive these transmissions and measure the time difference between each received signal. From these time differences, ships are able to calculate the TDoA hyperbolas. To simplify the process, the TDoA hyperbolas have been converted into published charts so that a navigator can directly look up the time differences for each transmitter pair and find an intersection of two time difference pairs to perform a location at sea.

TV Broadcast

Recently, Rosum Corporation (Mountain View, California) announced that they had developed a technique to recognize the ghost canceling reference (GCR) signal chirp that is included on the 19th line of the vertical retrace [9, 10]. By measuring the exact time that this chirp is received for several TV stations, and comparing the time delay between the measured arrival and the precalculated transmit time for that pulse, they can estimate a ToA for each detectable TV signal. TV transmitter locations are known with high accuracy, and they transmit with high power, so this TDoA system can work in urban areas where satellite signals fail. Since the bandwidth is 4.5 MHz versus the 1 MHz bandwidth of the standard GPS signal, it may also improve spatial resolution by a factor of approximately 4. Finally, because networks include precision time information as a data component during vertical retrace, all the information necessary for a sophisticated TDoA system is present in urban areas. The GCR chirp described above can be used to recognize and suppress multipath components, and therefore can be used in urban canyons.

Timing Estimates

Time can be derived from a number of sources, including atomic clocks, standard clocks, GPS time, disciplined GPS, phase estimation techniques, and correlation techniques in wideband transmission environments. The accuracy of the time estimate directly influences the ranging estimate accuracy at approximately $1\,\mathrm{ns}\,30\,\mathrm{cm}^{-1}$.

A popular way to obtain TDoA information is through cross correlation. At a fixed point in time (maybe on the GPS one pulse per second boundary), two stations digitize their received signals and pass the measurements to a common processing location. The processor calculates the cross correlation of the two signals. The peak of the cross correlation reveals the TDoA.

8.4.2 AOA Approach

The AOA approach requires an antenna array at the receivers. Multiple receivers estimate the AOA of a signal. Combining the bearing to the signal with the known location of multiple receivers yields an intersection point of the transmitter. This is simple triangulation.

Geometry of AOA Approach

Geometry, of course, affects how well any time or angle measurements work, including the AOA approach (Figure 8.5).[3] If the "baseline" of receivers does not have a sufficiently large angle of observation, the result is poor. If the accuracy of the AOA sensors is poor and the range to the receiver is long, the "angular dispersion" impacts the measurement. However, the computations are simple if the AOA is known to high accuracy and the interprocessor communications data volume is low.

Figure 8.5: AOA: The geometry affects the accuracy of AOA-based geolocation estimates (Rx: receiver; Tx: transmitter).

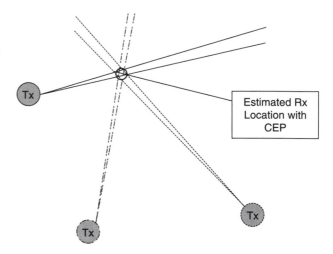

[3] Geometric dilution of precision (GDOP) is one way of relating approximately the ranging errors to position errors. Even though the distance measurement to each transmitter and its corresponding time may be measured quite accurately, the geometry of the position of the transmitter as observed by the receiver, may find that the lines of possible position intersect at shallow angles. This will cause the intersection to have much larger positional error than the error from the range estimate alone.

VHF VOR

Aircraft navigate using a number of transmitters located around the globe, frequently located near airports. These transmitters transmit an amplitude modulated (AM) signal with a constant frequency. In addition, they transmit this signal through a series of many antennas, each selected sequentially around a circle of radius approximately 5 m. The energy of the transmission is thus electronically swept around in a circle such that it introduces frequency modulated (FM) Doppler onto the received signal. Each receiver will perceive the relative phase angle between AM and FM as a different phase, reflecting the relative location of the receiver relative to the transmitter. VOR receivers convert this modulation phase angle into the angle bearing to the VOR. By locating two such VOR transmitters, the receiver is able to estimate its location.

8.4.3 RSS Approach

If the transmit power on a signal is accurately known, the patterns of the transmit and receive antenna gains are known accurately, and the receiver is able to measure RSS accurately (all generally difficult assumptions to assure), then a propagation model can be used to estimate the distance to the transmitter as a function of RSS. But propagation channels are very dynamic, so this approach is problematic. The greatest source of error in this approach is multipath fading and shadowing, and the effect can easily be a 40 dB impairment or a 6 dB increment over the direct path LOS in as little as a half of a wavelength. Mobility allows a receiver to average out these effects to some degree for FM music, but multipath represents a significant impairment to data services.

This location approach is similar to the ToA approach. Four estimates of range (upper and lower bounds of range) to four known transmitter locations are used to calculate the intersection of iso-range spheres. This yields an estimate of position and position error in three dimensions.

Simple propagation channel models predict path loss attenuation as an exponent ranging from 2 to 4.5, depending on terrain, foliage, and building shadowing and blockage losses. Since the physical environment is such a key component to other more realistic channel models, some averaging over position is required and some setting of the attenuation exponent must be assumed. The setting may be a table lookup as a function of rural, suburban, or urban environments and frequency.

If a correlation process based on a licensed transmitter's database is undertaken, an RSS-based receiver application could determine in which regulatory region it is located. For example, if a CR is receiving certain TV channels and certain FM stations and certain AM stations all at the same time, it can infer its city location. If the location of the transmitters is included in the database along with transmission levels, the RSS process could improve this estimate due to the fairly large number of measurements.

The quality of RSS-based geolocation estimates is fairly low. It is useful to CRs for some applications but not for others. For example, it may be close enough to regionalize an estimate to a city or a part of a city, and thus to a regional accuracy of 30 km.

8.5 Network-Based Approaches

Wired networks have already developed a database that allows a table lookup translation from IP addresses to geolocation. This database provides to the subscriber component the immediate ability to determine a geolocation to the accuracy of a service region. For WLAN devices, this allows the device to know its location to within a radius of 100 m or so. Services offered from longer-range connectivity, such as fiber-optic or wired service, may be able to infer regional positional accuracy from such a table lookup. At the present time, the database does not track its geographic uncertainty, nor does the database guarantee that it is currently up to date. Chapter 11 provides more information on how infrastructure-supported data services can provide geopositional support and timing to wireless subscribers.

8.6 Boundary Decisions

The ability to determine the location of a CR to a geographic region enables at least one beneficial capability. The benefit of automatic compliance with spatially variant regulations has been discussed extensively in the CR community. Other applications may develop from this capability as well.

8.6.1 Regulatory Region Selection

The capability of a CR to determine in which geopolitical region it is operating enables spatially variant policy selection, and therefore worldwide mobility and radio compliance with local regulatory rules. One approach for this capability is a map database that contains a boundary set for each regulatory region. A hierarchal search starting with continent, focusing into country (i.e., regulatory authority), and finally selecting a specific policy region would minimize the time and computational load required to determine in which region a radio is located. The regulatory region returned is then used to select a set of policies. If the radio does not have the correct policy, it may acquire it from locally available infrastructure, such as a policy database server, or apply a default set applicable to the largest number of regions in the last known continent.

Border Database Representation Analysis

This section explores several methods to determine the applicable geographic region, and therefore the associated geopolitical region. Various methods are feasible for defining a geopolitical region based on current GPS coordinates of a radio node. The objective is to find methods that minimize memory resources to represent the regions a radio will experience during its product lifetime or mission lifetime, so that it can choose corresponding policies for the associated region.

We assume in this analysis that the usual preference is for the mechanism that consumes the least memory resource to represent each country's borders. (There are currently 192 countries in the world [11].) Specifically, we explore three methods to determine whether the current GPS location is inside a given complex polygon defining the borders of the region:

A. Successive tiling using latitude and longitude (east–west and north–south) boundaries, in which the aggregated tiles define the geopolitical region.

B. A list of endpoints of successive line segments defining the borders.

C. A set of K nearest neighbor (KNN) position points.

Method A

In method A, we define a large rectangular region in the center of a country, and then smaller rectangular tiles to fill in the shape of the irregular border regions. If we assume one central region and an average of 100 smaller tiles, and four coordinates per tile, then each country would require, on average, $A = 101 \times 4 = 404$ points.

Searching this database can be quite efficient, particularly if we assume an outer constraint rectangle as well. So, for example, if Figure 8.6 defines the maximum northern and southern extent as well as the maximum eastern and western extent of each country, then we can perform a subset analysis for any given GPS point that can determine within which countries it may possibly lie. This is likely to include a maximum of six countries. We then search the interior subtiles for each of those possible countries, performing east, west, north, and south boundary comparisons,

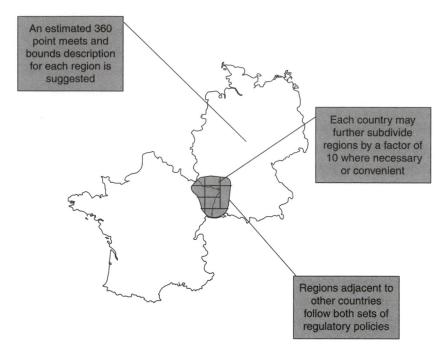

An estimated 360 point meets and bounds description for each region is suggested

Each country may further subdivide regions by a factor of 10 where necessary or convenient

Regions adjacent to other countries follow both sets of regulatory policies

Figure 8.6: Defining geopolitical regions.

and stop as soon as we determine that a point is within a specific subtile. The resulting subtile is then associated to a country and the country to a policy set.

Method B

In method B, countries may define their borders as a linked list of straight-line segments using GPS coordinates. We assume that sufficient resolution requires a maximum of 360 2-D GPS locations, plus a country geographic centroid. Therefore, $B = 2 \times 361 = 722$ points.

Although it may appear that this method is more frugal than method A in data that must be stored, the search problem is more complex because the search algorithm must determine whether a point is enclosed inside or outside the region defined by many line segments. To perform this calculation, we can begin, as in method A, by determining which countries must be searched by comparing the putative GPS coordinate against the extrema of all countries and only searching those that are relevant. Again, we assume a maximum search size of six countries. Next we calculate the perpendicular vector to each line segment. Along that vector, we calculate the minimum distance from the point to the line, and we save the identity of the two closest line segments.

For the sake of a simple explanation, we generalize that there are very few instances of spiral borders. We can then determine whether most of the country lies to the east, west, north, or south of the two line segments, and whether the putative point is likewise to the east, west, north, or south of the two line segments. Failing that, we use the borders of each country until we find a case in which the radio node location lies in the same direction as the centroid of the country relative to the two nearest line segments. Note that this is equivalent to dividing each country into a set of 100 triangular (pie-shaped) regions and determining whether a point is within the triangle.

Method C

In method C, we define a KNN approach to determine whether a point is in a country. In this case, all countries must define a set of points approximately equidistant from the border that result in a KNN border that smoothly approximates the true border.

We assume that the countries agree to place their KNN points at 10-km distance from the border and at approximately 10-km spacings. Furthermore, to assure fairness, these points are placed on each side of the border along the same perpendicular bisectors of border boundary line segments. This has the result of smoothing fine-grain border detail to approximately 3-km feature sizes. The database size is now proportional to the length of the border. If we assume 100 sample points for the average border, we must store $C = 100 \times 2 = 200$ points.

Furthermore, let us assume that $K = 2$ in our KNN search. Now, for each country, we calculate the Euclidean distance to all KNN locations, and we save the final distance as the average of the two smallest distances. We compute this calculation for all countries, and select the correct country as the one with the smallest Euclidean distance. As with the other search processes, the search computation can be reduced to a subset of local neighbors by selecting to search only those countries whose maximal extents cover the range where our point of interest is currently located.

Anomalies

There may be countries that have anomalous shapes that may not behave correctly when using the algorithms presented in section *Border Database Representation Analysis*, such as spiral arms brought about by the circuitous path of rivers through mountain ranges. In addition, islands are anomalies, but an island can be handled as a second country with the same name and same policy.

More likely sources of trouble, however, are disagreements about what the actual border really is, such as overlapped regions claimed by more than one country. Methods A and B can be extended to report that two countries both satisfy the criteria, defining the location of a radio. Under these conditions, it may be possible to define a common subset policy or to report to the user that there are no applicable rules in disputed zones.

The format for the latitude and longitude is negotiable. One suggestion is a 32-bit floating point number (4 bytes) representing degrees, minutes, and seconds of latitude or longitude. This would result in a worst-case resolution along the equator of about 5 m.

For each of the representation databases, and the conservative assumption of 200 countries, we can estimate the size of the database to be as shown in Table 8.2. Note that at any one time only a subset of such a database is needed and the meets and bounds database can be loaded into a CR similar to loading a working set into a cache memory.

Table 8.2: Database sizes for three representation methods to define whether a position is inside of a country.

Representation type	Database size (bytes)
A	323,200
B	577,600
C	160,000

8.6.2 Policy Servers and Regions

Possible sources of policy are: (1) a periodic broadcast (this protocol is not defined at this time); (2) a load from a certified source (this could be realized in the form of kiosks); or (3) a wired database server, which the device must proactively acquire, interrogate, and capture.

A policy region is not a defined entity at this time. An example of a policy region might be the greater Washington, DC, area. In a specific regulatory region, legacy transmitters are licensed, thus defining other broadcast bands that could be harvested (only under specific technical circumstances). Another example of something defined in a regulatory region is a licensed spectrum holder and a registry of frequency sublease server(s) (frequency and protocol) that may be contacted for leasing a channel for a short period of time. Both of these examples are "reach-out" capabilities and have not been approved in the United States. These types of technologies have been discussed only in limited circles (see, e.g., Chapter 2).

In the short term, a small set of policy regions could be defined as boxes in space and a subset of a small experimental band (there have been suggestions for an experimental CR band) where demonstrations are conducted. Field demonstrations of these capabilities could show the degree of spatial and frequency band reuse achievable. Demonstrations also will show the ability of CRs to share the space on a non-interference basis and the ability to comply with machine-readable policies.

8.6.3 Other Uses of Boundary Decisions

Radio policy selection is the frontrunner of CR applications that are dependent on knowing in which geopolitical region a radio is located. Other potential applications also exist. Given natural language processing capability, a CR that knows it is in Germany might offer to translate an English-speaking user's word from English into German when transmitting and could translate German words into English when receiving. Although this could be accomplished through language recognition more efficiently than through political boundary decisions, a boundary decision could be used to improve the language recognition search by prioritizing by local languages.

If a tourist is using the CR, the boundary decision engine could be used to display local points of interest such as historical markers. On a more personal note, the boundary decision engine could be correlated with the user's address book to locate friends and family in the vicinity, and the user can apply this information appropriately.

8.7 Example of Cellular Telephone 911 Geolocation for First Responders

The productivity of ubiquitous cellular telephone and handheld communications devices such as pagers or BlackBerry™-type devices have dramatically increased. Additional problems, however, have been introduced. For example, with wired telephones emergency calls were traditionally easily located to the phone where the call was placed and the location was very reliable. At the time of the initial deployment of cellular telephones, a call placed to an emergency response center did not come with a location attached for the emergency operator. The problem prompted the US Federal Communications Commission (FCC), in 1996, to mandate geolocation services in the cellular network infrastructure. The FCC mandate was for 125 m accuracy in 67 percent of all measurements by October 31, 2001.

Multiple cellular telephone interfaces exist including Advanced Mobile Phone System (AMPS), code division multiple access (CDMA) EIA/TIA IS-95, time division multiple access (TDMA) EIA/TIA IS-136, the older TDMA IS-54, and the Global System for Mobile Communications (GSM). The common characteristic of these systems is a set of base stations that communicate directly with mobile stations or handsets. The frequencies, bandwidths, modulations, and protocols vary from standard to standard.

The two broad categories of geolocation techniques for cellular telephones are: (1) network- or infrastructure-based approaches and (2) handset-based approaches. The advantage of the network-based approach is there are no requirements placed on the owners of the cell phones. The advantage of the handset-based approach is the precision available.

Infrastructure-based approaches include ToA, TDoA, and AOA. The characteristics and techniques for these approaches have been discussed in Section 8.4.1. The geolocation application executes on cooperating base stations, measures one or more of the essential physical properties, exchanges information, and processes the signals to produce a location estimate.

The best example of a handset-based approach is putting a GPS receiver into the handset and interrogating the handset for its location when an emergency call is placed.

8.8 Interfaces to Other Cognitive Technologies

This section discusses interfaces between the geolocation engine and other entities in a CR. The policy engine, networking functions, planning engines, and user, at least, will interface with the geolocation engine.

8.8.1 Interface to Policy Engines

One possible interface of the geolocation engine (server) and the policy engine (client) is a client–server model. For example, the client requests from the server a location in terms of latitude, longitude, and altitude. Another request would be for a coded geopolitical regulatory region. These two requests would spawn different signal processing steps to answer the inquiry.

Another interface to the policy engine is an "interference analysis" request. This requires the relative location of other users to process the "message." The policy engine may request relative position directly to estimate the possibility of interference. However, this does not address hidden node issues that are inherent in the harvesting of spectrum.

8.8.2 Interface to Networking Functions

The networking functions may use a similar client–server model to request relative position of other CR and non-CR devices. The resulting information may be used to direct steerable antenna beams and nulls or to select next-hop destinations in energy-efficient routing algorithms.

Networking functions may also request absolute position and use this in a search for local services, such as access points. The database of services is contained in the networking engine in this case. If the database of services is contained in the geolocation engine, the networking function could request location of the closest service provider, and the engine would return that. The partitioning of these functions to the two engines has not yet been discussed in the literature.

8.8.3 Interface to Planning Engines

A new capability for the geolocation engine is a distance, as the crow flies, from the CR's current location to a designated position. An address to coordinate system function could be included. A variation of this request is from position A to position B. The positions would need to be provided in a variety of coordinate systems including ECF, latitude–longitude, or addresses. A planning engine executing a traveling salesman algorithm uses the geolocation engine for metrics. The distances returned could be straight line or driving distances and paths, which may be obtained from the Internet.

8.8.4 Interface to User

A user interface to the geolocation engine may be a simple "where am I?" that is used to select a digital map segment. The map is overlaid with a "you are here" marker. If relative positions of other emitters are requested, their positions may be overlaid as well. This capability is very useful for navigation applications and may be integrated with time and space management functions.

8.9 Summary

A CR that is aware of its position is enabled to demonstrate spatially variant behavior. This is a critical capability for new functions, such as spatially variant regulatory policy or spatially variant networking functions. If the radio knows where it is located, it can self-report for such important functions as 911 responder interfaces.

Sections 8.2 and 8.4 covered a number of ways to determine where a system is located. An inertial navigation system can be used to integrate a current position relative to a known starting point, but this approach is fraught with unreliability and excessive expense. A better alternative is a GPS receiver. This inexpensive subsystem provides 3-D position and current time. The system is based on a constellation of satellites and it has the capability for two resolutions, precise positioning system and standard positioning system. Very precise GPS location can be obtained using differential approaches.

Non-infrastructure-based approaches include ToA, TDoA, AOA, and even RSS. The first two approaches use spatial diversity and inter-radio communications to estimate a position for an emitter. A CEP may be calculated a priori and is accurate with the exceptions introduced by high multipath channels. The AOA approach uses spatial diversity and inter-radio communications to make a geometric interpretation of the location of an emitter. The last approach, RSS, yields a regional estimate of position, but also uses spatial diversity and inter-radio communication to complete its estimate.

Section 8.3 touched on the value of geolocation knowledge to enable spatially aware networking functions. Routing may take into account position for energy savings or other purposes. Services may be accessed as a function of location. New novelty functions, such as spatially variant advertising, is possible when a CR knows its location.

Section 8.5 briefly looked at how networking can provide a coarse degree of localization, and Section 8.6 explored how geolocation is extended to become boundary analysis. One of the key interfaces to a geolocation engine is a boundary decision. A meets and bounds database may be employed for a CR to obtain its current geopolitical region. This approach may be used for a variety of other regions of interest decisions as well.

Section 8.7 provided examples of geolocation in the context of cellular emergency location, and Section 8.8 reviewed the many interfaces to other supporting subsystems a CR will need to build well-integrated, systems-level functionality. These interfaces will be the areas for significant standards development as CR technology evolves.

References

[1] http://www.colorado.edu/geography/gcraft/notes/gps/gps_f.htm
[2] http://tf.nist.gov/timefreq/general/pdf/1383.pdf
[3] http://tf.nist.gov/timefreq/stations/wwvb.htm
[4] http://tf.nist.gov/timefreq/stations/wwvh.htm
[5] http://www.navfltsm.addr.com/vor-nav.htm

[6] http://www.lucent.com/press/0699/990630.bla.html

[7] http://www.linuxjournal.com/article/7856

[8] http://www.catchoday.com/archives/39006.html

[9] http://www.uspto.gov patent search on patent #6,879,286

[10] R. Gold, "Optimal Binary Sequences for Spread Spectrum Multiplexing," *IEEE Transactions on Information Theory* IT-13, 1967, pp. 619–621.

[11] www.google.com

CHAPTER 9

Cognitive Techniques: Network Awareness

Jonathan M. Smith

CIS Department, University of Pennsylvania
Philadelphia, PA, USA

9.1 Introduction

Users see the network through the window of distributed applications, which carry out some combination of communication and computation to meet user needs. Familiar examples include web applications for shopping and banking, interactive applications such as video teleconferencing, Voice over Internet Protocol (VoIP) and chat, as well as music sharing and massive multiplayer games. What each of these applications has in common is a model of the interactions between the user and the system, and how the system supports interaction among sets of users.

The variety of applications listed above suggests that these models might differ substantially, but in the abstract these models share the need for a set of communications protocols that deliver network services required by the applications (Figure 9.1). Section 9.2 discusses specific requirements of typical applications, but as the understanding of networks and the variety of services they can provide has improved with experience, the commonality among many application requirements has been exploited to build network protocols that can be used by many applications. The general-purpose nature ("one size fits all") of such protocols is extremely attractive, but may sacrifice the ability to support specific applications as well as they might be supported by a dedicated and optimized protocol.

Note: Approved for public release, distribution unlimited.

Figure 9.1: Protocols provide interfaces and services to applications.

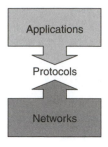

9.2 Applications and their Requirements

There is an increasing trend toward mobile applications in the commercial world, but mobility has always been a requirement for the military. Unlike those of the commercial world, military communications challenges include lack of predeployed infrastructure, long latencies, higher bit error rates (BERs) than commercial software systems assume, and higher standards for security. For example, the scenario illustrated in Figure 9.2 depicts a mobile "reachback" scenario, in which a mobile satellite terminal is used to request data over a wireless satellite communications channel. Such a method may be slow and certainly represents substantial propagation delay. If a conventional protocol architecture such as Hyper Text Transfer Protocol (HTTP) 1.0 running over Transmission Control Protocol with Internet Protocol (TCP/IP) is deployed end-to-end (i.e., between the mobile satellite terminal and the continental US ground data source) then the situation illustrated in Figure 9.3 can occur.

What occurs is that the application makes a request for data, such as an image of a map, and the request is forwarded over the satellite channel. The data source responds with the map, which can be substantial in size and will certainly require multiple packets for transport. The application requirements here are reliable delivery *and* minimal latency. When TCP is used to obtain a reliable byte stream to support the reliable delivery requirement, it also provides congestion control, a feature desired by the network operator but generally not by an individual user.[1] The TCP congestion control scheme interprets packet loss as indicative of congestion and therefore paces the number of packets in flight during a round-trip time

---

[1]TCP provides two services: flow control, which ensures that the sender does not transmit more packets than the receiver can handle, and congestion control, which mitigates the effect of an overflowed buffer in intermediate devices. The basic interaction between these two services is that the flow control variable forms an upper bound to the size of the congestion control window.

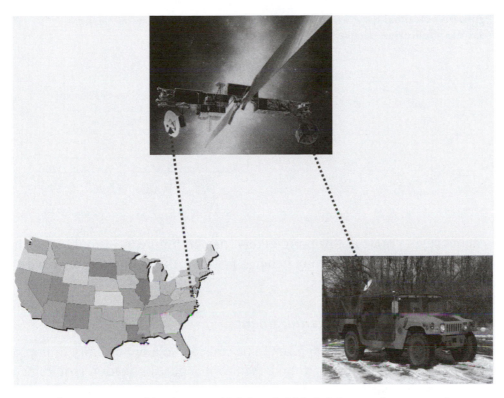

Figure 9.2: Reachback over a high bandwidth * delay product network.

Figure 9.3: Conceptual illustration of how TCP congestion control adapts to packet loss.

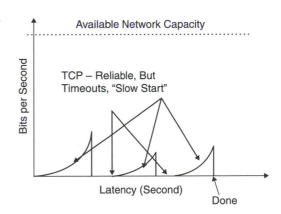

(RTT) (the "window size") according to its perception of whether the network is congested or not. Thus, even when capacity is available, as indicated by the horizontal dashed line in Figure 9.3, the conventional protocol architecture cannot exploit it.

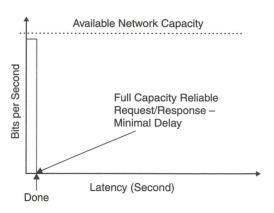

Figure 9.4: Optimal network behavior from the application perspective.

Without saying more at this point in this chapter, Figure 9.4 should tease the reader into thinking that something better is possible, the nature of which we will explore in the following sections.

9.3 Network Solutions to Requirements

As shown in Section 9.2, there appears to be considerable opportunity to improve the performance of applications if more is known about application requirements *and* this knowledge can be brought to bear in the network. Before pursuing this, it is worthwhile to examine the engineering issues at play in the protocol architecture implementations at work. Again, a convenient example for this analysis is the widely used IP suite, which is often represented conceptually as an "hourglass" of layers, as shown in Figure 9.5, where the hourglass is meant to indicate that the interoperability layer, here labeled IP, is used to coerce multiple incompatible subnets into a "virtual" network, the inter-network [1], whose properties can then be assumed to be present by end-to-end protocols. The inter-network is overlaid on multiple sub-networks, and routes through the inter-network are used to provide end-to-end connectivity to protocols such as TCP, which, as mentioned above, provides a virtual-circuit-like reliable byte-stream model. HTTP is shown as a user of TCP.

This architectural model has consequences for network awareness. The layering model employed reflects the software engineering mindset of the time: "modular programming" and "information hiding." The idea of modular programming is that more understandable (and therefore more robust) software can be written when the software structure is decomposed into modules of small size, with localized concerns, such as a certain class of operation or a particular data object. Information hiding is a discipline that limits the flow of information among modules,

302

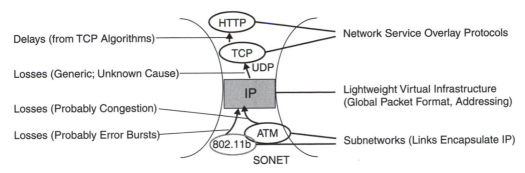

Figure 9.5: The IP hourglass—an interoperability solution (ATM: asynchronous transfer mode; SONET: synchronous optical network; UDP: User Datagram Protocol).

for example, by limiting data sharing to a small number of carefully type-checked parameters.

In the case examined here, the concerns are localized in the *layers*, and the information hiding discipline is enforced by the implementation, which passes information either in the data structure for representing the packets (e.g., the memory buffer (mbuf) in UNIX implementations) or in the limited additional information used in procedure calls between layers. It is clear that such a discipline enhances interoperability, a major goal achieved by the IP architecture. However, like most architectural choices, it represents a particular choice in a rich space of trade-offs, and as a consequence may introduce some undesirable properties.

As one example, we can consider the issue of packet loss. In many cases, if all the information available to the end host was employed, an intelligent statistical model for causes of packet loss could be employed. For example, the failure of link layer checksums computed in the device driver for the particular line card employed to connect the host to the network would be indicative of error bursts on the link, whereas a recent trend of increasing delays in communication to a particular host or set of hosts might be indicative of congestion.

However, the IP protocol does not indicate why packets are lost (as above, it could be burst errors, either causing whole packet loss or damage to a checksum, or it could be congestion), so TCP must make an estimate with *no information*. On commercial networks, the reasonable assumption as to the cause of loss is congestion. Thus, TCP slows down (sends less, in the decision matrix of Figure 9.6), using a sophisticated algorithm that is clocked in units of RTTs.[2] If, in fact, the

---

[2]Because everything is clocked in RTTs, TCP's "discovery" cycle is much slower in long-latency networks such as those that incorporate satellite communications links.

cause of loss is *not* congestion but some other cause, as is more common in mobile military communications systems found in tactical scenarios, then havoc ensues and performance plummets.

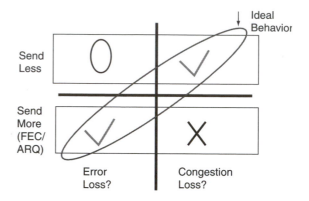

Figure 9.6: Matrix of loss causality versus congestion control actions (ARQ: automatic repeat request; FEC: forward error correction).

9.4 Coping with the Complex Trade-Space

One possible approach to overcoming certain limitations of information hiding, as reflected in protocol architectures, is shown in Figure 9.7. The left-hand side illustrates that the limits are imposed by narrow service interfaces, as discussed in Section 9.3. These service interfaces may be, for a variety of reasons, immutable (e.g., because a large body of existing software assumes the interfaces and the economic consequences of changing the interface would be unacceptable). In such circumstances, a protocol architecture could be adjusted [2], as shown in the right-hand side of Figure 9.7, by inserting a "shim" module into the protocol stack which is compliant with the interfaces defined above and below it in the layered architecture. In the example illustrated in Figure 9.7, an additional protocol function, automatic repeat request (ARQ), has been inserted beneath the IP layer to

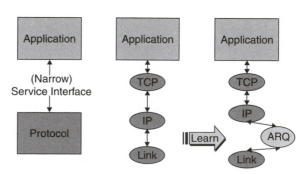

Figure 9.7: Protocols built with service interfaces. Service interfaces are often narrow and opaque. In spite of this, structural adaptation can occur through dynamic protocol composition.

reduce the observed losses due to errors. The addition of this protocol function reduces the probability of burst errors being observed by the IP layer. Thus, the assumption by the TCP layer that packet loss is due to congestion is made more probable, and the congestion control strategy is therefore much more likely to respond correctly to the packet loss events that it observes post-forward error correction (FEC).

Since the ARQ protocol seems useful, why not always employ it at the link layer? This is, in fact, done for certain types of links. For example, the 802.11 medium access control (MAC) layer employs an ARQ-like strategy to emulate the reliability of a wired local area network (LAN). Some satellite communications links employ performance-enhancing proxies (PEPs) at the link endpoints to provide a reliable substrate to end-to-end protocols. Such link protocols are not universally employed, however, due to the performance consequences, for example, increased latency (the 802.11 MAC may account for much more measured latency than propagation delay). It should be clear, then, that an adaptive protocol architecture that employs the error compensation mechanism, as needed, is desirable. In the illustration on the right-hand side of Figure 9.7, we might learn that there has recently been an increase in the checksum failures on link layer frames, and deploy ARQ or an FEC scheme. Over time, the wisdom of this decision can be balanced against the cost of the deployment, likely duration of error bursts, hysteresis models for the module removal decision, and so forth. Section 9.6 further discusses the robust use of system measurements on the Defense Advanced Research Projects Agency (DARPA) SAPIENT program (see Section 9.6).

Generalizing from this example, we can see that structural adaptation of protocols [2, 3] is a powerful technique with which we can provide application performance in spite of an extremely wide range of application requirements and network conditions, by adding and removing protocol elements in response to changes in the situation. Although this problem might be solved in specific circumstances by event-driven adaptations, such as the insertion of ARQ discussed in this section, there is a wide variety of additional information that must be available to fully exploit structural adaptation in network protocols.

Consider an application that requires data privacy and integrity. For such an application, it may be desirable to employ an encryption algorithm, which protects data privacy through a secret-controlled data transformation, and as a consequence of this transformation, detects data tampering as a failure to successfully reverse the transformation at the receiver. Redundancy present in the application data and limited network throughput might suggest that a data compression protocol be employed.

Consider the two possibilities for ordering compression and encryption protocol elements illustrated in Figure 9.8. In the compress/encrypt composition of two protocol elements, as might occur in the course of a structural adaptation process such as that discussed earlier, if compression is performed first, the data size is reduced by removing redundancy, and then an encryption transformation is performed to hide the bits of the compressed data. If, in contrast, the encryption is performed first, then the structure exploited by a compression algorithm is obscured, meaning that no data size reduction will be accomplished in spite of the considerable computation typically required for protocol elements of this type. So the first composition is desirable and meets the goals of "compressed + encrypted," whereas the second composition meets only the "encrypted" goal.

Figure 9.8: Protocol composition possibilities. Protocol composition must be aware of protocol element interactions.

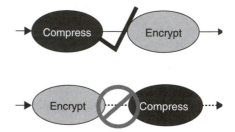

This example illustrates the point that several kinds of knowledge must be available in a general structural adaptation scheme. First is knowledge of what external conditions the protocol element is responsive to, such as network conditions and application requirements. Second is knowledge of the assumptions inherent in particular protocol elements responsive to external conditions, which may affect its suitability and role in a complete protocol architecture resulting from a multiplicity of structural adaptations. In the past, the daunting complexity of managing the many possible adaptations and their possible interactions has inhibited adaptations for which simple robust models (e.g., the control laws employed for TCP/IP's adaptation to network congestion, or the definition of a "spanning tree" [3]) have not been devised.

9.5 Cognition to the Rescue

A new approach, which employs what can broadly be viewed as an approach to automating the management of complexity, now seems possible. Automating the management of complexity does not seek to eliminate or obscure the complexity inherent in a system (as the information hiding approach might be seen to do).

Rather, automation seeks to remove the detailed management of complexity from the purview of the application programmer or protocol designer. Because many protocol design issues are quite subtle, as has been illustrated in this discussion, any such protocol automation strategy must be able to represent many kinds of knowledge. Thus, empirical knowledge (such as observed performance) should be able to be "fed back" into the knowledge employed by the system to create protocol structures suitable for particular applications and network conditions.

The opportunity afforded by the approach is considerable. Figure 9.9 illustrates the possibilities for structural adaptations that closely couple application requirements to the encountered network conditions (which, of course, as in the packet loss example, may vary considerably with time). When the network conditions are very attractive, as might be the case on the interconnection network of a parallel computer or a set of hosts resident on a common LAN, very little protocol support may be necessary for typical desiderata such as high "goodput" with low latency. Where conditions degrade, more and more protocol mechanisms may be employed to adapt the application to the degraded network, with these compensatory mechanisms permitting the application to operate in domains unreachable by traditional architectures (the limitations of which are illustrated by the demarcation of the vertical dotted line in Figure 9.9).

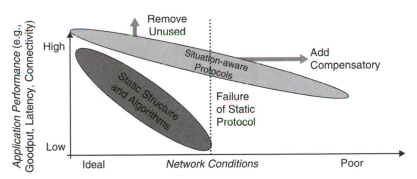

Figure 9.9: Performance of network conditions. Situation-Aware Protocols occupy a new region of a conceptual "trade-space."

A detailed example of the possibilities can be given by using the TCP/IP congestion control algorithm's response to packet loss. Simulations and laboratory experiments have both shown that this protocol's response to error conditions above a certain threshold (e.g., a BER greater than 10^{-6}) causes the TCP/IP algorithm to close its congestion window. As a consequence, its performance is reduced to the level of a "stop-and-wait" protocol, which is extremely sensitive to the delay inherent in the link. If error compensation is inserted (as in Figure 9.7), the windowing strategy is made effective, and relatively high throughput can be achieved in operating regimes with high BERs. Hadzic and Smith [4], for example, have shown experimentally that a protocol stack incorporating TCP/IP and adaptive compensation can operate with BERs as large as 10^{-4}.

Cognitive systems are systems that apply human-style reasoning and capabilities. In the case of the structural adaptations discussed here, the cognitive system would be required to identify the combination of application requirements and network conditions, select or devise an appropriate composition of protocol elements, and deploy that composition. Notions such as similarity, trial-and-error, and improvement by learning appropriate responses over time help to make a cognitive system both less brittle and more evolvable than a conventional protocol design. It provides a clean *global* separation of policy and mechanism, which is otherwise done only locally in modules, if at all.

9.6 The DARPA SAPIENT Program

The clear utility of such an approach in military networks, particularly those at the tactical "edge," has stimulated DARPA to support a research effort to investigate the effectiveness of cognitive approaches to rapid adaptive composition and adaptation of protocol structures. This effort is called SAPIENT, for Situation-Aware Protocols In Edge Network Technologies, and has identified three core challenges that must be addressed in a working system:

1. Knowledge representation
2. Learning
3. Selection and composition.

Presuming that these challenges are addressed, a conceptual architecture for an SAPIENT system might be structured similarly to that shown in Figure 9.10. An application that employs the network to carry out its intended task (such as those discussed in Section 9.2), is shown in the upper left of the illustration.

The requirements of the application might be characterized explicitly (by a designer or user), although it is far more attractive if the SAPIENT system deduces or infers these requirements automatically. An example of explicit characterization might be the quality of service (QoS) trade-offs specified in Nahrstedt and Smith's [5] OMEGA system, whereas automatic inference might be done by

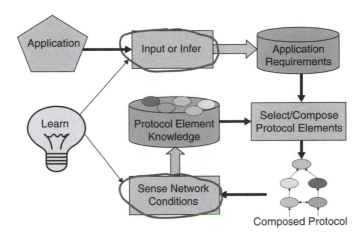

Figure 9.10: One possible structure for a Situation-Aware Protocol System.

technologies such as statistical machine learning operating on the packet stream emitted by the application.

As a simple example of automatic inference, consider observing packet size and a time stamp. A relatively constant packet size and inter-arrival delay might indicate the packets are part of a media stream, and the protocol support might be adjusted to support this inference. Conversely, great variability in timing, small sent packets, and highly variable sizes in received packets might indicate a transactional application such as that discussed at the end of Section 9.2 and the desirable protocol behavior illustrated in Figure 9.4.

The goal is to produce a knowledge base of application requirements, as shown in the upper right side of Figure 9.10. This knowledge is intended to be combined with other knowledge, namely, that about network conditions and available protocol elements, to produce a responsive SAPIENT system. The protocol element knowledge, illustrated at the center of Figure 9.10, is used to react to network conditions. It is brought together with the application information in the process of selecting and composing protocol elements, which results in the composed protocol illustrated in the lower right side of Figure 9.10.

Many opportunities for learning remain. Learning to automatically identify applications and their requirements has already been discussed, but the sensing of network conditions (losses, timing, load, etc.) is central to understanding whether a new protocol is needed, and if so, what network conditions it must address.

Sensing network conditions can involve a variety of sensors, including those that monitor packet checksums, packet ordering, packet sequencing, packet loss, congestion indications, latencies, and relative application usage of link capacity, as well as alerts such as might occur on the loss of a link or a dramatic reduction in signal strength on a wireless link. Sophisticated sensors might include hysteresis or damping algorithms to compensate for stochastic "noise" that might otherwise induce unwanted responses. Input from multiple low-level sensors might be fused to gain a better overall estimate of the state of the network; such sensory information could include information obtained through active probes as well as from remote systems, whether they employ an SAPIENT architecture or not.

The sensing of network conditions is a central element of an SAPIENT system's "situation awareness" because it provides the measurement basis for maintaining the current state or pursuing an adaptation. When the system is initialized, it might start by using a composition equivalent to TCP/IP as a baseline, and then modify the system while a positive performance gradient is realized. If a protocol composition is not working, then the network sensing process should deduce this and stimulate production of a more appropriate protocol. When a performance "plateau" is achieved, the SAPIENT system would preserve the status quo until a significant change in network conditions occurred.

Over time, the protocol element knowledge base incorporates this sensed information in its learned responses to detected network conditions. Here, statistical machine learning can prove powerful, for example, in classifying situations in which the operating protocol should be left untouched, and those in which it should be adjusted. One property possible in such a system is the ability to recover from protocol elements with errors, presuming that the sensed data is correct and that it is correctly interpreted.

The use of these varieties of knowledge in creating situation awareness, resulting in application-specialized protocols, offers a new way to support networked applications.

9.7 Summary

This chapter has presented a new form of network awareness—situation awareness—and its application in the construction of application-specific protocols. Use in the fashion described creates an evolvable system, as new protocol elements and knowledge about them can be inserted directly into the system. This would permit, for example, data from an "experienced" SAPIENT system to be loaded into a new system, in essence sharing the learned knowledge. The structure illustrated in Figure 9.10 permits the system, after this initial load, to continue learning as it encounters novel combinations of network situations local to its operating environment.

The SAPIENT effort and its approach to managing complexity represent a pathfinder to new methods of building robust systems that can operate successfully in extreme environments and unforeseen conditions, such as those faced by military tactical communications systems on a daily basis.

References

[1] V.G. Cerf and R.E. Kahn, "A Protocol for Packet Network Intercommunication," *IEEE Transactions on Communications* COM-22, May 1974, pp. 637–648.

[2] D.C. Feldmeier, A.J. McAuley, J.M. Smith, D.S. Bakin, W.S. Marcus and T.M. Raleigh, "Protocol Boosters," *IEEE Journal on Selected Areas in Communications*, Vol. 16, No. 3, April 1998, pp. 437–444.

[3] D.S. Alexander, M.S. Shaw, S.M. Nettles and J.M. Smith, "Active Bridging," in *Proceedings of the ACM SIGCOMM*, Cannes, FR, 1997, pp. 101–111.

[4] I. Hadzic and J.M. Smith, "Balancing Performance and Flexibility with Hardware support for Network Architectures," *ACM Transactions on Computer Systems*, Vol. 21, No. 4, November 2003, pp. 375–411.

[5] K. Nahrstedt and J.M. Smith, "Design, Implementations and Experiences of the OMEGA Endpoint Architecture," *IEEE Journal on Selected Areas in Communications*, Vol. 14, No. 7, September 1996, pp. 1263–1279.

Cognitive Services for the User

Joseph P. Campbell, William M. Campbell,
Scott M. Lewandowski and Clifford J. Weinstein
Information Systems Technology Group, MIT Lincoln Laboratory,
Lexington, MA, USA

10.1 Introduction

Software-defined cognitive radios (CRs) use voice as a primary input/output modality and are expected to have substantial computational resources capable of supporting advanced speech- and audio-processing applications. This chapter extends previous work on military speech applications (see, e.g., Ref. [1]) to cognitive-like services that enhance military mission capability by capitalizing on automatic processes, such as speech-information extraction and understanding the environment. Such capabilities go beyond interaction with the intended user of the software-defined radio (SDR)—they extend to speech and audio applications that can be applied to information that has been extracted from voice and acoustic noise, gathered from other users and entities in the environment. For example, in a military environment, situational awareness and understanding could be enhanced by informing users based on processing voice and noise from both friendly and hostile forces operating in a given battle space. This chapter provides a survey of a number of speech- and audio-processing technologies and their potential applications to CR, including:

- A description of the technology and its current state of practice.

- An explanation of how the technology is currently being applied, or could be applied, to CR.

Note: This work was sponsored by the Department of Defense under Air Force contract FA8721-05-C-0002. Opinions, interpretations, conclusions, and recommendations are those of the authors and are not necessarily endorsed by the US Government.

- Descriptions and concepts of operations for how the technology can be applied to benefit users of CRs.

- A description of relevant future research directions for both the speech and audio technologies and their applications to CR.

A pictorial overview of many of the core technologies with some applications presented in the following sections is shown in Figure 10.1. Also shown are some overlapping components between the technologies. For example, Gaussian mixture models (GMM) and support vector machines (SVM) are used in both speaker and language recognition technologies [2]. These technologies and components are described in further detail in the following sections of this chapter.

Speech and concierge cognitive services and their corresponding applications are covered in the following sections. The services covered include speaker recognition, language identification (LID), text-to-speech (TTS) conversion, speech-to-text (STT) conversion, machine translation (MT), background noise suppression, speech coding, speaker characterization, noise management, noise characterization, and concierge services. These technologies and their potential applications to CR are discussed at varying levels of detail commensurate with their innovation and utility.

10.2 Speech and Language Processing

The following speech- and language-processing technologies begin with the acoustic speech signal collected from a single microphone (multisensor collection using a variety of sensors is also shown). All the speech and language technologies here can be viewed as being abstractly related to the anatomy and dynamics of the vocal apparatus and behaviors expressed via speech and language and/or viewed as statistical modeling methods.

10.2.1 Speaker Recognition

Speaker recognition can enable a CR to authenticate users for access control, identify communicating parties, personalize the device, and adapt the device and its applications to individuals. Speaker recognition technologies allow systems to automatically determine who is talking or, to be precise, whether the incoming voice compares favorably with an enrolled user's voice. This determination can then be used to provide user authentication for access control, identification of communicating parties, and personalization and adaptation of the device and its applications. Figure 10.2 shows the basic operations of a speaker recognition system and its two phases of operation: enrollment and verification.

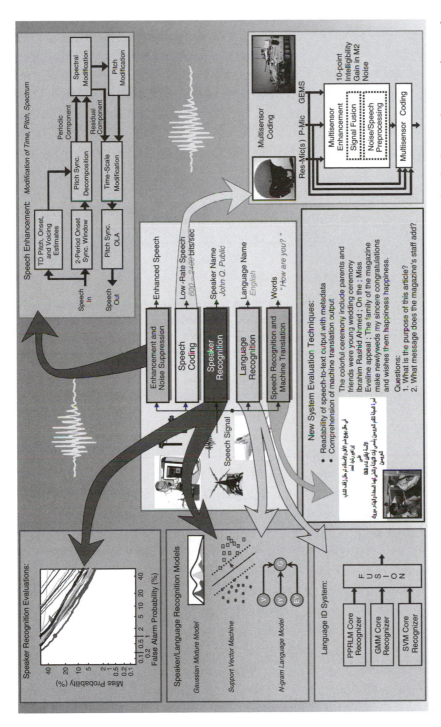

Figure 10.1: Speech and language technologies. A wide variety of speech and language technologies, such as speaker and language recognition and MT can be used to enable cognitive-like services for software-defined radios. In addition, multisensor speech coding and enhancement technologies can aid users, especially in harsh noise environments.

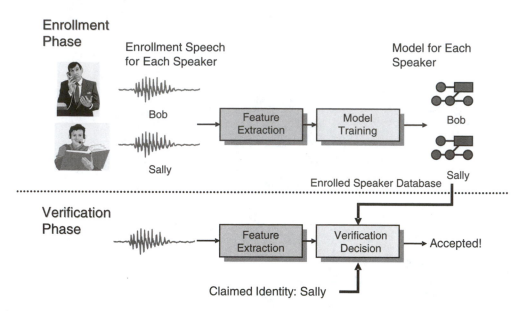

Figure 10.2: Speaker recognition. Speaker identification is based on extracting features from speech that best characterize the anatomical and behavioral differences of each individual when compared to the distributions of those features among the entire speaker population. Those differences are used to identify speakers or verify the claim of a speaker's identity.

Enrollment and Verification

In the enrollment phase, voice samples from the subject are used to create a model or template, which is sometimes improperly referred to as a voiceprint, for the specific speaker. In the verification phase, the unknown voice is compared with the model of the claimed identity. A verification decision can be made to accept or reject the identity claim.

Speaker recognition is imperfect and is characterized by two types of errors: false alarm (FA—meaning false recognition) and miss (failure to recognize the claimed enrolled speaker). These systems are characterized by whether the speech they use is text dependent (e.g., phrase prompted or pass phrases) or text independent (e.g., conversational speech). The performance of these systems is quantified by a plot of the miss rate versus the FA rate, which vary as a parametric tradeoff of a control threshold selected by the system designer called the operating point. The operating point is adjusted by varying an acceptance threshold that shifts tradeoffs between the two error types up or down the curve of a given system. Often, a combined measure is cited to provide an approximate representation of overall system accuracy; this measure, known as the equal-error rate (EER), indicates the operating point at which the miss and FA rates are equal. The state-of-the-art text-independent speaker recognition performance for conversational telephone speech of a few minutes in duration is in the range of 7–12 percent EER [3]. Given an extended duration of enrollment speech, even better performance can be achieved, as shown in Figure 10.3.

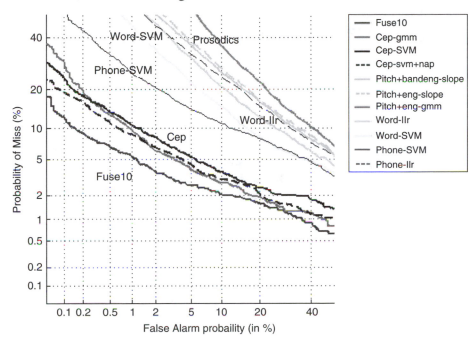

Figure 10.3: Performance curves of speaker recognition systems. Shown are the performance tradeoffs between probability of miss and false alarm of ten algorithms and their fusion. The speaker models were trained on approximately 20 minutes of speech and tested on about 2 minutes of speech.

As introduced in Campbell et al. [4], voice biometrics[1] are well suited to radios that already incorporate microphones and speech coders[2] (necessary for encrypted voice communications). Additionally, voice biometrics can be combined with other biometrics, such as face recognition, as shown in Figure 10.4, for increased security and/or backup operating modes (e.g., face recognition could be more reliable than speaker recognition in high-noise environments or vice versa in

---

[1] Biometrics is the automatic recognition of individuals based on biological and behavioral traits.
[2] Low-rate speech-coding vocoders parameterize the speech signal for transmission. This parameterization can be shared with speaker recognition engines, typically with some loss in performance, to conserve processing.

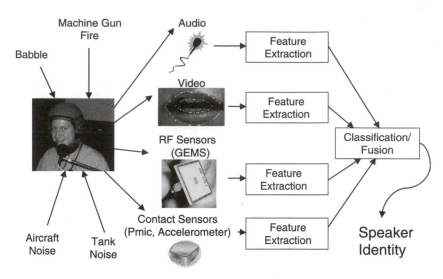

Figure 10.4: Speaker identification combined with facial and other biometrics. This combination enhances the accuracy, reliability, and survivability of the biometric system. Additionally, the suite of sensors shown enables robust low-rate speech coding in extremely harsh military noise environments [5].

adverse lighting conditions). Given the proliferation and improving quality of cameras in cellular phones, CRs are also likely to have cameras that might be suitable for face recognition.

Voice biometrics can provide access control via biometric logins and screen locks. This includes guarding against unauthorized use of lost CRs, such as disabling an idle radio that has been left behind. Voice biometrics can also enable user conveniences, such as recalling preferences and adapting to users (as do modern operating systems and application software). Conventional biometrics are generalized here to incorporate an additional authentication factor by learning and recognizing the users' distinctive behaviors. Future research directions for speaker recognition focus on making it more robust to mismatched channel conditions and applying high-level features that resemble those used by humans [6–8].

User Authentication

This section presents various means of user authentication, introduces generalized biometrics, and illustrates continuous and confidence-based user authentication. As shown in Figure 10.5, the four pillars of user authentication are *knowledge* (e.g., PIN or password), *tokens* (e.g., key or badge), *behaviors* (e.g., usage

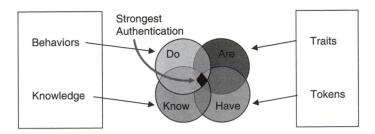

Figure 10.5: The four pillars of user authentication. Recognizing people based on what they do, what they know, who they are, and/or what they have. Using multiple attributes can increase reliability and survivability. Using all four attributes can provide the strongest authentication [4].

patterns or outcomes), and *traits* (e.g., voice or fingerprint). The proper combination of all four pillars provides the strongest user authentication. Biometrics are used to authenticate users,[3] as opposed to authenticating something they know (such as a password, which can be forgotten or compromised) or possess (such as an identification card, which can be lost or stolen). Unlike knowledge- and token-based authenticators, however, the inability of users to transfer biometrics can lead to difficulties; as in, for example, an emergency transfer of operation of a radio with biometric access control to an unenrolled user. To solve this difficulty, knowledge- and token-based authenticators can be used to authenticate users in these situations.

Popular biometrics includes voice, face, fingerprint, and iris.[4] Voice and face biometrics (possibly in combination) are well suited to radios that already incorporate microphones and cameras. Some biometrics lend themselves to continuous user authentication (e.g., to guard against lost or stolen radios being misused) and allow a system to assess varying levels of trust. For example, voice verification can be used to continuously authenticate a user while the user is talking; this can be useful if the voice quality makes it difficult for the interlocutor to detect a change in talkers. Figure 10.6 shows an example of an authentication process over time with varying levels of trust [4]. This example begins in a state of provisional trust and, over time, proceeds in continued states of provisional trust and then to a

---

[3] Strictly speaking, biometric verification is a binary hypothesis test. Here, the hypothesis is that the live voice matches an enrolled and stored voice model. The biometric system decides to accept or reject this hypothesis.

[4] See http://www.biometrics.org/ for additional biometrics.

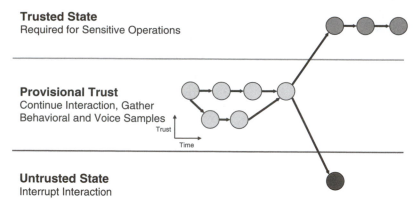

Figure 10.6: Continuous user authentication and trust. Varying levels of trust assessed over time via continuous authentication of a user are represented as state transitions. With sufficiently long samples of speech, the system is generally able to make a confident decision about the speaker's identity and, therefore, whether or not to trust the user [4].

trusted or untrusted state. While in a state of provisional trust, benign operations (e.g., adjusting radio volume) can be performed, whereas sensitive operations (e.g., downloading an SDR waveform) require a trusted state.

Behavior-based user authentication recognizes users via their actions, interests, tendencies, preferences, and other patterns. Examples of distinctive behaviors include:

1. *How* a user does something (e.g., speed and pattern of typing, stylus angle and intensity, use of menus versus keyboard shortcuts).

2. What a user *does* (e.g., patterns of applications use, program features used, patterns of collaboration).

3. What a user *causes to happen* (e.g., sequences of system calls, patterns of resource access).

These behaviors include not only a user's local actions, but also network interactions and outcomes. Behavior-based user authentication, like voice verification, has minimal adverse impact on a user's normal activities. The authentication is inherent and transparent; there is continuous mode operation and modest resource utilization, and user acceptance is likely to be high. Monitoring these behaviors can be combined with situational awareness to fuse multiple factors into the authentication process.

A cognitive approach allows for many interesting possibilities. First, the threshold to reach the trusted state of user authentication can be adapted based on situational, environmental, and mission awareness and the risk of the requested operation (e.g., benign volume adjustment to sensitive security operations). Second, authentication can be performed over time by combining available information—voice communication, mouse/stylus movement, dialogue structure, etc.

Some issues and questions in biometric deployments involve:

1. Whether to use remote versus distributed versus network enrollment and verification.
2. Where user models are created and stored.
3. How models are maintained and updated.
4. How enrollment is conducted.
5. How models are bound to users.
6. What the tolerable verification time or rate is.
7. How models of new users are distributed and their integrity assured.
8. Whether there are accuracy or policy requirements.
9. What the architecture to support the biometrics is.

Biometric Sensors

There are many approaches to biometric-based user authentication and all require some form of hardware input device to gather the required information about the user to be authenticated. For example, fingerprint recognition requires a fingerprint scanner, user voice recognition requires a microphone, and user behavior monitoring requires a traditional user input device (e.g., a keyboard or mouse). These hardware devices must be an integral part of the CR platform and must communicate with their software counterparts over a secure channel. Many of these devices are high bandwidth, although the utilization is often in bursts. The channel connecting the hardware and software must be capable of supporting the data transfer requirements without an undue performance impact on the device's core functionality.

Security Architecture with Biometric Processing

Once data has been gathered from a biometric sensor, it must be processed to determine user identity (or other user characteristics that the sensor has been designed to assess). Such processing can occur either in software, as shown in Figure 10.7, or in specialized hardware. Although the use of specialized hardware provides the advantages of increased tamper-proofing and higher performance, the complexity of managing updates and modifications to the functionality of the hardware often outweighs these benefits. Implementing the biometric processor in software does not fundamentally diminish the overall security of the system, given that the overall security of the radios (and the network services provided by radios) is also built on software components. Although not shown in Figure 10.7, most biometric processors require access to a database that contains information required to authenticate users, such as biometric models or templates, profiles, or logs of past behavior.

Figure 10.7: Notional radio security architecture. This architecture supports continuous-biometric processing and trust assessment over time for CR security [4].

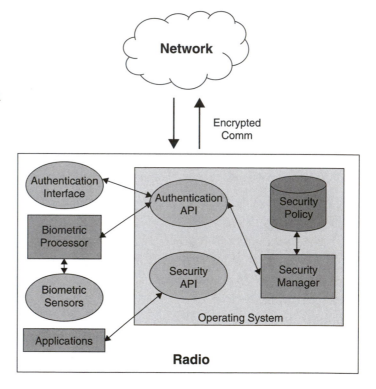

Applications

Although the platform has been designed to accommodate existing applications without modification, new applications can be designed to leverage the capabilities exposed by the security application programming interface to improve the user experience and to improve overall system performance. Designing applications to leverage this architecture is critical to its success [4].

Legacy applications will automatically benefit from continuous authentication, but will be completely unaware of confidence-based authentication. Therefore, the platform will need to define a confidence level at which the user is considered authenticated, and any level of confidence below that will cause the user to be considered completely unauthenticated by legacy applications. Applications that are aware of confidence-based authentication, in contrast, can enable functionality or access to data based on the confidence of the user's identity.

10.2.2 Language Identification

LID can enable a CR to infer whether its user or a remote communicator is a friend, a neutral party, or a foe based on identifying the language transmitted over a radio channel. LID technologies allow systems to automatically determine the language of the user from a list of possibilities. These technologies are available for over a dozen languages. These systems usually require about 30 seconds of speech to obtain good spoken LID performance.

Methods for spoken language recognition have traditionally been based on phonetic transcription of different languages [9]. By discovering the relation between occurrences of phones[5] in different languages (i.e., phonotactics[6]), a statistical model can be constructed of a particular language for later use in identification.

An emerging class of recent methods for language recognition is based on novel features [10]. These new features, shifted-delta cepstral coefficients,[7] measure changes in the speech spectrum over multiple frames[8] of speech to model long-term language characteristics. These methods need only a speech corpus labeled by language for training in order to achieve good results.

As mentioned previously, current system performance [10] is measured in terms of FA rate and target miss rate for detectors of individual languages. State-of-the-art error rates for speech from telephone environments are shown in

---

[5] Phones are subunits of the basic sound units of speech (e.g., "ah" or "t.").

[6] The set of allowed arrangements or sequences of speech sounds in a given language.

[7] The cepstral coefficients are common features used in speech-signal processing that are derived from a speech signal's spectral content. The *cepstrum*, pronounced kepstrum and an anagram of spectrum, is the result of taking the Fourier transform of the decibel spectrum, as if it were a signal. The mel-frequency cepstrum coefficients are a common variation and popular features used in speech and speaker recognition.

[8] Each frame of speech is on the order of 20 milliseconds in duration.

Figure 10.8. This plot shows average results for the following 12 languages: Arabic, English, Farsi, French, German, Hindi, Japanese, Korean, Mandarin, Spanish, Tamil, and Vietnamese. Results are shown for males (m), females (f), and both males and females (b) speakers for 3-second, 10-second, and 30-second utterances. EERs are less than 3 percent for 30-second utterances.

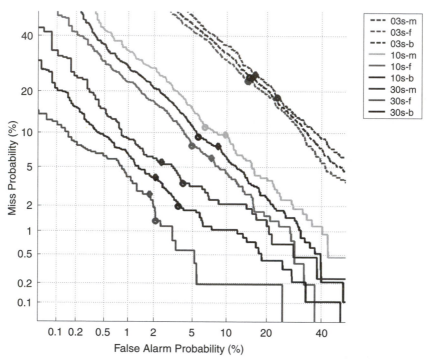

Figure 10.8: Typical performance of a language recognition system. Results are taken from the 2003 National Institute of Standards and Technology (NIST) Language Recognition Evaluation. Performance improves (curves shift toward origin) as the test speech duration increases and performance is fairly independent of the talker's sex.

LID has many potential applications in CR. First, LID could be used as a defense against system overrun; that is, the system could allow only certain languages to be used for radio communications. A more experimental strategy may be to look for *shibboleths*[9] to recognize the actual dialect or accent of the speaker; for example, whether the speaker has a foreign accent. A second application of LID is in situational awareness. If speech communications can be intercepted (and

---

[9]This biblical term has come into modern usage as a linguistic test to determine members of one group versus outsiders.

decrypted), their language could be identified to aid in the recognition of friends and foes and to alert soldiers of changes in the languages spoken by nearby forces.

10.2.3 Text-to-Speech Conversion

TTS conversion can provide status information to an eyes-busy CR user. TTS technology automatically speaks textual information. Textual information could originate from text-based communications or equipment display readouts. Text-based examples of communications that could be spoken via TTS includes e-mail, news, web, rich site summary (RSS), weblogs (blogs), instant messaging (IM), Internet relay chat (IRC), and short message service (SMS). Traditional equipment display readouts could also be spoken via TTS, such as radio frequency, battery power, signal strength, network speed, time, speed, location, and bearing.

By providing status information to an eyes-busy user, TTS would enable soldiers to focus on their mission while hearing an explanation of their battle space and status. Different synthesized voice types (e.g., male and female) are usually employed to convey different types of information. For example, routine or urgent information could be conveyed in male or female voices, respectively.

The current state of TTS technology produces mostly reasonable sounding speech; however, it does not yet sound quite human. Future research directions in TTS are focusing on improving the quality of voice synthesis, pronunciation of named entities, conveyance of expression, and integration with MT and STT conversion.

10.2.4 Speech-to-Text Conversion

STT conversion can recognize a CR user's voice commands, take dictation, and compose text messages from voice. STT technology attempts to automatically convert speech into a form that can be read by a user, produce entire transcripts of a conversation (continuous speech recognition), perform word spotting (looking for particular words), and provide command and control functions via voice. As CR keypads and displays shrink, the voice control modality becomes even more useful.

Recently, speech recognition has developed along several paths. One path is work on large vocabulary, continuous speech recognition for conversational situations. This work has been funded through projects such as the Defense Advanced Research Projects Agency's (DARPA) Effective Affordable Reusable Speech Recognition (EARS) program; work in this area can be found in, for example,

Schwartz et al. [11]. Progress in STT has brought error rates down to less than 12 percent word error rate for telephone speech. Another recent path for STT work is in noise robustness. An overview of some of these methods can be found in Junqua and Haton [12]. Noise robustness has been studied extensively for standardization by the European Telecommunications Standards Institute (ETSI) for distributed speech recognition (DSR), as exemplified by Parihar and Picone [13]. DSR's goal is to make STT a client–server application, in which the client uses the DSR front-end to parameterize the speech, while recognition is done on the server.

STT has many possible applications in CR:

First, STT can be used for gisting—rather than having a user listen to the complete conversation, a summarized version of the output can be produced.

Second, STT can be used to route certain conversations to appropriate users (see Riccardi and Gorin [14] and related references).

Third, STT can be used for data mining speech. If radio communication is processed by STT and stored, then text-retrieval techniques (such as those used to search for documents on the Internet) can be a quick and efficient way of searching for content.

Fourth, STT can be used for command and control of a CR, as described by Brown and Campbell [15]. In this scenario, a speech interface frees up tactile and visual modalities so that the user can more effectively multitask. The speech interface can be used to control various aspects of the CR: radio modes, sensor interfaces, sensor analysis, etc.

10.2.5 Machine Translation

MT automatically converts words or phrases from one language into another. This is generally done on text; however, MT can be combined with STT and/or TTS conversions to provide mixed mode translation [16].

MT technology could help a soldier during operations in foreign-language environments. For example, foreign-language signs, news, and radio intercepts could be roughly translated to the soldier's language to aid in understanding the battle space.

Current MT technology, as typified by various web-based systems, can be helpful for extracting some of the key words and phrases from foreign-language material, but such translations are by no means transparent, as they generally contain many errors. Transcription problems are frequent and are often, but not always, easily detectable by users (it could be argued that it is more problematic when users are unable to detect transcription problems). Future MT-related research will likely be aimed at improving basic MT performance, automatically extracting meaning, gisting, and summarization.

10.2.6 Background Noise Suppression

Background noise suppression enhances communication in the presence of noise, but extracts useful background noises to enhance situational awareness. Background noise suppression is primarily used in conjunction with STT conversion and voice communication. For voice communication, many new technologies have become available over the last few years.

Noise suppression can be used in voice communication to enhance the effectiveness of a speech coder. In this case, a noise suppression system attempts to improve both the quality and the intelligibility of coded speech. These methods fall into several categories. First, methods that attempt to "subtract out" the noise spectrum have achieved considerable success (see, e.g., Ref. [17]). A sophisticated form of spectral subtraction [18, 19] is integrated in the enhanced mixed-excitation linear-predictive (MELPe) 1200/2400 bps speech coder [20]. A second class of noise suppression algorithms is based on computational auditory scene analysis (CASA) [21]. The idea, in this case, is to use algorithms inspired by human processing; people effectively separate a sound field into multiple components such as music, voice, noise, among others. CASA methods use techniques such as independent component analysis and array processing to achieve noise suppression. A third class of noise suppression methods is based on multimodality. A well-known phenomenon for humans is that visual processing and audio processing of speech are fused, as evidenced by the McGurk effect[10] [22]. Several systems have been developed to take advantage of the visual component (see, e.g., Ref. [23]). Alternate non-acoustic modalities have also been explored; these include electromagnetic, accelerometer, and electroglottogram (EGG) sensors. Significant improvement in noise suppression has been achieved with these approaches [5].

Active noise suppression is another technology that is being incorporated into radio systems. Active noise suppression reduces the noise that a user perceives by emitting sound to cancel the undesired noise field. Active noise suppression can

---

[10] The McGurk effect is a phenomenon that indicates an interaction between hearing and vision in speech perception.

be used to decrease fatigue caused by exposure to high noise levels and reduce the Lombard effect.[11] Consumer headphones incorporating active (and passive) noise suppression are commercially available.

Noise suppression is a critical component of a CR with a speech user interface. Although not usually perceived as a cognitive capability, noise suppression is ultimately a test of a system's ability to deal with real-world conditions. Techniques such as multimodality and CASA show the sophistication and the challenge of matching human processing in this task.

10.2.7 Speech Coding

Speech coding is the effective use of varying or limited capacity channels for voice communications, as is widely experienced via cellular telephony. Speech coding seeks to provide communicability, intelligible speech, quality speech, talker and state (e.g., stress) recognizability, low delay for two-way communications (the total system one-way delay must be under 300 milliseconds for normal conversation), talker and language independence, naturalness, robustness in acoustic noise (including background talkers), insensitivity to transmission errors, tandem (synchronous and asynchronous) coding capability, ability to transmit signaling/information tones, near minimum rate coding (variable bit rates can complicates encryption and conferencing), and minimal computational and memory complexities to maximize battery life. Speech coding designers trade off many of these objectives when designing or choosing from among the many coders for real-world applications.

The current generation of speech coding standards capitalizes on the fact that people listen to speech communications systems and, thus, the systems attempt to minimize perceptual distortion. Future research will likely focus on optimizing coding for speech, as opposed to other types of signals; taking advantage of the language being spoken; widening the analysis bandwidths; and fusing multiple sensor streams. New adaptive speech coders will not only provide good communications-quality speech under noisy conditions, but will operate at dramatically reduced bit rates to conserve battery life and/or provide high processing gain to decrease the probability of a communication being intercepted, detected, or jammed [24]. This will yield improved and safer voice communications for the soldier.

---

[11] The Lombard effect is the tendency to increase one's vocal effort, which alters speech, in noise.

10.2.8 Speaker Stress Characterization

Speaker characterization is the process of automatically determining the state of a user by using speech-processing techniques. Typically, this has meant trying to determine a person's emotional state, which is often related to the stress a user is experiencing.

Speaker characterization is still a developing science. One of the difficulties is elicitation of an emotional state for corpus collection—how can an experimenter truly ensure that a participant is stressed and still comply with Institutional Review Board human subject testing requirements? Another difficulty is the definition of emotional states. For example, stress can take many forms: physical stress, emotional stress, task-based stress, noise-induced stress, and so on. Should all of these be separate categories of stress? Regardless of the experimental difficulties, several practical techniques for stress recognition and compensation have been examined; for examples, see the earlier work [25, 26] and work on the Speech Under Simulated and Actual Stress (SUSAS) corpus [27].

Speaker characterization is related to CR in various ways. Speaker characterization can be part of a broader strategy of affective computing [28]. Some examples include:

- Knowing the stress state of the local user as well as other users in the field to improve situational awareness.
- Knowing if a user is irritated by a particular feature by relying on the user's voice characteristics.
- Using stress level to determine appropriate modality (e.g., visual versus audio) for response to a query.
- Using verbal cues to determine if the CR made a correct decision.

Ultimately, if a CR is able to perceive and exploit a user's emotional state, it will make better decisions and be a more effective device.

10.2.9 Noise Characterization

Although noise has been considered a nuisance up to this point, it can be a useful source of information. Noise can be exploited in several ways. First, noise characterization can provide situational awareness; it can infer the location and situation of friends and foes by characterizing the acoustic environment in which the radio is operating. The CR, or a user, could catalog and track noise types in an environment to recognize anomalies that might indicate the presence of friend or foe. In this case, a noise characterization system would have to find features and provide recognition of different types of noise sources: vehicles, guns, planes, and so on. Also, the directionality of noise sources in the acoustic and radio-frequency environments would be a critical property to assess. Second, noise characterization can provide diagnostics. Noise analysis can be used to detect imminent mechanical failure of common military equipment [29] or provide a quick diagnosis of mechanical problems.

10.3 Concierge Services

With CR, there is an opportunity to provide sophisticated human interfaces. Broadly speaking, these interfaces fall into two different categories—transparent and concierge. For a transparent interface, the goal is to provide services to the user without the user explicitly invoking the interface. A simple example of this service would be automatic detection of input modality. A more advanced interface example is an augmented reality interface where the user sees information overlaid on real objects through a heads-up display. The other interface type, concierge, requires the user to explicitly invoke and respond to cognitive services. An example of this type of service is an agent that searches for information on a topic based on a verbal request. Hybrid architectures between these two interfaces are also a possibility—we may not want services automatically invoked, but we may want the CR to transparently know the user.

A unique opportunity in CR is the combination of advanced interfaces with a portable radio. A simple example that illustrates some of the ideas involved is shown in Figure 10.9. Suppose two users are communicating, and that they want to decide on a rendezvous for lunch (time, place). Both CRs have knowledge of their users' preferences and current status (money, schedule, etc.). Both radios also have a common ontology for communicating about the users' preferences. When one user asks to have lunch, the two radios can communicate and reason about a rendezvous place. This problem has many facets and can be quite sophisticated; the system will have to reason about:

- *Location*: The location of the two users and how long it would take to go to a common restaurant.
- *Schedule*: The available schedules of the two users, and how traveling to lunch would impact these schedules.
- *Resources*: How much money do the two users have? Is the restaurant available?

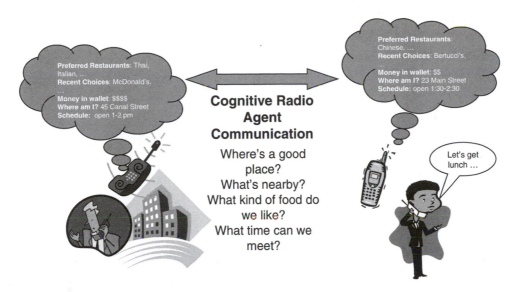

Figure 10.9: An example concierge service for a CR. The radios' concierge services make arrangements for a rendezvous for lunch. In a cognitive-like manner, the radio might anticipate the users' needs for lunch and sense nearby associates to invite.

- *User preferences*: What type of restaurant is reasonable for both users? What prior choices have the users made?

- *Social network*: If they invite other people, how does this impact the choices?

- *Dynamic rescheduling*: The CR could suggest an alternative rendezvous, if new information was discovered en route to the first choice.

- *Network resources*: The CR will access a comprehensive list of restaurants stored on the web.

- *User intent*: Is this meeting business or personal? Can the radio anticipate a lunch date from the user's schedule?

Ideally, the CR would find a list of prioritized restaurant choices. The users could then verbally negotiate the remaining details.

The example highlights the need for several technologies for cognitive interfaces. First, a common method of describing the user state and preferences is needed. For many situations, a language such as Web Ontology Language (OWL) would be appropriate. A second technology of importance would be modeling user intent. Several options exist in this area—Hidden Markov Models or Dynamic Bayesian Network models, for example Ref. [30] and [31], and plan/goal recognition [32]. These technologies attempt to predict what task the user is trying to accomplish and the current state of the task. Third, technologies that mimic a human interface may be appropriate—avatars, speech technologies (as previously discussed), visual interfaces, haptic interfaces, etc. Finally, sophisticated reasoning technologies must be able to handle time, geography, and uncertainty.

Reasoning and learning in a CR to support the user interface consists of many different types; three significant examples are reactive, adaptive, and network-based. Reactive reasoning is quick and does not involve explicit planning. A cognitive system would be expected to respond immediately to user voice commands, user preferences, etc.; that is, a reactive reasoner would be local to the CR. An adaptive CR would learn from its past experiences by adapting user preferences, speech recognition models, user intent models, etc. It could also pass this knowledge on to other CRs. Finally, one of the large potentials for CR would be in network-based reasoning. A CR could access a large network of services to respond to user requests. For instance, in the concierge example, the agent representing the user would be expected to obtain information about location, restaurants, travel time, etc. The benefit of network-based reasoning is flexibility and minimization of local resource usage; the drawback is latency and protocol standardization (a common communication ontology).

10.4 Summary

This chapter has provided an overview of several speech- and audio-processing technologies exploiting the voice and acoustic noise streams that are likely to be available to a software-defined CR.

An integrated approach to user authentication and architecture to enhance trusted radio communications networks has been presented. User authentication, via generalized biometrics, can be combined with other authenticators to provide a continuous, flexible, and strong system. This biometrically enhanced authentication system approach can be extended to become part of a CR system that learns about users, situations, and surroundings and takes appropriate proactive or reactive actions. One kind of learning presented here was generalized biometric authentication, where the users' distinctive behaviors and traits are learned and recognized. Cognitive-like applications to CR were given using speaker recognition, LID, TTS conversion, STT conversion, MT, background noise suppression, speech coding, speaker characterization, noise management, noise characterization, and concierge services technologies. An advanced CR will capitalize on these technologies to learn about and take action based on user preferences, availability of resources, and other elements of the situation and environment.

These technologies leverage the significant computational capabilities of future CRs to improve the capability, effectiveness, and efficiency of users of CRs. A high-impact implication of this technology is the ability to drastically reduce the complexity of the man-machine interface with the radio. As these technologies mature and become more robust, they will provide significant efficiency for the user, which will better enable the user to accomplish his most important tasks.

References

[1] C.J. Weinstein, "Opportunities for Advanced Speech Processing in Military Computer-Based Systems," *Proceedings of the IEEE*, Vol. 79, No. 11, 1991, pp. 1626–1641.

[2] W.M. Campbell, J.P. Campbell, D.A. Reynolds, E. Singer and P.A. Torres-Carrasquillo, "Support Vector Machines for Speaker and Language Recognition," *Computer Speech and Language*, Vol. 20, No. 2–3, 2006, pp. 210–229.

[3] M.A. Przybocki and A.F. Martin, "NIST Speaker Recognition Evaluation Chronicles," in *Proceedings of Odyssey: The Speaker and Language Recognition Workshop*, Toledo, Spain, 2004, pp. 15–22.

[4] J.P. Campbell, W.M. Campbell, D.A. Jones, S.M. Lewandowski, D.A. Reynolds and C.J. Weinstein, "Biometrically Enhanced Software-Defined Radios," in *Proceedings of Software Defined Radio Technical Conference (SDR '03)*, Orlando, Florida, 2003.

[5] T.F. Quatieri, K. Brady, D. Messing, J.P. Campbell, W.M. Campbell, M.S. Brandstein, C.J. Weinstein, J.D. Tardelli and P.D. Gatewood, "Exploiting Nonacoustic Sensors for Speech Encoding," *IEEE Transactions on Audio, Speech, and Language Processing*, Vol. 14, No. 2, 2006, pp. 533–544.

[6] J.P. Campbell, D.A. Reynolds and R.B. Dunn, "Fusing High- and Low-Level Features for Speaker Recognition," in *Proceedings of Eurospeech*, Geneva, Switzerland, 2003, pp. 2665–2668.

[7] W.M. Campbell, J.P. Campbell, D.A. Reynolds, D.A. Jones and T.R. Leek, "High-Level Speaker Verification with Support Vector Machines," in *Proceedings of International Conference on Acoustics, Speech, and Signal Processing*, Hong Kong, China, 2003, pp. 73–76.

[8] W.M. Campbell, "Compensating for Mismatch in High-Level Speaker Recognition," to appear in *Proceedings of IEEE Odyssey: The Speaker and Language Recognition Workshop*, San Juan, Puerto Rico, 2006.

[9] M.A. Zissman, "Comparison of Four Approaches to Automatic Language Identification of Telephone Speech," *IEEE Transactions on Speech and Audio Processing*, Vol. 4, No. 1, 1996, pp. 31–44.

[10] E. Singer, P.A. Torres-Carrasquillo, T.P. Gleason, W.M. Campbell and D.A. Reynolds, "Acoustic, Phonetic, and Discriminative Approaches to Automatic Language Recognition," in *Proceedings of Eurospeech*, Geneva, Switzerland, 2003, pp. 1345–1348.

[11] R. Schwartz, T. Colthurst, H. Gish, R. Iyer, C.-L. Kao, D. Liu, O. Kimball, J. Makhoul, S. Matsouka, L. Nguyen, M. Noamany, R. Prasad, B. Xiang, D. Xu, J.-L. Gauvain, L. Lamel, H. Schwenk, G. Adda, L. Chen and J. Ma, "Speech Recognition in Multiple Languages and Domains: The 2003 BBN/LIMSI EARS System," in *Proceedings of International Conference on Acoustics, Speech, and Signal Processing*, Montréal, Québec, Canada, 2004, pp. 753–756.

[12] J.-C. Junqua and J.-P. Haton, *Robustness in Automatic Speech Recognition*, Kluwer Academic Publishers, Boston, Massachusetts, 1996.

[13] N. Parihar and J. Picone, "Analysis of the Aurora Large Vocabulary Evaluations," in *Proceedings of Eurospeech*, Geneva, Switzerland, 2003, pp. 337–340.

[14] G. Riccardi and A.L. Gorin, "Stochastic Language Adaptation over Time and State in Natural Spoken Dialog Systems," *IEEE Transactions on Speech and Audio Processing*, Vol. 8, No. 1, 2000, pp. 3–10.

[15] C. Broun and W.M. Campbell, "Force XXI Land Warrior: A Systems Approach to Speech Recognition," in *Proceedings of International Conference on Acoustics, Speech, and Signal Processing*, Salt Lake City, Utah, 2001, pp. 973–976.

[16] W. Shen, B. Delaney and T.R. Anderson, "The MIT-LL/AFRL MT System," in *Proceedings of International Workshop on Spoken Language Translation*, Pittsburgh, Pennsylvania, 2005.

[17] Y. Ephraim and D. Malah, "Speech Enhancement Using a Minimum Mean-Square Error Log-Spectral Amplitude Estimator," *IEEE Transactions on Acoustics, Speech, and Signal Processing*, Vol. 33, No. 2, 1985, pp. 443–445.

[18] R. Martin, I. Wittke and P. Jax, "Optimized Estimation of Spectral Parameters for the Coding of Noisy Speech," in *Proceedings of International Conference on Acoustics, Speech, and Signal Processing*, Istanbul, Turkey, 2000, pp. 1479–1482.

[19] R. Martin, D. Malah, R.V. Cox and A.J. Accardi, "A Noise Reduction Preprocessor for Mobile Voice Communication," *EURASIP Journal on Applied Signal Processing*, Vol. 2004, No. 8, 2004, pp. 1046–1058.

[20] T. Wang, K. Koishida, V. Cuperman, A. Gersho and J.S. Collura, "A 1200/2400 bps Coding Suite Based on MELP," in *Proceedings of 2002 IEEE Workshop on Speech Coding*, Tsukuba, Ibaraki, Japan, 2002, pp. 90–92.

[21] M. Cooke and D. Ellis, "The Auditory Organization of Speech and Other Sources in Listeners and Computational Models," *Speech Communication*, Vol. 35, No. 3–4, 2001, pp. 141–177.

[22] H. McGurk and J. MacDonald, "Hearing Lips and Seeing Voices," in *Nature*, Vol. 264, 1976, pp. 746–748.

[23] X. Zhang, C.C. Broun, R.M. Mersereau and M.A. Clements, "Automatic Speechreading with Applications to Human-Computer Interfaces," *EURASIP Journal on Applied Signal Processing*, Vol. 2002, No. 11, 2002, pp. 1228–1247.

[24] A. McCree, "A Scalable Phonetic Vocoder Framework Using Joint Predictive Vector Quantization of MELP Parameters," in *Proceedings of International Conference on Acoustics, Speech, and Signal Processing*, Toulouse, France, 2006.

[25] K. Cummings and M.A. Clements, "Analysis of the Glottal Excitation of Emotionally Stressed Speech," *Journal of the Acoustical Society of America*, Vol. 98, No. 1, 1995, pp. 88–98.

[26] J. Hansen and M. Clements, "Source Generation Equalization and Enhancement of Spectral Properties for Robust Speech Recognition in Noise and Stress," *IEEE Transactions of Speech and Audio Processing*, Vol. 3, No. 5, 1995, pp. 407–415.

[27] S.E. Bou-Ghazale and J.H.L. Hansen, "Speech Feature Modeling for Robust Stressed Speech Recognition," in *Proceedings of International Conference of Spoken Language Processing*, Sydney, Australia, 1998, pp. 887–890.

[28] R.W. Picard, *Affective Computing*, MIT Press, Cambridge, Massachusetts, 2000.

[29] C. Chen and C. Mo, "A Method for Intelligent Fault Diagnosis of Rotating Machinery," *Digital Signal Processing*, Vol. 14, 2004, pp. 203–217.

[30] N. Oliver, E. Horvitz and A. Garg, "Layered Representations for Human Activity Recognition," in *Proceedings of Fourth IEEE International Conference on Multimodal Interfaces*, Pittsburg, Pennsylvania, 2002, pp. 3–8.

[31] D.W. Albrecht, I. Zukerman and A.E. Nicholson, "Bayesian Models for Keyhole Plan Recognition in an Adventure Game," *User Modeling and User-Adapted Interaction*, Vol. 8, No. 1–2, 1998, pp. 5–47.

[32] N. Blaylock and J. Allen, "Corpus-Based Statistical Goal Recognition," in *Proceedings of Eighteenth International Conference on Artificial Intelligence*, Acapulco, Mexico, 2003, pp. 1303–1308.

Website

http://www.biometrics.org/

CHAPTER 11

Network Support: The Radio Environment Map

Youping Zhao[1], Bin Le[2] and Jeffrey H. Reed[1]

[1] Mobile and Portable Radio Research Group, [2] Center for Wireless Telecommunications, Wireless@Virginia Tech, Bradley Department of Electrical and Computer Engineering, Virginia Tech, Blacksburg, VA, USA

11.1 Introduction

The motivations for network support for cognitive radio are threefold:

1. With powerful network support, the requirements on cognitive radio user equipment could be significantly relaxed because many computation-intensive cognition functionalities can be realized at the network side. Distributed and collaborative information processing over the network can reduce the workload of single user's equipment and speed up the adaptation process of cognitive radio. This is an important strategy to facilitate the commercialization of cognitive radio technology, considering the many constraints imposed on cost-sensitive user equipment, such as limited battery power, signal processing capability, and memory footprint.

2. Key cognitive functionality, such as incumbent primary user (PU) detection, cannot be reliably accomplished by user equipment itself due to the shadowing

Note: Youping Zhao is supported through funding from the Electronics and Telecommunications Research Institute (ETRI) and is advised by Jeffrey H. Reed. Bin Le is supported by the National Science Foundation (NSF) under Grant No. CNS-0519959 and advised by Charles W. Bostian. Any opinions, findings, and conclusions or recommendations expressed in this material are those of the author(s) and do not necessarily reflect the views of ETRI or NSF.

or fading effects of radio propagation and the practical system limitations of the sensitivity, dynamic range, and the noise floor [1, 2]. The radio has to resort to network support for many situations to solve the hidden node problem and achieve the operational goals of cognitive radio.

3. Network support is critically important to the evolution of wireless communications from legacy radios to cognitive radios, and from the coexistence of various disparate radio networks to converged cooperative networks. As explained in this chapter, network-enabled cognitive radio offers maximal flexibility to the government, regulator, and service provider by supporting dynamic policies on spectrum access and utilization.

The radio environment map (REM) has been proposed as a vehicle of network support for cognitive radio [3, 4]. The REM is an abstraction of real-world radio scenarios; it characterizes the radio environment of cognitive radios in multiple domains, such as geographical features, regulation, policy, radio equipment capability profile, and radio frequency (RF) emissions. The REM, which is essentially an integrated spatiotemporal database, can be exploited to support cognitive functionality of the user equipment, such as situation awareness (SA), reasoning, learning, and planning, even if the subscriber unit is relatively simple. The REM can also be viewed as an extension to the available resource map (ARM), which is proposed to be a real-time map of all radio activities in the network for cognitive radio applications in unlicensed wide area networks (UWANs) [2, 5].

This chapter discusses the motivation and the important role of network support in cognitive radios, then introduces the REM and explains how the REM can provide network support to cognitive radios for various applications. The main concerns of network support and the related research issues are also addressed in this chapter.

11.2 Internal and External Network Support

From the cognitive radio user's point of view, the network support to the cognitive radio can be classified into two categories: internal network support and external network support. The internal network refers to the radio network with which the cognitive radio is associated. Along with various communication services, the internal network can provide some cognitive functionality as well. For example, the cognitive radio network can provide location information and location-based services to the user; it can also characterize the usage pattern of other users in the neighborhood. The external network refers to any other networks that can provide meaningful knowledge to support the cognitive functionalities of the radio. For example, a separate sensor network could be dedicated to gather information for cognitive radio networks [6]. The external network could be a legacy network or other cognitive radio networks.

Both internal and external networks can contribute to building up the REM and can be employed in a collaborative way. For instance, location information needed for a cognitive radio can be obtained either from internal network support through a network-based positioning method for indoor scenarios, or from external network support through the global positioning system (GPS) for outdoor scenarios.

As depicted in Figure 11.1, network support can be realized through a global REM and local REMs. In this figure, the cognitive radio is symbolized as a "brain"—empowered radio. The global REM maintained on the network keeps an overview of the radio environment, and the local REMs stored at the user equipment only present more specific views to reduce the memory footprint and communication overhead. The local REMs and global REM may exchange information in a timely manner so as to keep the information stored at different entities current. In this figure, a regional REM can be aggregated from the combined experiences of several local REMs.

Figure 11.1: REM. This provides cognitive services to both the associated internal networks and a useful awareness of external networks such as noncognitive legacy systems.

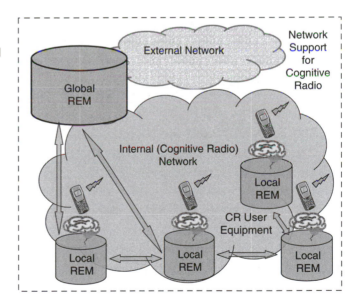

11.3 Introduction to the REM

This section starts with an insightful analogy about cognitive radio, and then introduces the REM as a cost-effective navigator for cognitive radios. Essentially,

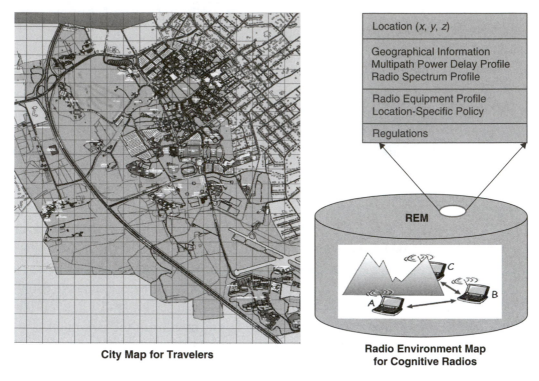

| Location (x, y, z) |
| Geographical Information
Multipath Power Delay Profile
Radio Spectrum Profile |
| Radio Equipment Profile
Location-Specific Policy |
| Regulations |

City Map for Travelers

Radio Environment Map
for Cognitive Radios

Figure 11.2: City map versus REM. A REM provides services to local cognitive radios similar to the way a city map aids a driver with local navigation. In contrast to the city map, the REM database must be current to be useful.

the REM is a comprehensive spatiotemporal database and an abstraction of real-world radio scenarios [3, 4].

Similar to how a city map helps a traveler, the REM can help the cognitive radio to know the radio environment by providing information on, for example, spectral regulatory rules and user-defined policies to which the cognitive radio should conform; spectrum opportunities; where the radio is now and where it is heading; the appropriate channel model to use; the expected path loss and signal-to-noise ratio (SNR); hidden nodes present in the neighborhood; usage patterns of PUs[1] and/or secondary users (SUs); and interference or jamming sources. Figure 11.2 shows such an insightful analogy. In fact, there are many commonalities between transportation and telecommunication, such as the concept of throughput, channel capacity, routing, signaling, rules, etiquette, and utilization efficiency of

---

[1]PUs usually hold the license to use certain spectrum. Therefore, PUs have higher priority over SUs. For more discussion on PUs and SUs, please refer to http://www.wireless-world-research.org/index.html.

resources. Enlightened by this analogy, some lessons may be drawn from the field of transportation and employed in the field of cognitive radio.

The REM can provide both current and historic radio environment information, so that most cognitive functionalities in the cognitive radio network can be realized in a cost-efficient way. By leveraging the REM, cognitive radios can conduct spectrum sensing with prior knowledge rather than continually blindly scanning over the whole spectrum. Thus, observation time and energy consumption in the radio front-end can be significantly reduced. By incorporating collaborative information processing techniques, the costs of a cognitive radio system can be further reduced by relaxing the requirements on transmission power, dynamic range, and sensitivity of the individual radio device. By combining reasoning and learning with data mining in a REM, network intelligence directly enables cognitive capabilities for its network nodes whether the subscriber radios are cognitive or not. In this sense, even legacy networks can become cognitive by resorting to a REM. The REM also supports a system-level solution to the fundamental cognitive radio technical challenges, such as SA, cross-layer optimization, the hidden node and exposed node problem, load balancing across the network, opportunistic spectrum access, and dynamic spectrum regulations and policy.

11.4 REM Infrastructure Support to Cognitive Radios

11.4.1 The Role of the REM in Cognitive Radio

The REM plays an important role in the cognition cycle of cognitive radio, as illustrated in Figure 11.3. Both direct observations from the radio and knowledge derived from network support can contribute to the global and/or local REM. The radio's environment awareness can be obtained from direct observation, such as spectrum sensing, and/or from the REM. Reasoning and learning help the cognitive radio to identify the specific radio scenario, learn from past experience and observations, and make decisions and plans to meet its goals. The global REM and/or the local REM should be updated once action is taken or scheduled by the radio to keep the REM's information current.

The underlying techniques to support a REM include, but are not limited to, database design, database management, database transactions (such as query, search, and update), and data mining.

11.4.2 REM Design

The REM contains information at multiple layers, as illustrated in Figure 11.4. By integrating various databases, the REM enables or supports cognitive functionality

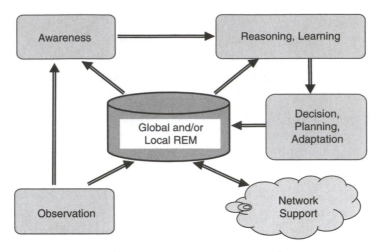

Figure 11.3: Role of REM for cognitive radio. The REM provides the infrastructure to support the fusion of multiple cognitive services to the local subscribers.

for radios with different levels of intelligence. The REM helps a cognitive radio to be aware of situations and make optimal adaptations according to its goals; for legacy or hardware reconfigurable radios, the REM facilitates smart network operations by providing cognitive strategies to the network radio resource management control. Just like the city map that is informative to every traveler, no matter whether driving a car or taking the bus, the REM is transparent to the specific radio access technology (RAT) to be employed regardless of whether the subscriber radio is cognitive or not.

With the help of the REM, a radio can become cognitive of performance metrics, the application, topology, and network (routing), as well as the medium access control (MAC) and physical (PHY) layers of communication stacks under different and varying radio environments. For example, if the radio is used on the battlefield, reliability and security are of high importance. Therefore, special source coding, encryption, anti-jamming channel coding, frequency planning, and routing algorithms could be employed accordingly. The REM can support various network architectures: centralized, distributed, or heterogeneous networks, or even point-to-point communications. It can also support collaborative information processing among multiple nodes for obtaining comprehensive awareness. With the REM, a cognitive radio can choose an access network based on cost, data rate, spectral efficiency, and many other performance metrics. The optimal adaptation is subject to the constraints of various radio scenarios, such as

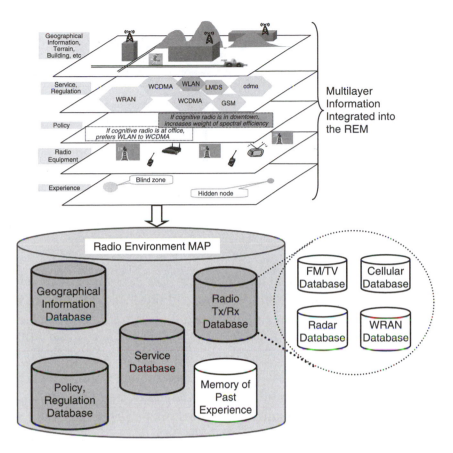

Figure 11.4: Integrating various databases for building up the REM.

available services, available spectrum, user policy, and the capability of the radio equipment.

11.4.3 Enabling Techniques for Implementing REM

Figure 11.5 shows the interdisciplinary research nature of the REM design for cognitive radios. To implement and exploit the REM, various technologies must be employed, such as artificial intelligence (AI), detection and estimation, pattern classification, cross-layer optimization, database management and data mining, site-specific propagation prediction, and network-based ontology.

The next generation of cognitive radios and their required infrastructure are enabled by a collection of many new technologies: data storage technology, very large scale integration (VLSI), RF integrated circuits (RFICs), geographical information systems (GIS) (e.g., three-dimensional (3-D) digital map), terrain-based

Figure 11.5:
Interdisciplinary research
for developing and
exploiting REM.

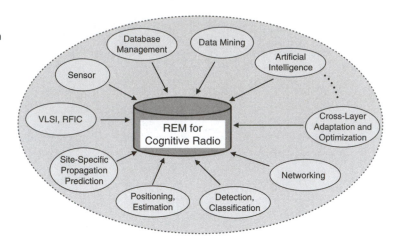

radio propagation modeling, and GPS receivers. For example, several electronics manufacturers have recently developed handheld products with Gigabyte storage capacity, and many cellular phones have already been equipped with GPS receivers. Therefore, it is quite possible that future cognitive radios will have enough memory to contain a comprehensive database such as the REM.

Radio propagation simulation tools can predict many important parameters such as path loss, SNR or signal-to-interference and noise ratio (SINR), as long as the geographical environment information is available [7]. Compared to the empirical channel model-based prediction, the advanced site-specific radio propagation techniques and software tools using 3-D terrain maps have been successfully employed to provide much more reliable predictions for radio propagation prediction and system planning and management [8]. This makes it possible to embed several contours into the REM, such as the service contour, blind zone, and interference region, which will enable cognitive radios to make decisions and adaptations that overcome the most common user complaints.

Information to populate the REM can be obtained by:

1. Integrating and/or correlating various existing databases, such as GIS databases or radio equipment databases.

2. Sensing with collaboration among distributed nodes (which may or may not be cognitive radios).

3. Observing from a dedicated sensor network and/or other external networks.

4. Probing the radio environment.

5. Estimating the radio propagation characteristics with software tools.

11.5 Obtaining Awareness with the REM

This section explains the meaning and importance of SA to cognitive radios. It also shows how radios become situation aware with the help of the REM. As discussed in the previous section, the REM can present multidimensional information for cognitive radios, such as geographical environment, location and activities of radios, regulations, and policy of the user or service provider. One of the most important features of the REM is that it is transparent to the specific application of cognitive radios.

11.5.1 Awareness: Prerequisite for Cognitive Radio

The REM presents comprehensive system-level knowledge for a cognitive engine to exploit. The cognitive engine, which is the cognition core of the cognitive radio, is typically implemented as a software system consisting of learning and adaptation algorithms.

An interesting analogy between two intelligent agents—a taxi driver and a cognitive engine—makes this clear (Figure 11.6). The comparisons of cognition components between these two intelligent agents are listed in Tables 11.1 and 11.2. The SA and performance measure for a taxi driver or pilot have been extensively discussed [9, 10]. The capability of SA is one of the most important features that differentiates a cognitive radio from an adaptive radio.

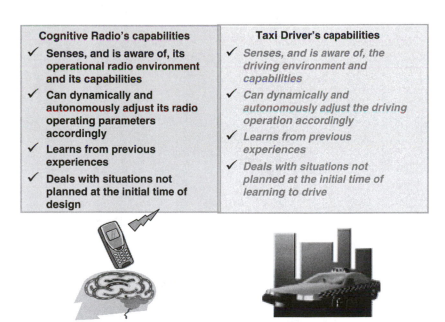

Figure 11.6: Capabilities of two intelligent agents: cognitive radio engine and taxi driver.

Table 11.1: Comparison of two intelligent agents: taxi driver and cognitive radio engine.

Agent type	Environment	Performance measure	Sensors	Actuators
Taxi driver [9]	Roads, other traffic, pedestrians, customers	Safe, fast, legal, comfortable trip, maximize profits, minimize collisions	Cameras, sonar, speedometer, GPS, odometer, accelerometer, engine sensors, keyboard	Steering, accelerator, brake, signal, horn, display
Cognitive radio engine	Radio spectrum, other traffic by PU and SU, jammer, RF noise and interference	Spectrum utilization, reliable, fast, legal, cost-efficient, low power consumption, minimize interference	GPS, antenna, BER/ PER/FER, interference temperature, quality of service	Transmission power control, MAC, beamforming

BER: bit error rate; FER: frame error rate; PER: packet error rate.

Table 11.2: Three-level situation awareness.

Agent type	Level 1 awareness	Level 2 awareness	Level 3 awareness
Taxi driver (or Pilot) [10]	*Looks* and *perceives* basic information	*Thinks* about and *understands* the meanings of that information	Uses the environmental understanding to *anticipate* what will happen ahead in time and space
Cognitive radio engine	*Observes* the RF spectrum, and waveform and surrounding activities, with focused attention	Performs calculation, estimation, reasoning, and *understands* the radioactivity and what it represents	Performs prediction and planning to *anticipate* radio performance, network requirements, and user needs

It is apparent that both intelligent agents observe the environment and become aware of their situations; make in situ decisions according to their observations, anticipations, and experiences; and then execute intelligent adaptations to reach their goals. They evolve by a spiral learning process (also known as the "cognition cycle") throughout their lifetimes.

Like the taxi driver, the cognitive radio may also have three levels of SA, as shown in Table 11.2.

To appreciate the significance of three-level awareness, consider the following scenario: suppose you are driving from your office to a restaurant for lunch. You may make many "cognitive" decisions and adaptations according to your current situation, observations, and previous experience:

- If you have tight schedule for lunch, you may prefer to drive through a fast-food restaurant nearby.
- If it is a sunny Friday, and you want to enjoy the lunch with your buddies, you may carpool with your friends to a nice restaurant in another town.
- If it is a rainy day, and the road is curvy, you may turn on the headlights and drive slowly to avoid accidents.
- As you drive, you still need to be aware of the traffic lights, road signs, and speed limits.
- You may look at a map for directions, or recall from your memory the most convenient way (short-cut) to the restaurant.
- When you learn from the radio news broadcast that there is an accident ahead, you may anticipate that it will result in slow traffic and choose an alternate route.

This list could be extensive, considering the many other possible situations and their adaptations for a simple automotive errand. Obviously, the cognitive engine has both awareness and learning capability, but with an REM the ability to predict radio performance is facilitated as readily as predicting the path distance for the car trip.

11.5.2 Classification of Awareness

Awareness means the understanding of the situation [11]. Both geographical and RF environment-related information (such as radio propagation characteristics, waveform, and spectral regulations) play major parts in cognitive radio knowledge [12]. In addition, policy, goals, and contexts are also important issues for the cognitive radio.

SA here means that the cognitive radio knows its current radio scenario, the intent of the user, and the regulations with which it must comply. In summary, SA may include, but is not limited to, the following key types of awareness:

(a) *Location awareness*: The cognitive radio knows where it is, in the form of latitude, longitude, and altitude, or relative location to some reference nodes.

(b) *Geographical environment awareness*: The cognitive radio knows the terrain and geographical information related to the radio propagation and channel characteristics. This awareness is critically important for a cognitive radio to choose the appropriate spectrum, channel model, or RAT, antenna configuration, and networking techniques.

(c) *RF environment and waveform awareness*: The cognitive radio knows the spectrum utilization, the existence of PUs and/or SUs, the topology of the user group, the interference profile, and other RF characteristics that may be of concern.

(d) *Mobility and trajectory awareness*: The cognitive radio knows its moving speed and direction. For example, in conjunction with geographical awareness, the cognitive radio can know it is moving south along Main Street at a speed of 45 miles per hour, and it can "foresee" the radio environment ahead, such as the available channel after the user passes over the next hill or the radio standards supported along the route.

(e) *Power supply and energy efficiency awareness*: The cognitive radio knows the source of its power supply, the remaining battery life, and the energy efficiency of alternative adaptation schemes.

(f) *Regulation awareness*: The cognitive radio knows the spectrum allocation and emission masks at specific locations and frequency bands, which are regulated by government authorities such as the US Federal Communications Commission (FCC).

(g) *Policy awareness*: The cognitive radio knows the policy defined by the user and/or the service provider. For example, the user may prefer to use the wireless local area network (WLAN) from a specific service provider at some locations for quality of service (QoS) or security reasons.

(h) *Capability awareness*: The cognitive radio knows its own capabilities as well as those of its team members and/or the network. Such awareness may include knowing which waveforms are supported, the maximum transmit power, and the sensitivity of the cognitive radio.

(i) *Mission, context, and background awareness*: The cognitive radio understands the intent of the user, and knows what mode and volume of traffic it is going to generate and what the impact of that traffic will be to the local networks. The cognitive radio understands the QoS requirements, and how overhead activities may trigger additional network traffic and latency.

(j) *Priority awareness*: The cognitive radio knows the user's priorities and habits. For example, the user may prefer to use low-cost services whenever possible (e.g., to switch to WLAN from a third-generation (3G) system when entering a Wi-Fi® zone), or may prefer reliability over cost.

(k) *Language awareness*: The cognitive radio knows the signs, ontologies, and etiquette used among cognitive radios to communicate with each other.

(l) *Past experience awareness*: The cognitive radio remembers the past experience and learns from it.

Note that all of these items can be interrelated and employed together in various ways. For example, the cognitive radio may adapt network topology, and make dynamic spectrum access (DSA) decisions based on the user's location, mission goal, RF environment, regulations, and service priority. Table 11.3 summarizes the various types of awareness a cognitive radio may have, the relative significance of each type of awareness, and how to obtain such awareness.

11.5.3 Obtaining SA

The cognitive radio can obtain SA through different approaches:

1. direct observation (e.g., through field measurement);
2. inference from network support (e.g., through a network-maintained REM); or
3. analysis of local terrain propagation models combined with existing database structures defining known communication systems.

Using the previous analogy, suppose you are driving a car. You cannot just rely on your own vision when you drive, especially under unfavorable weather conditions, such as a snowstorm, or at night. The map complements your limited local vision and helps you know the road conditions ahead and how far to the destination. You can make an informed decision whether you need to stop at a gas station. You may take extra caution because you know the road ahead will be winding. You may schedule to have a dinner at the next rest area 10 miles ahead. In summary,

Table 11.3: Summary of situation awareness for cognitive radio.

Type	Significance	Approaches to obtain awareness	Current status
Location	High	• GPS (or assisted GPS) • Network-based positioning • Landmark or RF fingerprint matching • Combination of inertial navigation and GPS	Many kinds of positioning techniques are commercially available
Geographical environment	Low to medium	• Query and exploit GIS database • Terrain recognition • Site-specific propagation prediction	GIS database is available (e.g., from US Geological Survey [13]). Many site-specific propagation tools can predict path loss, delay spread, and service coverage
RF environment/ waveform	Medium to high	• Radio transceiver database • Collective observations by cognitive radios • Sensor network • Field measurement	Microwave point-to-point radio, FM, and TV station databases are available and maintained by the FCC, which provides radio station information, such as site location and transmission power [14]
Mobility and trajectory	Low to medium	Estimate moving speed and trajectory of the user by analyzing the change of locations over a period of time and correlating with GIS	Can be addressed with current technologies together with the REM
Power supply and energy efficiency	Battery: high AC: low	Measure the voltage and/or current of power supply	Mature technique (e.g., the cellular phone knows the source of its power supply and the remaining power)

Policy	High	Defined by the service provider and/or the user	Can be addressed with the policy database in the REM
Regulation	High	Defined by the government authorities	Can be addressed with the regulation database in the REM
Capability	High	Provided by the cognitive radios and/or networks	Can be addressed with the capability database in the REM
Mission/ Context/Back-ground environment	Low to medium	• Using machine intelligence, various sensors • Applying speech and/or image recognition and understanding techniques	Common industry software tools demonstrate some context-awareness, and even interact with the user occasionally
Priority	Low to medium	Defined by the service provider and/or the user	Can be addressed with the priority database in the REM
Language	Medium to high	Standardizing cognitive radio languages and etiquettes	This is under development (e.g., by the DARPA XG program [15])
Past experience	Medium to high	Long- and short-term memory of experience for recall	Can be realized with case-based decision theory (CBDT) and other technologies

DARPA XG program: Defense Advanced Research Projects Agency NeXt Generation program; FM: frequency modulation.

you easily obtain helpful context information. Therefore, when you drive, you can examine the map, refresh your past experience if you had been there before, or just take a "trial and error" learning approach if the map or previous experience is insufficient.

For the cognitive radio, shadowing, fading, and Doppler shift are the most common degradation or distortions imparted by the channel. REM can provide the cognitive radio with channel characteristics associated with the location and direction of a mobile user. Channel information can be obtained through observing instantaneous measurements of the environment as well as long-term measurements and learning of the general characteristics of the environment. Once these channel measurements are available, models can be created to predict the performance of the link. Of course, these models are stochastic and produce outputs that are random variables. The variance of model outputs can be incorporated into the decision process of cognitive radio. Furthermore, cognitive radio can take advantage of channel awareness for planning. With the awareness of shadowing and fading characteristics, cognitive radio may adopt the appropriate waveform (i.e., PHY and MAC layer) to adapt to or take advantage of the propagation characteristics. For example, in a multipath-intensive environment, cognitive radio may choose to apply multiple input, multiple output (MIMO) techniques to improve its performance. Cognitive radio can also anticipate emerging call drop due to multipath or shadowing and take pre-emptive measures, such as switching on the backup power amplifier, increasing the number of RAKE fingers, leveraging smart antenna resources or spreading gain, altering the power-control policy, or making an intersystem handoff.

The REM can exist at the user's terminal equipment (local REM) and/or at the network level (global REM). The local REM may be unique to each user's device. Each cognitive radio may use its own local REM to memorize its past experience as well as its current status. For example, the experience of a blind zone stored at the REM of a high-quality radio could be different from that of a low-quality, less-sophisticated radio.

As shown in Figure 11.7,[2] by using a global REM, it is possible that even the legacy radio network can be upgraded to support some cognitive functionality and behave as if it were a cognitive radio. For instance, through a software upgrade to the network-level radio resource management system, the legacy network can know the subscriber's location, and the interference environment, and

---

[2] In Figure 11.7, the term "radio" may refer to any type of radio device, even a cognitive RF identification (RFID) tag.

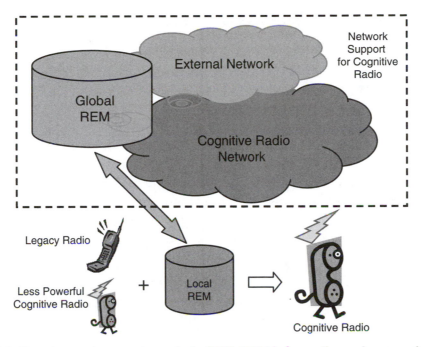

Figure 11.7: Situation awareness through the REM. REM helps radios to become situation aware by bootstrapping them with local accumulated experience, customized to each radio's functionality requirements.

instruct the radio to use the most effective PHY and MAC layers supported by the user's device. A simple radio with limited cognition capability can become more capable by leveraging the REM-based network support. The local REM may exchange information with the Global REM, for example, through some common control channel.

11.6 Network Support Scenarios and Applications

This section illustrates the implementation of the REM for cognitive radio network support through the analyses of various application scenarios requiring different network structures and services [16]. It shows that REM-based network support has the following fundamental features:

- It fits into the core of cognitive radio functional architecture.
- It is independent from specific network topology.
- It is compatible with hybrid node technology and intelligence.

353

11.6.1 Infrastructure-Based Network and Centralized Global REM

The centralized global REM can play an important role in many infrastructure-based radio systems, such as the IEEE 802.22 wireless regional area network (WRAN) and cellular radio systems. For example, the 802.22 WRAN is the first worldwide wireless standard based on cognitive radios. It is composed of WRAN base stations (BS), repeaters, and consumer premise equipment (CPE). 802.22 systems are primarily targeted at rural and remote areas offering fixed wireless access services. Figure 11.8 shows a typical WRAN scenario, in which TV stations, TV receivers, wireless microphones, and public safety systems operating on certain TV channels are PUs, and 802.22 system subscribers are SUs [17, 18]. Coexistence is the key requirement for 802.22 systems because the SUs must avoid generating harmful interference to the PUs and/or other collocated SUs [19, 20].

With a global REM maintained at a WRAN infrastructure, the WRAN BS can know the location, antenna height, and transmit power of nearby TV stations; the local terrain; the Grade B service contours of TV stations; the forbidden spectrum

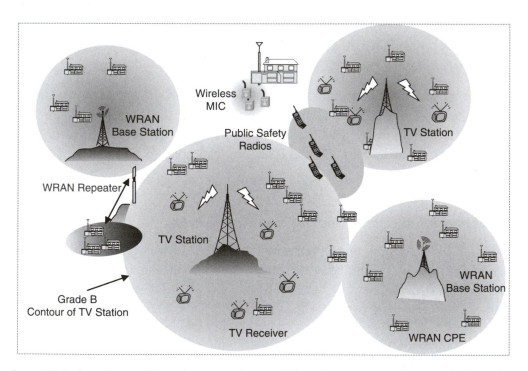

Figure 11.8: A typical radio environment for cognitive WRANs. It aggregates the knowledge of all local wireless activity.

used by public safety radios and satellite communications; the demographical distribution of TV receivers; and the available TV channels for use. In addition, it knows the distribution and usage patterns of other WRAN service subscribers. Such information helps the WRAN BS choose the best spectrum opportunity to use at the optimal transmission power. Smart antenna techniques can be more efficiently employed at the BS and/or the CPE with the radio environment information from the REM.

Another example application of the global REM is cognitive WLAN supporting network-based interference management. Figure 11.9 shows the typical radio environment of 802.11 WLAN in an office building, where various industrial, scientific, and medical (ISM) band interferences (such as microwave oven leakage, cordless phone, Bluetooth devices, and ZigBee devices) co-exist [21]. A global REM can be built up at the network infrastructure and used as a WLAN interference management tool. WLAN interference is detected, classified, and located by the cognitive subscribers or access points (APs). Such information is stored or updated in the global REM. In this way, the WLANs can cooperate to effectively identify and/or report sources of interference, and make adaptations at the AP or the subscriber device to avoid or mitigate the interference, such as changing the channel, managing the diversity mechanisms, selecting proper MIMO schemes, or switching *off* some APs. Note that no changes to the current WLAN PHY/MAC standards are required; however, flexibility in the APs and/or subscriber devices may be needed.

11.6.2 Ad hoc Mesh Networks and Distributed Local REMs

Local REMs can be used in ad hoc mesh networks that consist of cognitive radios. The Wireless World Research Forum (WWRF) has established a work group (WG3) to study cooperative and ad hoc networks as an integral and evolving part of the future communication infrastructure, taking into account both wireless and fixed aspects [22]. In such networks, the topology can be self-organized and optimized; various modulation and coding schemes can be adapted; smart antenna and MIMO techniques can be selected dynamically; and information sharing and collaboration methods, such as maximal throughput, improved reliability, or extended transmit range, can be employed to achieve the goal(s) defined by the network or participating radios. As pointed out in Chapter 7, for ad hoc cognitive radio networks, distributed learning is another effective way of learning that can increase the power of cognitive radios greatly. By sharing or exchanging the local REMs, cognitive radios may learn to match the right routing protocol, either

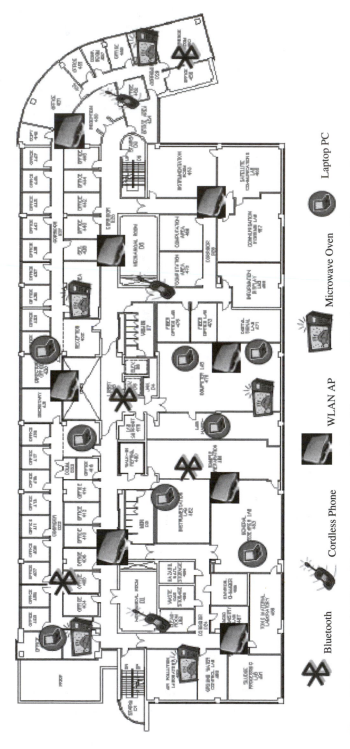

Figure 11.9: A typical REM of 2.4 GHz WLAN activity. This map provides interference avoidance and management services.

Bluetooth   Cordless Phone   WLAN AP

Laptop PC   Microwave Oven

proactively or reactively. Master nodes may be selected to be responsible for collecting the distributed local REMs and combining them into a complete REM for the network. Such a complete map can be accessed by each individual node, in a role similar to the routing table for an ad hoc network.

The Defense Advanced Research Projects Agency (DARPA) Adaptive Cognition Enhanced Radio Teams (ACERT) program intends to create a "distributed radio" greater than the sum of its parts [23], as illustrated in Figure 11.10. This scenario involves two teams of people forming two ad hoc networks. Both intrateam and interteam communications may be needed. Under an ad hoc mesh network scenario, the network support can be accomplished through distributed local REMs, where the multipath profile and path loss experienced by each user can be shared. Node and channel awareness can be realized by sharing and exchanging local REMs.

11.7 Supporting Elements to the REM

The REM is essentially a database that stores information utilized in the decision process of a cognitive network. As such, various elements in conjunction with the REM are needed to create the cognitive network. These elements include the learning, reasoning, and decision algorithms. Separating the REM from the learning, reasoning, and decision mechanisms helps to facilitate a system that can serve a variety of radio types and networks and support a variety of applications.

Reasoning, data mining, and query capability to external databases can furnish information for the REM; this composite capability is called the self-informing system (SIS), and is shown in Figure 11.11. Decision-making outside of the SIS is the key to integrating heterogeneous networks or even dissimilar radios within a homogenous network. Given the information provided by the SIS, a radio with its own decision-making capability can make decisions that are relevant to the radio and/or network capabilities and the supported application. Even for non-cognitive radios, a cognitive network can be achieved by interfacing the SIS with the network radio resource management system. Note that the SIS does not need to reside at a central location. It can, in fact, be distributed among the nodes in the network.

More complicated learning and reasoning techniques can be practically implemented on the network than on an individual cognitive radio. Data mining can be used for extracting hidden information and predicting future events. For example, data mining the Global REM can be used to find the usage pattern of PUs, SUs, and other interferers from large quantities of data (i.e., observations over an

Figure 11.10: DARPA distributed radio scenario. Two ad hoc networks form around two teams of users with intrateam and interteam command requirements, supported by a distributed REM database for both teams.

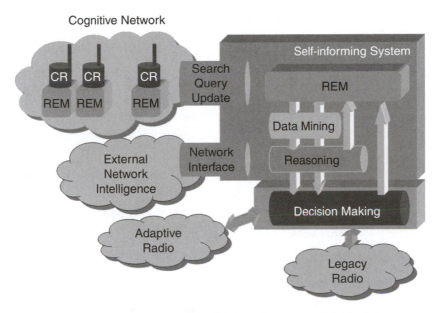

Figure 11.11: Cognitive radio self-informing system block diagram.

extended period of time), using such techniques as decision trees, neural networks, and neural-fuzzy systems.

As discussed in more detail in Chapter 12, the cognitive radio can employ many learning methods, such as statistical learning, instance-based learning, reinforcement learning, and decision-tree induction learning. Two basic optimization approaches may be employed by cognitive radios to optimize complex parameters: (1) the classical optimization based on the properties of the objective function; and (2) heuristic optimization. Common classical optimization routines include the cyclic-coordinate method, the steepest descent method, and the quasi-Newton and conjugate-gradient methods. Heuristic techniques include genetic algorithm, simulated annealing, ant colony optimization, Tabu search, and neural network, among others [4, 24]. Classical optimization is more analytically tractable compared to the heuristic approach; however, neither of them can be guaranteed to yield optimal solutions. As explained in Chapter 7, the cognitive radio environment is typically characterized by multi-objective decision-making (MODM). Without a single objective function measurement, it is difficult to apply classic optimization theory to adapt the radio parameters. Genetic algorithms are suitable for searching the optimal MAC/PHY/NET combinations and other parameters for cognitive radios [4, 25]. Hidden Markov Models (HMMs) can be used for radio scene classification, case recognition, and for making meaningful predictions based on past experience captured into the training data.

AI techniques are useful to fully exploit the REM to obtain SA and to enable reasoning, decision-making, and self-learning. For example, a cognitive radio that learns to foresee where and when a spectrum hole or interference will appear will do better than one that does not. Just as evolution provides animals with enough built-in reflexes so that they can survive long enough to learn for themselves, it would be reasonable to provide cognitive radio with some initial knowledge as well as an ability to learn. After sufficient experience in its environment, the behavior of the cognitive radio can become effectively independent of its prior knowledge. The incorporation of learning allows one to design a cognitive radio that will succeed in a vast variety of environments.

11.8 Summary and Open Issues

This chapter has addressed the important role of network support in developing cognitive radios for various application scenarios, including infrastructure-based as well as infrastructure-less ad hoc networks. As a vehicle to providing network support to cognitive radios, the REM is proposed to be an integrated database that consists of multidomain information, such as geographical features, available services, spectral regulations, locations and activities of radios, policies of the user and/or service provider, and past experience. A REM can be exploited by a cognitive engine to enhance or achieve most cognitive functionalities, such as SA, reasoning, learning, planning, and decision support. Leveraging both internal and external network support through global and local REMs presents a sensible approach to implement cognitive radios in a reliable, flexible, and cost-effective way.

Network support can dramatically relax the requirements on a cognitive radio device and improve the reliability of the whole cognitive radio network. Considering the dynamic nature of spectral regulation and operation policy, the REM-based cognitive radio is flexible and future-proof in the sense that it allows regulators or service providers to modify or change their rules or policies simply by updating the REMs accordingly. Because the REM is a comprehensive database that contains multidomain information needed for cognitive radios, even a low-cost/low-complexity radio device can obtain basic cognitive functionalities by referring to the REM.

As a system-level solution of the cognitive radio, the REM not only makes cognitive applications feasible, but, more importantly, it facilitates the evolution and convergence of wireless communication networks with cost-efficient database development. The REM presents a smooth evolutionary path from the legacy radio to the cognitive radio. The REM can also be viewed as a natural, but major evolution of radio resource management used in today's commercial wireless networks.

The REM also has great potential in bridging or converging different wireless communication networks, and thus facilitating the integration of various radio network architectures and access technologies, such as WRAN, Worldwide Interoperability for Microwave Access (WiMAX), WLAN, wireless personal area network (WPAN), Beyond 3G (B3G), and 4G systems, to support heterogenous cognitive wireless communications [26].

Although the network-enabled approach for cognitive radio looks promising, many technical issues currently remain open. Some important open questions include:

1. How can the global REM and/or the local REM keep its information current?

2. How current and at what level of granularity does the information contained in the global and local REMs need to be in order to provide desired performance?

3. How much additional overhead is needed to keep the REM current, and what is the constraint on network latency?

4. How well can the distributed local REMs serve an ad hoc mesh network?

5. How can the integrity, security, privacy, and reliability of the REM be assured?

Certainly there are many challenges ahead. Although these technical challenges seem achievable, business, regulatory, and political challenges can be much more difficult to predict and address.

References

[1]   J.H. Reed and C.W. Bostian, "Understanding the Issues in Software Defined Cognitive Radio," in *Tutorial for 2005 1st IEEE International Symposium on New Frontiers in Dynamic Spectrum Access Network*, Baltimore, November 2005.

[2]   W. Krenik and A. Batra, "Cognitive Radio Techniques for Wide Area Networks," in *42nd Design Automation Conference*, Anaheim, CA, June 13–17, 2005.

[3]   Y. Zhao and J.H. Reed, "Radio Environment Map Design and Exploitation," *MPRG Technical Report*, Virginia Tech, Blacksburg, VA, December 2005.

[4]   J.H. Reed, C. Dietrich, J. Gaeddert, K. Kim, R. Menon, L. Morales and Y. Zhao, "Development of a Cognitive Engine and Analysis of WRAN Cognitive Radio Algorithms," *MPRG Technical Report*, Virginia Tech, December 2005.

[5]   B. Krenik and C. Panasik, "The Potential for Unlicensed Wide Area Networks," Wireless Advanced Architectures Group, *Texas Instruments White Paper*, October 2004.

[6]   N. Sai Shankar, C. Corderio and K. Challapali, "Spectrum Agile Radios: Utilization and Sensing Architectures," in *2005 1st IEEE International Symposium on New Frontiers in Dynamic Spectrum Access Network*, Baltimore, November 2005.

[7]   T.S. Rappaport, *Wireless Communications Principles and Practice*, Prentice Hall, Upper Saddle River, NJ, 2002.

[8]   G. Bauer, R. Bose and R. Jakoby, "Three-Dimensional Interference Investigations for LMDS Networks Using an Urban Database," *IEEE Transactions on Antennas and Propagation*, Vol. 53, No. 8, Part 1, August 2005.

[9]   J.S. Russell and P. Norvig, *Artificial Intelligence: A Modern Approach*, 2nd Edition, Pearson Education, Upper Saddle River, NJ, 2003.

[10]   http://www.2pass.co.uk/awareness.htm#SAdefinition

[11]   J. Mitola III and G.Q. Maguire Jr., "Cognitive Radio: Making Software Radios More Personal," *IEEE Personal Communications*, Vol. 6, No. 4, August 1999.

[12]   B. Fette, "The Promise and the Challenge of Cognitive Radio," March 2004; http://www.sdrforum.org/MTGS/mtg_40_sep04/fette_cognitive_radio.pdf

[13]   http://www.usgs.gov/

[14]   http://www.fcc.gov/mb/databases/cdbs/

[15]   http://www.darpa.mil/ato/programs/xg/

[16]   J. Polson, "Cognitive Radio Applications in Software Defined Radio," *Software Defined Radio Technical Conference—SDR04*, Phoenix, November 2004.

[17]   C. Corderio, K. Challapali, D. Birru and N.S. Shankar, "IEEE 802.22: The First Worldwide Wireless Standard Based on Cognitive Radio," in *2005 1st IEEE International Symposium on New Frontiers in Dynamic Spectrum Access Network*, Baltimore, November 2005.

[18]   C.-J. Kim et al., "WRAN PHY/MAC Proposal for TDD/FDD," *Doc.: IEEE 802.22-05/0109r0*, November 7, 2005.

[19]   Federal Communication Committee (FCC), "In the Matter of Facilitating Opportunities for Flexible, Efficient, and Reliable Spectrum Use Employing Cognitive Radio Technologies, Authorization and Use of Software Defined Radios," *FCC NPRM 03-322*, Washington, DC, December 30, 2003.

[20]   Federal Communication Committee (FCC), "Facilitating Opportunities for Flexible, Efficient, and Reliable Spectrum Use Employing Cognitive Radio Technologies," *FCC NAO 05-57*, March 11, 2005.

[21]   Y. Zhao, B.G. Agee and J.H. Reed, "Simulation and Measurement of Microwave Oven Leakage for 802.11 WLAN Interference Management," *IEEE International Symposium on Microwave, Antenna, Propagation and EMC Technologies for Wireless Communications*, Beijing, China, August 8–12, 2005.

[22]   http://www.wireless-world-research.org/index.html

[23]   J.M. Smith, "Adaptive Cognition Enhanced Radio Teams (ACERT) Presentation," May 2005; http://www.darpa.mil/ipto/solicitations/open/05-37_PIP.htm

[24] T. Wang, *Global Optimization for Constrained Nonlinear Programming*, Ph.D. Thesis, Department of Computer Science, University of Illinois, Urbana, IL, December 2000.

[25] T.W. Rondeau, B. Le, C.J. Rieser and C.W. Bostian, "Cognitive Radios with Genetic Algorithms: Intelligent Control of Software Defined Radios," in *Software Defined Radio Technical Conference and Product Exposition*, November 15–18, 2004.

[26] Wireless World Research Forum Working Group 6 (WWRF-WG6), "Cognitive Radio, Spectrum and Radio Resource Management," *WWRF-WG6 White Paper*, 2004.

CHAPTER 12

Cognitive Research: Knowledge Representation and Learning

Vincent J. Kovarik Jr.
Harris Corporation, Melbourne, FL, USA

12.1 Introduction

Generally speaking, a cognitive radio is a radio system that can intelligently adapt its behavior or operational characteristics in response to changes in the radio's internal state or external environment. This ability to adapt promises a range of functional capabilities, including making the radio systems more "personal," as suggested by Mitola [1]. Some examples of internal states that may alter the operational behavior of the radio are:

- *Low Battery Power*: A reduction in available power may result in the inability to support multiple waveforms simultaneously. The radio may select and disable a low-priority waveform in order to conserve power or maintain operation of a mission-critical waveform.

- *Component Failure*: The radio may reroute around a failure point; it may provide an alternative service possibly at a lower data rate, or it may terminate a lower priority service in favor of a mission-critical communication.

Some examples of external influences that can affect the operational behavior include:

- *Co-Site Interference*: Use of the same radio frequency (RF) by multiple radios or a radio attempting to exercise multiple frequencies through a single set of RF equipment will diminish or negate successful communications. The radio may select an alternate operating frequency or change other characteristics of the waveform.

- *Background Noise*: Normal RF background noise can impede the effectiveness of communications on a given channel or frequency. Again, the radio may sense the level of background noise and alter the frequency, initiate frequency-hopping, or take other adaptive transmission actions to enable end-to-end communications.

Although, a cognitive radio system may exhibit intelligent behavior by adapting to changes in its environment, it is, nonetheless, following a fixed set of behavioral guidelines. As long as the environmental and internal states remain within the boundaries of its established guidelines, the radio system will continue to successfully modify its operational behavior in response to those changes. However, if the internal state or the environmental conditions fall outside the range of the precoded patterns and responses, the radio system cannot identify a feasible solution path to follow. Without a prescribed set of actions, the system typically ceases to operate or vacillates between states in an attempt to find a stable operational mode.

Within a cognitive radio system are several common architectural components:

1. Sensors or other methods for gathering information about the external environment and internal state of the system.
2. A corpus of knowledge that represents a set of behaviors to be performed in response to some set or pattern of inputs and/or states.
3. A reasoning engine or algorithm that applies the knowledge to the current state of the system and reaches one or more conclusions.
4. Some element that enables the system to change its operational characteristics or, in some way, act upon the conclusions.

The reasoning engine modifies the operation of the system based on the application of knowledge to the combined state information. Reasoning is the process by which the system has an existing set of knowledge, applies it to a current situation, and identifies a course of action.

Figure 12.1 illustrates the basic operational control of a cognitive radio. The radio is performing some communications function represented by the operational function box in the figure. The waveform implementation and general radio hardware information provide the internal state data of the radio system. One or more sensors provide information regarding the environment in which the radio is operating. This state information provides input data to the reasoning engine. The reasoning engine applies the collection of knowledge to the current state of the system. Based on the current combined state and the knowledge, the reasoning system changes one or more operational parameters for the function being performed.

As illustrated in Figure 12.1, a cognitive radio can intelligently adapt its operational behavior in response to external and internal influences. However, such a system has a critical-limiting factor: the system can adapt based solely on predefined behaviors as represented in the knowledge base. Thus, even though the radio can react intelligently to external environment and internal state changes, it can adapt its behavior only within the bounds of the previously defined knowledge—it cannot adapt to new situations. In order to modify its behavior in response to new situations, the radio must *learn*.

Learning entails not only the ability to sense and adapt based on precoded algorithms or heuristics but also the ability to analyze sensory input, recognize patterns, and modify internal behavioral specifications based on the resultant analysis of the new situation.

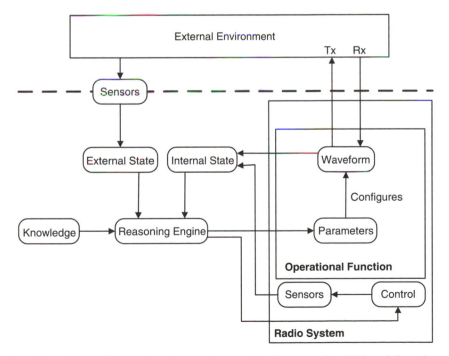

Figure 12.1: Cognitive radio monitoring and control adds several additional functions to the basic radio, labeled "operational function" in this figure.

The basic machine-learning architecture is illustrated in Figure 12.2. The architecture has a set of control parameters (e.g., rules, functions, algorithms, etc.) that provide some control output to the external environment. The essential concept is that the machine-learning system has some perception of the environment, gained through sensors, external data input, and other means, that provides the baseline truth on which the system asserts some conclusion or action. Coupled with the action is some predictive assertion regarding the anticipated impact or change to the environment as a result of the action. The action, in turn, alters or modifies the environment in some fashion. As the environment is modified, the changes are received by the learning component and compared against the expected changes. If the resultant changes are the same as the anticipated changes, then, based on the proximity of the actual changes to the expected changes, the system reinforces the parameters that led to the decision. If the resultant changes are different than the predicted values, then the learning system will alter the parameters of the decision process to more closely fit the actual results.

Figure 12.2: General machine learning process is built on a reasoning system and a decision-making system.

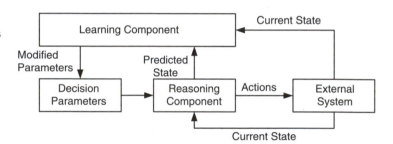

So, how must the architecture shown in Figure 12.1 be modified in order to support learning? Several changes must be made to the system, as illustrated in Figure 12.3. In order to learn, there must be some method for assessing the operational performance of the reasoning engine and, based on that assessment, a method for modifying the set of knowledge that is applied by the reasoning engine. So, as illustrated in Figure 12.3, a learning algorithm is integrated into the radio system. The learning algorithm, although potentially complex, has a simple mission: observe the state of the radio system and the actions selected by the reasoning engine. Compare the resultant state after the action selected by the reasoning engine is performed. Then alter the knowledge base to reflect the success (or failure) of the selected action. Alteration of the knowledge base can take multiple forms, from modifying existing knowledge, to changing the action selection policy, to generating entirely new knowledge entries for the system to apply in subsequent actions.

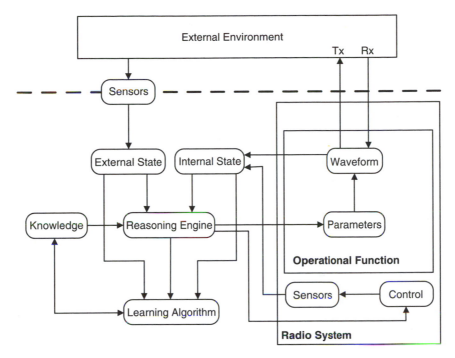

Figure 12.3: Learning integrated into cognitive radio architecture.

The balance of this chapter is organized into three key sections and a summarization. Section 12.2 provides a brief overview of knowledge representation and reasoning paradigms. Understanding the different representation mechanisms is an important prerequisite to discussing learning algorithms because the algorithm may be inextricably tied to the underlying representation and reasoning mechanism. Section 12.3 presents several approaches to machine learning and their application within a cognitive radio system. Section 12.4 presents some of the issues and trade-offs related to the implementation aspects of the learning systems. Finally, Section 12.5 summarizes the chapter.

This chapter provides a representative cross section of machine learning with an emphasis on application within the radio communications domain. It does not, however, provide an exhaustive overview of the multiple areas of machine learning, related research, and approaches; rather, some approaches are discussed in depth whereas others may be discussed at a superficial level or not at all.

12.2 Knowledge Representation and Reasoning

A common theme throughout all of the machine-learning methods discussed in this chapter—and, to a great extent, throughout all of machine-learning research—is the underlying tenet to enable a machine to react intelligently to some future situation based on knowledge it has gained through past experiences. Knowledge can have multiple forms based on its type and the selected reasoning mechanism. How the reasoning system represents and applies the knowledge underlying its decision process has an impact on the learning process applied to improve the system. This section looks at several common implementation methods and their potential uses within a cognitive radio system.

In humans, knowledge is represented as a complex set of neural connections that are activated in an associative manner. The set and pattern of the activation represents a concept or memory, a contextual relationship, or an action or event. Thus, when we are faced with a situation that we remember encountering before, the set of neural connections activated by the current state also activates a set of connections representing the action or response performed when last encountered.

The different sensory inputs interact to form a multidimensional perspective of a concept or object in the physical world. For example, waking up in the morning to the smell of bacon cooking can evoke a set of associated references. Based on the smell of the food, we can visualize the bacon cooking in the frying pan and remember the taste of the bacon we have eaten.

More significant than simply remembering the situation, however, is the ability to recall actions associated with a particular situation. Continuing with our breakfast scenario, we may also remember the last time we cooked bacon and the steps involved: getting the bacon out of the refrigerator, getting the frying pan from the cabinet, getting tongs or another instrument to manipulate the bacon, placing the bacon in the pan, flipping it over as it cooks, placing the cooked bacon on a plate, and then placing the plate on the table. The key concept here is that, in addition to remembering the *objects* or concepts, we also remember the *process* or action sequence associated with achieving a particular goal.

Supporting a reasoning mechanism requires two critical capabilities. First, there must be some form of capturing, categorizing, and storing the knowledge. Second, the stored knowledge must be activated and applied to a given set of parameters or conditions. Learning provides a method for extending the set of knowledge structures without explicit programming on the part of a human developer. The types of learning that can be supported depend on the underlying structure of the knowledge. Representing knowledge can take many forms, and in some cases, it may take multiple forms within a single system. For the objectives of this chapter, the knowledge representation forms are limited to two basic categories: *declarative* and *behavioral*.

Declarative knowledge and reasoning refers to any method of representation that asserts and reasons over descriptive factual knowledge about an object or entity. For example, the statement, "The car is red," provides a simple declarative fact—that the color of the car is red. In this example, the fact describes a property or attribute about the car. This type of declarative knowledge is different than that provided by the statement, "The car is owned by John." In this statement, declarative knowledge is also provided, but, whereas the first example provides information that describes a property of the car, the second asserts a relationship between two entities, the car and John. Thus, the second assertion of declarative knowledge provides contextual or semantic knowledge about the car and its relationship to other entities within the knowledge space. The key distinction is that the car's color is independent of contextual relationships and is a primitive property of the car entity. The "owned" relationship, on the other hand, describes semantic connections between two objects.

Behavioral knowledge and reasoning is concerned with actions and the effect they have on the knowledge system. The actions may be initiated by some agent or actor within the system or they may be external events that alter the system's environment. In both cases, the event has some initial state or set of assertions at the start of the action, some state at the end or completion of the action, and some temporal extent over which the action occurs.

The remainder of this section briefly discusses several different representation and reasoning approaches that intersect the above knowledge types. It should be noted that a particular approach may encompass reasoning, representation, or both. Furthermore, one reasoning approach may utilize different representation approaches, depending on the situation.

12.2.1 Symbolic Representation

In a symbolic representation and reasoning system, extensible data structures capture the salient facts, descriptions, and properties associated with a concept. This encapsulation of multiple descriptive information forms the *symbol*, or a unit of knowledge. Symbolic knowledge representation can have any of multiple forms. For example, knowledge may be represented symbolically as a conceptual entity that has a set of properties describing the entity, or it may represent a semantic association between two entities. For example, Figure 12.4 illustrates a simple assertion in symbolic form expressing a *transmits* relationship between Radio-1 and Radio-2. This concise pictorial view captures the fact that Radio-1 transmits some information or signal to Radio-2. The assertion may be represented in basic data structures within a high-level programming language such as C++, Java, or list processing (LISP), to name a few.

Each of the radios in Figure 12.4 can be represented by a data structure that may have additional descriptive information or properties about each of the radios. This is illustrated in Figure 12.5, in which Radio-1 now has additional information associated with it that describes its operational frequency and the power level of its transmitter amplifier.

Figure 12.4: Symbolic knowledge assertion: radio transmission represented as a from-to "transmits to" graph.

Figure 12.5: Properties associated with radio-1.

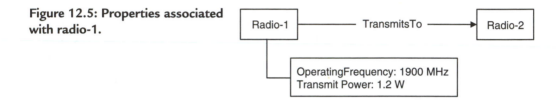

The benefits of a symbolic knowledge representation approach include that it is relatively easy for humans to understand and it is easily captured in graphical and/or natural language notation. Symbolic knowledge representation also provides a more natural means of conveyance of the underlying information for human understanding and interpretation.

Knowledge can be represented symbolically by using a variety of approaches, including semantic nets, rules, frames, and objects, among others. The key underlying concept is that the representation formalism utilizes a symbolic form of representing declarative knowledge.

As a cognitive radio system functions, part of the process, as described in the prior section, is to observe the actual response or result of its interaction with the environment, to analyze the result with respect to its expectations, and then to adapt based on some learning algorithm. One of the applications of learning is to extend existing knowledge based on newly observed data or patterns.

12.2.2 Ontologies and Frame Systems

Declarative knowledge is of little value when it is simply a disjoint collection of facts and assertions. The data must be organized into a useful form in order to support reasoning and learning. This organization is typically referred to as an ontology. Ontology-based knowledge representation and reasoning for software radios have been investigated by a number of individuals, including Wang et al. [2, 3], and in this text by Kokar et al. [4].

The classical *is-a-kind-of* relationship organizes classes or types of entities into a type/subtype hierarchy similar to object-oriented programming languages. The differences between the type and subtype can be structural or behavioral. For example, a pager and a cell phone may be both classified as a *"kind-of"* personal communications device.

Ontology-based representation can be traced back to early frame systems [5], and both have several aspects or viewpoints that can be applied to the organization of a corpus of knowledge into an ontology. The Web Ontology Language (OWL) by Bechhofer et al. [6] has been utilized in several efforts. For example, Baclawski et al. [7] explore the use of ontology-based reasoning for communications protocol interoperability.

Figure 12.6 illustrates a possible ontology of communications devices according to a type hierarchy. Two common types of devices, a cell phone and pager, are described within this contextual organization.

The value of this type of organization from a learning system perspective is that, as new concepts and situations are encountered, the existing knowledge ontology can be searched to identify an appropriate classification for the new

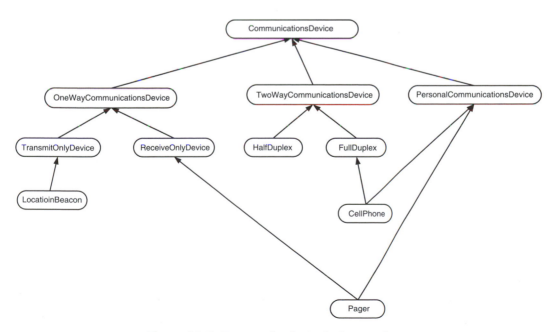

Figure 12.6: Communications device ontology.

situation or concept. For example, if a Family Radio Service (FRS) walkie-talkie is introduced into the system, it may be described as being used by an individual capable of either sending or receiving at any given point in time. Using this knowledge to guide the traversal of the knowledge hierarchy, the system would follow the hierarchical links based on the known attributes of the new entity and reach the conclusion that the cell phone radio system should be classified as a *PersonalCommunicationsDevice* and a *FullDuplex* device.

Learning within an ontology-based system is performed by incorporating and integrating new concepts into the existing set of entries. New concepts are incorporated into the ontology based on their similarity to existing entities; that is, new information is processed by a classifier. The classifier analyzes the properties and values associated with the new concept and compares it against the existing set of concepts ontology within the ontology. Wherever there is a similarity, the similar concept or concepts are candidates for associating the new concept into the ontology.

This is similar to the method by which we incorporate new concepts within the framework of our knowledge and experience. Using this approach, new entities can be linked into an ontology, enabling the system to extend its set of knowledge. Note that, although the example shown describes physical entities, the ontology can also represent related actions, events, or abstract entities such as waveforms and the attributes associated with each of these types of entities. Thus, organizing new knowledge can be performed across a range of conceptual domains.

Also note, however, that learning based on ontology extensions is predicated on the existence of an initial set of entities already organized into a hierarchical network of concepts.

12.2.3 Behavioral Representation

Behavioral knowledge refers to the embodiment or description of actions that an object or entity can perform or has performed. This may be behavior that is executed in response to a simple stimulus, as in the case of object-oriented programming languages. This may be described within a neural net or genetic algorithm (GA) as set of patterns and actions, or it may be a more complex behavior embodied within a set of rules or heuristics that allow additional options or degrees of flexibility in the entity's response to a given stimulus.

Although each type of behavioral knowledge may vary significantly in both flexibility and the types of behavioral patterns it supports, each has the same fundamental objective. Each describes behavioral responses that an intelligent system can exhibit in response to a stimulus that it receives.

Autonomously extending behavioral knowledge requires the system to incorporate a learning algorithm implementation that monitors the results of the system's actions and adapts or modifies the decision parameters based on the specific approach implemented by the learning algorithm. Typically, this is accomplished by strengthening those decisions or actions that lead to achieving a system goal, weakening those choices that did not lead to the attainment of the goal, or both.

12.2.4 Case-Based Reasoning

People faced with a new situation or problem typically attempt to find some prior experience or situation that is similar and recall what actions were taken in response to the situation. In some instances, there is a high degree of similarity between the current situation and a prior experience in memory. When there is a high degree of similarity, then the actions previously taken can usually be applied to the current situation. In those instances where there is some similarity to prior experience, then typically only some subset of the prior actions may be applicable, or some may be applicable but require modification. This process of comparing current situations to prior experience and then applying, potentially with adaptations, prior actions to the current situation is the fundamental approach of case-based reasoning (CBR). The origin of CBR is generally attributed to Schank [8].

Figure 12.7 illustrates the general architecture of a CBR system. Performing CBR consists of several steps:

1. The current situation or state must be assessed and quantified. This entails collecting the set of external and internal sensory input, whether obtained autonomously or through external input, and coercing it into a canonical form that is consistent with the set of previously stored cases.

Figure 12.7: CBR applies prior experience to the current situation.

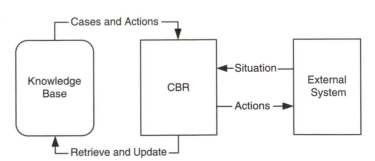

2. The current case is compared against the set of stored cases to identify previous experience that is similar to the current state.

3. The actions taken for one or more of the most similar cases are retrieved and applied to the current situation. If there are no matching cases, then a set of actions must be adapted from a similar case or new actions must be proposed [9].

4. The case and the actions taken are updated or stored. If the case performed is new, then it must be classified and inserted into the library of cases with the appropriate semantic relationships to the other cases.

Implementations of CBR systems may vary as widely as the systems to which they are applied. The underlying representation may be frame-based, organized into ontologies, or based on symbolic knowledge representations. Regardless of the underlying representation, the essential reasoning mechanism is to perform a pattern-matching process that codifies the current state of the system, both internally and externally, and compares that against stored representations of previously encountered or codified states entered as a baseline set of knowledge.

This last point raises a key aspect of CBR. The accuracy and value of the reasoning process increases proportionately to the breadth and depth of the cases that are available. In other words, to be effective, a case-based system must have a significant set of prior experiences from which it can search and identify similar situations that it can apply to the current state.

The second key aspect of an effective CBR system is its ability to adapt the actions associated with a prior case that is close to but not a clear match for the current situation. The ability to identify specific action steps requiring modification, to change those actions identified in a meaningful way, and to perform the actions is a central CBR capability.

Depending on the underlying representation of the cases and the number of cases stored in the knowledge base, the memory requirements for a comprehensive case-based system can be fairly significant. Another performance consideration involves the computational requirements to (1) perform the pattern matching required to identify potential cases that are similar to the current state, and (2) perform the modifications necessary to the stored set of actions, when there is not a highly similar match, in order to apply them to the current situation.

CBR would typically not be a candidate for low-end, resource-limited systems due to its memory and processing requirements. It is more suitable for larger systems with adequate computing resources, such as fixed systems in vehicles, larger aircraft, and ships.

12.2.5 Rule-Based Systems

Rule-based (production) systems have a long history [10] and have been applied to a variety of applications. A rule-based system has a knowledge base represented as a collection of "rules" that are typically expressed as "if-then" clauses. The set of rules forms the knowledge base that is applied to the current set of facts. Rule-based systems provide a method for representing inferential knowledge by using a simple if-then form, which is relatively easy to state and understand. The rule paradigm is naturally understood by humans. The basic architecture of a production rule system is shown in Figure 12.8.

As seen in Figure 12.8, the current state of the system is represented as a set of facts or assertions in working memory. The corpus of knowledge is stored within a set of production rules that form the knowledge base. The inference engine performs a pattern match of the antecedents or conditional portion of the production rule against the set of assertions in the working memory. When there is a match, the rule is tagged for possible execution. Because more than one rule may match the current state in working memory, some mechanism is usually provided for resolving the conflict and deciding which rule should be executed. The selected rule is then executed, or "fired," resulting in some action being performed or one or more facts being asserted (added to) or retracted (removed from) the working memory. Any external data that is represented in working memory is updated and the cycle continues.

Figure 12.8: Basic architecture of a rule-based system. The production rule matches the current state in working memory to one or more rules in the knowledge base.

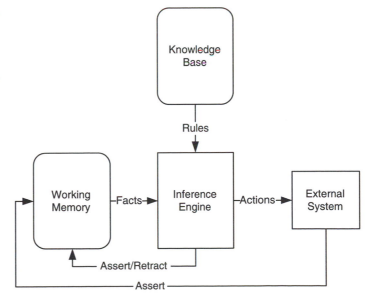

Figure 12.9 illustrates a simple power conservation rule that states that if the radio has a waveform instantiated and the waveform is not actively transmitting, then the system may conserve power by reducing the power level on the power amplifier. The syntax shown is a common form for production systems. The antecedents are expressed as a set of tuples in parentheses. The implied relationship between the tuples is usually AND. So, for the example rule in Figure 12.9, there are two antecedent expressions, and both must be true for the rule to fire.

Figure 12.9: An example of rule-based reasoning.

```
If (?waveform is-instantiated)
   (not (?waveform is transmitting))
=>
   (set PA-power conserve)
```

The use of the question mark (?) in front of waveform denotes that the item is a variable. So, any two-value tuple in the working memory for which the second value equals "is-instantiated" is matched to the antecedent expression with the first value of the tuple being bound to the ?waveform variable. Then, when the second antecedent expression is evaluated, the ?waveform variable is set to the value bound in the first expression and the working memory is searched for the pattern.

One of the shortcomings of the rule-based paradigm, however, is that it typically has no means to introspect the knowledge within the system. In other words, the system's knowledge provides guidance regarding the domain of the system and responds based on the set of knowledge and the current state of the environment. However, the rule engine cannot scan through the rules to adjust them, add rules, or delete rules. To accomplish these activities, an additional set of reasoning must be implemented. Thus the learning algorithm would be applied to monitor which rules were applied in a particular situation, assess the success of the decision process based on the actual outcome versus the predicted outcome, and then have at its disposal access to the rules in a form that the learning algorithm can understand and process.

12.2.6 Temporal Knowledge

Temporal reasoning provides the ability for a system to reason about its operational characteristics within the context of time. More than simply a discrete set of points, temporal reasoning that captures relationships between temporal intervals provides a basis for qualitative temporal reasoning that does not require assignment of specific time values. The notion of temporal interval logic proposed by

Allen [11] comprises a set of 13 relationships that capture the relationship between any two temporal intervals. For example, interval A may overlap interval B, implying that A starts before B and A ends sometime between the start and end of B. A number of researchers have developed additional aspects of temporal interval logic, including Kovarik and Gonzalez [12], resulting in a rich variety of temporal representation and reasoning approaches, as described by Allen [13].

Dynamic spectrum utilization can be mapped into a temporal representation paradigm by mapping the times associated with observed frequency usage. Once a map of spectrum usage over time has been built, intervals of time can be identified when a particular portion of the spectrum is underutilized. The temporal "holes" can then be used to more intelligently allocate spectrum in a collaborative fashion across a number of devices. This concept is illustrated in Figure 12.10.

Temporal reasoning also comes into play when considering aspects such as co-site interference and concurrent requests for physical and processing resources.

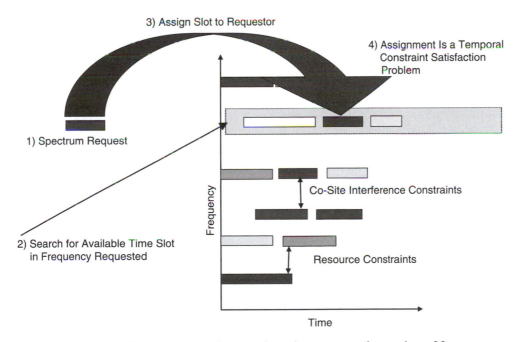

Figure 12.10: Dynamic spectrum assignment based on temporal mapping of frequency usage.

12.2.7 Knowledge Representation Summary

Section 12.2 has thus far presented a brief overview of representation and reasoning paradigms. Each of the different implementation approaches has benefits and limitations. Although there are multiple reasoning paradigms from which to choose, several key tenets can be asserted regarding the use of a reasoning system within a cognitive radio system.

First, even though any one of the approaches described, as well as those not covered in this section, will provide a measurable degree of intelligent behavior within a cognitive radio, no single representation and reasoning system is capable of embodying all of the types of knowledge and reasoning expressiveness required for the range of operational situations that a radio system must be capable of handling.

Second, although each reasoning mechanism provides a degree of adaptability to operational situations not explicitly defined, when any of the mechanisms encounters a totally new situation or scenario, the ability of the system to reason and decide on an appropriate behavior or response drops off sharply. This brittleness at the edge conditions leads to the pattern of oscillation between stable and unstable states or termination of operation because the system has encountered a situation for which it does not have any precoded knowledge.

Finally, the performance of each of the reasoning systems discussed is highly dependent on the breadth and depth of the knowledge codified within the system. The more knowledge that is encoded within a system, the better it performs. This axiom has its consequences, of course, in that the cost in terms of memory and computational resources required to support a specific reasoning system may be significantly more than is available on some resource-limited systems.

The next section discusses different learning mechanisms. Each of the methods can typically be applied to different representation and reasoning implementations, as discussed in this section.

12.3 Machine Learning

Research in machine learning has grown dramatically in the past decade, with a significant amount of progress along several research paths. Multiple strategies exist for implementing machine learning, each with some degree of benefits as well as drawbacks both in general and in the context of a cognitive radio system. This section addresses common learning approaches and provides some analysis of their applicability to cognitive radio systems.

One of the important aspects of the learning mechanism is whether the learning performed is supervised or unsupervised. In a supervised learning system, as the name implies, learning is performed through a set of predetermined conditions, and the system is trained to select the right choice or outcome. In effect, the learning mechanism learns to associate a particular set of input stimuli with a learned response or output. In the context of a cognitive radio system, either technique may be applied as the radio's method of learning. Supervised learning may take the form of a dialog between the operator and the radio, in which the radio may develop some new assertion or operational behavior model based on the learning mechanism within the radio and then ask for confirmation from the operator that its conclusion is correct.

Conversely, the radio system may extend its knowledge through the learning algorithm and simply add the new knowledge to its existing base of knowledge assertions and behaviors. In operational radio systems, there will likely be some level of protection or guard to limit the degree of behavioral modifications that a cognitive radio may incorporate without external verification and validation.

The potential problem with autonomous or unsupervised learning is that the learning process may result in many wrong choices before a feasible decision path is developed. This can be computationally intensive and time consuming. Learning algorithms, such as reinforcement and temporal difference, which are discussed in Sections 12.3.5 and 12.3.6, respectively, are applicable to this type of learning.

The following sections provide an overview of several types of learning mechanisms. As previously noted, the focus is on those learning mechanisms that are applicable to cognitive radio systems. Consequently, not all learning methods are covered within this chapter.

12.3.1 Memorization

One of the most basic learning mechanisms is memorization, or rote learning. This approach captures a sequence of steps or a response to a specific set of conditions and then, when the same task or set of conditions is again encountered, the memorized responses are applied. This can be an effective method of learning if the range of situations encountered by the radio system is limited and well defined. Incomplete and partial data and situations, however, are not amenable to rote learning.

Rote learning forms a set of steps that represent the observed or described sequence of actions for a given set of conditions. Applying the sequence learned requires the system to take a set of measurements, sensor data, or descriptive input data provided by an external source and then use the set of information as an associative key to look up the information within an internal database of patterns. When the matching set of descriptive data is found, then the set of associated actions are activated and applied to the current state.

An example of memorization would be the selection of a particular waveform or specific adjustments to the operating parameters of a waveform in response to measured interference. Although the memorization would allow for multiple response sequences, each associated with a specific interface value, each sequence of actions would be explicitly tied to the measured values. Memorization does not allow for generalization of responses based on similar responses to different measured values. This capability could be thought of as a computational implementation of conditioned response, as pioneered by Ivan Pavlov.[1] When the radio senses some specific combination of stimuli, it performs some specific conditioned response based on the association of the response to the stimuli over time. The problem is, of course, that if the set of stimuli that may be sensed by the radio are insufficient to delineate two distinct conditions that may exist given the same set of stimuli, the radio will execute a learned behavior that may not be correct for the actual condition.

Due to the explicit and inflexible nature of memorization learning, it is best suited for relatively fixed sequences of actions in which there is not a great deal of variation in the set of choices. Examples would be the ordered set of steps associated with radio start-up and initialization.

12.3.2 Classifiers

Classifiers refer to algorithms that analyze two or more units of data or knowledge and identify similarities or patterns in the structure and content of the data. Classifiers are applied to areas such as data mining and, in the case of declarative knowledge structures, provide a learning mechanism by extending an ontology with new concepts based on the similarity of the new concepts to those already within the system.

Similarities can be discovered between two distinct pieces of declarative knowledge by a number of methods. A classifier may compare the set of properties associated with each of the concepts and, based on the similarities of the two concepts, item B may be deemed to be *subtype* of item A. This would result in the insertion of item B into the ontology as a subtype or subclass of item A.

The process of quantitatively assessing the similarity between two concepts can be attributed to Tversky in his "Features of Similarity" paper [15]. A computational

---

[1]Pavlov's research on conditional reflexes greatly influenced not only science, but also popular culture. The phrase "Pavlov's dog" is often used to describe someone who merely reacts to a situation rather than using critical thinking [14].

implementation of a common method for assessing the similarity between two objects is to take the set of properties associated with each of the objects and create an *n*-dimensional space, with each dimension representing a property, and then assign a number between 0 and 1 representing the value of each property for each object.

Once this structure has been built for each object, the center-of-gravity point can be calculated as the sum of the distances between each of the property values in the *n*-dimensional space. Then, the relative similarity between the two concepts is the distance (i.e., difference in magnitude) between the two center-of-gravity points. The shorter the distance, the more similar the concepts.

In addition to classifiers being applied to declarative knowledge structures in an ontology, they can also be applied to identifying similarities and patterns in declarative structures representing temporal relationships. Learning as applied to temporal knowledge enables the radio system to enhance or extend its repertoire of known patterns of behavior and then apply those patterns to new situations. For example, the system may observe a particular pattern of interference on a regular set of frequencies at a particular time of day. Or, a system may observe that a particular pattern of activities and events precedes the initiation of a communications activity that is of interest from a signal analysis perspective. Repeated observation of the temporal patterns reinforces the validity of the set of observed temporal events. The stored temporal pattern can then be applied by matching observed events against the collection of known temporal patterns to predict a specific activity and initiate an appropriate action or countermeasure based on the temporal prediction.

12.3.3 Bayesian Logic

Statistical learning methods employ some method of probability of a given outcome for a given set of input stimuli. The system matches a set of active input stimuli to one or more sets of statistical functions having the same input parameters, and then applies the function to the input values, thus generating an expected outcome, course of action, or classification assignment. The probabilities applied to the calculation affect the result.

Learning is accomplished through the system observing the *actual* outcome as opposed to the *expected* outcome and adjusting the weights accordingly. Here the range of statistical functions that can be applied is significantly large.

One of the fundamental behaviors of a reasoning system is the tendency to alter the prediction or expectation of the outcome of an event based on prior history. This concept is the fundamental idea of Bayesian logic. The probability behind Bayes' theorem is shown in Eq. (12.1):

$$P(\phi|x) = \frac{P(x|\phi)P(\phi)}{P(x|\phi)P(\phi) + P(x|{\sim}\phi)P({\sim}\phi)} \tag{12.1}$$

The event or occurrence of interest is represented as ϕ, and x is the observed phenomenon or evidence of the occurrence. $P(\phi|x)$ is the probability that the event ϕ has occurred given the observation x or, in other words, a measure of the reliability of the assertion that ϕ has actually occurred given the observation x. The probability of the observation for a specified set of the occurrence of interest is $P(x|\phi)$, and $P(\phi)$ is the overall probability of the occurrence within a sample population. Conversely, $P(x|{\sim}\phi)$ is the probability that the observation will manifest itself in a sample population that does not contain the event or occurrence of interest, and $P({\sim}\phi)$ represents the percentage of the set that does not contain the event or occurrence of interest. An example in the context of a cognitive radio illustrates how Bayes' theorem might be applied as a learning mechanism.

Suppose the problem domain is predicting whether there will be interference, ϕ, encountered for a given geographical region. Further, assume that the radio has one or more sensors that can detect the interference. When the interference is detected, a memory location is set to a specified value, flagging the detection that provides the evidence, x, of the interference phenomenon.

As a starting point, assign a value, $P(\phi)$, representing the probability that there actually is interference in a given region. The probability that the interference is not present would thus be $P({\sim}\phi)$. However, because detection equipment is not perfectly accurate all the time, there is a finite probability that it will detect interference when it is present, $P(x|\phi)$, and also a finite probability that it will detect interference when it is not present, $P(x|{\sim}\phi)$, which is also called a false positive.

Assume that the detection equipment is calibrated in the lab and is shown to be accurate 80 percent of the time. So, for every 100 times that interference is present, 80 will be detected and 20 will not. Assume also, however, that for every 100 readings when there is no interference, the detector registers interference (false positive) 10 percent of the time. Now assume a total population of 10,000 samples of the environment and that the probability of interference is 30 percent. So, of the 10,000 samples, 3000 are likely to be samples containing interference and 7000 are not.

Inserting these values into Eq. (12.1) yields Eq. (12.2):

$$P(\phi/x) = \frac{0.8 \times 0.3}{0.8 \times 0.3 + 0.1 \times 0.7}$$

$$= 0.774 \tag{12.2}$$

Therefore, based on an observed interference phenomenon, it can be asserted with roughly a 77 percent degree of confidence that the interference is actually present.

Bayesian logic can be adapted as part of the learning mechanism through adjustment of the weights based on observed phenomenon. The limiting factor in general use is the required verification of the prediction in order to adjust the probabilities. Thus, the learning mechanism must have either external input or some internal method of verification of the predicted value in order to adjust the weighted values. Nonetheless, Bayesian logic can provide valuable reasoning support within a cognitive radio system by tempering the predicted outcome based on prior experience.

12.3.4 Decision Trees

A decision tree, as the name implies, is a directed graph consisting of a hierarchical set of nodes connected via arcs, where each node represents a choice or decision, and the arcs leading from that node to the next decision node represent the set of possible choices for a given node. A decision tree can be visualized as a sequence of choices in which the path taken through the choices from the starting point to an endpoint is governed by the choices made at the starting node and each successive node. The root node provides the starting point and the leaf nodes contain the decisions or actions to be taken.

Decision trees may be simplistic in nature, representing a fixed procedural process, such as problem diagnosis, or may be more complex, applying Bayesian reasoning to the decision process governing the transition between the nodes of the graph.

This type of reasoning approach is referred to as a solution space search problem. That is, the system generates a set of possible alternative actions for a given state. Each of the generated alternatives is then explored, alternative actions are generated for each of those states, and so on. As each alternative is generated, its weight is calculated, subsequent alternatives are generated, and a total weight or probability is calculated. Thus, for any given node in a decision tree, the total weight for the path leading to that point is maintained.

The calculation of the weights can be dynamically adjusted based on an observed state compared to an anticipated state. Thus, the system can learn to respond differently to different scenarios and states. The learning aspect can be extended further by enabling the system to generate wholly new actions for a given node. This enables the radio system to not only learn new behavioral responses based on performance observations, but also to extend the range of choices.

12.3.5 Reinforcement-Based Learning

Reinforcement-based learning provides a method for assessing the success of a particular action, where success is defined as the actual outcome of the event being the same as (or near to) the anticipated or desired outcome. Based on the degree of success, a reward weight is assigned to the action, thereby reinforcing the selection of that particular action if the system finds itself in the same state at some future time.

There have been a number of advances in reinforcement learning [16]. Reinforcement learning is represented by a series of states as a directed graph. As the system moves through the sequence of states by applying some policy for selecting a path or alternative, the degree of success attained in achieving the system's goals, both short term and long term, are measured. Based on the degree of success, the policy is reinforced.

At each point in time, t, the system is in some state, s. The system has a set of actions, A, from which it selects a specific action, a, where $a \in A$ to perform in the given situation with the objective of transitioning to some goal state, s', at the completion of the action. In order to select an action, the system has a set of policies or rules for deciding which action to apply. The selection of a specific action, a, given a state, s, by a policy, π, can be represented as $a = \pi(s)$.

Based on the degree of success, the system assigns a reward to the action performed, which, in effect, is a reinforcement of the policy selection rules. The reward function, $R(s,a,s')$, provides immediate feedback to the selection process. So, as the system selects actions that achieve the goal state, the policy rules that contributed to that selection process are reinforced and the probability of selecting the same action again, if the same set of conditions is present, is increased.

Each correct selection of an action provides an immediate reward. Over the course of several actions, a cumulative reward value is calculated. This provides the mechanism for strengthening the selection of a sequence of actions. In essence, it is reinforcing the ordered set of actions for the attainment of a final objective or goal. Therefore, two methods influence the selection of an action.

One is the selection of the current action based on the reinforcement of that action successfully achieving the desired objectives for the current situation, and the other is the influence of the successful attainment of the end objective providing reinforcement of the action as a part of the ordered set of actions.

This attainment of a long-term goal is typically represented within the reinforcement paradigm as a cumulative value that is the sum of each of the immediate reward values. Each of the immediate reward values that are part of the sum is discounted by applying a factor, γ^t, to the reward value. The cumulative reward is then expressed as shown in Eq. (12.3).

$$\text{CumulativeReward} = \sum_{t=0}^{\infty} \gamma^t R(s, a, s') \qquad (12.3)$$

where $0 \leq \gamma^t < 1$.

The effect of the discount factor is to balance the reinforcement value between realization of a short-term goal (i.e., whether the selected action yielded the expected state, given the current state) versus the long-term goal (i.e., whether the selected action contributed positively to the attainment of the long-term goal).

Thus, for any given state and action pair, there is a computed value function for the policy based on the current state, denoted as $V^{\pi}(s)$. Given that there may be multiple states, s', that may be the result of performing the selected action, there is a probability assignment that is made for each of the possible states. The probability of ending in state s', given a current state of s and that action $\pi(s)$ is taken, is represented as $P(s'|s, \pi'(s))$. This probability is then combined with the sum of the short-term reward associated with the current action, $R(s, \pi(s), s')$, and the cumulative discounted reward, $\gamma V^{\pi}(s')$. This computed value for each possible path is summed together, as shown in Eq. (12.4):

$$V^{\pi}(s) = \sum_{s'} P(s'|s, \pi(s)) \cdot [R(s, \pi(s), s') + \gamma V^{\pi}(s')] \qquad (12.4)$$

The overall objective of reinforcement learning is to compute the optimal policy, π^*, for each iteration of the process that maximizes the total reward shown in Eq. (12.4).

Note that the term policy rule can be realized within the reinforcement-learning system in a wide variety of ways. The policy may start out as nothing more than a random selection process which, over time, evolves to select the appropriate action based solely on matching the pattern of current state variables to those patterns that were previously encountered, stored, and reinforced because the system chose the right action.

Figure 12.11 illustrates the application of this approach. The radio system is in some current state, s, and a policy, a, has been selected. There are four possible states that may be entered upon the completion of the action. For each of these possible states, a probability, $P(s'|s, \pi(s))$, has been assigned. The immediate reward, $R(s|s, \pi(s), s')$, for each of the subsequent states is shown, along with the estimated value of each of the states, $V^{\pi}(s')$.

Figure 12.11:
Reinforcement learning
applied to policy
selection.

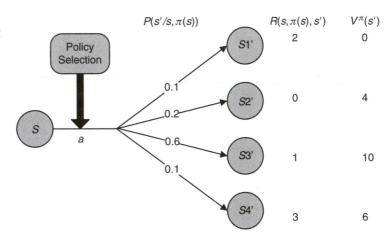

Assume that the discount factor, γ, is equal to 0.6. This would yield a new estimated value of $V^{\pi}(s)$ as:

$$V^{\pi}(s) = 0.1(2 + 0.6(0)) + 0.2(0 + 0.6(4)) +$$
$$0.6(1 + 0.6(10)) + 0.1(3 + 0.6(6)) \quad \text{(12.5)}$$
$$V^{\pi}(s) = 5.5$$

This value becomes the new value of the selected policy for the current node $V^{\pi}(s)$. This process of modifying the policy value of the current node based on the probable path and subsequent nodes is referred to as backing up the policy value.

Problems arise with reinforcement learning because, in order to function, a complete model of the system is required that provides the transition probability between states and the reward function. Defining the model to the level of detail required may not be feasible for a cognitive radio system.

Finally, in addition to the infeasibility of specifying the complete model description, there is the issue of the algorithm performance. Performing the policy iteration calculations requires $O(n^3)$ time where n is the number of states in the system.

12.3.6 Temporal Difference

Temporal difference also applies a reinforcement algorithm to the policy selected based on the degree of success. However, the temporal difference algorithm does not require a priori model of the sequence of possible states, as the standard reinforcement algorithm does. Thus the temporal difference algorithm builds the state representation "on the fly" and does not require the back-propagation used in the reinforcement algorithm to update policy weights.

Introduced by Sutton [17], the general temporal difference algorithm is referred to as TD(λ). There is a variant of temporal difference learning, TD(0), that performs the computation of simple backups as the system transitions through the states. As with reinforcement learning, as described in Section 12.3.5, the system is in state s and, based on the selection of an action based on a policy associated for state s, $a = \pi(s)$, performs a probabilistic transition to a new state, s'. The transition produces the immediate reward $R(s, a, s')$. The TD(0) algorithm updates the value function by using Eq. (12.6):

$$V^{\pi}(s) := (1 - \alpha)V^{\pi}(s) + \alpha[R(s, a, s') + \gamma V^{\pi}(s')] \qquad \text{(12.6)}$$

In Eq. (12.6), α is a *learning rate parameter*. The initial value of this parameter is usually between 0.01 and 0.5 and it must, over time, reduce down to zero to allow the TD(0) algorithm to converge. As can be seen in the equation, as the learning parameter reaches zero, the latter portion of the function that incorporates the immediate reward and the discounted value of the next state is algebraically eliminated, and the system converges to the state value $V^{\pi}(s)$.

Thus, as a node is visited multiple times, the effect is the same as performing a simple backup, as with the reinforcement-learning algorithm in Section 12.3.5. The benefit of the temporal difference algorithm is that the policy value can be calculated without an explicit model. The set of states traversed as a result of the actions and the environment become the model. This can be accomplished because the empirical data gathered as the system traverses the set of states provides the basic data required to compute a probability function for each node traversed.

A second advancement of the temporal difference algorithm introduced by Sutton [17] over the reinforcement algorithm is that, rather than maintaining a separate value, $V^\pi(s)$, for each state s, the value function can be stored as a neural network or some other method for providing a differential approximation. Thus, the value function can be represented as $V(s, W)$, where W is a vector of weights that are modified based on observed state versus expected state. This prevents the ability to directly assign a value to a state. However, the weights specified in W may be adjusted such that $V(s, W)$ fits closely to the desired value through the use of an error function. The temporal difference error function is shown in Eq. (12.7):

$$J(W) = \tfrac{1}{2}(V^\pi(s, W) - [R(s, a, s') + \gamma V^\pi(s', W)])^2 \qquad \text{(12.7)}$$

This equation captures the temporal difference error as one-half the squared difference between the current estimated value of state s and the backed-up value. Thus, the objective is to modify the set of weights, W, such that the temporal difference error, $J(W)$, is reduced. By differentiating the equation with respect to W and limiting only the first occurrence as being adjustable, the general weight assignment function is obtained. This is shown in Eq. (12.8):

$$W = W - \alpha \nabla_w V(s, W)(V^\pi(s, W) - [R(s, a, s') + \gamma V^\pi(s', W)]) \qquad \text{(12.8)}$$

The term ∇ represents the gradient of V with respect to the weights W. The term α represents the step size to be taken to reduce the error term calculated as the temporal difference error.

12.3.7 Neural Networks

Neural networks and GAs also rely on the reinforcement of a decision or selection based on the actual result or outcome of a decision. (See Section 12.3.8 and Chapter 7 for a discussion on GAs.) The neural net typically has a vector of input values and a vector of output values. An intermediate layer couples the input and output values together, propagating the input values along a set of connections from the input to the output.

Essentially, each node in the neural network takes an input vector, A, and applies a weight vector, W, to perform the propagation of the input value along the link to the next layer in the network. This propagation is typically tempered by a constant or bias value, b. Thus, for an individual node, the propagation output, A, of a given node, j, can be expressed as shown in Eq. (12.9):

$$A_j = \left[\sum_{i=0}^{n} a_i w_i + b_j \right]$$

(12.9)

When the output value, A_j, is greater than some threshold value, T_j, the value is propagated along the output, as illustrated in Figure 12.12. Learning, within the context of a cognitive radio, involves the adjustment of the threshold value, the bias value, or the weight associated with a node.

Figure 12.12: Neural network node illustration. Shown are the input vector A (with elements $a_0 \cdots a_n$); weight matrix W (with elements $w_{0j} \cdots w_{nj}$); bias value, b; and output.

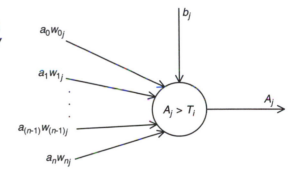

A neural network is formed when a collection of individual nodes is organized together in a multilayer fashion, as illustrated in Figure 12.13. The neural network has a set of input points. The values sensed or input through these points is propagated forward, through the middle layer (also referred to as the hidden layer), to a set of output points. The output points activated by the forward-propagation are then compared with the actual value (i.e., anticipated versus actual). If there is a match, the path followed to arrive at the output point is followed through back-propagation, and the intervening nodes and paths are reinforced for that particular output point.

Learning within a neural network requires feedback that allows the network to compare the expected output value associated with a set of input data against the conclusion reached by the neural network. This forms the back-propagation illustrated in Figure 12.13. As the vectors of input values are applied to the system, the data propagates through the intermediate layers to the output. The expected output is mapped onto the output vectors. Those elements in an output

Figure 12.13: Neural network. Shown are forward-propagation of input and back-propagation learning.

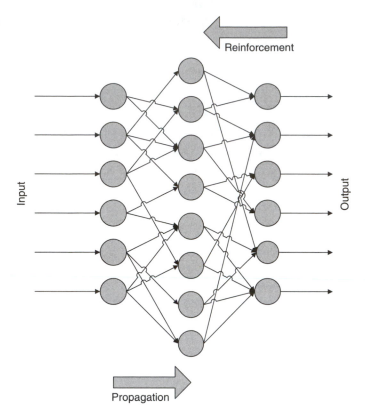

vector whose value matches the applied expected value are reinforced by back-propagation. Thus, the intermediate layers that contributed to the propagation of data resulting in the correct (i.e., expected) output values are reinforced. Reinforcement may be performed through increasing weights, of the nodes involved in the propagation from the input data to the correct output or conclusion.

This reinforcement of neural links increases the probability that the same input values will result in the same propagation to the correct output values. Those output values that did not match the expected output value are weakened, thereby decreasing the probability that they would be applied again given the same set of conditions.

12.3.8 Genetic Algorithms

GAs utilize a vector that represents a given condition or input stimuli and action. Based on the success of the action taken, the input vector is modified. In the case of GAs, however, the vector may undergo alteration, or genetic mutation, in a random fashion. Even though this may introduce potentially large vector sets, GAs may introduce new solutions through the random mutation process. This enables a system to generate new knowledge and evaluate its effectiveness by using empirical data collected through the operational environment.

The initial implementations of GAs utilized a bit sequence as the representation of the *gene*. The bit pattern essentially consists of two parts. One part forms a pattern that is used to match a given set of conditions. The second part forms a bit pattern representing the action or actions to be taken as a result of the pattern match.

As the process of matching the input part and performing the corresponding actions is performed, two modifications of the "gene" occur. The first is a strengthening of those genes that contributed to a successful set of actions (and a weakening of those genes that did not). This serves to enhance the survival rate of those genes that contributed positively, and effectively causes those genes that did not contribute positively to die off (i.e., they are removed from the set of bit sequences).

The second action is a random mutation of the bit sequence. This provides a mechanism for generating new alternatives within the set of bit sequences. Thus, not only will genes live or die off through a form of natural selection, but wholly new strains can be introduced that may result in more efficient reasoning.

Implementation of GAs has continued to evolve to include the use of a vector of multivalued data items. In effect, each item in the sequence is represented as a discrete variable that can be modified as part of the process. See Chapter 7 for a detailed discussion of GAs.

12.3.9 Simulation and Gaming

The application of simulation and game theory to machine learning provides a method for developing and evaluating alternative actions in a given scenario, thereby proposing new paths or actions. At a fundamental level, game-theoretic learning utilizes a reinforcement mechanism similar to those previously discussed. Unlike reinforcement and temporal learning, which require a comprehensive model of states, however, game theory—due to its incorporation of strategy and probability—supports learning in systems that are not well defined, or are changing. Game-theoretic learning utilizes game theory as a means for proposing actions to be taken, given a particular situation.

Game-theoretic learning introduces some features that are unique to this approach. The player chooses from a set of actions for any given state. The action may be selected based on a numerical probability, random generation, or any other selection method. As the player applies the selected strategy, the success of the strategy is recorded for future reference. Thus, as in the reinforcement and temporal difference algorithms, there is a bias value applied to a set of actions for a given situation. Chapter 15 provides a comprehensive discussion of the application of game theory to cognitive radio systems.

12.4 Implementation Considerations

The previous sections discussed different approaches to knowledge representation as well as reasoning and learning approaches. This section discusses implementation and deployment considerations related to the use of this technology within a radio system. Several areas are related to the technological requirements necessary to realize a cognitive radio that learns, and some of the key issues that need to be addressed are related to operational and sociological issues that are raised when a learning system is deployed.

12.4.1 Computational Requirements

As with any type of work, some effort must be expended in order to produce the desired artifact or product. Cognitive radio systems must expend effort in the form of computational resources and the power to drive them in order to reach conclusions or decisions concerning the operation of the system. Similarly, computational effort must be expended by a learning system in order to monitor and analyze the results of the actions performed against the expected outcome, perform analysis on the differences between the two, and generate new or changed knowledge to be applied to the next iteration of the decision process.

It can be argued that whereas the cognitive decision process is an integral part of the operational behavior of the radio system, the learning process is not. Consequently, relegating the learning process to non-critical performance times would be a reasonable operational assumption. However, the primary driver is the availability of computational resources that can be applied to the learning process without affecting the mission-critical performance of the radio system.

12.4.2 Brittleness and Edge Conditions

Learning has the potential to overcome the brittleness of cognitive implementations when faced with a new situation or scenario. However, the resources and time required to generate additional or new knowledge that can be applied to the situation may not be sufficient to result in a feasible alternative in a time frame that can be applied to the situation. Nonetheless, the ability of the system to extend its behavior provides unique opportunities for radio systems. Systems that are to be deployed for long-duration missions in locations that cannot be easily serviced would benefit from the capabilities afforded by learning algorithms.

12.4.3 Predictable Behavior

One of the critical aspects of any automated system is the tacit assumption that the system is deterministic in its behavior. That is, given a set of inputs, the output, or action taken by the system, can be determined. The problem that arises in any complex system is that each step taken to reach a decision may be individually predictable, but the end decision or action reached may not be what was expected on the part of the human operating or interacting with the system. This trait of individually predictable steps leading to an unexpected or unforeseen outcome is one of the founding premises of chaos theory. The unpredictability of the endpoint action is further exacerbated by the introduction of learning to the system.

Enabling learning within a system is a powerful two-edged sword. On the one hand, it enables the radio system to adapt autonomously to new situations, change operational characteristics, make operational decisions based on newly learned knowledge, and provide more reliable and dependable service. On the other hand, as systems learn, there is no guarantee that the observed data used by the learning system to adapt or form new knowledge is accurate, nor is it assured that the methodology applied to form some new chunk of knowledge is without flaw.

Just as in the case of humans, the ability to extend the knowledge and behavior of the system is not assured to be predictable. Yet, it is predictability that underlies current methodologies and approaches to radio system certification. Regulatory bodies, such as the US Federal Communications Commission (FCC), are just beginning to consider the ramifications of a radio system that is largely or exclusively software based and may change its operational characteristics through changes in the software. Regulatory methods and approaches, as discussed in previous chapters, will be stretched, as cognitive radio technology becomes an integral part of radio systems, allowing them to autonomously change operational characteristics. Finally, as learning technologies are integrated within cognitive radio systems, the concept of radio system certification will again need to evolve as cognitive radio systems gain the ability not only to change operational behavior based on the operational environment but also to modify the method by which the decisions to change operational behavior is reached.

12.5 Summary

This chapter has reviewed several methods of knowledge representation and reasoning, approaches to computational learning, and possible applications of these methods within the domain of cognitive radio systems. As shown, there is no single approach to knowledge representation and reasoning or learning that will address all aspects of cognitive radio systems. The level of intelligence exhibited by a cognitive radio system will be dynamic and varied, depending on the implementation platform, the computational resources available, and the specific mission of the system.

Also, the type of knowledge representation and reasoning system employed as well as the type of knowledge may "prefer" a particular learning method. Radio systems may therefore employ a particular representation mechanism, and hence a specific learning implementation based on the operational constraints and objectives imposed. Some systems may host two or more methods, or may interact with other radios in a networked fashion to implement a larger intelligent entity.

Learning within the context of a cognitive radio must encompass multiple representation paradigms, integrate multiple reasoning mechanisms, provide hybrid-learning mechanisms, and span multiple functional layers within the cognitive radio. This concept is illustrated in Figure 12.14.

The cognitive radio can be viewed through three essential aspects or perspectives:

- At the lowest layer is the physical set of hardware. This provides the physical infrastructure for the radio system.

- The infrastructure aspect provides an abstraction layer that implements a logical set of interfaces to manage, control, configure, and operate the physical components and provide software views of these resources to the waveform implementation. This layer also manages the suite of components that implement a waveform.

- At the application and services layer, waveforms and system services are viewed and manipulated as a logical entity.

A fourth perspective is the user's aspect. From that viewpoint, the user is either interacting with the physical radio (e.g., power up, diagnostic checks, etc.) or interacting with an instantiation of a waveform as a logical entity.

As the right-hand side of Figure 12.14 illustrates, knowledge representation, reasoning, and learning must be implemented within and across each of these

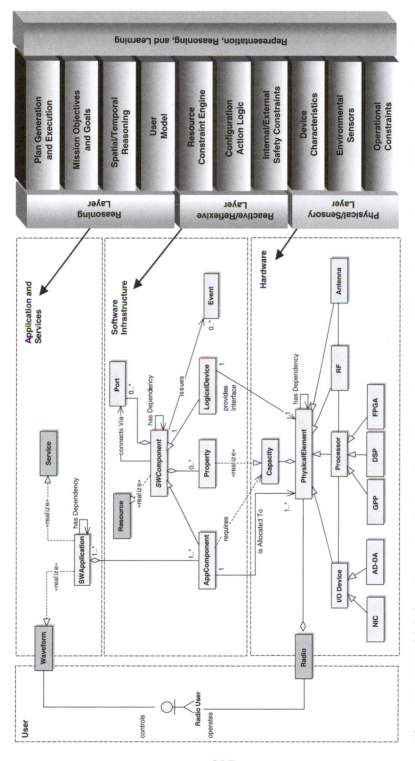

Figure 12.14: Multiple learning methods. Learning and reasoning span multiple architectural layers within the cognitive radio.

397

layers, yielding a hybrid system. Physical hardware will become smarter with low-level implementation of simple learning algorithms to adaptively modify the power usage or RF components. The software infrastructure will provide a basic reflexive behavioral and learning capability that provides, among other capabilities, safety limitations and operational constraints. At the upper levels of the cognitive radio, reasoning and learning become more abstract, addressing mission requirements and user needs.

Future research and development of computational learning applications for radio systems will need to address the hybrid nature of reasoning and learning within cognitive radio systems.

References

[1] J. Mitola III and G.Q. Maguire Jr., "Cognitive Radio: Making Software Radios More Personal," *IEEE Personal Communications*, Vol. 6, No. 4, August 1999, pp. 13–18.

[2] J. Wang, D. Brady, K. Baclawski and M. Kokar, "The Use of Ontologies for the Self-Awareness of the Communications Nodes," in *Proceedings of the Software Defined Radio Technical Conference—SDR03*, Orlando, FL, November 2003.

[3] J. Wang, M. Kokar, K. Baclawski and D. Brady, "Achieving Self-Awareness of SD Nodes Through Ontology-Based Reasoning and Reflection," in *Proceedings of the Software Defined Radio Technical Conference—SDR04*, Phoenix, AZ, November 2004.

[4] M. Kokar, D. Brady and K. Baclawski, "Roles of Ontologies in Cognitive Radios," in *Cognitive Radio Technology*, B. Fette (Ed.), Elsevier, 2006, pp. 401–434.

[5] M. Minsky, "A Framework for Representing Knowledge," Massachusetts Institute of Technology, Technical Report, UMI Order Number: AIM-306, 1974.

[6] S. Bechhofer, et al., "OWL Web Ontology Language Reference," http://www.w3.org/TR/owl-ref/, M. Dean, G. Schreiber (Eds.), February 2004.

[7] K. Baclawski, D. Brady and M. Kokar, "Achieving Dynamic Interoperability of Communication at the Data Link Layer Through Ontology Based Reasoning," in *Proceedings of the Software Defined Radio Technical Conference—SDR05*, Garden Grove, CA, November 2005.

[8] R. Schank, *Dynamic Memory: A Theory of Learning in Computers and People*, Cambridge University Press, Cambridge, England, 1982.

[9] J. Kolodner, "Improving Human Decision Making Through Case-Based Decision Aiding," *The AI Magazine*, Vol. 12, No. 2, 1991, pp. 52–68.

[10] C. Forgy and J. McDermott, "OPS, A Domain-Independent Production System Language," *International Joint Conference on Artificial Intelligence*, 1977, pp. 933–939.

[11] J.F. Allen, "Maintaining Knowledge about Temporal Intervals," *Communications of the ACM*, Vol. 26, No. 11, November 1983, pp. 832–843.

[12] V. Kovarik and A. Gonzalez, "An Interval-Based Temporal Algebra Based on Binary Encoding of Point Relations," *International Journal of Intelligent Systems*, Vol. 15, No. 6, June 2000, pp. 495–523.

[13] J.F. Allen, "Time and Time Again: The Many Ways to Represent time," *International Journal of Intelligent Systems*, Vol. 6, No. 4, July 1991, pp. 341–356.

[14] http://en.wikipedia.ord/wiki/Ivan_Pavlov

[15] A. Tversky, "Features of Similarity," *Psychological Review*, Vol. 84, No. 4, July 1977, pp. 327–352.

[16] R. Sutton and A.G. Barto, *Reinforcement Learning: An Introduction*, MIT Press, Cambridge, MA, 1998.

[17] R.S. Sutton, "Learning to Predict by the Methods of Temporal Differences," *Machine Learning*, Vol. 3, No. 1, 1988, pp. 9–44.

Roles of Ontologies in Cognitive Radios

Mieczyslaw M. Kokar, David Brady and Kenneth Baclawski
Northeastern University
Boston, MA, USA

13.1 Introduction to Ontology-Based Radio

This chapter discusses the role of knowledge representation (ontologies) in cognitive radio. The emphasis is on those capabilities of cognitive radio that are practical and amenable to ontological treatment. After a brief introduction to ontologies, this chapter develops an ontology for cognitive radio functionality that is organized by using the same layers as communication networks. Specific examples of the use of ontologies in cognitive radio are then developed in some detail. The examples show how ontologies can be the basis for achieving interoperability at the physical (PHY) and data link (DL) layers. This chapter then outlines some of the major research issues for ontology-based cognitive radio.

13.2 Knowledge-Intensive Characteristics of Cognitive Radio

The term *cognitive radio* is interpreted differently by different people. This chapter uses an interpretation that includes a *cognitive agent*, as well as standard radio functionality. Toward this aim we first discuss the features that a cognitive agent is expected to have, and then propose some features that a cognitive radio could have.

To put the discussion in context, a cognitive radio is viewed as being part of a larger functionality, that is, of an *intelligent agent* or *intelligent personal digital assistant* (PDA) that can support a mobile user. Such a PDA would not only have to advise the user, but it would also have to be connected essentially all the time. Although there are many definitions of a cognitive agent, all of them revolve around similar ideas. For instance, the Defense Advanced Research Projects Agency (DARPA) [1] provides the following definition of a cognitive system:

A cognitive system is one that:

- *can reason, using substantial amounts of appropriately represented knowledge*
- *can learn from its experience so that it performs better tomorrow than it did today*
- *can explain itself and be told what to do*
- *can be aware of its own capabilities and reflect on its own behavior*
- *can respond robustly to surprise.*

The functionalities of a cognitive agent are often viewed according to Boyd's so-called OODA (Observe, Orient, Decide, Act) loop [2]. This approach is especially popular in the information fusion community where the OODA loop is used to model human behavior, which then serves as a pattern to be followed by a fusion system. In the intelligent control community, this loop is presented in a somewhat simpler form, called the perception-reasoning-action triad [3], used to represent an intelligent controller.

Following roughly the same line of reasoning, this chapter identifies the following basic functionalities that a cognitive radio should include:

- information collection and fusion;
- self-awareness;
- awareness of constraints and requirements;
- query by user, self or other radio;
- command execution;
- dynamic interoperability at any stack layer;
- situation awareness and advise;
- negotiation for resources.

Each of these capabilities is discussed in this chapter from two points of view: (1) the user's point of view and (2) the radio's point of view. This reasoning is illustrated and explained in Figure 13.1.

As shown in Figure 13.1, multiple conversations are taking place at several layers. At a higher layer, the two speakers pose and respond to queries through their radios; simultaneously, at a lower layer the radios converse in a similar

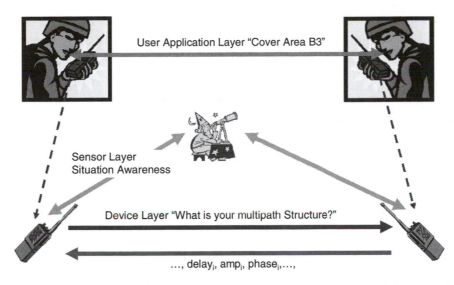

Figure 13.1: Three layers of cognitive radio conversation: user layer, sensor layer, and device layer. Cognitive radios permit conversations at all three layers. Conversation at user layer is a standard feature of all radios. Communication at the sensor layer is required for self-awareness, situation awareness, and information collection. Communication at the radio layer enables resource negotiation, query, and response.

fashion. In this case, one radio is trying to establish the multipath characteristics observed by the other with the intent of improving performance. In order to achieve this goal, the radios need to monitor the communication (indicated in the figure as dashed arrows between the radios and their users) as well as possibly some environmental characteristics, for instance, obtained from other sensors (indicated by the arrows from the radios to the "ancient astronomer" icon). The notion of simultaneous communications at different levels may be applied to each layer in the protocol stack, as discussed in Section 13.2.8. These conversations will be interrelated by the requirement that the "layer below" is supporting the "layer above".

13.2.1 Knowledge of Constraints and Requirements

At the base of any cognitive activity is knowledge. A cognitive agent must possess knowledge and must be able to make use of the knowledge for deriving its decisions and actions. Any software system contains some knowledge, but the real issue is how knowledge can be represented so it can be used in the most flexible way. This issue is related to the requirement that a cognitive system "respond robustly to surprise." When knowledge is encoded in a purely procedural way, the use of that knowledge must be prescribed (encoded) during system development. Thus, it is the responsibility of the system developer to not only encode all the necessary *procedures*, but also to encode the *invocations* of the procedures. In contrast to this approach, the *declarative* programming approach requires only that knowledge fragments be represented in the system's data structures (e.g., rules); the selection of the knowledge is left to the system. Such decisions are made by the system at run-time based upon *pattern matching*; that is, specific knowledge fragments are used when specific patterns are recognized by the system. This approach gives more flexibility to the use of stored knowledge and at least partially satisfies the requirement of the system responding to surprise.

13.2.2 Information Collection and Fusion

In addition to knowledge encoded in the cognitive agent, the agent must also have access to current information. Whereas knowledge is encoded by either the system designer or the user, the current information is collected from various sources at run-time. For instance, a PDA could have access to a global positioning system (GPS) receiver and thus obtain information of its current location. Additionally, a PDA could (potentially) have access to traffic reports, including the status of roads and bridges, as well as accident reports, weather conditions such as blizzards and black ice, and other pertinent details. Based on such information, the PDA could advise the user on efficient routing. Some of this information could be collected by subscribing to a service, whereas other information would require explicit requests. Because such advice would depend on various types of information from multiple sources, the information would have to be integrated and fused appropriately in order to be useful to the user.

13.2.3 Situation Awareness and Advice

Having knowledge and current information stored in the PDA's memory does not automatically guarantee that the PDA is aware of the current situation of the user (and, in effect, the PDA). According to Endsley and Garland [4], "*Situation Awareness* is the perception of the elements in the environment within a volume of time and space, the comprehension of their meaning, and the projection of their status in the near future." In other words, in order to understand a situation, the agent needs to know not only about all the objects of interest, but also about their relations to other objects and also possible future states of the objects and the relations. This, in turn, requires that knowledge of models of objects, including their dynamics, be known, so that future states could be predicted and rules for deter-mining the relationships could be derived. With respect to the PDA's task as a travel advisor, an example of a relevant relation might be "on-the-path-from-to," meaning that a specific geographic feature, such as a bridge or a river, is on the path that leads "from-to" specific destinations.

13.2.4 Self-awareness

To plan and schedule a task for execution, a PDA must know whether the task is within its own capabilities. It needs to understand what it does and does not know, as well as the limits of its capabilities. This is referred to as *self-awareness*. For instance, the radio should know its current performance, such as bit error rate (BER), signal-to-interference and noise ratio (SINR), multipath, and others. In a more advanced case, the agent might need to reflect on its previous actions and their results. For instance, for the radio to assess its travel speed a fortnight ago between locations A and B, it might be able to extract parameters from its log file and do the calculation. For the radio to decide whether it should search for the specific entries in the log and then perform appropriate calculations (or simply guess), it needs to know the effort required to perform such a task and the required accuracy of the estimate to its current task.

13.2.5 Query by User, Self, or Other Radio

Similar to humans, intelligent agents need to be able to answer queries from their users and other agents. Moreover, they should be able to query the user and other agents about relevant information whenever they cannot infer it from their own knowledge. For agents to formulate such queries, they must understand a common language, formally defined in syntax and semantics. Moreover, to formulate queries, agents must have a planning capability. Finally, they should be able to wait for an answer and incorporate the answer into their own knowledge struc-tures. For instance, the PDA might ask the user, "Where do you go next?" or ask another agent, "What is the temperature at your location?" Another radio could ask, "What is your multipath structure?" or "What is your BER?"

13.2.6 Query Responsiveness and Command Execution

If two humans agree on a plan to perform a specific task, they then (normally) execute the plan by performing the actions they agreed to. For instance, if two persons are driving around the neighborhood searching for a runaway dog, they communicate over cell phones and decide who will search which street and when. The implementation of such a feature in a software radio is not a simple matter. For instance, if two radios decide that one of them should adjust its communication protocol after exchanging information about the multipath structure, by selecting a different modulation scheme, the radio must be able to understand how to implement such a command. A simple way would seem to be to provide a number of procedures and invoke one of them using an if–then–else programming construct. However, this would limit the flexibility of reacting to different multipath structures to only the procedures encoded at the development time of the software radio. Moreover, the invocation of the procedures would have to be decided at development time. At the other extreme, one might envision an approach in which code is automatically generated at run-time and then executed. A middle ground solution would be to use the declarative programming approach in which knowledge fragments are developed and represented as rules at design time and then the run-time system selects, combines, and executes the rules as needed. This approach would provide more flexibility in terms of possible behaviors in response to various external conditions.

13.2.7 Negotiation for Resources

The human activity of planning involves, among others, negotiation for resources. For instance, if two people decide that one of them will drive to a grocery store to get drinks and the other will drive to the hardware store to buy a ladder, and if they have a sedan and a SUV at their disposal, it is likely that through a rational negotiation process they will use the SUV for the hardware store and the sedan for the grocery store. Similarly, resource negotiation is very useful in software radios. Negotiation of the most important resource, the spectrum, is being investigated in the DARPA NeXt Generation (XG) program [5]. According to the XG vision, software radios would be able to request unused spectrum that is allocated to another radio. This approach should lead to the achievement of a higher utilization of the spectrum resource.

13.2.8 Dynamic Interoperability at Any Stack Layer

The flexibility of ontology-based radios (OBRs) postulated in the above discussion can be extended to protocols at any layer of the protocol stack. This extension would enable intercommunication and negotiation among layers. Query and response among layers invites the possibility of cross-layer optimization of communication efficiency while maintaining the strict functional division between layers. For example, the medium access control (MAC) layer, which handles channel access and routing, might query the local PHY layer about the residual error in an equalizer, with the goal of preempting an outage. If the reported error is sufficiently large, the MAC might avoid an outage by seeking an alternative channel and route to the destination. As another example, MAC and transport layers at different nodes may also use query and response to reduce the frequency of broadcasting routing tables. Through the process of query and response, MAC layers could also provide a much more meaningful routing metric than hop count or aggregate delay because much more PHY layer information would be available. DL layers, which manage automatic repeat request (ARQ) protocols, may avoid some of the rigidness of fixed protocols such as selective repeat go-back-n or stop-and-wait. Through the use of reasoning, ARQs may be tailored on the fly for individual links as the nodes learn about the link performance.

This chapter focuses on ontological presentations for the PHY and DL layers because these layers are the ones that were implemented in hardware or firmware prior to the introduction of software-defined radio. Higher layers are more likely to be already implemented in software. Furthermore, the lower layers present challenges that are unique to radio communication.

13.3 Ontologies and Their Roles in Cognitive Radio

13.3.1 Introduction

An *ontology* is an explicit mechanism for capturing the basic terminology and knowledge (the concepts) of a domain of interest as well as the relationships among the concepts [6]. Ontologies are an increasingly important mechanism for the integration of disparate software systems. Indeed, a shared ontology is a fundamental prerequisite for meaningful communication between systems. The ontology can be hard-coded via shared data formats, database schemas, and procedures, but there are significant advantages for expressing the shared ontology by using a formal declarative ontology language that is either difficult to achieve or even impossible without it. The advantages include support for interoperability, flexible querying, run-time modifiability, validation against specifications, and consistency checking.

In the case of software-defined radio, ontologies offer the additional advantage of *self-awareness*: communication nodes can understand their own structure and can modify their functioning at run-time. Furthermore, nodes can query the capabilities and current state of other nodes, allowing them to modify the processing of packets during a communication session both at the source and the destination. The ontology specifies not only the structure of communication packets but also the processing of those packets according to the communication protocol. The use of ontologies adds flexibility, inferencing, and reasoning features that are not available with ad hoc data structures or database schemas [7, 8].

Systems that do not initially share an ontology might still be able to interoperate by mapping or merging ontologies in order to synthesize a shared ontology. It may be possible to construct systems that not only use ontologies but also modify them or even learn them dynamically.

Basics

An ontology specifies the concepts of a domain, attributes of the concepts, and relationships among the concepts. Each concept is expressed by using a *class*, which may be interpreted as a set of things. Anything that belongs to a class is called an *instance* of that class. Waveforms, packets, and symbol alphabets are all fundamental concepts in radio communication. In the ontology for software radio, these concepts are expressed as classes. For example, Waveform is a class whose instances are particular waveforms. Classes are organized into a *hierarchy* of classes by the *subclass* relationship. For example, binary phase shift keying (BPSK) and quadrature phase shift keying (QPSK) are both special cases of *M*-ary phase shift keying (MPSK). This is expressed by specifying that BPSK and QPSK are subclasses of MPSK.

An *attribute* is a property that something has, such as the number of symbols in an alphabet or the carrier frequency of a waveform. An attribute is a characteristic of a single entity, where that characteristic is a data value such as a number. A *relationship* is an association among various entities. For example, a waveform is used to represent a sequence of symbols from an alphabet. This is expressed by linking the waveform to the symbol sequence. An ontology will generally have many different kinds of attributes and relationships. The number of symbols in an alphabet might be called *numberOfSymbols*, and the relationship of a waveform with the sequence of symbols it represents might be called the *usedToRepresent* relationship. The term *property* is used for either an attribute or a relationship. As with classes, one kind of property may be regarded as a set of instances, called *facts*. For example, when a particular waveform *w* is being used to represent a particular sequence *S*, this fact is the triple (*w*, *usedToRepresent*, *S*). Properties can be organized in a hierarchy by the *subproperty* relationship. For example, the *usedToRepresent* property is a subproperty of the more generic *used* property.

Ontology Languages

The rapid expansion of the World Wide Web (WWW) has had a profound impact on communication. One consequence of this expansion is the emergence of the eXtensible Markup Language (XML) as the most commonly supported format for data interchange [9]. This trend has also affected ontology and knowledge representation languages. In order to be interoperable, it is becoming essential that such languages be web based and expressible in XML. Being web based means that the ontology is concerned with *web resources*. A web resource is identified by a universal resource indicator (URI). In other words, anything being described by a web-based ontology is a web resource.

There are three major web-based ontology languages: XML Topic Maps (XTM) [10], the Resource Description Framework (RDF) [11] and the Web Ontology Language (OWL) [12]. XTM is an International Organization for Standardization (ISO) standard (ISO13250), whereas RDF and OWL are standards of the World Wide Web Consortium (W3C). XTM allows one to specify relationships among any number of web resources. By contrast, RDF and OWL restrict all relationships to be binary. Restricting relationships to binary simplifies the language and processors, but it makes it much more awkward to deal with relationships that involve more than two resources. It also restricts the portability of knowledge structures to hardware platforms other than those on which the structure was created.

When one restricts relationships to be binary, all facts are *triples*, consisting of the two entities being related (called the *subject* and *object*) and the relationship between them (called the *predicate*). RDF is an XML-based language for representing triples. The RDF Schema (RDFS) language is an extension of RDF that allows one to specify subclass and subproperty relationships. It also allows one to specify the domain and range of a property.

The OWL language has three levels: OWL Lite, OWL-DL, and OWL Full. They differ in the constructs that are allowed, with Lite being the most restrictive and Full being the least restrictive. OWL adds many new capabilities to RDF and RDFS, such as cardinality constraints, disjointness constraints, enumerations, and inverse properties. However, the most significant new feature is the ability to construct classes from other classes, such as defining a class to be the intersection or union of two or more other classes. For example, a BPSK is the special case of

MPSK for which there are exactly two symbols in the waveform alphabet (each one being 180 degrees away from the other). Class constructors are the basis for a form of reasoning known as *description logic* [13].

OWL Lite differs from OWL-DL in the class constructors that are allowed. For example, in OWL-DL, one can specify the *complement* of a class (i.e., all instances that are *not* in the class), but this specification is not allowed in OWL Lite. Although OWL-DL allows all class constructors, it does not allow one to cross *metalevels.* For example, in OWL-DL a class cannot be an instance of another class. By contrast, in OWL Full, a class can be an instance of another class, which itself is an instance of yet another class, and so on to any number of levels. Although OWL Full is a very rich ontology language, it is still not as rich as arbitrary first order predicate logic. Table 13.1 summarizes the features and differences between the various web-based ontology languages.

Table 13.1: Existing web-based ontology languages and their characteristics. The last column describes the level of reasoning as well as the level of processing complexity as a function of the number of triples.

Language/ organization	Features	Reasoning/complexity
XTM/ISO	Higher order relationships	None/linear
RDF/W3C	Binary relationships	None/linear
RDFS/W3C	RDF plus subclass, subproperty, domain, and range	Subsumption/polynomial
OWL Lite/ W3C	RDFS plus some class constructors, but no crossing of metalevels	Limited form of description logic/exponential
OWL-DL/ W3C	All class constructors, but no crossing of metalevels	General description logic/ decidable
OWL Full/ W3C	No restrictions	Limited form of first-order predicate logic/undecidable

One might think that the richer the ontology language the better. However, richer ontology languages are also more difficult to process. RDF and RDFS are relatively easy to process. The time to process RDF is linear in the number of triples. RDFS is a little more difficult, being polynomial in the number of triples and NP-complete in the size of the query in the worst case. OWL Lite is much more difficult, requiring exponential time in the number of triples for the worst case. OWL-DL is *decidable*, meaning that the processing will finish in a finite amount of time, but the amount of time can be more than exponential. Finally, OWL Full is *undecidable*, which means that ontologies cannot be processed, and some queries cannot be answered, in a finite amount of time.

Querying

Given a database, one can extract information by using a *query*. Many query languages have been proposed for XTM, RDF, and OWL. The query language that has been proposed for RDF is called RDQL [14]. As is the case with most query languages, RDQL is syntactically and semantically very similar to the Structured Query Language (SQL) for relational database systems. RDQL differs primarily by allowing one to specify patterns. A *pattern* is a fact in which some of the components can be variables.

For example, the query in Figure 13.2 retrieves all fields of all DL frames. The query patterns in Figure 13.2 specify both variables, such as ?x, and constants, such as <http://ontoradio.org/contains>. The constants are the URIs of web resources. The ontoradio.org domain is the OBR domain where the OBR ontology resources are defined. The query language that has been proposed for OWL is called OWL-QL [15].

```
1.   SELECT ?x
2.   WHERE (?x, <http://www.w3.org/1999/02/22-rdf-syntax-ns#type>,
                  <http://ontoradio.org/2005/datalink#DataLinkFrame>)
3.   AND (?y, <http://www.w3.org/1999/02/22-rdf-syntax-ns#type>,
                <http://ontoradio.org/2005/datalink#DataLinkField>)
4.   AND (?x, <http://ontoradio.org/2005/datalink#contains>,?y)
```

Figure 13.2: An example of RDQL, a query language for RDF. The first line enables the retrieval of attributes of variable ?x that match patterns. Patterns are expressed as triples (subject, predicate, object) following the WHERE delimiter. The second line limits all ?x to be of type *DataLinkFrame*. The third line describes those attributes that are of interest (those ?y of type *DataLinkField*). The last line restricts the objects to be retrieved to be *DataLinkFields* contained in *DataLinkFrames*.

The RDQL and OWL-QL query languages differ from database query languages in several important ways. One important difference is that they are web based. Whereas databases are generally restricted to a single server or at least one site, RDF and OWL effectively regard the entire web as being a single database.

OWL-QL has the additional feature of specifying a protocol for query requests and answers to support its use by web services.

The most important difference between database queries and query languages such as RDQL and OWL-QL is their support for reasoning. In addition to facts that have been explicitly asserted, a query can also retrieve facts that have been inferred. It is for this reason that RDF and OWL databases are said to be knowledge bases. A *knowledge base* is the set of all currently known or inferred facts in a particular context. The reasoning capability of OWL is especially powerful. We elaborate on this feature in section *Reasoning*.

Radio communication introduces an additional requirement on query languages. Unlike databases, which have explicitly stored data, and knowledge bases, which have a combination of explicitly asserted facts and inferred facts, an OBR knowledge base includes data that is embedded in the software that implements the communication protocols. Extracting such data requires a new software capability known as *self-awareness*, or *reflection*.

Reflection is a property that enables software to understand its own *run-time structure*. Reflection is a key feature of any software system that is expected to respond to unanticipated queries. Run-time structure is significantly different from program structure, which is an artifact of the language itself, because it also is a function of the compiler and the run-time platform. Run-time structure includes memory pointers for variables, for their subfields, for methods, and for procedures. This is essential for the software to be able to answer queries and execute methods dynamically.

Reasoning

One of the important features of ontology languages that distinguishes them from databases is the ability to make logic deductions. In other words, one can *reason* about the information in a knowledge base. A fact is *deduced* if one can infer that it is true even though the fact has never been explicitly asserted to be true. One of the most important examples of deduction is *subsumption.* For example, if one knows that an analog signal uses BPSK, then one can deduce that it is also of type MPSK. In general, whenever something is an instance of a subclass, then it is also an instance of all superclasses of the subclass. Subsumption is the basis for reasoning in description logic. For example, all features and axioms applicable to MPSK signaling also apply to BPSK signaling.

While subsumption reasoning is useful, it is not sufficient for all reasoning tasks. When reasoning involves several linked facts, one cannot express the inference using subsumption alone. When database records are linked by common attributes, they are said to be *joined*. To express reasoning involving *joins* of facts, it is necessary to introduce *rules*. A rule is knowledge in the form of an *if-then* statement. If a *hypothesis* holds, then a *conclusion* must also hold. The hypothesis is also known as the *antecedent*, and the conclusion is also known as the *consequent*. The rule language that has been proposed for OWL is the Semantic Web Rule Language (SWRL) [16]. An example of a SWRL rule is shown in Section 13.4.2.

13.3.2 Role of Ontology in Knowledge-Intensive Applications

Two-way radio communication introduces a number of challenges not shared by most other ontology-based applications:

1. Real-time processing demands higher performance for inference and reasoning than an interactive application.
2. The knowledge base of a node includes state information that is continually varying, in contrast with the static knowledge bases required by most reasoning systems.
3. The facts are not explicitly stored in a knowledge base but rather are embedded in the software that implements the communication protocol [2].
4. The radios may not have access to the WWW or the Semantic Web for broad support.
5. Most radios are continually moving, and the link performance is bandwidth restrictive and time varying.

If these challenges can be overcome, ontologies can play a number of important roles in radio communication, such as the following:

1. *Interoperability*: Radios can use ontologies to deduce important information, such as the protocol being used by other radios.
2. *Flexible querying*: Information, such as multipath structures, can be queried. Furthermore, such queries can be answered without having any explicit pre-programmed monitoring capability.
3. *Run-time modifiability*: Protocols, packet structures, and even waveforms can be modified at run-time in response to environmental conditions and application requirements.

4. *Validation*: Formalization allows one to check the consistency of protocols and to validate the correctness of algorithms that implement the protocols.

5. *Self-awareness*: Communication nodes can understand their own structure and modify their functioning at run-time based on this understanding.

13.4 A Layered Ontology and Reference Model

This section discusses two (partial) ontologies, for the PHY layer and for the DL layer. These ontologies are then used for the discussion of the realization of the cognitive aspects discussed in Section 13.5.

13.4.1 Physical Layer Ontology

A piece of the ontology for the PHY layer is shown in Figure 13.3. It is represented in the Unified Modeling Language (UML) notation. UML is used primarily in software engineering to represent software and systems. The boxes in UML represent *classes*, and the connecting lines represent *associations*. UML classes correspond to the classes used in object-oriented programming languages such as C++, Java, or C#. For programming languages such as C, which are not object oriented, classes are implemented by using *structs*. Associations can be implemented in a variety of ways, depending on the kind of association. In OWL, associations are called *properties*.

Some associations have a predefined meaning. For instance, the hollow arrow at the end of the line in Figure 13.3 represents the *subclass* relation. In the figure, M-FSK, QAM, and M-PSK are all subclasses of Alphabet. The subclass relation is supported by all object-oriented programming languages. In Java and C# one specifies that a class is a subclass of another by using the "extends" keyword. C++ is more succinct, specifying a subclass by using the colon character (":"). Diamonds represent *aggregation*. Thus Alphabet is an aggregate of symbols (instances of class Symbol).

Aggregations are normally implemented by using arrays or linked lists, but other implementations are also used, depending on the number of elements in the aggregate and whether the aggregation has a variable number of elements. Because a particular Alphabet has a fixed number of Symbol instances, it would be implemented as an array. Non-object-oriented programming languages, such as C, implement the subclass relation by using aggregation. Other associations are identified either by the names placed in the middle of the association line, such as encodes, or by the roles that elements of the related classes play (roles are

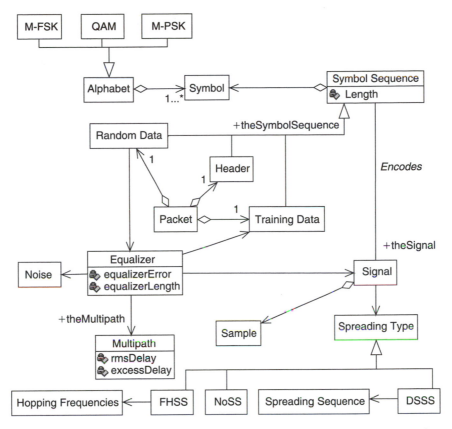

Figure 13.3: A partial ontology for the PHY layer in UML [8]. The boxes denote classes, and the lines connecting them represent associations. The hollow arrowhead denotes the subclass relation, and the diamond denotes aggregation.

attached to the line ends and are distinguished by the "+" symbol). So the association encodes in Figure 13.3 has two roles, +theSymbolSequence and +theSignal. Associations such as these are implemented by using pointers or references to the associated object. In some cases, associations have a prespecified multiplicity. For instance, an Alphabet must have at least one instance of Symbol. Multiplicity specifications affect how an association can be implemented. Because an Alphabet can have more than one instance of Symbol, it cannot be implemented using a single pointer or reference to a Symbol.

The ontology consists of two groups of classes. The upper part represents classes and properties related to symbols. We can see that Packet consists of one Header, one Training Data, and one Random Data. Each of them is an instance of Symbol Sequence.

The bottom part of Figure 13.3 represents classes and properties related to the signal domain. The main association between the two parts is encodes. It relates a Symbol Sequence to a Signal. Signal is an aggregate of instances of Sample. A signal can be modulated in different ways. This ontology example shows two of them, DSSS and FHSS. Another connection between the symbol and the signal domain occurs through Equalizer, which estimates the Multipath using the Training Sequence and then uses the result for signal interpretation.

13.4.2 Data Link Layer Ontology

The DL layer is responsible for transmitting frames and for error detection and correction in communication links. There are many *DL protocols*. Each protocol specifies the frame types and structure as well as how the communication link is controlled. Many of the DL protocols specialize and extend other protocols. Consequently, the DL protocols form a hierarchy as shown in Figure 13.4. Because of the diversity and complexity of DL protocols, not all of the details of the DL ontology are shown in the figure.

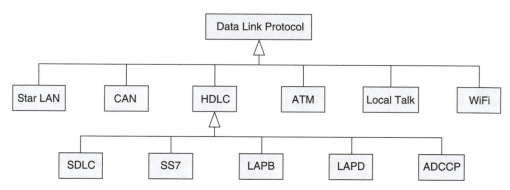

Figure 13.4: A partial hierarchy of data link protocols in UML.

There are a large variety of frame types. The names and semantics of the frame types depend on the protocol, but there is considerable overlap among the protocols. Consequently, although each protocol has its own frame type hierarchy, the frame types in different hierarchies can be related by the OWL *sameAs* relationship. The frame type hierarchy for High-Level Data Link Control (HDLC) protocols is shown in Figure 13.5 and the hierarchy for WiFi® protocols is shown in Figure 13.6.

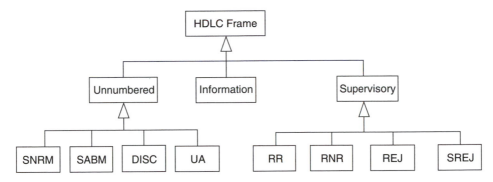

Figure 13.5: HDLC frame hierarchy.

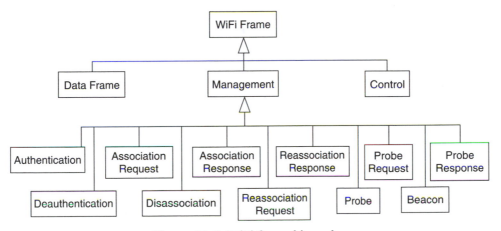

Figure 13.6: WiFi frame hierarchy.

A frame consists of a sequence of *fields*. The frame structure is defined by the order of the field types, the number of bits allowed in each field, and the values (bit sequences) allowed in each field. The HDLC protocol has the field types shown in Figure 13.7.

The WiFi protocol has many of the same field types, except for changes in terminology. For example, the WiFi Frame Control Field is called the Control Field in the OWL syntax. Similarly, the WiFi Checksum Field is called the CRC Field in the OWL syntax. In the OWL syntax, this is written as: Frame Control Field *owl:sameAs* Control Field, and Checksum Field *owl:sameAs* CRC Field.

The DL classes are related to one another as shown in Figure 13.8.

A DL frame contains an ordered sequence of fields, each of which has a size (in bits) and a value. A DL field belongs to a protocol and also has a *mode*. For example, the HDLC protocol has three operational modes: NR (Normal

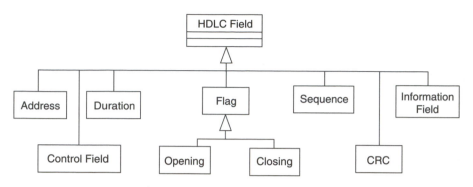

Figure 13.7: HDLC field hierarchy.

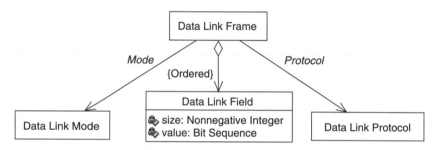

Figure 13.8: DL layer ontology.

Response), AR (Asynchronous Response), and AB (Asynchronous Balanced). These modes are not shown in Figure 13.8.

In addition to the hierarchies and relationships already described, the DL ontology specifies a large number of rules that constrain DL fields within the same frame and in related frames. Examples of such rules shown informally (and implemented in Figures 13.9 and 13.10) include:

1. If a frame belongs to the HDLC protocol, then its opening flag field has value $0 \times 7E$ (i.e., the bit sequence 01111110).

2. If a frame belongs to the SDLC protocol, then the address field has 8 bits.

3. If a frame belongs to the SS7 protocol, then the address field has 0 bits.

4. If a frame belongs to the HDLC protocol, then the control field has 8 or 16 bits.

5. If a frame belongs to the HDLC protocol and the mode is available resource map (ARM), then the address field has 0 bits.

6. If a frame belongs to the WiFi protocol then the first two bits of the control field are zeroes. This subfield represents the version number.

```
 1 <owl:Class>
 2   <owl:intersectionOf rdf:parseType="Collection">
 3     <owl:Class rdf:about="#OpeningFlagField"/>
 4     <owl:Restriction>
 5       <owl:onProperty rdf:resource="#containedIn"/>
 6       <owl:allValuesFrom>
 7         <owl:Restriction>
 8           <owl:onProperty rdf:resource="#protocol"/>
 9           <owl:allValuesFrom rdf:resource="#HDLCProtocol"/>
10         </owl:Restriction>
11       </owl:allValuesFrom>
12     </owl:Restriction>
13   </owl:intersectionOf>
14   <rdfs:subClassOf>
15     <owl:Restriction>
16       <owl:onProperty rdf:resource="#value"/>
17       <owl:hasValue rdf:datatype="&xsd;hexBinary">7E</owl:hasValue>
18     </owl:Restriction>
19   </rdfs:subClassOf>
20 </owl:Class>
```

Figure 13.9: An OWL implementation of the DL layer protocol, which categorizes the opening flag field of DL layer frames.

All of these rules can be represented either in SWRL or in OWL. For instance, the first rule may be represented in OWL, as shown in Figure 13.9.

In XML, element tags, attribute names and attribute values can belong to various *namespaces*. The namespace of a tag or an attribute name is specified with a colon. For example, rdf: specifies the RDF namespace; rdfs: specifies the RDFS namespace; and owl: specifies the OWL namespace. Within an attribute value, a namespace is specified with an XML *entity*. For example, &xsd; specifies the XML Schema namespace. If no namespace is specified, the namespace is the current default namespace. In this case, the default namespace is obr, the namespace of OBR. For example, OpeningFlagField is in the obr namespace.

At the highest level, the rule in Figure 13.9 says that an intersection of two classes is a subclass of another class (lines 1, 2, 13, 14, 19, 20). This is expressed using the owl:intersectionOf (lines 2, 13) and the rdfs:subClassOf (lines 14, 19) properties. The intersection is used to represent the Boolean AND operator, and subclass is used to present the logic IMPLIES or IF–THEN operator. The two classes being intersected are the class of opening flag fields (line 3) and the class of HDLC fields (lines 4–12). The former class is part of the ontology and has a name (line 3). The latter class does not have a name. It is specified by two relationships in the ontology: containedIn (line 5) and protocol (line 8). An HDLC field is one that is contained in a frame whose protocol is of type HDLC. The owl:Restriction is used for constructing classes of instances that satisfy a constraint for a particular property. The first restriction (line 4) specifies "a field that is contained in" (lines 5, 6), and the second restriction (line 7) specifies "a frame whose protocol has type HDLC" (lines 8, 9). The last restriction (line 15) specifies a value that must be taken by the field (lines 16, 17). Putting all of these together, the rule can be expressed as "IF something is an opening flag field AND is a field that is contained in a frame whose protocol has type HDLC THEN it has value 7E. Alternatively, the same rule could be represented in SWRL, as shown in Figure 13.10.

```
 1  <ruleml:Imp>
 2    <ruleml:body>
 3      <swrlx:classAtom>
 4        <owlx:Class owlx:name="#OpeningFlagField"/>
 5        <ruleml:var>x</ruleml:var>
 6      </swrlx:classAtom>
 7      <swrlx:individualPropertyAtom swrlx:property="#containedIn">
 8        <ruleml:var>x</ruleml:var>
 9        <ruleml:var>y</ruleml:var>
10      </swrlx:individualPropertyAtom>
11      <swrlx:individualPropertyAtom swrlx:property="#protocol">
12        <ruleml:var>y</ruleml:var>
13        <ruleml:var>z</ruleml:var>
14      </swrlx:individualPropertyAtom>
15      <swrlx:classAtom>
16        <owlx:Class owlx:name="#HDLCProtocol"/>
17        <ruleml:var>z</ruleml:var>
18      </swrlx:classAtom>
19    </ruleml:body>
20    <ruleml:head>
21      <swrlx:datavaluedPropertyAtomsw rlx:property="#value">
22        <ruleml:var>x</ruleml:var>
23        <owlx:DataValue
24          owlx:datatype="&xsd;hexBinary">7E</owlx:DataValue>
25      </swrlx:datavaluedPropertyAtom>
26    </ruleml:head>
27  </ruleml:imp>
```

Figure 13.10: A SWRL implementation of the rule from Figure 13.9. This rule is used to characterize the opening flag field of the DL layer frames.

At the highest level of the rule in Figure 13.10, there is an IF–THEN operator (lines 1 and 27). An IF–THEN operator has two parts: the *hypothesis* or *body* (lines 2–19), and the *conclusion* or *head* (lines 20–26). Within either body or head, one specifies a sequence of *atoms*. A *class atom* specifies an instance of a class. In this case the class atoms specify that x is an instance of the class OpeningFlagField (lines 3–6), and z is an instance of HDLCProtocol (lines 15–18). A *property atom* specifies a triple. Individual property atoms are for properties whose values are *individuals* (lines 7–10 and 11–14). Data-valued property atoms are for properties whose values are data values (lines 21–25). The second atom in the body specifies the triple (x obr:containedIn y) (lines 7–10) and the third atom in the body specifies the triple (y obr:protocol z) (lines 11–14). The head consists of a single atom, which specifies the triple (x obr:value "7E").

The full specification of all DL ontology rules in OWL is very large [17]. Rules dealing with multiple frames (such as a request and response to it) are especially complex because they are specifying functionality rather than just formats.

The DL ontology provides for the following capabilities:

1. *Self-awareness of DL layer functionality*: The ontology specifies the protocols. More precisely, it specifies the frame structure and the functionality associated with the various fields.

2. *Dynamic interoperability at the DL layer*: An OBR can infer the DL protocol being used by another radio. A simple example of such a deduction was shown by the list of properties 1 through 6 earlier in this section.

3. *Command capability*: A radio can remotely request the use of a different protocol or modify the features of an existing protocol.

13.5 Examples

Some examples of how two radios can exchange information about their communications characteristics and parameters and then adapt the communications protocol so that the quality of communication is maintained or improved demonstrate the cognitive aspects described in the preceding sections.

13.5.1 Responding to Delays and Errors

Wireless transmission requires a robust and efficient communication protocol. When the channel has been estimated and the estimation has been sent back to the transmitter, then the transmission can be adapted according to the channel characteristics. The basic idea behind adaptive transmission is to maintain a constant signal-to-noise ratio (SNR) level, (E_b/N_0), by varying the transmission power level, symbol transmission rate, constellation size, and coding rate/scheme, or any combination of these parameters [7]. In experiments with protocol adaptation [7], radios monitored their excess delay (multipath delay spread), rms delay (root mean square delay spread) of the multipath structure of the channel, and the mean square root error of the equalizer. Here the mean square root error of the equalizer represents the average distance between the equalized data (the input data of the equalizer multiplied by the equalizer chips) and the output of the equalizer (the estimated symbol).

OBRs are able to query each other by using a query expressed in an appropriate query language (e.g., OWL-QL). An example of such a query regarding the rms delay, excess delay, and equalizer error is shown in Figures 13.11 and 13.12. In this example, radio A first sends the following query to radio B (Figure 13.11).

```
<owl-ql:query ID="query1">
 <owl-ql:queryPattern>
  <rdf:RDF>
     <obr:Multipath rdf:about="&obr;multipath">
       <obr:rmsDelay rdf:resource="&var;x"/>
 <obr:excessDelay rdf:resource="&var;y"/>
     </obr:Multipath>
     <obr:Equalizer rdf:about="&obr;equalizer">
       <obr:equalizerError rdf:resource="&var;z"/>
     </obr:Equalizer>
   </rdf:RDF>
 </owl-ql:queryPattern>
 </owl-ql:query>
```

Figure 13.11: An OWL-QL query. The query to radio B from radio A is requesting information concerning multipath parameters (rms delay and mean excess delay) as well as the mean-squared decision error at the receiver B. The first line identifies the fragment as a query, and creates the string identifier for it. The fourth line restricts attention to multipath parameters, and the fifth and sixth lines request the particular variables rmsDelay (query variable x) and excessDelay (query variable y) from that group. The eighth line refers to equalizer parameters, and the ninth line requests the value of equalizerError (query variable z).

As was shown in Figure 13.2 for RDQL, a query consists of a series of triples in which some of the slots contain variables. In OWL-QL, the query pattern is specified by using RDF syntax. In RDF, the subject of a triple is specified by using rdf:about. The first line of the query in Figure 13.2 specifies the triple (obr:multipath rdf:type obr:Multipath) (i.e., obr:multipath is an instance of the obr:Multipath class). The second and third lines specify triples in which the subject is obr:multipath, the predicate is the XML element tag, and the object is specified by using rdf:resource=. Thus the third line specifies the triple (obr:multipath obr:rmsDelay var:x). In other words, the variable x should be assigned to the rms delay of the multipath. The var namespace contains the variables used by queries. Similarly, in the fifth line obr:equalizer is an instance of the obr:Equalizer class, and the sixth line specifies the triple (obr:equalizer obr:equalizerError var:z).

When radio B receives this query, it invokes its reasoner to derive an answer. The query was formulated in the ontology that the two radios share, so the other radio understands what obr:rmsDelay, obr:excessDelay, and obr:equalizer Error mean. For instance, the ontology would include information on the units of measure for each of the parameters. To answer the query, the radio first searches its *annotation* (or *markup*) file to find or to infer facts that can match the query pattern. However, this file contains only facts that have been explicitly asserted. It does not contain facts that are embedded in the software. Consequently, the reflection mechanism is invoked to extract facts that can be used to match the query pattern, either directly or by means of inference. The answers to the query are then transmitted to radio A. Such a response would look like the fragment in Figure 13.12.

```
<owl-ql:answerBundle about="query1">
  <owl-ql:answer>
    <owl-ql:binding-set>
      <var:x>1.0078370372505556</var:x>
      <var:y>1.062759005498691</var:y>
      <var:z>0.025987243652343</var:z>
    </owl-ql:binding-set>
  </owl-ql:answer>
</owl-ql:answerBundle>
```

Figure 13.12: An OWL-QL response to radio A from radio B. The first line specifies the fragment as a response, and identifies the corresponding query identifier (see Figure 13.11). The fourth through sixth lines specify the values for query variables x, y, and z, which were requested in the *query* query1, namely rmsDelay, excessDelay, and equalizerError.

Upon receiving this answer, radio A invokes its reasoner in order to decide how to adjust its protocol so that the communications quality is improved. For instance, the radio could do one of the following: select a different alphabet (e.g., 2-QAM, 4-QAM or 16-QAM), increase the equalizer length, or increase the transmit power.

13.5.2 Adaptation of Training Sequence Length

The following example considers the case of negotiating the length of the training data according to the channel dynamics and noise level [8]. The goal is to have a low packet overhead, when possible, and increase it in the case of high-noise or high-delay spread. In these experiments, the transmitter used the DSSS spreading type DSSS(2,7) for transmitting the header (i.e., each symbol is mapped to 127 chips), and a sequence of DSSS(2,5) symbols (31 chips for each symbol) for the training data. The number of symbols in the training data was changed gradually in response to the changing characteristics of the channel.

In the example described in Wang et al. [8], negotiation of the length of the training data was accomplished in six transmissions. First, node A sends data to node B. This data comes to node B in an ontology-defined envelope (as is done in Figure 13.11) as *Data*. Node B then checks performance. For this purpose, nodes invoke their rules for performance checking. If performance is satisfactory, then node B returns a *Confirm* message with content "*Continue*" and node A continues to send data to B. If the performance is not satisfactory, a *Confirm* message with content "*CommandReq*" (request a command from the other node) is returned. Node A then generates a *Command* to change the communication protocol. The command generation rules first send a *Query* from node A to node B requesting the channel condition and the current protocol parameters (in a similar way as discussed in the previous example). When node B receives this query, it infers the answer and sends the *Answer* back to node A. When node A receives the answer, it generates the command. After the *Command* is generated, it is sent to node B, and executed on node A, thus changing the protocol at node A. When node B receives the *Command*, it executes the command, thus changing the protocol at node B. A *Confirm* message with content "*Continue*" is then sent to node A. The negotiation sequence involves five message types: data, confirmation, query, answer, and command, as shown in the inheritance diagram in Figure 13.13. These message types each have message contents, as described next.

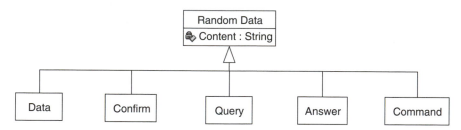

Figure 13.13: Classification of message types.

To realize this kind of negotiation, the radios must not only share an ontology but also have means for inferring ways of behaving in particular situations. In pragmatic terms, this means that radios must have some domain knowledge of radio communication. The knowledge may be represented in many different ways, but as discussed in Section 13.2.1, the declarative representation is preferred to the procedural. Consequently, this knowledge should be encoded in the form of *rules*. For instance, the rule that is used in deriving a command concerning the number of symbols in the training sequence in response to this query might look like this:

IF
- the difference between the rmsDelay and the previous equalizer feed-back coefficient vector length is larger than half of the previous equalizer feedback coefficient vector length

THEN
- construct a new equalizer with the length *3(integer(rmsDelay) + 1)* and construct new training data of length *3 · rmsDelay · 20 MOD symbol length*

ELSE IF
- the equalizer error is smaller than a predefined performance threshold

THEN
- decrease the length of the training data by a predefined value AND use the old equalizer

ELSE
- increase the length of the training data by a predefined value and use the old equalizer

An informal textual representation of the rule is used here. The experiments reported by Wang et al. [8] used a version of Prolog (Kernel Prolog) for this purpose. In the future, the use of SWRL [16], is envisioned instead.

The experiments reported by Wang et al. [8] considered three different inter-node distances under different multipath conditions. It was shown that for the worst condition, about 20 symbols were selected according to the negotiation rules, whereas in other cases, only five symbols were needed to maintain a desired level of quality of communication.

13.5.3 Data Link Layer Protocol Consistency and Selection

The DL ontology specifies the formats and functionality of the DL protocols. This section presents a few examples of how the DL ontology might be used. The first example illustrates how the formal specification of protocols can uncover inconsistencies. This kind of reasoning would normally be performed offline. The second example shows a simple example of how the ontology can be used for interoperability.

The first example uses the following two rules (2 and 5) from the DL ontology list presented in Section 13.4.2:

- If a frame belongs to the SDLC protocol, then the address field has 8 bits.
- If a frame belongs to the HDLC protocol and the mode is ARM, then the address field has 0 bits.

Because the SDLC protocol is a subclass of the HDLC protocol, the two rules imply that an ARM frame belonging to the SDLC protocol has an address field with both 8 and 0 bits. It follows that the SDLC protocol cannot have ARM frames.

We can address this inconsistency in several ways. We could remove the subclass relationship between the SDLC protocol and the HDLC protocol classes from the ontology. This has the disadvantage that the SDLC protocol would now have to be completely specified from scratch rather than being derived from a more general class. Alternatively, we could modify the first of these two rules to recognize that the address field for the SDLC protocol will have only 8 bits for modes other than ARM.

The second example concerns the issue of interoperability. Suppose that a radio receives a stream of packets from another radio with which it has not previously communicated. Can the receiving radio deduce the protocol being used by the transmitter? In the case of HDLC versus WiFi, the following two rules (1 and 6 from the list in Section 13.4.2) allow one to disambiguate the packets:

- If a frame belongs to the HDLC protocol, then its opening flag field has value $0 \times 7E$ (i.e., the bit sequence 01111110).
- If a frame belongs to the WiFi protocol then the first two bits of the control field are zeroes. This subfield represents the version number.

From these two rules, we can deduce that the first two bits of the packet will distinguish these two protocols. If the first two bits are 01, then it is an HDLC frame. If the first two bits are 00, then it is a WiFi frame. If the bits have some other value, then the frame belongs to neither protocol. This example is somewhat artificial, and we could argue that it is easy to write a small procedure that could make this same determination. However, this is only one example among many possible inferences. The software would quickly become extremely unwieldy if we needed a separate procedure for every possible deduction. Indeed, we would need infinitely many procedures because we can make infinitely many deductions, and every time a new feature or protocol variation was added, we would need to revise all of the procedures.

13.6 Open Research Issues

Even though the concept of OBR can be implemented as discussed in this chapter, a number of open issues still need to be investigated. Additionally, some of the engineering issues need to be addressed before this concept finds its way into real radio applications. Several of the most important open issues are discussed in this section.

13.6.1 Ontology Development and Consensus

As discussed in this chapter, ontologies can bring many advantages to software radio communication. However, to use an ontology, one must first develop it. Building high-quality ontologies is a substantial task. It is further complicated by the diversity of web-based ontology languages.

The first step in an ontology development project is to agree upon the purpose and motivation for embarking on this activity. This step includes determining the following:

1. *Why* the ontology is being developed.
2. *What* will be covered by the ontology.
3. *Who* will be using the ontology.
4. *When* and for how long the ontology will be used.
5. *How* the ontology is intended to be used.

Once there is a clear understanding of the purpose of the ontology, four major activities must be undertaken: (1) choosing an ontology language, (2) obtaining a development tool, (3) acquiring domain knowledge, and (4) reusing existing ontologies. None of these is especially easy, and it is important to reach consensus because the ontology forms the basis for communication within the community. Generally speaking, the larger the community, the more difficult it is to reach consensus. One can mitigate this difficulty to some degree by establishing a precise statement of the purpose of the ontology before starting the project. However, one cannot entirely eliminate it.

Ultimately, the success of ontologies for software and cognitive radio communication will depend not only on the quality of the ontology design but also on the extent to which the community accepts the ontology. The task is a challenging one, but experience in other communities suggests that it is achievable, nonetheless.

13.6.2 Ontology Mapping

At every level of the communication protocol stack are many protocols currently in use. Each protocol has its own terminology and therefore its own ontology. Generally speaking, the higher in the stack, the more diverse the ontologies. To achieve interoperability between protocols and other applications, it is necessary to have mappings between the ontologies. Such mappings can be specified by subject matter experts, but when the ontologies use the same concepts, it should be possible to automate the process of transforming from one ontology to the other. This problem has been studied for many years. Most of the work has been devoted to mapping relational database schemas, but there has also been some recent work on this problem for XML DTDs and even for the more sophisticated ontologies of the Semantic Web.

A survey of relational schema integration tools is presented by Rahm and Bernstein [18]. When the data from a variety of relational database sources are transformed to a single target database, the process is called data warehousing. Many data warehousing companies now also support XML. If a query using one vocabulary is rewritten so as to retrieve data from various sources, each of which uses its own vocabulary, it is called virtual data integration [19].

Ontology mapping depends on identifying semantically corresponding elements. This is a difficult problem to solve because different sources may use very different structural and naming conventions for the same entities. In addition, the same name can be used for elements having totally different meanings, such as different units, precision, resolution, measurement protocol, and so on. It is usually necessary to annotate an ontology with auxiliary information to assist one in determining the meaning of elements, but ontology mapping is difficult to automate even with this additional information. Furthermore, a single element in one ontology may correspond to multiple elements in another. In general, the correspondence between elements is many-to-many: many elements correspond to many elements.

Tools for automating ontology mapping have been proposed, and some research prototypes exist. However, these tools mainly help to discover simple one-to-one matches, and they do not consider the meaning of the data or how the transformation will be used. Using such a tool requires significant manual effort to correct wrong matches and add missing matches. One of the few tools that can handle many-to-many mappings was developed by Goguen and his students [20].

In practice, ontology mapping is done manually by domain experts and is very time-consuming when there are many data sources, or when ontologies are large or complex. Automated ontology-mapping systems are designed to reduce manual effort. However, such systems require a substantial amount of time to prepare input to the system as well as to guide the matching process. This amount of time may easily swamp time saved by using the system. Unfortunately, existing ontology-mapping prototypes focus on measuring accuracy and completeness rather than on whether they provide a net gain. In practice, the best that one can hope for from current systems is that they can help to record and to manage the ontology matches that have been detected, by whatever means.

Many ontology-mapping projects exist, and some have developed prototypes. Examples are PROMPT [21], from the Stanford Medical Informatics Laboratory, and the Semantic Knowledge Articulation Tool (SKAT) [22], also from Stanford. None of these prototypes are available for public use, however, either via open source software or commercial software. But even if the tools were available, they are only prototypes, not developed to deal with the issues that are specific to radio communication, such as reflection and high-performance reasoning. Consequently, ontology mapping remains a challenging field for development.

13.6.3 Learning

Ontology learning is an emerging field aimed at automated assistance in the construction of ontologies and semantic page annotation through the use of machine learning techniques [23]. The relevance of this field is tied to the prohibitively large amount of manual ontological construction required to represent the vast amount of information in communication systems. The implementation of a fixed ontological system presumes that the knowledge engineer anticipated the extent of the language involved in queries, responses, and reasoning. Although this goal could be attained at some time instant, the continual evolution of radio standards and implementations mandates a requisite evolution in the ontology. For example, WiFi was not a recognized radio standard 15 years ago, IEEE802.11g was not available prior to 2000, and the pre-release version of 802.11n was not available until 2004. Additionally, new variants and derivatives (subclasses) of the HDLC protocol class will continue to appear for some time. Without automated ontological learning, a cognitive radio that is expected to communicate using these emerging standards would require continuous updating of its ontological database system. Ontology learning is one mechanism to automate this evolution.

13.6.4 Efficiency of Reasoning

The implementation of the concept of OBR discussed in this chapter relies heavily on OWL, a language for representing ontologies. Another component of the OBR architecture is a generic reasoning engine. Even though a generic reasoning mechanism satisfies the requirement of being able to respond to surprise by a cognitive agent, it also poses very high demand on the reasoning engine. In general, logic reasoning is undecidable. In simple terms, this means that an inference engine might never terminate its process of answering a specific query.

With this concern in mind, the designers of OWL provided the language in three levels: OWL Full, OWL-DL, and OWL Lite, as discussed in section *Ontology Languages*. The three levels differ in both expressive power and computational complexity. OWL Lite is the simplest of the three, with the least expressive power of the three OWL languages. In spite of its relative simplicity, the computational complexity of OWL Lite is very high; it is in the EXPTIME complexity class, meaning that in the worst case a deduction will require time equal to $O(2^{p(n)})$, where $p(n)$ is a polynomial in the size n of the ontology. OWL-DL is more expressive than OWL Lite, but its complexity is even worse; it is in the NEXPTIME complexity class. Although the preside complexity of this class is not known, it is at least as high as EXPTIME, and it may be much higher still, possibly an exponential of an exponential of a polynomial (i.e., $O(2^{f(n)})$, where $f(n) = 2^{p(n)}$. Furthermore, there is no bound on how much memory may be required. OWL Full is the most expressive of the three OWL languages, but its complexity is the worst possible: it is undecidable. Undecidability means that in the worst case there is no bound on the amount of time or space that may be needed to perform a deduction. In practice, even OWL Full is not sufficiently expressive, and it is necessary to use rules to express some of the axioms of the ontology. Unfortunately, if rules are added to any of the three languages, including OWL Lite, then the computational complexity of deduction is undecidable.

In order to make systems based on such generic reasoners scalable, the issues of complexity of reasoning must be resolved so that reasoning conclusions are derived within the constraints of the radio domain. Various approaches to such a problem are known in the literature. In general, such approaches are based on a trade-off between the quality of the solution and the computation time. For instance, for algorithms known as *any time algorithms*, the problem-solving activity can be stopped at any time, and the result will be the best one that has been found up to that time. Another approach is to combine logic reasoning with an uncertainty handling mechanism, like Bayesian networks, fuzzy logic, and others.

13.7 Summary

This chapter presented an ontology-based approach to address some of the requirements of cognitive radio. Throughout, this chapter illustrated how we can capture knowledge about the domain of radio communication using ontologies. All of the examples used a purely declarative representation of domain knowledge. The objective was to show that such a knowledge representation can be used by generic reasoning tools both to formulate queries and to infer answers to such queries. Such a query-answering mechanism is generic and thus satisfies the requirement of being capable of "responding to surprise," one of the key features of a cognitive agent.

Another feature of a cognitive agent is the ability to collect and use information about the environment. The example in Section 13.5.1 showed how radios can request multipath information and then use it to modify the transmission protocol. Moreover, such radios are situation-aware because they know not only transmission conditions at other nodes, but also how such conditions relate to their own transmission protocol parameters. Section on *Querying* also discussed how self-awareness can be implemented so that radios can know the values of their internal variables and how to change those values on demand. For instance, a radio can know its rms delay, excess delay, or equalizer error. Another example of this capability, shown in Section 13.5.2, is the knowledge of such values as the BER and the multipath structure.

This chapter also showed examples of querying of radios by other radios. Although it did not show any queries by end users, such queries can be implemented in the same manner as queries by other radios. This chapter has also shown that self-aware radios can execute commands. For instance, if two radios agree to use a specific length of the training sequence, they first negotiate a contract, and then both update their internal structures. In general, any information expressible in terms of the ontology can be retrieved and modified.

References

[1] R. Brachman, "A DARPA Information Processing Technology Renaissance: Developing Cognitive Systems." http://www.darpa.mil/ipto/presentations/brachman.ppt

[2] J. Boyd, *A Discourse on Winning and Losing*, Technical Report, Maxwell AFB, 1987.

[3] K.M. Passino and P.J. Antsaklis (Eds.), *Introduction to Intelligent and Autonomous Control*, Kluwer Academic Publishers, Boston, Dodrech, /London, 1992.

[4] M. Endsley and D. Garland, *Situation Awareness, Analysis and Measurement*, Lawrence Erlbaum Associates, Publishers, Mahway, New Jersey, 2000.

[5] XG Working Group. The XG Vision. Request For Comments, Version 2.0.

[6] N. Guarino, "Formal Ontology in Information Systems," in *Proceedings of Formal Ontology in Information Systems*, IOS Press, Amsterdam, 1998, pp. 3–15.

[7] J. Wang, D. Brady, K. Baclawski, M. Kokar and L. Lechowicz, "The Use of Ontologies for the Self-Awareness of the Communication Nodes," in *Proceedings of the Software Defined Radio Technical Conference SDR '03*, Orlando, Florida, USA, 2003.

[8] J. Wang, M.M. Kokar, K. Baclawski and D. Brady, "Achieving Self-Awareness of SDR Nodes Through Ontology-Based Reasoning and Reflection," in *Proceedings of the Software Defined Radio Technical Conference SDR '04*, Phoenix, Arizona, November 2004.

[9] World Wide Web Consortium, "eXtensible Markup Language," www.w3.org/XML/, 2001.

[10] TopicMaps.org, The XTM website, topicmaps.org, 2000.

[11] O. Lassila and R. Swick, "Resource Description Framework (RDF) Model and Syntax Specification," www.w3.org/TR/REC-rdf-syntax, February 1999.

[12] F. van Harmelen, J. Hendler, I. Horrocks, D. McGuinness, P. Patel-Schneider and L. Stein, "OWL Web Ontology Language Reference," www.w3.org/TR/owl-ref/ M. Dean and G. Schreiber (Eds.), March 2003 .

[13] F. Baader, D. Nardi, D. McGuinness, P. Patel-Schneider and D. Calvanese. *The Description Logic Handbook: Theory, Implementation and Applications*, Cambridge University Press, Cambridge, England, 2003.

[14] A. Seaborne. "RDQL—A Query Language for RDF," www.w3.org/Submission/RDQL, January 2004.

[15] R. Fikes, P. Hayes and I. Horrocks, "OWL-QL: A Language for Deductive Query Answering on the Semantic Web," *Journal of Web Semantics*, Vol. 2, No. 2, 2005.

[16] I. Horrocks, P. Patel-Schneider, H. Boley, S. Tabet, B. Grosof and M. Dean, "SWRL: A Semantic Web Rule Language Combining OWL and RuleML," http://www.daml.org/2003/11/swrl, 2003.

[17] D. Brady, K. Baclawski and M. Kokar, "Achieving Dynamic Interoperability of Communication at the DataLink Layer through Ontology Based Reasoning," in *Proceedings of SDR Forum Tech Conference SDR05*, Anaheim, CA, November 2005.

[18] E. Rahm and P. Bernstein, "On Matching Schemas Automatically," *VLDB Journal*, Vol. 10, No. 4, 334–350, 2001.

[19] D. Embley, et al., "Multifaceted Exploitation of Metadata for Attribute Match Discovery in Information Integration," In *Proceedings of the International Workshop on Information Integration on the Web (WIIW'01)*, pages 110--117, Rio de Janeiro, Brazil, April 2001.

[20] Y. Nam, J. Goguen and G. Wang, " A Metadata Integration Assistant Generator for Heterogeneous Distributed Databases," in *Proceedings of International Conference Ontologies, Databases, and Applications of Semantics for Large Scale Information Systems, Lecture Notes in Computer Science* 2519, Springer-Verlag, Berlin, Heidelberg, New York, 2002, 1332–1344.

[21] N. Noy and M. Musen, *PROMPT: Algorithm and Tool for Automated Intelligence*, AAAI-2000, Austin TX, 2000.

[22] P. Mitra, G. Weiderhold and J. Jannink, "Semi-automatic Integration of Knowledge Sources," in *Proceedings of 2nd International Conference on Information Fusion*, Sunnyvale, CA, 1999.

[23] B. Omelayenko, "Learning of Ontologies for the Web: The Analysis of Existent Approaches," in *Proceedings of International Conference on Database Theory*, London, UK, January, 2001.

Cognitive Radio Architecture

Joseph Mitola III
The MITRE Corporation
Tampa, FL, USA

14.1 Introduction

Architecture is a comprehensive, consistent set of *design rules* by which a specified set of *components* achieves a specified set of *functions* in products and services that evolve through multiple design points over time [1]. This section introduces the fundamental design rules by which software-defined radio (SDR), sensors, perception, and automated machine learning (AML) may be integrated to create aware, adaptive, and cognitive radios (AACRs). These SDRs will have better quality of information (QoI) through capabilities to observe (sense, perceive), orient, plan, decide, act, and learn (the so-called OOPDAL loop) in radio frequency (RF) and in the user domains. By performing this integration, we will transition from merely adaptive to a demonstrably cognitive radio (CR).

This section develops five complementary perspectives of CR architecture (CRA), called CRA I through CRA V. The CRA I perspective defines six functional components, black boxes to which are ascribed a first-level decomposition of AACR functions and among which important interfaces are defined. One of these boxes is SDR, a proper subset of AACR. One of these boxes performs

---

Note: The MITRE Corporation is provided for identification purposes only and should not be interpreted as the endorsement of the material by the MITRE Corporation or any of its sponsors. Adapted from *Cognitive Radio Architecture: The Engineering Foundations of Radio XML*, Wiley, 2006.

cognition via the <Self/>,[1] a self-referential subsystem that strictly embodies finite computing (e.g., no While or Until loops) avoiding the Gödel-Turing paradox.[2]

The CRA II perspective examines the flow of inference through a cognition cycle that arranges the core capabilities of *ideal* CR (iCR) in temporal sequence for a logical flow and circadian rhythm for the CRA. The CRA III perspective examines the related levels of abstraction for AACR to sense elementary sensory stimuli and to perceive QoI relevant aspects of a <Scene/> consisting of the <User/> in an <Environment/> that includes <RF/>.[3] The CRA IV perspective examines the mathematical structure of this architecture, identifying mappings among topological spaces represented and manipulated to preserve set-theoretic properties. Finally, the CRA V perspective reviews SDR architecture, sketching an evolutionary path from the Software Communications Architecture/Software Radio Architecture (SCA/SRA) to the CRA. The CRA is expressed in Radio eXtensible Markup Language (RXML). The CRA is introduced in this chapter and developed in [12].

14.2 CRA I: Functions, Components, and Design Rules

The *functions* of AACR exceed those of SDR. Reformulating the AACR <Self/> as a *peer* of its own <User/> establishes the need for added functions by which the <Self/> accurately perceives the local scene including the <User/> and autonomously learns to tailor the information services to the specific <User/> in the current RF and physical <Scene/>.

14.2.1 AACR Functional Component Architecture

The SDR components and the related cognitive components of iCR appear in Figure 14.1. The cognition components describe the SDR in RXML so that the <Self/> can know that it is a radio and that its goal is to achieve high QoI tailored to its own users. RXML intelligence includes a priori radio background and user stereotypes as well as knowledge of RF and space–time <Scenes/> perceived and experienced. This includes both structured reasoning with iCR peers and cognitive

---

[1] Note that <Self/> is how the CR will refer to itself, so "<Self/>" can also be read as "the radio."
[2] This refers to the classic Turing machine, which can fall into unbounded analysis when attempting to reason about itself. One solution to this problem is a timer that kills any task that isn't completed within a specified time.
[3] The use of Semantic Web technology (e.g., referring to the <User/> in an <Environment/>) is addressed by Mahonen [2].

wireless networks (CWNs), and ad hoc reasoning with users, all the while learning from experience.

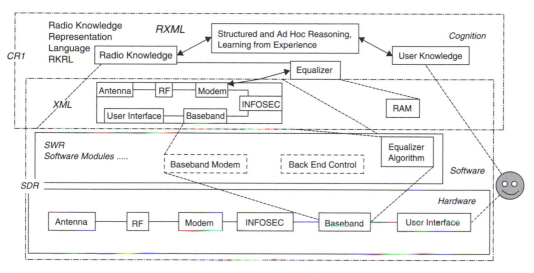

Figure 14.1: The CRA augments SDR with computational intelligence and learning capacity (© Dr. Joseph Mitola III, used with permission).

The detailed allocation of functions to components with interfaces among the components requires closer consideration of the SDR component as the foundation of CRA.

SDR Components

SDRs include a hardware platform with RF access and computational resources, plus at least one software-defined personality. The SDR Forum has defined its SCA [3] and the Object Management Group (OMG) has defined its SRA [4]. These are similar fine-grained architecture constructs enabling reduced-cost wireless connectivity with next-generation plug-and-play. These SDR architectures are defined in Unified Modeling Language (UML) object models [5], Common Object Request Broker Architecture (CORBA) Interface Design Language (IDL) [7], and XML descriptions of the UML models. The SDR Forum and OMG standards describe the technical details of SDR both for radio engineering and for an initial level of wireless air interface ("waveform") plug-and-play. The SCA/SRA was sketched in 1996 at the first US Department of Defense (DoD) inspired modular multifunctional information transfer system (MMITS) Forum, was developed by the DoD in the 1990s and the architecture is now in use by the US military [7]. This architecture emphasizes plug-and-play wireless personalities on computationally capable mobile nodes where network connectivity is often intermittent at best.

The commercial wireless community [8], in contrast, led by cell phone giants Motorola, Ericsson, and Nokia, envisions a much simpler architecture for mobile wireless devices, consisting of two application programming interfaces (APIs)—one for the service provider and another for the network operator. Those users define a knowledge plane in the future intelligent wireless networks that is not dissimilar from a distributed CWN. That community promotes the business model of the user → service provider → network operator → large manufacturer → device, in which the user buys mobile devices consistent with services from a service provider, and the technical emphasis is on *intelligence in the network*. This perspective no doubt will yield computationally intelligent networks in the near- to mid-term.

The CRA developed in this text, however, envisions the computational intelligence to create ad hoc and flexible networks with the *intelligence in the mobile device*. This technical perspective enables the business model of user → device → heterogeneous networks, typical of the Internet model in which the user buys a device (e.g., a wireless laptop) that can connect to the Internet via any available Internet service provider (ISP). The CRA builds on both the SCA/SRA and the commercial API model, but integrates Semantic Web intelligence in RXML for more of an Internet business model. This chapter describes how SDR, AACR, and iCR form a continuum facilitated by RXML.

AACR Node Functional Components

A simple CRA includes the functional components shown in Figure 14.2. A functional component is a black box to which functions have been allocated, but for which implementation is not specified. Thus, while the applications component is likely to be primarily software, the nature of those software components is yet to be determined. User interface functions, however, may include optimized hardware (e.g., for computing video flow vectors in real time to assist scene perception). At the level of abstraction of this figure, the components are functional, not physical.

These functional components are as follows:

1. The *user sensory perception* (SP), which includes haptic, acoustic, and video sensing and perception functions.

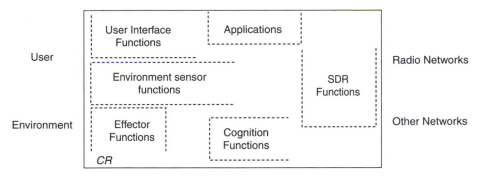

Figure 14.2: Minimal AACR node architecture (© Dr. Joseph Mitola III, used with permission).

2. The local *environment* sensors (location, temperature, accelerometer, compass, etc.).

3. The *system applications* (sys apps) media-independent services such as playing a network game.

4. The *SDR* functions which include RF sensing and SDR applications.

5. The *cognition* functions (symbol grounding for system control, planning, and learning).

6. The *local effector* functions (speech synthesis, text, graphics, and multimedia displays).

These functional components are embodied on an iCR platform, a hardware realization of the six functions. To support the capabilities described in the prior chapters, these components go beyond SDR in critical ways. First, the user interface goes well beyond buttons and displays. The traditional user interface has been partitioned into a substantial user sensory subsystem and a set of local effectors. The user sensory interface includes buttons (the haptic interface) and microphones (the audio interface) to include acoustic sensing that is directional, capable of handling multiple speakers simultaneously, and able to include full motion video with visual scene perception. In addition, the audio subsystem does not just encode audio for (possible) transmission; it also parses and interprets the audio from designated speakers, such as the <User/>, for a high-performance spoken natural language (NL) interface. Similarly, the text subsystem parses and interprets the language to track the user's information states, detecting plans and potential communications and information needs unobtrusively as the user conducts normal activities. The local effectors synthesize speech along with traditional text, graphics, and multimedia displays.

Sys apps are those *information services* that define value for the user. Historically, voice communications with a phone book, text messaging, and the exchange of images or video clips comprised the core value proposition of sys apps for SDR. These applications were generally integral to the SDR application, such as data services via general packet radio service (GPRS), which is really a wireless SDR personality more than an information service. AACR sys apps break the service out of the SDR waveform, so that the user need not be limited by details of wireless connectivity unless that is of particular interest. Should the user care whether he or she plays the distributed video game via 802.11 or Bluetooth over the last 3 m? Probably not. The typical user might care if the AACR wants to switch to third generation (3G) at $5 per minute, but a particularly affluent user might not care and would leave all that up to the AACR.

The cognition component provides all the cognition functions—from the semantic grounding of entities in the perception system to the control of the overall system through planning and initiating actions—learning user preferences and RF situations in the process.

Each of these subsystems contains its own processing, local memory, integral power conversion, built-in test (BIT), and related technical features.

The Ontological <Self/>

AACR consists of six functional components: user SP, environment, effectors, SDR, sys apps, and cognition. Those components of the <Self/> enable external communications and internal reasoning about the <Self/> by using the RXML syntax. Given the top-level outline of these functional components along with the requirement that they be embodied in physical hardware and software (the "platform"), the six functional components are defined ontologically in the equation in Figure 14.3. In part, this equation states that the hardware–software platform and

```
<iCR-Platform/>
<Functional-Components>
      <User SP/><Environment/><Effectors/><SDR/><Sys Apps/><Cognition/>
</Functional-Components>
</Self>
```

Figure 14.3: Components of the AACR <Self/>. The AACR <Self/> is defined to be an iCR platform, consisting of six functional components using the RXML syntax.

the functional components of the AACR are independent. Platform-independent computer languages such as Java are well understood. This ontological perspective envisions platform independence as an architecture design principle for AACR. In other words, the burden is on the (software) functional components to adapt to whatever RF–hardware–operating system platform might be available.

14.2.2 Design Rules Include Functional Component Interfaces

The six functional components (see Tables 14.1(a) and 14.1(b)) imply associated functional interfaces. In architecture, design rules may include a list of the quantities and types of components as well as the interfaces among those components. This section addresses the interfaces among the functional components.

The AACR *N*-squared diagram of Table 14.1(a) characterizes AACR interfaces. These constitute an initial set of AACR APIs. In some ways, these APIs augment the established SDR APIs. This is entirely new and much needed in order for basic AACRs to accommodate even the basic ideas of the Defense Advanced Research Projects Agency (DARPA) NeXt-Generation (XG) radio communications program.

In other ways, these APIs supersede the existing SDR APIs. In particular, the SDR user interface becomes the user sensory and effector API. User sensory APIs include acoustics, voice, and video, and the effector APIs include speech synthesis

Table 14.1(a): AACR *N*-squared diagram. This matrix characterizes internal interfaces between functional processes. Interface notes 1–36 are explained in Table 14.1(b).

From\to	User SP	Environment	Sys apps	SDR	Cognition	Effectors
User SP	1	7	13 PA[a]	19	25 PA[b]	31
Environment	2	8	14 SA[a]	20	26 PA[b]	32
Sys apps	3	9	15 SCM[a]	21 SD[a]	27 PDC[a,b]	33 PEM[a]
SDR	4	10	16 PD[a]	22 SD	28 PC[b]	34 SD
Cognition	5 PEC[b]	11 PEC[b]	17 PC[a,b]	23 PAE[b]	29 SC[b]	35 PE[b]
Effectors	6 SC	12	18[a]	24	30 PCD[b]	36

P: primary; A: afferent; E: efferent; C: control; M: multimedia; D: data; S: secondary; others not designated P or S are ancillary.
[a]Information services API; [b]CAPI.

Table 14.1(b): Explanations of interface notes for functional processes shown in Table 14.1(a).

Note number	Process interface	Explanation
1	User SP– User SP	Cross-media correlation interfaces (video-acoustic, haptic-speech, etc.) to limit search and reduce uncertainty (e.g., if video indicates user is not talking, acoustics may be ignored or processed less aggressively for command inputs than if user is speaking).
2	Environment– User SP	Environment sensors parameterize user sensor-perception. Temperature below freezing may limit video.
3	Sys apps– User SP	Sys apps may focus scene perception by identifying entities, range, expected sounds for video, audio, and spatial perception processing.
4	SDR–User SP	SDR applications may provide expectations of user input to the perception system to improve probability of detection and correct classification of perceived inputs.
5	Cognition– User SP	This is the *primary control efferent* path from cognition to the control of the user SP subsystem, controlling speech recognition, acoustic signal processing, video processing, and related SP. Plans from cognition may set expectations for user scene perception, improving perception.
6	Effectors– User SP	Effectors may supply a replica of the effect to user perception so that self-generated effects (e.g., synthesized speech) may be accurately attributed to the <Self/>, validated as having been expressed, and/or canceled from the scene perception to limit search.
7	User SP– Environment	Perception of rain, buildings, indoor/outdoor can set GPS integration parameters.
8	Environment– Environment	Environment sensors would consist of location sensing such as GPS or GLONASS; ambient temperature; light level to detect inside versus outside locations; possibly

		smell sensors to detect spoiled food, fire, etc. There seems to be little benefit in enabling interfaces among these elements directly.
9	Sys apps–Environment	Data from the sys apps to environment sensors would also be minimal.
10	SDR–Environment	Data from the SDR personalities to the environment sensors would be minimal.
11	Cognition–Environment (primary control path)	Data from the cognition system to the environment sensors controls those sensors, turning them on and off, setting control parameters, and establishing internal paths from the environment sensors.
12	Effectors–Environment	Data from effectors directly to environment sensors would be minimal.
13	UserSP–Sys apps	Data from the user SP system to sys apps is a *primary afferent path* for multimedia streams and entity states that effect information services implemented as sys apps. Speech, images, and video to be transmitted move along this path for delivery by the relevant sys apps or information service to the relevant wired or SDR communications path. Sys apps overcomes the limitations of individual paths by maintaining continuity of conversations, data integrity, and application coherence (e.g., for multimedia games). Whereas the cognition function sets up, tears down, and orchestrates the sys apps, the primary API between the user scene and the information service consists of this interface and its companions—the environment afferent path, the effector efferent path, and the SDR afferent and efferent paths.
14	Environment–Sys apps	Data on this path assists sys apps in providing location awareness to services.

(Continued)

Table 14.1(b): Explanations of interface notes for functional processes shown in Table 14.1(a). (*Continued*)

Note number	Process interface	Explanation
15	Sys apps– Sys apps	Different information services interoperate by passing control information through the cognition interfaces and by passing domain multimedia flows through this interface. The cognition system sets up and tears down these interfaces.
16	SDR–Sys apps	This is the primary afferent path from external communications to the AACR. It includes control and multimedia information flows for all the information services. Following the SDR Forum's SCA, this path embraces wired as well as wireless interfaces.
17	Cognition– Sys apps	Through this path, the AACR <Self/> exerts control over the information services provided to the <User/>.
18	Effectors– Sys apps	Effectors may provide incidental feedback to information services through this afferent path, but the use of this path is deprecated. Information services are supposed to control and obtain feedback through the mediation of the cognition subsystem.
19	User SP–SDR	Although the SP system may send data directly to the SDR subsystem (e.g., to satisfy security rules that user biometrics be provided directly to the wireless security subsystem), the use of this path is deprecated. Perception subsystem information is supposed to be interpreted by the cognition system so that accurate information, not raw data, can be conveyed to other subsystems.
20	Environment– SDR	Environment sensors such as GPS historically have accessed SDR waveforms directly (e.g., providing timing data for air interface signal generation). The cognition system may establish such paths in cases where cognition provides little or no value added, such as providing a precise timing reference from GPS to an SDR waveform. The use of this path is deprecated because all of the environment sensors,

		including GPS, are unreliable. Cognition has the capability to "de-glitch" GPS (e.g., recognize from video that the <Self/> is in an urban canyon and therefore not allow GPS to report directly, but report to the GPS subscribers, on behalf of GPS, location estimates based perhaps on landmark correlation, dead reckoning, etc.).
21	Sys apps–SDR	This is the primary efferent path from information services to SDR through the services API.
22	SDR–SDR	The linking of different wireless services directly to each other is deprecated. If an incoming voice service needs to be connected to an outgoing voice service, there should be a bridging service in sys apps through which the SDR waveforms communicate with each other. That service should be set up and taken down by the cognition system.
23	Cognition–SDR	This is the primary control interface, replacing the control interface of the SDR SCA and the OMG SRA.
24	Effectors–SDR	Effectors such as speech synthesis and displays should not need to provide state information directly to SDR waveforms, but if needed, the cognition function should set up and tear down these interfaces.
25	User SP–Cognition	This is the primary afferent flow for the results from acoustics, speech, images, video, video flow, and other sensor-perception subsystems. The primary results passed across this interface should be the specific states of <Entities/> in the scene, which would include scene characteristics such as the recognition of landmarks, known vehicles, furniture, and the like. In other words, this is the interface by which the presence of <Entities/> in the local scene is established and their characteristics are made known to the cognition system.
26	Environment–Cognition	This is the primary afferent flow for environment sensors.

(Continued)

Table 14.1(b): Explanations of interface notes for functional processes shown in Table 14.1(a). (*Continued*)

Note number	Process interface	Explanation
27	Sys apps–Cognition	This is the interface through which information services request services and receive support from the AACR platform. This is also the control interface by which cognition sets up, monitors, and tears down information services.
28	SDR–Cognition	This is the primary afferent interface by which the state of waveforms, including a distinguished RF-sensor waveform, is made known to the cognition system. The cognition system can establish primary and backup waveforms for information services, enabling the services to select paths in real time for low-latency services. Those paths are set up and monitored for quality and validity (e.g., obeying XG rules) by the cognition system, however.
29	Cognition–Cognition	The cognition system as defined in this six-component architecture entails (1) orienting to information from <RF/> sensors in the SDR subsystem and from scene sensors in the user SP and environment sensors; (2) planning; (3) making decisions; and (4) initiating actions, including the control over all of the cognition resources of the <Self/>. The <User/> may directly control any of the elements of the systems via paths through the cognition system that enable it to monitor what the user is doing in order to learn from a user's direct actions, such as manually tuning in the user's favorite radio station when the <Self/> either failed to do so properly or was not asked.
30	Effectors–Cognition	This is the primary afferent flow for status information from the effector subsystem, including speech synthesis, displays, and the like.
31	User SP–Effectors	In general, the user SP system should not interface directly to the effectors, but should be routed through the cognition system for observation.

446

32	Environment–Effectors	The environment system should not interface directly to the effectors. This path is deprecated.
33	Sys apps–Effectors	sys apps may display streams, generate speech, and otherwise directly control any effectors once the paths and constraints have been established by the cognition subsystem.
34	SDR–Effectors	This path may be used if the cognition system establishes a path, such as from an SDR's voice track to a speaker. Generally, however, the SDR should provide streams to the information services of the sys apps. This path may be necessary for legacy compatibility during the migration from SDR through AACR to iCR, but it is deprecated.
35	Cognition–Effectors	This is the primary efferent path for the control of effectors. Information services provide the streams to the effectors, but cognition sets them up, establishes paths, and monitors the information flows for support to the user's <Need/> or intent.
36	Effectors–Effectors	These paths are deprecated, but may be needed for legacy compatibility.

to give the AACR <Self/> its own voice. In addition, wireless applications are growing rapidly. Voice and short-message service provide an ability to exchange images and video clips with ontological tags among wireless users. The distinctions between cell phone, personal digital assistant (PDA), and game box continue to disappear.

These interface changes enable the AACR to sense the situation represented in the environment, to interact with the user, and to access radio networks on behalf of the user in a situation-aware way.

The above information flows aggregated into an initial set of AACR APIs define an information services API (ISAPI) by which an information service accesses the other five components (interfaces 13–18, 21, 27, and 33 in Table 14.1(a)). They would also define a CAPI by which the cognition system obtains status and exerts control over the rest of the system (interfaces 25–30, 5, 11, 17, 23, 29, and 35 in Table 14.1(a)). Although the constituent interfaces of these APIs are suggested in this table, it would be premature to define these APIs without first developing detailed information flows and interdependencies. We will define and analyze these APIs in this chapter. It would also be premature to develop such APIs without a clear idea of the kinds of RF and user domain knowledge and performance expected of the AACR architecture over time. These aspects are developed in the balance of this chapter, enabling one to draw some conclusions about these APIs in the final part of this chapter.

A fully defined set of interfaces and APIs would be circumscribed in RXML.

14.2.3 Near-Term Implementations

One way to implement this set of functions is to embed into an SDR a reasoning engine, such as a rule base with an associated inference engine, as the cognition function. If the effector functions control parts of the radio, then we have the simplest AACR based on the simple six-component architecture of Figure 14.2. Such an approach may be sufficient to expand the control paradigm from today's state machines with limited flexibility to tomorrow's AACR control based on reasoning over more complex RF states and user situations. Such simple approaches may well be the next practical steps in AACR evolution from SDR toward iCR.

This incremental step doesn't suggest how to mediate the interfaces between multisensory perception, situation-sensitive prior experience, and a priori knowledge to achieve situation-dependent radio control. Such radio control enables more sophisticated information services. A simple architecture does not proactively allocate machine-learning (ML) functions to fully understood components. For example, will AML require an embedded radio propagation modeling tool? If so, then what is the division of function between a rule base that knows about radio propagation and a propagation tool that can predict values such as the received signal-strength indicator (RSSI)? Similarly, in the user domain, some aspects of user behavior, such as movement by foot and in vehicles, may be modeled in detail based on physics. Will movement modeling be a separate subsystem based on physics and global positioning system (GPS)? How will that work inside of buildings? How is the knowledge and skill in tracking user movements divided between physics-based computational modeling and the symbolic inference of a rule base or set of Horn clauses[4] [34] with a PROLOG engine? For that matter, how will the learning architecture accommodate a variety of learning methods such as neural networks, PROLOG, forward chaining, or support vector machines (SVMs) if learning occurs entirely in a cognition subsystem?

Although hiding such details may be a good thing for AACR in the near term, it may severely limit the mass customization needed for AACRs to learn user patterns and thus to deliver RF services dramatically better than mere SDRs. Thus, we need to go further "inside" the cognition and perception subsystems to establish more of a fine-grained architecture. This enables one to structure the data sets and functions that mediate multisensory domain perception of complex scenes and related learning technologies that can autonomously adapt to user needs and preferences. The sequel thus proactively addresses the embedding of ML technology into the radio architecture.

Next, consider the networks. Network-independent SDRs retain multiple personalities in local storage, whereas network-dependent SDRs receive alternate personalities from a supporting network infrastructure—CWNs. High-end SDRs both retain alternate personalities locally and have the ability to validate and accept personalities by download from trusted sources. Whatever architecture emerges must be consistent with the distribution of RXML knowledge aggregated in a variety of networks from a tightly coupled CWN to the Internet, with a degree of <Authority/> and trust reflecting the pragmatics of such different repositories.

---

[4] A Horn clause is a Boolean expression in which no more than one of the Boolean variables is positive (not negated). Horn clauses are used in artificial intelligence systems to prove theorems [9].

Thus, the stage is set for the development of CRA. The following sections address the cognition cycle, the inference hierarchies, and the SDR architecture embedded into the CRA.

14.2.4 The Cognition Components

Figure 14.1 shows three computational intelligence aspects of CR:

1. Radio knowledge—RXML:RF
2. User knowledge—RXML:User
3. The capacity to learn

The minimalist architecture of Figure 14.2 and the functional interfaces of Tables 14.1(a) and 14.1(b) do not assist the radio engineer in structuring knowledge, nor do they assist much in integrating ML into the system. Rather, the fine-grained architecture developed in this chapter is derived from the functional requirements to fully develop these three core capabilities.

Radio Knowledge in the Architecture

Radio knowledge has to be translated from the classroom and engineering teams into a body of computationally accessible, structured technical knowledge about radio. RXML is the primary enabler and product of this foray into formalization of radio knowledge. This text starts a process of RXML definition and development that can be brought to fruition only by industry over time. This process is similar to the evolution of the SCA of the SDR Forum [3]. The SCA structures the technical knowledge of the radio components into UML and XML. RXML will enable the structuring of sufficient RF and user world knowledge to build advanced wireless-enabled or enhanced information services. Thus, whereas the SRA and SCA focus on building radios, RXML focuses on using radios.

The World Wide Web (WWW) is now sprouting with computational ontologies some of which are nontechnical but include radio, such as the open Cyc[5]

---

[5] Cyc is an artificial intelligence (AI) project that attempts to assemble an encyclopedic comprehensive ontology and database of everyday common sense knowledge, with the goal of enabling AI applications to perform human-like reasoning [9].

ontology. They bring the radio domain into the Semantic Web, which helps people know about radio. This informal knowledge lacks the technical scope, precision, and accuracy of authoritative radio references such as the European Telecommunications Standards Institute (ETSI) documents defining the Global System for Mobile Communications (GSM) and the International Telecommunication Union (ITU) definitions of, for example 3GPP.[6]

Not only must radio knowledge be precise, it must be stated at a useful level of abstraction, yet with the level of detail appropriate to the use case. Thus, ETSI GSM in most cases would over-kill the level of detail without providing sufficient knowledge of the user-centric functionality of GSM. In addition, AACR is multi-band, multimode radio (MBMMR), so the knowledge must be comprehensive, addressing the majority of radio bands and modes available to an MBMMR. This knowledge is formalized with precision that should be acceptable to ETSI, the ITU, and regulatory authorities (RAs), yet also be at a level of abstraction appropriate to internal reasoning, formal dialog with a CWN, or informal dialog with users.

The capabilities required for an AACR node to be a cognitive entity are to sense, perceive, orient, plan, decide, act, and learn. To relate ITU standards to these required capabilities is a process of extracting content from highly formalized knowledge bases that exist in a unique place and that bear substantial authority, encapsulating that knowledge in less complete and therefore somewhat approximate form that can be reasoned with on the AACR node and in real time to support RF-related use cases. Table 14.2 illustrates this process.

Table 14.2 is illustrative and not comprehensive, but it characterizes the technical issues that drive an information-oriented AACR node architecture. Where ITU, ETSI, … (meaning other regional and local standards bodies), and CWN supply source knowledge, the CWN is the repository for authoritative knowledge derived from the standards bodies and RAs, the <Authorities/>. A user-oriented AACR may note differences in the interpretation of source knowledge from <Authorities/> between alternate CWNs, precipitating further knowledge exchanges.

---

[6]3GPP is a collaboration agreement for the 3G portable phone among ETSI (Europe), the Association of Radio Industries and Businesses/Telecommunication Technology Committee (ARIB/TTC; Japan), the China Communications Standards Association (CCSA; China), the Alliance for Telecommunications Industry Solutions (ATIS; North America), and the Telecommunications Technology Association (TTA; South Korea).

Table 14.2: Radio knowledge in the node architecture.

Need	*Source knowledge*	*AACR internalization*
Sense RF	RF platform	Calibration of RF, noise floor, antennas, direction
Perceive RF	ITU, ETSI, ARIB, RAs	Location-based table of radio spectrum allocation
Observe RF (sense and perceive)	Unknown RF	RF sensor measurements and knowledge of basic types (AM, FM, simple digital channel symbols, typical TDMA, FDMA, CDMA signal structures)
Orient	XG-like policy	Receive, parse, and interpret policy language
	Known waveform	Measure parameters in RF, space, and time
Plan	Known waveform	Enable SDR for which licensing is current
	Restrictive policy	Optimize transmitted waveform, space–time plan
Decide	Legacy waveform, policy	Defer spectrum use to legacy users per policy
Act	Applications layer	Query for available services (white/yellow pages)
	ITU, ETSI, … CWN	Obtain new skills encapsulated as download
Learn	Unknown RF	Remember space–time–RF signatures; discover spectrum use norms and exceptions
	ITU, ETSI, … CWN	Extract relevant aspects such as new feature

User Knowledge in the Architecture

Next, user knowledge is formalized at the level of abstraction and degree of detail necessary to give the CR the ability to acquire, from its owner and other designated users, the user knowledge relevant to information services incrementally. Incremental knowledge acquisition was motivated in the introduction to ML by describing how frequent occurrences with similar activity sequences identifies learning opportunities. ML machines may recognize these opportunities for learning through joint probability statistics <Histogram/>. Effective use cases clearly identify the classes of user and the specific knowledge learned to customize envisioned services. Use cases may also supply sufficient initial knowledge to render incremental ML not only effective, but also—if possible—enjoyable to the user.

This knowledge is defined in RXML:User. As with RF knowledge, the capabilities required for an AACR node to be a cognitive entity are to OOPDAL. To relate a use case to these capabilities, one extracts specific and easily recognizable <Anchors/> for stereotypical situations observable in diverse times, places, and situations. One expresses the anchor knowledge in RXML for use on the AACR node.

Cross-domain Grounding for Flexible Information Services

The knowledge about radio and about user needs for wireless services must be expressed internally in a consistent form so that information services relationships may be autonomously discovered and maintained by the <Self/> on behalf of the <User/>. Figure 14.4 shows relationships among user and RF domains.

Staying better connected requires the normalization of knowledge between <User/> and <RF/> domains. If, for example, the <User/> says, "What's on one

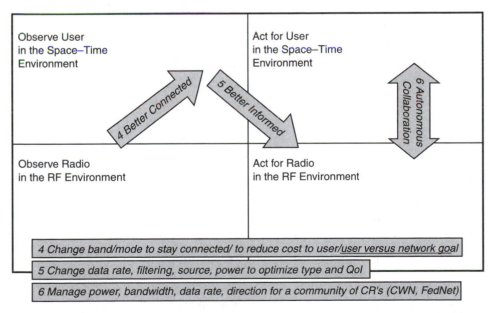

Figure 14.4: Discovering and maintaining services (© Dr. Joseph Mitola III, used with permission).

oh seven, seven," near the Washington, DC, area, then the dynamic <User/> ontology should enable the CR to infer that the user is talking about the current frequency modulation (FM) radio broadcast, the units are in MHz, and the user wants to know what is on WTOP. If it can't infer this, then it should ask the user or discover by first dialing a reasonable default, such as 107.7 FM, a broadcast radio station, and asking, "Is this the radio station you want?" Steps 4, 5, and 6 in Figure 14.4 all benefit from agreement across domains on how to refer to radio services. Optimizing behavior to best support the user requires continually adapting the <User/> ontology with repeated regrounding of terms in the <User/> domain to conceptual primitives and actions in the <RF/> domain.

The CRA facilitates this by seeding the speech recognition subsystem with the most likely expressions a particular <User/> employs when referring to information services. These would be acquired from the specific users via text and speech recognition, with dialogs oriented toward continual grounding by posing "yes/no" questions, either verbally or in displays or both, and obtaining reinforcement either verbally or via haptic interaction or both. The required degree of mutual grounding would benefit from specific grounding-oriented features in the AACR information architecture [12].

The process of linking user expressions of interest to the appropriate radio technical operations sometimes may be extremely difficult. Military radios, for example, have many technical parameters: a "channel" in SINCGARS[7] consists of dehopped digital voice in one context (voice communications) or a 25 kHz band of spectrum in another context (and that may be either an FM channel into which its frequency hop waveform has hopped or a frequency division multiple access (FDMA) channel when in single-channel mode). If the user says, "Give me the commander's channel," the SINCGARS user is talking about a "dehopped CVSD voice stream."[8] If the same user a few seconds later says, "This sounds awful. Who else is in this channel?" the user is referring to interference with a collection of hop sets. If the CR observes, "There is strong interference in almost half of your assigned channels," then the CR is referring to a related set of 25 kHz channels. If the user then says, "OK, notch the strongest three interference channels," the user is talking about a different subset of the channels. If in the next breath the user says, "Is anything on our emergency channel?" then the user has switched from SINCGARS context to <Self/> context, asking about one of the cognitive

---

[7] SINCGARS is the Single-Channel Ground and Airborne Radio System used by the US DoD (see Chapter 4).

[8] CVSD stands for continuously variable slope delta modulation, a voice coding method.

military radio's physical RF access channels. The complexity of such exchanges demands cross-domain grounding; and the necessity of communicating accurately but quickly under stress motivates a structured NL and rich radio ontology aspects of the architecture, developed further in this section.

Thus, both commercial and military information services entail cross-domain grounding with ontology oriented to NL in the <User/> domain and oriented to RXML formalized a priori knowledge in the <RF/> domain. Specific methods of cross-domain grounding with associated architectural features include:

1. *<RF/> to <User/> shaping dialog* to express precise <RF/> concepts to non-expert users in an intuitive way, such as:
   (a) Grounding: "If you move the speaker box a little bit, it can make a big difference in how well the remote speaker is connected to the wireless transmitter on the television (TV)."
   (b) AACR information architecture: Include facility for *rich set of synonyms* to mediate cognition–NL–synthesis interface (<Antenna> ≅ <Wireless-remote-speaker> ≅ "Speaker box").

2. *<RF/> to <User/> learning jargon* to express <RF> connectivity opportunities in <User/> terms:
   (a) Grounding: "Tee oh pee" for "WTOP," "Hot ninety-two" for "FM 92.3."
   (b) AACR information architecture: NL-visual facility for single-instance update of user jargon.

3. *<User/> to <RF/> relating values to actions*: Relate <User> expression of values ("low-cost") to features of situations ("normal") that are computable (<NOT> (<CONTAINS> <Situation> <Unusual/></..>)) and that relate directly to <RF> domain decisions:
   (a) Grounding: Normally wait for free wireless local area network (WLAN) for big attachment; if situation is <Unusual/>, ask if user wants to pay for 3G.
   (b) AACR information architecture: Associative inference hierarchy that relates observable features of a <Scene/> to user sensitivities, such as <Late-for-work/> => <Unusual/>; "The President of the company needs this" => <Unusual/> because "President" => <VIP/> and <VIP/> is not in most scenes.

14.2.5 Self-referential Components

The cognition component must to assess, manage, and control all of its own resources, including validating downloads. Thus, in addition to <RF> and <User>

domains, RXML must describe the <Self/>, defining the AACR architecture to the AACR itself in RXML.

Self-referential Inconsistency

This class of self-referential reasoning is well known in the theory of computing to be a potential black hole for computational resources. Specifically, any Turing-capable (TC) computational entity that reasons about itself can encounter unexpected Gödel-Turing situations from which it cannot recover. Thus, TC systems are known to be "partial"—only partially defined because the result obtained when attempting to execute certain classes of procedure are not definable (the computing procedure will never terminate).

To avoid this paradox, CR architecture mandates the use of only "total" functions, typically restricted to bounded minimalization [10]. Watchdog "step-counting" functions [11] or timers must be in place in all its self-referential reasoning and radio functions. The timer and related computationally indivisible control construct are equivalent to the computer-theoretic construct of a step-counting function over "finite minimalization." It has been proven that computations that are limited with certain classes of reliable watchdog timers on finite computing resources can avoid the Gödel-Turing paradox or at least reduce it to the reliability of the timer. This proof is the fundamental theorem for practical self-modifying systems.

In brief, if a system can compute in advance the amount of time or the number of instructions that any given computation should take, then if that time or step-count is exceeded, the procedure returns a fixed result such as "Unreachable in Time T." As long as the algorithm does not explicitly or implicitly restart itself on the same problem, then with the associated invocation of a tightly time- and computationally-constrained alternative tantamount to giving up, it

(a) is not TC, but

(b) is sufficiently computationally capable to perform real-time communications tasks such as transmitting and receiving data as well as bounded user interface functions, and

(c) is not susceptible to the Gödel-Turing incompleteness dilemma, and thus

(d) will not crash because of consuming unbounded or unpredictable resources in unpredictable self-referential loops.

This is not a general result. This is a highly radio domain-specific result that has been established only for isochronous communications domains in which

(a) processes are defined in terms of a priori tightly bounded time epochs such as code division multiple access (CDMA) frames and Signaling System 7 (SS7) time-outs;[9] and

(b) for every situation, there is a default action that has been identified in advance that consumes $O(1)$ resources; and

(c) the watchdog timer or step-counting function is reliable.

Because radio air interfaces transmit and receive data, there are always defaults such as "repeat the last packed" or "clear the buffer" that may degrade the performance of the overall communications system. A default has $O(1)$ complexity and the layers of the protocol stack can implement the default without using unbounded computing resources.

Watchdog Timer

Without the reliable watchdog timer in the architecture and without this proof to establish the rules for acceptable computing constructs on CRs, engineers and computer programmers would build CRs that would crash in extremely unpredictable ways as their adaptation algorithms got trapped in unpredictable unbounded self-referential loops. Because planning problems exist that cannot be solved with algorithms so constrained, either an unbounded community of CRs must cooperatively work on the more general problems or the cognitive network (CN) must employ a TC algorithm to solve the more difficult problems (e.g., *NP*-hard with large *N*) offline. There is also the interesting possibility of trading off space and time by remembering partial solutions and restarting *NP*-hard problems with these subproblems already solved. Although it doesn't actually avoid any necessary calculations, with $O(N)$ pattern matching for solved subproblems, it may reduce the total computational burden, somewhat.[10] This class of approach to parallel problem-solving is similar to the use of pheromones by ants to solve the traveling

---

[9] SS7 is the seventh and most stable software update of the long-distance telephony software. Time-out here means that the software expected tasks to complete by a certain time-out; if not, it determined that something had gone awry, and either tried a different route or delivered a tone indicating that a system fault had occurred.

[10] For example, the fast Fourier transform (FFT) converts $O(N^2)$ steps to $O(N \log N)$ by avoiding the recomputation of already computed partial products.

salesman problem in less than $(2^N)/M$ time with M ants. But this is an engineering text, not a text on the theory of computing, so these aspects are not developed further here, but it suffices to show the predictable finiteness and proof that the approach is boundable and hence compatible with the real-time performance needs of CR.

This timer-based finite computing regime also works for user interfaces because users will not wait forever before changing the situation (e.g., by shutting off the radio or hitting another key); and the CR can always kind of throw up its hands and ask the user to take over.

Thus, with a proof of stability based on the theory of computing, the CRA structures systems that not only can modify themselves, but also can do it in such a way that they are not likely to induce nonrecoverable crashes from the "partial" property of self-referential computing.

14.2.6 Flexible Functions of the Component Architecture

Although this chapter develops the six-element component architecture of one particular information architecture and one reference implementation, many possible architectures exist. The purpose is not to try to sell a particular architecture, but to illustrate the architecture principles. The CRA and research implementation CR1[11] therefore offer open source licensing for noncommercial educational purposes. Table 14.3 further differentiates architectural features.

These functions of the architecture shown in Table 14.3 are not different from those of the six-component architecture, but represent varying degrees of instantiation of the six components. Consider the following degrees of architecture instantiations:

- *Cognition functions of radio* entail the monitoring and structuring knowledge of the behavior patterns of the <Self/>, the <User>, and the environment (physical, user situation, and radio) to provide information services, learning from experience to tailor services to user preferences and differing radio environments.

- *Adaptation functions of radio* respond to a changing environment, but can be achieved without learning if the adaptation is preprogrammed.

- *Awareness functions of radio* extract usable information from a sensor domain. Awareness stops short of perception. Awareness is required for adaptation, but awareness does not guarantee adaptation. For example, embedding a GPS

---

[11]CR1 is a CR research prototype developed by Joseph Mitola III [12].

Table 14.3: Features of AACR to be organized via architecture.

Feature	Function	Examples (RF; vision; speech; location; motion)
Cognition	Monitor and learn	Get to know user's daily patterns and model the local RF scene over space, time, and situations
Adaptation	Respond to changing environment	Use unused RF, protect owner's data
Awareness	Extract information from sensor domain	Sense or perceive
Perception	Continuously identify knowns, unknowns, and backgrounds in the sensor domain	TV channel; depth of visual scene, identity of objects; location of user, movement and speed of <Self/>
Sensing	Continuously sense and preprocess single-sensor field in single-sensory domain	RF FFT; binary vision; binaural acoustics; GPS; accelerometer; etc.

receiver into a cell phone makes the phone more location aware, but unless the value of the current location is actually used by the phone to do something that is location dependent, the phone is not location adaptive, only location aware. These functions are a subset of the CRA that enable adaptation.

- *Perception functions of radio* continuously identify and track knowns, unknowns, and backgrounds in a given sensor domain. Backgrounds are subsets of a sensory domain that share common features that entail no particular relevance to the functions of the radio. For a CR that learns initially to be a single-owner radio, in a crowd, the owner is the object that the radio continuously tracks in order to interact when needed. Worn from a belt as a cognitive wireless PDA (CWPDA), the iCR perception functions may track the entities in the scene. The nonowner entities comprise mostly irrelevant background because no matter what interactions may be offered by these entities, the CR will not obey them—only the interactions of the perceived owner. These functions are a subset of the CRA that enable cognition.

- *The sensory functions of radio* entail those hardware and/or software capabilities that enable a radio to measure features of a sensory domain. Sensory domains include anything that can be sensed, such as audio, video, vibration, temperature, time, power, fuel level, ambient light level, sun angle (e.g., through polarization), barometric pressure, smell, and anything else imaginable. Sensory domains for vehicular radios may be much richer, if less personal, than those of wearable radios. Sensory domains for fixed infrastructure could include weather features such as ultraviolet sunlight, wind direction and speed, humidity, traffic flow rate, or rain rate. These functions are a subset of the CRA that enable perception.

The platform-independent model (PIM) in the UML of SDR [13] provides a convenient, industry-standard computational model that an AACR can use to describe the SDR and computational-resource aspects of its own internal structure, as well as facilities that enable radio functions. The general structure of hardware and software by which a CR reasons about the <Self/> in its world is also part of its architecture defined in the SDR SCA/SRA as resources.

14.3 CRA II: The Cognition Cycle

The CRA comprises a set of design rules by which the cognitive level of information services may be achieved by a specified set of components in a way that supports the cost-effective evolution of increasingly capable implementations over time [1]. The cognition subsystem of the architecture includes an inference hierarchy and the temporal organization and flow of inferences and control states—the cognition cycle.

14.3.1 The Cognition Cycle

The cognition cycle developed for CR1 [14] is illustrated in Figure 14.5. This cycle implements the capabilities required of iCR in a reactive sequence. Stimuli enter the CR as sensory interrupts, dispatched to the cognition cycle for a response. Such an iCR continually observes (senses and perceives) the environment, orients itself, creates plans, decides, and then acts. In a single-processor inference system, the CR's flow of control may also move in the cycle from observation to action. In a multiprocessor system, temporal structures of sensing, preprocessing, reasoning, and acting may be parallel and complex. Special features synchronize the inferences of each phase. The tutorial code all works on a single processor in a rigid inference sequence defined in Figure 14.5. This process is called the "wake epoch" because the primary reasoning activities during this large epoch of time are reactive to the environment. We will refer to "sleep epochs" for power-down conditions, "dream epochs" for performing computationally intensive pattern recognition and learning, and "prayer epochs" for interacting with a

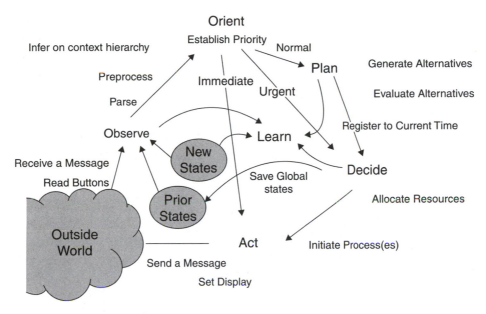

Figure 14.5: Simplified cognition cycle. The observe, orient, plan, decide, act (OOPDA) loop is a primary cycle; learning, planning, and sensing the outside world are crucial phases of the larger OOPDA-loop (© Dr. Joseph Mitola III, used with permission).

higher authority such as network infrastructure. See Section 14.5.4 for further discussion of behavioral epochs.

During the wake epoch, the receipt of a new stimulus on any of a CR's sensors or the completion of a prior cognition cycle initiates a new primary cognition cycle. The CR observes its environment by parsing incoming information streams. These can include monitoring and speech-to-text conversion of radio broadcasts (e.g., the Weather Channel, stock ticker tapes, etc.). Any RF-LAN or other short-range wireless broadcasts that provide services awareness information may be also parsed. In the observation phase, a CR also reads location, temperature, and light-level sensors, among other parameters, to infer the user's communications context.

14.3.2 Observe (Sense and Perceive)

The iCR senses and perceives the environment (via "observation phase" code) by accepting multiple stimuli in many dimensions simultaneously and by binding these stimuli—all together or more typically in subsets—to prior experience so that it can subsequently detect time-sensitive stimuli and ultimately generate plans for action.

Thus, iCR continuously aggregates experience and compares prior aggregates to the current situation. A CR may aggregate experience by remembering everything. This may not seem like a very smart thing to do until you calculate that all the audio, unique images, and e-mails the radio might experience in a year takes up only a few hundred gigabytes of memory, depending on image detail. So the computational architecture for remembering and rapidly correlating current experience against everything known previously is a core capability of the CRA. A *novelty* detector identifies new stimuli, using the new aspects of partially familiar stimuli to identify incremental-learning primitives.

In the six-component (user SP, environment, effectors, SDR, sys apps, and cognition) functional view of the architecture defined in section *AACR Node Functional Components*, the observe phase comprises both the user SP and the environment (RF and physical) sensor subsystems. The subsequent orient phase is part of the cognition component in this model of architecture.

14.3.3 Orient

The orient phase determines the significance of an observation by binding the observation to a previously known set of stimuli of a "scene." The orient phase contains the internal data structures that constitute the equivalent of the short-term memory (STM) that people use to engage in a dialog without necessarily remembering everything with the same degree of long-term memory (LTM). Typically people need repetition to retain information over the long term. The natural environment supplies the information redundancy needed to instigate transfer from STM to LTM. In the CRA, the transfer from STM to LTM is mediated by the sleep cycle in which the contents of STM since the last sleep cycle are analyzed both internally and with respect to existing LTM. How to do this robustly remains an important CR research topic, but the overall framework is defined in Section 14.5.4.

Matching of current stimuli to stored experience may be achieved by "stimulus recognition" or by "binding." The orient phase is the first collection of activity in the cognition component.

Stimulus Recognition

Stimulus recognition occurs when there is an exact match between a current stimulus and a prior experience. The CR1 prototype is continually recognizing exact matches and recording the number of exact matches that occurred along with the time measured in the number of cognition cycles between the last exact match. By default, the response to a given stimulus is to merely repeat that stimulus to the next layer up the inference hierarchy for aggregation of the raw stimuli. But if the system has been trained to respond to a location, a word, an RF condition, a signal on the power bus, or some other parameter, it may either react immediately or plan a task in reaction to the detected stimulus. If that reaction were in error, then it may be trained to ignore the stimulus, given the larger context, which consists of all the stimuli and relevant internal states, including time.

Sometimes, the orient phase causes an action to be initiated immediately as a "reactive" stimulus–response behavior. A power failure, for example, might directly invoke an act that saves the data (the "immediate" path to the act phase in Figure 14.5). A nonrecoverable loss of signal on a network might invoke reallocation of resources (e.g., from parsing input to searching for alternative RF channels. This may be accomplished via the path labeled "urgent" in Figure 14.5.

Binding

Binding occurs when there is a nearly exact match between a current stimulus and a prior experience and very general criteria for applying the prior experience to the current situation are met. One such criterion is the number of unmatched features of the current scene. If only one feature is unmatched and the scene occurs at a high level such as the phrase or dialog level of the inference hierarchy, then binding is the first step in generating a plan for behaving in the given state similar to the last occurrence of the stimuli. In addition to number of features that match exactly, which is a kind of hamming code, instance-based learning (IBL) supports inexact matching and binding. Binding also determines the priority associated with the stimuli. Better binding yields higher priority for autonomous learning, whereas less-effective binding yields lower priority for the incipient plan.

14.3.4 Plan

Most stimuli are dealt with "deliberatively" rather than "reactively." An incoming network message would normally be dealt with by generating a plan (in the plan phase, the "normal" path). Such planning includes plan generation. In research-quality or industrial-strength CRs, formal models of causality must be embedded into planning tools. The plan phase should also include reasoning about time. Typically, reactive responses are preprogrammed or defined by a network (i.e., the CR is "told" what to do), whereas other behaviors might be planned. A stimulus may be associated with a simple plan as a function of planning parameters with a simple planning system. Open source planning tools enable the embedding of planning subsystems into the CRA, enhancing the plan component. Such tools enable the synthesis of RF and information access behaviors in a goal-oriented way based on perceptions from the visual, audio, text, and RF domains as well as RA rules and previously learned user preferences.

14.3.5 Decide

The decide phase selects among the candidate plans. The radio might have the choice to alert the user to an incoming message (e.g., behave like a pager) or to defer the interruption until later (e.g., behave like a secretary who is screening calls during an important meeting).

14.3.6 Act

Acting initiates the selected processes using effector modules. Effectors may access the external world or the CR's internal states.

Externally Oriented Actions

Access to the external world consists primarily of composing messages to be spoken into the local environment or expressed in text form locally or to another CR or CN using the Knowledge Query and Manipulation Language (KQML), Radio Knowledge Representation Language (RKRL), Web Ontology Language (OWL), Radio eXtensible Markup Language (RXML), or some other appropriate knowledge interchange standard.

Internally Oriented Actions

Actions on internal states include controlling machine-controllable resources such as radio channels. The CR can also affect the contents of existing internal models, such as adding a model of stimulus–experience–response (serModel) to an existing internal model structure [12]. The new concept itself may assert-related concepts into the scene. Multiple independent sources of the same concept in a scene reinforce that concept for that scene. These models may be asserted by the <Self/> to encapsulate experience. The experience may be reactively integrated into RXML knowledge structures as well, provided the reactive response encodes them properly.

14.3.7 Learning

Learning is a function of perception, observations, decisions, and actions. Initial learning is mediated by the observe phase perception hierarchy in which all SP are continuously matched against all prior stimuli to continually count occurrences and to remember time since the last occurrence of the stimuli from primitives to aggregates.

Learning also occurs through the introduction of new internal models in response to existing models and case-based reasoning (CBR) bindings. In general, there are many opportunities to integrate ML into AACR. Each of the phases of the cognition cycle offers multiple opportunities for discovery processes, such as <Histogram/>, as well as many other ML approaches. The architecture includes internal reinforcement via counting occurrences and via serModels, so ML with uncertainty is also supported [12, 31].

Finally, a learning mechanism occurs when a new type of serModel is created in response to an action to instantiate an internally generated serModel. For example, prior and current internal states may be compared with expectations to learn about the effectiveness of a communications mode, instantiating a new mode-specific serModel.

14.3.8 Self-monitoring

Each of the prior phases must consist of computational structures for which the execution time may be computed in advance. In addition, each phase must restrict its computations to not consume more resources (time $\times$ allocated processing capacity) than the precomputed upper bound. Therefore, the architecture has some prohibitions and some data set requirements needed to obtain an acceptable degree of stability of behavior for CRs as self-referential self-modifying systems.

Since first-order predicate calculus (FOPC) used in some reasoning systems is not decidable, one cannot in general compute in advance how much time an FOPC expression will take to run to completion. There may be loops that will preclude this, and even with loop detection, the time to resolve an expression may be only loosely approximated as an exponential function of some parameters (such as the number of statements in the FOPC database of assertions and rules). Therefore, unrestricted FOPC is not allowed.

Similarly, unrestricted For, Until, and While loops are prohibited. In place of such loops are bounded iterations in which the time required for the loop to execute is computed or supplied independent of the computations that determine the iteration control of the loop. This seemingly unnatural act can be facilitated by next-generation compilers and computer-aided software engineering (CASE) tools. Because self-referential self-modifying code is prohibited by structured design and programming practices, no such tools are available on the market today. But CR is inherently self-referential and self-modifying, such tools most likely will emerge, perhaps assisted by the needs of CR and the architecture framework of the cognition cycle.

Finally, the cognition cycle itself cannot contain internal loops. Each iteration of the cycle must take a defined amount of time, just as each frame of a 3G air interface takes 10 milliseconds. As CR computational platforms continue to progress, the amount of computational work done within the cycle will increase, but under no conditions should explicit or implicit loops be introduced into the cognition cycle that would extend it beyond a given cycle time.

Retrospection

The assimilation of knowledge by ML can be computationally intensive, so, as previously stated in Section 14.3.1 and further discussed in Section 14.5.4, CR has sleep and prayer epochs that support ML. A sleep epoch is a relatively long period of time (e.g., minutes to hours) during which the radio will not be in use, but has sufficient electrical power for processing. During the sleep epoch, the radio can run ML algorithms without detracting from its ability to support its user's needs. ML algorithms may integrate experience by aggregating statistical parameters. The sleep epoch may re-run stimulus–response sequences with new learning parameters in the way that people dream. The sleep cycle could be less anthropomorphic, however, employing a genetic algorithm to explore a rugged fitness landscape, potentially improving the decision parameters from recent experience.

Reaching Out

Learning opportunities not resolved in the sleep epoch can be brought to the attention of the user, the host network, or a designer. We refer to elevating complex problems to an infrastructure support as a prayer epoch.

14.4 CRA III: The Inference Hierarchy

The phases of inference from observation to action show the flow of inference, a top-down view of how cognition is implemented algorithmically. The inference hierarchy is the part of the algorithm architecture that organizes the data structures. Inference hierarchies have been in use since Hearsay II in the 1970s [15], but the CR hierarchy is unique in its method of integrating ML with real-time performance during the wake epochs. An illustrative inference hierarchy includes layers from atomic stimuli at the bottom to information clusters that define action contexts, as shown in Figure 14.6.

Sequence	Level of Abstraction
Context Cluster	*Scenes* in a Play, Session
Sequence Clusters	*Dialogs,* Paragraphs, Protocol
Basic Sequences	*Phrases,* Video Clip, Message
Primitive Sequences	*Words,* Toke, Image
Atomic Symbols	*Raw Data,* Phoneme, Pixel
Atomic Stimuli	External Phenomena

Figure 14.6: Standard inference hierarchy (© Dr. Joseph Mitola III, used with permission).

The pattern of accumulating elements into sequences begins at the bottom of the hierarchy. Atomic stimuli originate in the external environment including RF, acoustic, image, and location domains, among others. The atomic symbols extracted from them are the most primitive symbolic units in the domain. In speech, the most primitive elements are the phonemes. In the exchange of textual data (e.g., in e-mail), the symbols are the typed characters. In images, the atomic symbols may be the individual picture elements (pixels) or they may be small groups of pixels with similar hue, intensity, texture, and so forth.

A related set of atomic symbols forms a primitive sequence. Words in text, tokens from a speech "tokenizer," and objects in images (or individual image regions in a video flow) are primitive sequences. Primitive sequences have spatial and/or temporal coincidence, standing out against the background (or noise), but there may be no particular meaning in that pattern of coincidence. Basic sequences, in contrast, are space–time–spectrum sequences that entail the communication of discrete messages.

These discrete messages (e.g., phrases) are typically defined with respect to an ontology of the primitive sequences (e.g., definitions of words). Sequences cluster together because of shared properties. For example, phrases that include words such as "hit," "pitch," "ball," and "out" may be associated with a discussion of a baseball game. Knowledge Discovery in Databases (KDD) and the Semantic Web offer approaches for defining, or inferring, the presence of such clusters from primitive and basic sequences.

A scene is a context cluster, a multidimensional space–time–frequency association, such as a discussion of a baseball game in the living room on a Sunday afternoon. Such clusters may be inferred from unsupervised ML (e.g., using statistical methods or nonlinear approaches such as SVMs).

Although presented here in a bottom-up fashion, there is no reason to limit multidimensional inference to the top layers of the inference hierarchy. The lower levels of the inference hierarchy may include correlated multisensor data. For example, a word may be characterized as a primitive acoustic sequence coupled to a primitive sequence of images of a person speaking that word. In fact, taking the cue that infants seem to thrive on multisensory stimulation, the key to reliable ML may be the use of multiple sensors with multisensor correlation at the lowest levels of abstraction.

Each of these levels of the inference hierarchy is now discussed in more detail.

14.4.1 Atomic Stimuli

Atomic stimuli originate in the external environment and are sensed and preprocessed by the sensory subsystems, which include sensors of the RF environment (e.g., radio receiver and related data and information processing) and of the local physical environment, including acoustic, video, and location sensors. Atomic symbols are the elementary stimuli extracted from the atomic stimuli. Atomic symbols may result from a simple noise-riding threshold algorithm, such as the squelch circuit in RF that differentiates signal from noise. Acoustic signals may be differentiated from simple background noise this way, but generally the result is the detection of a relatively large speech epoch that contains various kinds of speech energy. Thus, further signal processing is typically required in a preprocessing subsystem to isolate atomic symbols.

The transformation from atomic stimuli to atomic symbols is the job of the sensory preprocessing system. Thus, for example, acoustic signals may be transformed into phoneme hypotheses by an acoustic signal preprocessor. However, some software tools may not enable this level of interface via an API. To develop industrial-strength CR, advanced, and video-processing software tools are needed. Speech tools yield an errorful transcript in response to an acoustic signal. Thus, speech tool map from stimuli to basic sequences. One of the important contributions of architecture is to identify such maps and to define the role and the level of mapping tools.

There is nothing about the inference hierarchy, however, that forces data from a preprocessing system to be entered at the lowest level. In order for the more primitive symbolic abstractions such as atomic symbols to be related to more aggregate abstractions, one may either build up the aggregates from the primitive abstractions or derive the primitive abstractions from the aggregates. People are used to being exposed to "the whole thing" by immersion in the full experience of life—touch, sight, sound, taste, and balance—all at once; therefore, it seems possible (even likely) that the more primitive abstractions are somehow derived through the analysis of aggregates, perhaps by cross-correlation. This can be accomplished in a CRA sleep cycle. The idea is that the wake cycle is optimized for immediate reaction to stimuli, similar to what our ancestors needed to avoid predation, whereas the sleep cycle is optimized for introspection, for analyzing the day's stimuli to derive those objects that should be recognized and acted upon in the next cycle.

Stimuli are each counted. When an iCR that conforms to this architecture encounters a stimulus, it both counts how many such stimuli have been encountered and resets to zero a timer that keeps track of the time since the last occurrence of the stimulus.

14.4.2 Primitive Sequences: Words and Dead Time

The accumulation of sequences of atomic symbols forms primitive sequences. The key question at this level of the data structure hierarchy is the sequence boundary. The simplest situation is one in which a distinguished atomic symbol separates primitive sequences, which is exactly the case with white space between words in typed text. A text based ML system may be white space to separate a text stream into primitive sequences.

14.4.3 Basic Sequences

The pattern of aggregation is repeated vertically at the levels corresponding to words, phrases, dialogs, and scenes. The data structures generated by processing nodes create the concept hierarchy of Figure 14.6. These are the reinforced hierarchical sequences. They are reinforced by the inherent counting of the number of times each atomic or aggregated stimulus occurs. The phrase level typically contains or implies a verb (the verb "to be" may be implied if no other verb is implicit).

Unless digested (such as by a sleep process), the observation phase hierarchy accumulates all the sensor data, parsed and distributed among processing nodes for fast parallel retrieval. Because the hierarchy saves everything and compares new data to memories, it is a kind of memory-based learning approach, which takes a lot of space. When the stimuli retained are limited to atomic symbols and their aggregates, however, the total amount of data that needs to be stored is relatively modest. In addition, recent research shows the negative effects of discarding cases in word pronunciation. In word pronunciation, no example can be discarded even if it is "disruptive" to a well-developed model. Each exception has to be

469

followed. Thus, in the CR1 prototype, when multiple memories match partially, the most nearly exact match informs the orientation, planning, and action.

14.4.4 NL in the CRA Inference Hierarchy

In speech, words spoken in a phrase may be coarticulated with no distinct boundary between the primitive sequences in a basic sequence. Therefore, speech detection algorithms may extract a basic sequence, but the interpretation of that sequence as constituent primitive sequences may be much less reliable. Typically, the correct parse is within the top 10 candidates for contemporary speech tools. The flow of speech signal processing may be similar to the following:

1. Isolate a basic sequence (phrase) from background and noise by using an acoustic analysis to determine speech versus background.
2. Analyze the basic sequence to identify candidate primitive sequence boundaries (words).
3. Analyze the primitive sequences statistically for e.g. Hidden Markov Sequence.
4. Evaluate primitive and basic sequence hypotheses based on a statistical model of language to rank-order alternative interpretations of the basic sequence.

So a practical speech-processing algorithm may yield alternative strings of phonemes and candidate parses "all at once." NL-processing (NLP) tool sets may be embedded into the CRA inference hierarchy, as illustrated in Figure 14.7. Speech and/or text channels may be processed via such NL facilities with substantial a priori models of language and discourse. The use of those models entail mappings among the word, phrase, dialog, and scene levels of the observation phase hierarchy and the encapsulated component(s).

It is tempting to expect CR to integrate a commercial NLP system such as IBM's ViaVoice® or a derivative of an NLP research system such as SNePS [16], AGFL [17], or XTAG [18] perhaps using a morphological analyzer such as PCKIMMO [19]. These tools go too far and yet not far enough in the direction needed for CRA. One might like to employ existing tools by using a workable interface between the domain of radio engineering and some of these NL tool sets. The definition of such cross-discipline interfaces is in its infancy. A present, one cannot just express a radio ontology in Interlingua and plug it neatly into XTAG

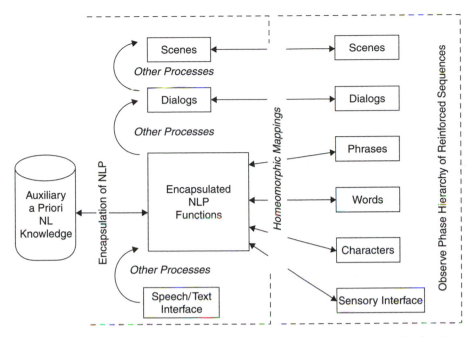

Figure 14.7: NL encapsulation in the observation hierarchy (© Dr. Joseph Mitola III, used with permission).

to get a working CR. The internal data structures that are used in radio mediate the performance of radio tasks (e.g., "transmit a waveform"). The data structures of XTAG, AGFL, and so forth mediate the conversion of language from one form to another. Thus, XTAG wants to know that "transmit" is a verb and "waveform" is a noun. The CR needs to know that if the user says "transmit" and a message has been defined, then the CR should call the SDR function *transmit()*. NLP systems also need scoping rules for transformations on the linguistic data structures. The way in which domain knowledge is integrated in linguistic structures of these tools tends to obscure the radio engineering aspects.

NLP systems work well on well-structured speech and text, such as the prepared text of a news anchor. But they do not yet work well on noisy, nongrammatical data structures encountered, for example, when a user is trying to order a cab in a crowded bar. Thus, less-linguistic or meta-linguistic data structures may be needed to integrate core CR reasoning with speech and/or text-processing frontends. The CRA has the flexibility illustrated in Figure 14.7 for the subsequent integration of evolved NLP tools. The emphasis of this version of the CRA is a structure of sets and maps required to create a viable CRA. Although introducing the issues required to integrate existing NLP tools, the discussion does not pretend to present a complete solution to this problem.

14.4.5 Observe–Orient Links for Scene Interpretation

CR may use an algorithm-generating language with which one may define self-similar inference processes. In one example, the first process (Proc1) partitions characters into words, detecting novel characters and phrase boundaries as well. Proc2 detects novel words and aggregates known words into phrases. Proc3 detects novel phrases, aggregating known phrases into dialogs. Proc4 aggregates dialogs into scenes, and Proc5 detects known scenes. In each case, a novel entity at level N will be bound in the context of the surrounding known entities at that level to the closest match at the next highest level, $N + 1$. For example, the word–phrase intersection of Proc2 would map the following phrases:

"Let me introduce Joe"

"Let me introduce Chip"

Since "Chip" is unknown, and "Joe" is known from a prior dialog, integrated CBR matches the phrases, binding <Chip/> = <Joe/>. In other words, it will try to act with respect to Chip in the way it was previously trained (at the dialog level) to interact with Joe. In response to the introduction, the system may say, "Hello, Chip. How are you?" mimicking the behavior that it had previously learned with respect to Joe. Not too bright, but not all that bad either for a relatively simple ML algorithm.

There is a particular kind of dialog that is characterized by reactive world knowledge in which there is some standard way of reacting to given speech-act inputs. For example, when someone says, "Hello," you may typically reply with "Hello" or some other greeting. The capability to generate such rote responses may be preprogrammed into a lateral component as Hearsay knowledge source (KS). The responses are not preprogrammed, but the general tendency to imitate phrase-level dialogs is a preprogrammed tendency that can be overruled by plan generation—but that is present in the orient phase, which is Proc6.

Words may evoke a similar tendency toward immediate action. What do you do when you hear the words "Help!!" or "Fire, fire! Get out, get out!!" The CR programmer, can capture reactive tendencies in a CR by preprogramming an ability to detect these kinds of situations in the word-sense KS, as implied by Figure 14.8. When confronted with such wording (which is preferred), CR should react appropriately if properly trained. To cheat, one can preprogram a wider array of stimulus–response pairs so that the CR has more a priori knowledge, but some of it may not be appropriate. Some responses are culturally conditioned. Will the CR be too rigid? If it has too much a priori knowledge, it will be perceived by its users as too rigid. If it doesn't have enough, it will be perceived as uninformed.

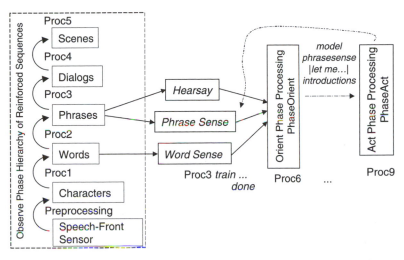

Figure 14.8: The inference hierarchy supports lateral KS (© Dr. Joseph Mitola III, used with permission).

14.4.6 Observe–Orient Links for Radio Skill Sets

Radio knowledge may be embodied in components called radio skills. Declarative radio knowledge is static, requiring interpretation by an algorithm such as an inference engine in order to accomplish anything. Radio skills, on the other hand, are knowledge embedded in serModels through the process of training or sleeping/ dreaming. This knowledge is continually pattern matched against all stimuli in parallel. That is, there are few a priori logical dependencies among knowledge components that mediate the application of the knowledge. With FOPC, the theorem prover must reach a defined state in the resolution of multiple axioms in order to initiate action. In contrast, serModels are continually compared to the level of the hierarchy to which they are attached, so their immediate responses are always cascading toward action. Organized as maps primarily among the wake-cycle phases "observe" and "orient," the radio procedure skill sets (SS's) control radio personalities, as illustrated in Figure 14.9.

These skill sets may either be reformatted into serModels directly from the a priori knowledge of an RKRL frame, or they may be acquired from training or sleep/dreaming. Each skill set may also save the knowledge it learns into an RKRL frame.

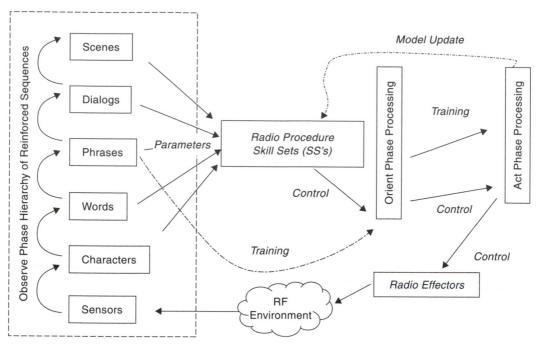

Figure 14.9: Radio skills respond to observations (© 1999 Dr. Joseph Mitola III, used with permission).

14.4.7 General World Knowledge

An AACR needs substantial knowledge embedded in the inference hierarchies. It needs both external RF knowledge and internal radio knowledge. Internal knowledge enables it to reason about itself as a radio. External radio knowledge enables it to reason about the role of the <Self/> in the world, such as respecting rights of other cognitive and not-so-CRs.

Figure 14.10 illustrates the classes of knowledge an AACR needs to employ in the inference hierarchies and cognition cycle. It is one thing to write down that the universe includes a physical world (there could also be a spiritual world, and that might be very important in some cultures)—it is quite another thing to express that knowledge in a way that the AACR will be able to use it effectively. Symbols such as "universe" take on meaning by their relationships to other symbols and to external stimuli. In this ontology, meta-level knowledge consists of *abstractions*, distinct from existential knowledge of the physical universe. In RXML, this ontological perspective includes all in a universe of discourse, <Universe>, expressed as shown by Figure 14.11.

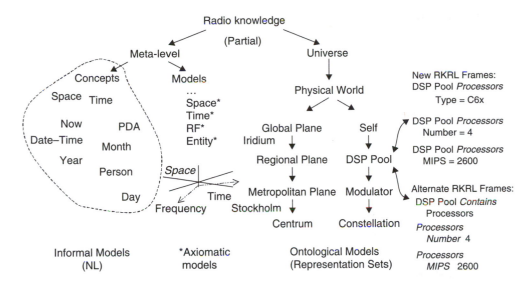

Figure 14.10: External radio knowledge includes concrete and abstract knowledge (© 1999, 2000 Dr. Joseph Mitola III, used with permission).

```
<Universe>
<Abstractions> <Time> <Now/> </Time> <Space> <Here/> </Space> ... <RF/> ...
<Intelligent-Entities/> ... </Abstractions>
<Physical-universe> ... <Instances/> of Abstraction ... </Physical-universe>
</Universe>
```

Figure 14.11: Equation for the Universe of discourse of AACR. This <Universe/> consists of abstractions plus the physical universe.

Abstractions include informal and formal meta-level knowledge from unstructured knowledge of concepts to the more mathematically structured models of space, time, RF, and entities that exist in space–time. To differentiate "now" as a temporal concept from "Now," which is the Chinese name of a plant, the CRA includes both the a priori knowledge of "now" as a space–time locus, <Now/>, as well as functions that access and manipulate instances of the concept <Now/>. <Now/> is axiomatic in the CRA, for temporal reference in planning actions. The architecture allows an algorithm to return the date-time code e.g. from Windows to define instances of <Now/>.

Definition-by-algorithm permits an inference system like the cognition subsystem to reason about whether a given event is in the past, present, or future. What is the present? The present is some region of time between "now" and the immediate past and future. If the user is a paleontologist, "now" may consist of the million-year epoch in which it is thought that man evolved from apes. To a rock star, "now" is probably a lot shorter than that. How will a CR learn its user's concept of now? The CRA design offers an axiomatic treatment of time, but the axioms do not reflect such subjective reality from the <User/> perspective. The CR1 [12] aggregates knowledge of time by a temporal CBR that illustrates the key principles. The CR1 [12] does not fix the definition of <Now/>, but enables the <Self/> to define the details in an <Instance/> in the physical world about which it can learn from the user, whether a paleontologist or a rock star.

Given the complexity of a system that includes both a multi-tiered inference hierarchy and the cognition cycle's observe, orient, plan, decide, act sequence with AML throughout, it is helpful to consider the mathematical structure of these information elements, processes, and flows. The mathematical treatment is the subject of the next section.

14.5 CRA IV: Architecture Maps

Cognition functions are implemented via cognition elements consisting of data structures, processes, and flows, which may be modeled as topological maps over the abstract domains identified in Figure 14.12.

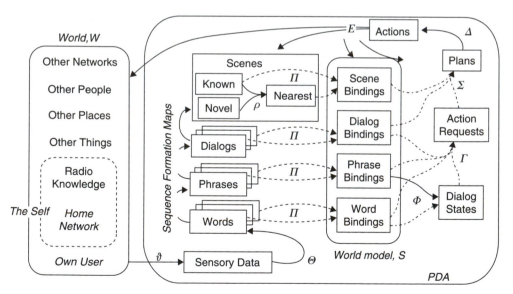

Figure 14.12: Architecture based on the cognition cycle (© 2000 Dr. Joseph Mitola III, used with permission).

The <Self/> is an entity in the world, whereas the internal organization of the <Self/> (annotated PDA in the figure) is an abstraction that models the <Self/>.

The hierarchy of words, phrases, and dialogs from sensory data to scenes is not inconsistent with visual perception. Words correspond to visual entities; phrases to detectable movement and juxtaposition of entities in a scene. Dialogs correspond to a coherent sequence of movement within the scope of a scene, such as walking across the room. Occlusion may be thought of as a dialog in which the room asserts itself in part of the scene while observable walking corresponds to assertion of the object. The model data structures may be read as generalized words, phrases, dialogs, and scenes that may be acoustic, visual, or perceived in other sensory domains (e.g., infrared). These structures refer to set-theoretic spaces consisting of a set X and a family of subsets Ox that contain {X} and { }, the null set, and that are closed under union and countable intersection. In other words, each is a topological space induced over the domain. Proceeding up the hierarchy, the scope of the space (X, Ox) increases. A <Scene/> is a subset of space–time that is circumscribed by the entity by sensory limits.

The cognition functions modeled in these spaces are topology-preserving maps (see Figure 14.12). Data- and knowledge-storage spaces are shown as rectangles (e.g., dialog states, plans), whereas processing elements that transform sets are modeled as homeomorphisms, or topology-preserving maps, shown as directed graphs (e.g., Π) in this figure.

14.5.1 CRA Topological Maps

The processing elements of the architecture are modeled topological maps, as shown in Figure 14.12:

1. The input map ϑ consists of components that transform external stimuli to the internal data structure sensory data.

2. The transformation Θ consists of entity recognition (via acoustic, optical, and other sensors), lower-level software radio (SWR) waveform interface components, and so forth, that create streams of primitive-reinforced sequences. The model includes maps that form successively higher-level sequences from the data on the immediately lower level.

3. Reasoning components include the map ρ that identifies the best match of known sequences to novel sequences. These are bound to scene variables by projection components, Π. The maps ϑ, Θ, ρ, and Π, constitute observe phase processing.

4. Generalized word- and phrase-level bindings are interpreted by the components Φ to form dialog states. Train, for example, is the dialog state of a training experience in the CRA.

5. The components of Γ create action requests from bindings and dialog states. The maps Φ and Γ constitute orient phase processing.

6. Scene bindings include user communications context. Context-sensitive plans are created by the component Σ that evaluates action requests in the plan phase.

7. The decision phase processing consists of map Δ that maps plans and scene context to actions.

8. Finally, the map *E* (consists of the effector components that change the PDA's internal states, change displays, synthesize speech, and transmit information on wireless networks using the SWR personalities.

14.5.2 CRA Identifies Self, Owner, and Home Network

The sets of entities in the world that are known to the CR are modeled graphically as rounded rectangles in Figure 14.12. These include the self-grounded in the outside world ("self"), as well as its knowledge of the self as self (e.g., as "PDA"). The critical entities are world, W, the PDA, and the PDA's World Model, S. (In the CRA, S includes the orient phase data structures and processes.) Entities in the world include the differentiated entities "Own User" or owner, and "Home Network." The architecture requires that the PDA be able to identify these entities so that it may treat them differentially. Other networks, people, places, and things may be identified in support of the primary cognition functions, but the architecture does not depend on such a capability.

14.5.3 CRA-Reinforced Hierarchical Sequences

The data structures for perception include the reinforced hierarchical sequences words, phrases, dialogs, and scenes of the observe phase. Within each of these sequences, the novel sequences represent the current stimulus–response cases of the cognitive behavior model. The known sequences represent the integrated knowledge of the cognitive behavior model. Known sequences may consist of a priori RXML statements embedded in the PDA or of knowledge acquired through independent ML. The nearest sequence is the known sequence that is closest in some sense to the novel sequence. The World Model, W, consists primarily of bindings between a priori data structures and the current scene. These associative structures are also associated with the observe phase. Dialog states, action requests, plans, and actions are additional data structures needed for the observe, orient, plan, and act phases, respectively. Each internal data structure maps to an RXML frame consisting of element (e.g., set or stimulus); model (e.g., embedded procedure, parameter values); content, typically a structure of elements terminating in either primitive concepts <concept/> (e.g., subset or response) or instance data; and associated resources. Context is defined as the RXML URL or root from <Universe>, to include source, time, and place of the <Scene>.

14.5.4 Behaviors in the CRA

CRA entails three modes of behavior: waking, sleeping, and praying. Behavior that lasts for a specific time interval is called a behavioral epoch. The axiomatic relationships among these behaviors are expressed in the topological maps of Figure 14.13.

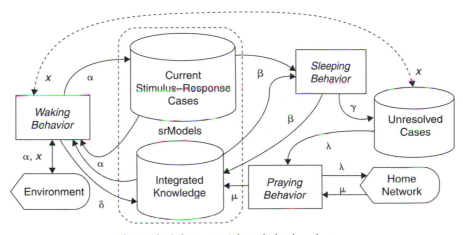

α: action cycle; δ: in cremental machnine learning;
β: nonincremental machine learning; γ: learning conflicts;
λ: RKRL/KQML requests for external assistance;
μ: authoritative assistance; x: attempt to resolve problems.

Figure 14.13: Cognitive behavior model consists of domains and topological maps (© 2000 Dr. Joseph Mitola III, used with permission).

Waking Behavior

Waking behavior is optimized for real-time interaction with the user, isochronous control of SWR assets, and real-time sensing of the environment. The conduct of the waking behavior is informally referred to as the awake-state, although it is not a specific system state, but a set of behaviors. Thus, referring to Figure 14.13, the awake-state cognition-actions (α) map the environment interactions to the current stimulus–response cases. These cases are the dynamic subset of the embedded serModels. Incremental ML (δ) maps these interactions to integrated knowledge, the persistent subset of the serModels.

Sleeping and Dreaming Behaviors

Cognitive PDAs (CPDAs) detect conditions that permit or require sleep and dreaming. For example, if the PDA predicts or becomes aware of a long epoch of low utilization (such as overnight hours), then the CPDA may autonomously initiate sleeping behavior. Sleep occurs during planned inactivity, for example, to recharge batteries. Dreaming behavior employs energy to retrospectively examine experience since the last period of sleep. In the CRA, all sleep includes dreaming. In some situations, the CPDA may request permission to enter sleeping/dreaming behavior from the user (e.g., if predefined limits of aggregate experience are reached). Regular sleeping/dreaming limits the combinatorial explosion of the process of assimilating aggregated experience into the serModels needed for real-time behavior during the waking behaviors. During the dreaming epochs, the CPDA processes experiences from the waking behavior using nonincremental ML algorithms. These algorithms map current cases and new knowledge into integrated knowledge (β).

A conflict is a context in which the user overrode a CPDA decision about which the CPDA had little or no uncertainty. Map β may resolve the conflict. If not, it will place the conflict on a list of unresolved conflicts (map γ).

Prayer Behavior

Attempts to resolve unresolved conflicts via the mediation of the PDA's home network may be called prayer behavior, referring the issue to a completely trusted source with substantially superior capabilities. The unresolved-conflicts list γ is mapped (λ) to RXML queries to the PDA's home CN expressed in XML, OWL, KQML, RKRL, RXML, or a mix of declared knowledge types. Successful resolution maps network responses to integrated knowledge (μ). Many research issues surround the successful download of such knowledge, including the set of support for referents in the unresolved-conflicts lists and the updating of knowledge in the CPDA needed for full assimilation of the new knowledge or procedural fix to the unresolved conflict. The prayer behavior may not be reducible to finite-resource introspection, and thus may be susceptible to the "partialness" of TC, even though the CPDA and CWN enforce watchdog timers.

Alternatively, the PDA may present the conflict sequence to the user, requesting the user's advice during the wake cycle (map π).

14.5.5 From Maps to APIs

Each of these maps has a domain and a range. Axiomatically, the domain is the set of subsets of internal data structures over which the map is defined, and the range is the set of subsets onto which the map projects its effects. Thus, for each map, $M:D => R$, there is an associated API or API component.

API-M: $\{m \in M: \{d \in D => r \in R\}\}$

In other words, the API for the map M specifies methods or attached procedures defined over subsets d of the domain D that map onto subsets r of the range R. So each map can be interpreted as a generalized API. Some APIs may entail more than one map. A planning API, for example, might include the maps that generate the plans and the maps that select among plan components and schedule plans for actions. In fact, APIs for many CR functions from perception to planning and action include more functionality than is needed for embedding into a CR, such as visualization tools and user interfaces. Therefore, the representation of API components as maps establishes the foundations of the API without over-constraining the definition of APIs for a given CR design. The evolution of the CRA from this set of maps to a set of APIs with broad industry support may be facilitated by the framework of the maps.

14.5.6 Industrial-Strength Inference Hierarchy

Although the CRA provides a framework for APIs, it doesn't specify the details of the data structures nor of the maps. The CRA research prototype emphasizes ubiquitous learning via serModels and CBR, but it doesn't implement critical features that would be required in deployable CRs. Other critical aspects of such industrial-strength architectures include more capable scene perception and situation interpretation, specifically addressing:

1. *Noise*: In utterances, images, objects, location estimates, and the like. Noise sources include thermal noise; conversion error introduced by the process of converting analog signals (audio, video, accelerometers, temperature, etc.) to digital form; error in converting from digital-to-analog form; preprocessing algorithm biases and random errors, such as the accumulation of error in a digital filter; or the truncation of a low-energy signal by threshold logic. Dealing

effectively with noise differentiates a tutorial demonstration from a useful product.

2. *Hypothesis management*: Keeping track of more than one possible binding of stimuli to response, dialog sense, scene, etc. Hypotheses may be managed by keeping the N-best hypotheses (with an associated degree of belief), by estimating the prior probability or other degree of belief in a hypothesis, and by keeping a sufficient number of hypotheses to exceed a threshold (e.g., 90 percent or 99 percent of all the possibilities) or by keeping multiple hypotheses until the probability for the next most likely (second) hypothesis is less than some threshold. The estimation of probability requires a measurable space, a sigma-algebra[12] that defines how to accumulate probability on that space, proof that the space obeys the axioms of probability, and a certainty calculus that defines how to combine degrees of belief in events as a function of the measures assigned to the probability of each event.

3. *Training interfaces*: The reverse flow of knowledge from the inference hierarchy back to the perception subsystems. The recognition of the user by a combination of face and voice could be more reliable than single-domain recognition either by voice or by vision. In addition, the location, temperature, and other aspects of the scene may influence object identification. Visual recognition of the owner outdoors in a snowstorm, for example, is more difficult than indoors in an office. Even though the CR might learn to recognize the user based on weaker cues outdoors, access to private data might be constrained until the quality of the recognition exceeds some learned threshold.

4. *Nonlinear flows*: Although the cognition cycle emphasizes the forward flow of perception enabling action, it is crucial to realize that actions may be internal, such as advising the vision subsystem that its recognition of the user is in error because the voice does not match and the location is wrong. Due to the way the cognition cycle operates on the self, these reverse flows from perception to training are implemented as forward flows from the perception system to the self, directed toward a specific subsystem such as vision or audition. There may also be direct interfaces from the CWN to the CR to upload data structures representing a priori knowledge integrated into the UCBR learning framework.

---

[12] In mathematics, a *sigma-algebra* over a set is a functional correspondence between subsets of that set and operations that measure, e.g. the size of the subset. The sigma-algebra is fundamental to probability theory [9].

14.6 CRA V: Building the CRA on SDR Architectures

A CR is an SWR or SDR with flexible formal semantics-based entity-to-entity messaging via RXML and integrated ML of the self, the user, the RF environment, and the "situation." This section reviews SWR, SDR, and the SCA, or SRA, as they relate to the SRA. Although it is not necessary for an AACR to use the SCA/SRA as its internal model of itself, it certainly must have some model, or it will be incapable of reasoning about its own internal structure and adapting or modifying its radio functionality autonomously.

14.6.1 Review of SWR and SDR Principles

Hardware-defined radios such as the typical amplitude/frequency modulation (AM/FM) broadcast receiver convert radio to audio using such radio hardware as antennas, filters, analog demodulators, and the like. SWR is the ideal digital radio in which the analog-to-digital converter (ADC) and digital-to-analog converter (DAC) convert digital signals to and from RF directly, and all RF channel modulation, demodulation, frequency translation, and filtering are accomplished digitally. For example, modulation may be accomplished digitally by multiplying sine and cosine components of a digitally sampled audio signal (called the "baseband" signal, to be transmitted) by the sampled digital values of a higher-frequency sine wave to upconvert it, ultimately to the RF spectrum.

Figure 14.14 shows how SDR principles apply to a cellular radio-base station. The ideal SWR would have essentially no RF conversion, just ADC/DAC blocks accessing the full RF spectrum available to the (wideband) antenna elements. Today's SDR-base stations approach this ideal by digital access (DAC and ADC) to a band of spectrum allocations, such as 75 MHz allocated to uplink and downlink frequencies for 3G services. In this architecture, RF conversion can be a substantial system component, sometimes 60 percent of the cost of the hardware, and not amenable to cost improvements through Moore's law. The ideal SDR would access more like 2.5 GHz from, say 30 MHz to around 2.5 GHz, supporting all kinds of services in TV bands, police bands, air traffic control bands, and other bands. Although this concept was considered radical when introduced in 1991 [20] and popularized in 1995 [21], recent regulatory rulings are encouraging the deployment of such "flexible spectrum" use architectures.

This ideal SWR may not be practical or affordable, so it is important for the radio engineer to understand the trade-offs (see Mitola [1] for SDR architecture trade-offs). In particular, the physics of RF devices (e.g., antennas, inductors, filters) makes it easier to synthesize narrowband RF and intervening analog RF

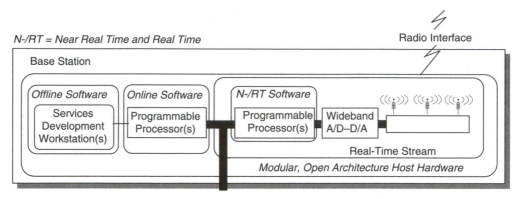

N-/RT = Near Real Time and Real Time

Figure 14.14: SWR principle applied to cellular-base station (© 1992 Dr. Joseph Mitola III, used with permission).

conversion and intermediate frequency (IF) conversion. Given narrowband RF, the hardware-defined radio might employ baseband (e.g., voice frequency) ADC, DAC, and digital signal processing. The programmable digital radios (PDRs) of the 1980s and 1990s used this approach. Historically, this approach has not been as expensive as wideband RF (i.e., the cost of antennas, conversion), ADCs, and DACs. Handsets are less amenable to SWR principles than the base station (Figure 14.15). Base stations access the power grid. Thus, the fact that wideband ADCs, DACs, and DSP (digital signal processor) consume many watts of power is not a major design driver. Conservation of battery life, however, *is* a major design driver in the handset.

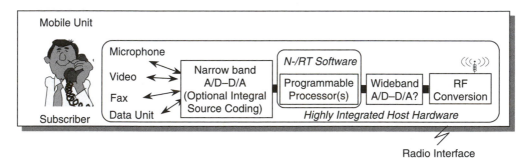

Figure 14.15: SWR principle: "ADC and DAC at the antenna" may not apply (© 1992 Dr. Joseph Mitola III, used with permission).

Thus, insertion of SWR technology into handsets has been relatively slow. Instead, the major handset manufacturers include multiple single-band RF chip sets into a given handset. This has been called the "Velcro" radio or "slice" radio.

The ideal SWR is not readily approached in many cases, so the SDR has comprised a sequence of practical steps from the baseband DSP of the 1990s toward the ideal SWR. As the economics of Moore's law and of increasingly wideband RF and IF devices allow, implementations move upward and to the right in the SDR design space (Figure 14.16).

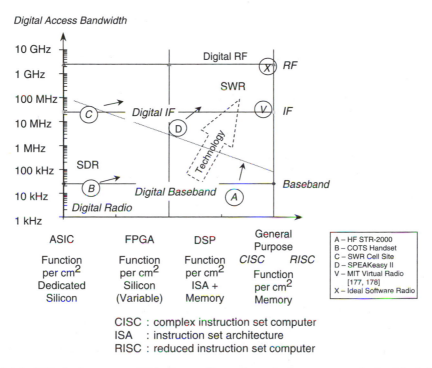

Figure 14.16: SDR design space. This figure shows how designs approach the ideal SWR (© 1996–2003 Dr. Joseph Mitola III, used with permission).

This space consists of the combination of digital access bandwidth and programmability. Access bandwidth consists of ADC/DAC sampling rates converted by the Nyquist criterion[13] and practice into effective bandwidth. Programmability

---

[13] The Nyquist criterion is that a signal must be sampled at more than twice the highest-frequency component that is present in the signal. Failure to do so will result in an alias, in which interference will be shifted from out of band to in the processing band. This is generally a distortion that most systems cannot tolerate. In other words, the Nyquist frequency is half the sample rate of the analog-to-digital (A/D) or digital-to-analog (D/A) converters.

of the digital subsystems is defined by the ease with which logic and interconnect may be changed after deployment. Application-specific integrated circuits (ASICs) cannot be changed at all, so the functions are "dedicated" in silicon. Field-programmable gate arrays (FPGAs) can be changed in the field, but if the new function exceeds some performance parameter of the chip, which is not uncommon, then one must upgrade the hardware to change the function, just like with ASICs. DSPs are typically easier or less expensive to program and are more efficient in power use than FPGAs. Memory limits and instruction set architecture (ISA) complexity can drive up costs of reprogramming the DSP. Finally, general-purpose processors, particularly reduced instruction set computers (RISCs), are most cost-effective to change in the field. To characterize a multiprocessor, such as a cell phone with a CDMA-ASIC, DSP speech codec, and RISC microcontroller, weight the point in the design space by equivalent-processing capacity.

Where should one place an SDR design within this space? The quick answer along a migration path of radio technology from the lower left toward the upper right, benefiting from lessons learned in the early migration projects captured in *SRA* [1].

14.6.2 Radio Architecture

The discussion of the SWR design space contains the first elements of radio architecture. It defines a mix of critical components for the radio. For SWR, the critical hardware components are the ADC, DAC, and processor suite. The critical software components are the user interface; the networking software; the information security (INFOSEC) capability (hardware and/or software); the RF media access software, including the physical (PHY) layer modulator and demodulator (modem) and media access control (MAC); and any antenna-related software, such as antenna selection, beamforming, pointing, and the like. INFOSEC consists of transmission security, such as the frequency-hopping spreading code selection, plus communications security encryption.

The SDR Forum defined a very simple, helpful model of radio in 1997, which is shown in Figure 14.17. This model highlights the relationships among radio functions at a tutorial level. The CR has to "know" about these functions, so this model is a good start because it shows both the relationships among the functions and the typical flow of signal transformations from analog RF to analog or (with SDR) digital modems, and on to other digital processing, including system control of which the user interface is a part.

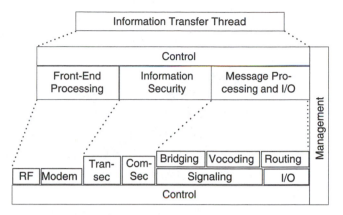

Figure 14.17: SDR Forum (formerly MMITS) information transfer thread architecture (© 1997 SDR Forum, used with permission).

This model, the techniques for implementing an SWR, and the various degrees of SDR capability are addressed in depth in the various texts on SDR [22–25].

14.6.3 The SCA

The US DoD developed the SCA for its Joint Tactical Radio System (JTRS) family of radios. The SCA identifies the components and interfaces shown in Figure 14.18. The APIs define access to the PHY layer, to the MAC layer, to the logical

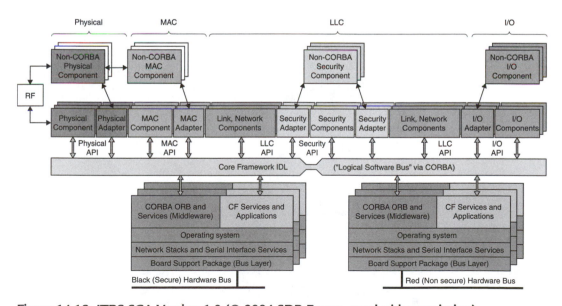

Figure 14.18: JTRS SCA Version 1.0 (© 2004 SDR Forum, used with permission).

link control (LLC) layer, to security features, and to the input/output of the physical radio device. The physical components consist of antennas and RF conversion hardware that are mostly analog and that typically lack the ability to declare or describe themselves to the system. Most other SCA-compliant components are capable of describing themselves to the system to enable and facilitate plug-and-play among hardware and software components. In addition, the SCA embraces the portable operating system interface (POSIX) and CORBA.

The model evolved through several stages of work in the SDR Forum and OMG into a UML-based object-oriented model of SDR (Figure 14.19). Waveforms are collections of load modules that provide wireless services, so from a radio designer's perspective, the waveform is the key application in a radio. From a user's perspective of a wireless PDA (WPDA), the radio waveform is just a means to an end, and the user doesn't want to know or to have to care about waveforms. Today, the cellular service providers hide this detail to some degree, but consumers sometimes know the difference between CDMA and GSM, for example, because first generation CDMA works in the United States, but not in Europe. With the deployment of the 3G of cellular technology, the amount of technical jargon consumers will need to know is increasing. So the CRA insulates the user from those details, unless the user really wants to know.

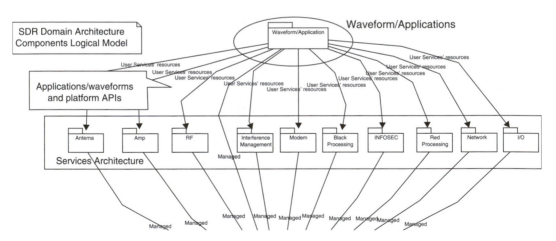

Figure 14.19: SDR Forum UML model of radio services (© 2004 SDR Forum, used with permission).

In the UML model shown in Figure 14.19, Amp refers to amplification services, RF refers to RF conversion, interference management refers to both avoiding interference and filtering it out of one's band of operation. In addition, the jargon for US military radios is that the "red" side contains the user's secret information, but when it is encrypted it becomes "black," or protected, so it can be transmitted. Black processing occurs between the antenna and the decryption process. Notice also that Figure 14.19 has no user interface. The UML model contains a sophisticated set of management facilities, illustrated further in Figure 14.20, to which the human–machine interface (HMI) or user interface is closely related.

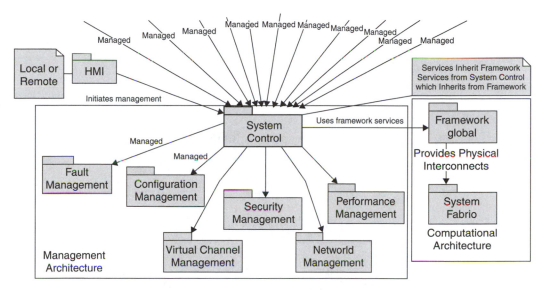

Figure 14.20: SDR Forum UML management and computational architectures (© 2004 SDR Forum, used with permission).

Systems control is based on a framework that includes very generic functions such as event logging, organized into a computational architecture, heavily influenced by CORBA. The management features are needed to control radios of the complexity of 3G and of the current generation of military radios. Although civil sector radios for police, fire, and aircraft lag these two sectors in complexity and are more cost-sensitive, baseband SDRs are beginning to insert themselves even into these historically less technology-driven markets.

Fault management features are needed to deal with the loss of a radio's processors, memory, or antenna channels. CR therefore interacts with fault management to determine what facilities may be available to the radio given recovery from hardware and/or software faults (e.g., error in a download). Security management is increasingly important in the protection of the user's data by the CR; balancing convenience and security can be very tedious and time consuming. The CR will direct virtual channel management (VCM) and will learn from the VCM function what radio resources are available, such as what bands the radio can listen to and transmit on and how many bands it can use at once. Network management does for the digital paths what VCM does for the radio paths. Finally, SDR performance depends on the availability of analog and digital resources, such as linearity in the antenna, millions of instructions per second (MIPS) in a processor, and the like.

14.6.4 Functions-Transforms Model of Radio

The CRA uses a self-referential model of a wireless device, the functions-transforms model, to define the RKRL and to train the CRA. In this model, illustrated in Figure 14.21, the radio knows about sources, source coding, networks, INFOSEC, and the collection of front-end services needed to access RF channels. Its knowledge also extends to the idea of multiple channels and their characteristics (the channel set), and the radio part may have many alternative personalities at a given point in time. Through evolution support, those alternatives change over time.

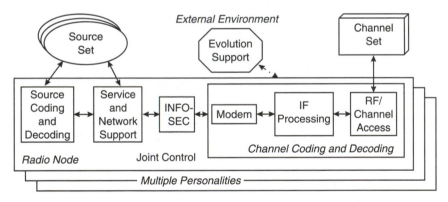

Figure 14.21: Functions-transforms model of a wireless node (© 1996 Dr. Joseph Mitola III, used with permission).

CR reasons about all of its internal resources via a computational model of analog and digital performance parameters, and how they are related to features it can measure or control. MIPS, for example, may be controlled by setting the clock speed. A high clock speed generally uses more total power than a lower clock speed, and this tends to reduce battery life. The same is true for the brightness of a display. The CR only "knows" this to the degree that it has a data structure that captures this information and algorithms, preprogrammed and/or learned, that deal with these relationships to the benefit of the user. Constraint languages may be used to express interdependencies, such as how many channels of a given personality are supported by a given hardware suite, particularly in failure modes. CR algorithms may employ this kind of structured reasoning as a specialized KS when using case-based learning to extend its ability to cope with internal changes.

The ontological structure of the above may be formalized as shown by the equation in Figure 14.22.

```
<SDR>
   <Sources/> <Channels/> <Personality> <HCI/>
      <Source-Coding-Decoding/> <Networking/> <INFOSEC/>
      <Channel-Codec><Modem/> <IF-Processing/> <RF-Access/> </Channel-Codec>
</Personality>
   <SDR-Platform> <Evolution-Support/>
   </SDR>
```

Figure 14.22: Equation that defines SDR subsystem components.

Although this text does not offer a comprehensive computational ontology of SDR, semantically based dialogs among AACRs about internal issues such as downloads may be mediated by RXML with the necessary ontological structures.

14.6.5 Architecture Migration: From SDR to AACR

Given the CRA and contemporary SDR architecture, one must address the transition of SDR through a phase of AACRs, toward the iCR. As the complexities of handheld, wearable, and vehicular wireless systems increase, the likelihood that the user or network will have the skill necessary to do the optimal thing in any given circumstance is reduced. Today's cellular networks manage the complexity of individual wireless protocols for the user, but the emergence of multiband multimode AACR moves the burden for complexity management toward the PDA. Likewise, the optimization of the choice of wireless service between the "free" home WLAN and the "for-sale" cellular equivalent moves the burden of radio-resource management from the network to the WPDA.

14.6.6 Cognitive Electronics

The increasing complexity of the PDA–user interface also accelerates the trend toward increasing the computational intelligence of personal electronics. AACR is in some sense just an example of a computationally intelligent personal electronics system. For example, using a laptop computer in the bright display mode uses up the battery power faster than when the display is set to minimum brightness.

A cognitive laptop could offer to further the brightness level when reduce only half charged in battery-powered mode. It would be even nicer if it would recognize operation aboard a commercial aircraft and know that the user's preference is to set the brightness low on an aircraft to conserve the battery, and automatically do so. A cognitive laptop shouldn't make a big deal over that, and it should let the user turn up the brightness without complaining. If it had an ambient light sensor or ambient light algorithm for an embedded camera, it also could tell that a window shade was open and that the user has to deal with the brightness. By sensing the brightness of the *on-board aircraft* scene and associating the user's control of the brightness of the display with the brightness of the environment, a hypothetical cognitive laptop could learn advice the user to do the right thing in the right situation (pull down the shade).

How does this relate to the CRA? For one thing, the CRA could be used as-is to increase the computational intelligence of the laptop. In this case, the self is the laptop and the PDA knows about itself as a laptop, not as a WPDA. It knows about its sensors suite, which includes at least a light-level sensor if not a camera through the data structures that define the <Self/>. It knows about the user by observing keystrokes and mouse action as well as by interpreting the images on the camera; it verifies that the user is still the owner because that is important to building user-specific models. It might build a space–time behavior model of any user or it might be a one-user laptop. Its actions include the setting of the display intensity level. In short, the CRA accommodates the cognitive laptop with suitable laptop knowledge and functions implemented in the CRA map sets.

14.6.7 When Should a Radio Transition toward Cognition?

If a wireless device accesses only a single-RF band and mode, then it is not a very good starting point for CR—it's just too simple. Even as complexity increases, as long as the user's needs are met by wireless devices managed by the network(s), then embedding computational intelligence in the device has limited benefits. In 1999, Mitsubishi and AT&T announced the first "four-mode handset." The T250 operated in Time Division Multiple Access (TDMA) mode on 850 or 1900 MHz, in first-generation Advanced Mobile Phone System (AMPS) mode on 850 MHz, and in Cellular Digital Packet Data (CDPD) mode on 1900 MHz. This illustrates the early development of multiband, multimode, multimedia (M3) wireless. These radios enhanced the service provider's ability to offer national roaming, but the complexity was not apparent to the user because the network managed the radio resources of the handset.

Even as device complexity increases in ways that the network does not manage, there may be no need for cognition. There are several examples of capabilities embedded in electronics typically are not heavily used. For example, how many people use their laptop's speech recognition system? What about its Infrared Data Association (IrDA) port? The typical users in 2004 didn't use either capability of their Windows XP laptop all that much. So complexity can increase without putting a burden on the user to manage that complexity if the capability isn't central to the way in which the user employs the system.

For radio, as the number of bands and modes increases, the SDR becomes a better candidate for the insertion of cognition technology. But it is not until the radio or the wireless part of the PDA has the capacity to access multiple RF bands that cognition technology begins to pay off. With the liberalization of RF spectrum use rules, the early evolution of AACR may be driven by RF spectrum-use etiquette for ad hoc bands such as the FCC use case. In the not-too-distant future, SDR PDAs could access satellite mobile services, cordless telephone, WLAN, GSM, and 3G bands. An ideal SDR device with these capabilities might affordably access three octave bands, from 0.4 to 0.96 GHz (skipping the air navigation and GPS band from 0.96 to 1.2 GHz), from 1.3 to 2.5 GHz, and from 2.5 to 5.9 GHz (Figure 14.23). Not counting satellite mobile and radio navigation bands, such radios would have access

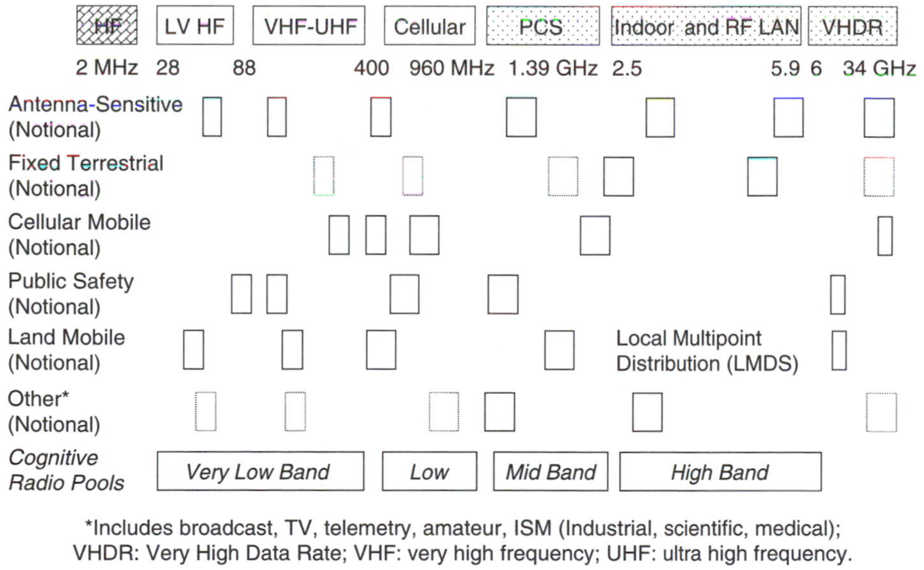

HF	LV HF	VHF-UHF	Cellular	PCS	Indoor and RF LAN	VHDR
2 MHz	28	88	400	960 MHz	1.39 GHz	2.5	5.9	6	34 GHz

Antenna-Sensitive (Notional)

Fixed Terrestrial (Notional)

Cellular Mobile (Notional)

Public Safety (Notional)

Land Mobile (Notional) — Local Multipoint Distribution (LMDS)

Other* (Notional)

Cognitive Radio Pools: Very Low Band | Low | Mid Band | High Band

*Includes broadcast, TV, telemetry, amateur, ISM (Industrial, scientific, medical); VHDR: Very High Data Rate; VHF: very high frequency; UHF: ultra high frequency.

Figure 14.23: Fixed spectrum allocations versus pooling with CR (© 1997 Dr. Joseph Mitola III, used with permission).

to more than 30 mobile subbands in 1463 MHz of potentially sharable outdoor mobile spectrum. The upper band provides another 1.07 GHz of sharable indoor and RF-LAN spectrum. This wideband radio technology will be affordable first for military applications, next for base station infrastructure, then for mobile vehicular radios, and later for handsets and PDAs. When a radio device accesses more RF bands than the host network controls, it is time for CR technology to mediate the dynamic sharing of spectrum. It is the well-heeled conformance to the radio etiquettes afforded by CR that makes such sharing practical.

14.6.8 Radio Evolution toward the CRA

Various protocols have been proposed by which radio devices may share the radio spectrum. The US FCC Part 15 rules permit low-power devices to operate in some bands. In 2003, a Report and Order (R&O) made unused TV spectrum available for low-power RF-LAN applications, making the manufacturer responsible for ensuring that the radios obey this simple constraint. DARPA's XG program developed a language for expressing spectrum use policy [26]. Other more general protocols based on peek-through to legacy users have also been proposed [33].

Does this mean that a radio must transition instantaneously from the SCA to the CRA? Probably not. The six-component AACR architecture may be implemented with minimal SP, minimal learning, and no autonomous ability to modify itself. Regulators hold manufacturers responsible for the behaviors of such radios. The simpler the architecture, the simpler the problem of explaining it to regulators and of getting concurrence among manufacturers regarding open architecture interfaces that facilitate technology insertion and teaming. Manufacturers who fully understand the level to which a highly autonomous CR might unintentionally reprogram itself to violate regulatory constraints may decide they want to field aware–adaptive (AA) radios, but may not want to take the risks associated with self-modifying CRs just yet.

Thus, one can envision a gradual evolution toward the CRA beginning initially with a minimal set of functions mutually agreeable among the growing community of AACR stakeholders. Subsequently, the introduction of new services will drive the introduction of new capabilities and additional APIs, perhaps informed by the CRA.

14.7 Cognition Architecture Research Topics

The cognition cycle and related inference hierarchy imply a large scope of hard research problems for CR. Parsing incoming messages requires NL text interpretation. Scanning the user's voice channels for content that further defines the communications context requires speech processing. Planning technology offers a wide range of alternatives in temporal calculus [27], constraint-based scheduling [28], task planning [29], causality modeling [30], and the like. Resource allocation includes algebraic methods for wait-free scheduling protocols [31], open distributed processing (ODP), and parallel virtual machines (PVMs). Finally, ML remains one of the core challenges in artificial intelligence research [32]. The focus of CRA research, then, is not on the development of any one of these technologies per se. Rather, it is on the organization of cognition tasks and on the development of cognition data structures needed to integrate contributions from these diverse disciplines for the context-sensitive delivery of wireless services by SDR.

Learning the difference between situations in which a reactive response is needed versus those in which deliberate planning is more appropriate is a key challenge in ML for CR. The CRA framed the issues. The CR1 goes further [12], providing useful KS and related ML, so that the CR designer can start there in developing good engineering solutions to this problem for evolving CR applications domains.

14.8 Industrial-Strength AACR Design Rules

The CRA allocates functions to components based on design rules. Typically design rules are captured in various interface specifications, including APIs, and object interfaces, such as Java's JINI/JADE structure of intelligent agents. This chapter so far has introduced the CRA; this section suggests additional design rules by which user domains, sensory domains, and radio RF band knowledge may be integrated into industrial-strength AACR products and systems.

The following design rules further circumscribe the integration of cognitive functions with the other components of a WPDA within the CRA:

1. The cognition function should maintain an explicit (topological) model of space–time of the user, the physical environment, the radio networks, and the internal states of the radio (the <Self/>).

2. The CRA requires each CR to predict, in advance, an upper bound on the amount of computational resources (e.g., time) required for each cognition cycle. The CR is must set a trusted (hardware) watchdog (e.g., a timer) before entering a cognition cycle. If the watchdog is violated, the system must detect

that event, log that event, and mark the components invoked in that event as nondeterministic.

3. The CRA should internalize knowledge as procedural skills (e.g., serModels):
   (a) The CRA requires each CR to maintain a trusted index to internal models and related experience.
   (b) Each CR must preclude cycles from its internal models and skills graph because a CRA conformance requires reliable detection of cycles to break cycles (e.g., via timer) to avoid Gödel-Turing unbounded resource use endemic to self-referential TC computational entities such as AACRs.

4. Context that references space, time, RF, the <User/>, and the <Self/> for every external and internal event shall be represented formally using a topologically valid and logically sound model of space–time–context.

5. Each CR conforming to the CRA shall include an explicit grounding map, M, that maps its internal data structures onto elements sensed in the external world represented in its sensory domains, including itself. If the CR cannot map a sensed entity to a space–time–context entity with specified time allocated to attempt that map, then the entity should be designated "ungroundable."

6. The model of the world shall follow a formal treatment of time, space, RF, radio propagation, and the grounding of entities in the environment.

7. Models shall be represented in an open architecture RKRL suited to the representation of radio knowledge (e.g., a Semantic Web derivative of RKRL). That language shall support topological properties and inference (e.g., forward chaining), but must not include unconstrained axiomatic FOPC, which per force violates the Gödel-Turing constraint.

8. The cognition functions shall maintain location awareness, including (a) the sensing of location from global positioning satellites, (b) sensing position from local wireless sensors and networks, and (c) sensing precise position visually:
   (a) Location shall be an element of all contexts.
   (b) The cognition functions shall estimate time to the accuracy necessary to support the user and radio functions.
   (c) The cognition functions shall maintain an awareness of the identity of the PDA, of its owner, of its primary user, and of other legitimate users designated by the owner or primary user.

9. The cognition functions shall reliably infer the user's communications context and apply that knowledge to the provisioning of wireless access by the SDR function.

10. The cognition functions shall model the propagation of the user's radio signals with sufficient fidelity to estimate interference to other spectrum users:
    (a) The cognition function shall also assure that interference is within limits specified by the spectrum use protocols in effect in its location (e.g., in spectrum rental protocols).
    (b) The cognitive function shall defer control of the <Self/> to the wireless network in contexts where a trusted network manages interference.

11. The cognition functions shall model the domain of applications running on the host platform, sufficient to infer the parameters needed to support the application. Parameters modeled include QoS, data rate, probability of link closure (grade of service), and the space–time–context domain within which wireless support is needed.

12. The cognition functions shall configure and manage the SDR assets to include hardware resources, software personalities, and functional capabilities as a function of network constraints and use context.

13. The cognition functions shall administer the computational resources of the platform. The management of SWR resources may be delegated to an appropriate SDR function (e.g., the SDR Forum domain manager). Constraints and parameters of those SDR assets shall be modeled by the cognition functions. The cognition functions shall assure that the computational resources allocated to applications, interfaces, cognition, and SDR functions are consistent with the user communications context.

14. The cognition functions shall represent the degree of certainty of understanding in external stimuli and in inferences. A certainty calculus shall be employed consistently in reasoning about uncertain information.

15. The cognition functions shall recognize preemptive actions taken by the network and/or the user. In case of conflict, the cognition functions shall defer the control of applications, interfaces, and/or SDR assets to the owner, to the network, or to the primary user, according to the appropriate priority and operations assurance protocol.

14.9 Summary and Future Directions

Often technical architectures of the kind presented in this chapter accelerate the state of practice by catalyzing work across the industry on plug-and-play, teaming, and collaboration. The thought is that to propel wireless technology from limited spectrum awareness toward valuable user awareness, an architecture such as the

CRA will be needed. In short, the CRA articulates the functions, components, and design rules of next-generation stand-alone and embedded wireless devices and networks. Each of the different aspects of the CRA contributes to the government, academic, and industry dialog:

1. The functional architecture identifies components and interfaces for CRs with sensory and perception capabilities in the user domain, not just the radio domain.

2. The cognition cycle identifies the processing structures for the integration of sensing and perception into radio: observe (sense and perceive), orient (react if necessary), plan, decide, act, and learn.

3. The inference hierarchies suggest levels of abstraction helpful in the integration of radio and user domains into the synthesis of services tailored to the specific user's current state of affairs given the corresponding state of affairs of the radio spectrum in space and time.

4. The introduction to ontology suggests an increasing role for Semantic Web technologies in making the radios smarter—initially, about radio; over time, about the user [see 12].

5. Although not strictly necessary for CR, SDR provides a very flexible platform for the regular enhancement of both computational intelligence and radio capability, particularly with each additional Moore's law cycle.

6. Finally, this chapter has introduced the CRA to the reader interested in the cutting edge, but it has not defined the CRA. The previous section suggested a few of the many aspects of the embryonic CRA that must be addressed by researchers, developers, and markets in the continuing evolution of SDR toward ubiquitous and "really fun" AACRs.

In conclusion, AACR seems headed for the Semantic Web, but the markets for services layered on practical radio networks will shape that evolution. Although many information-processing technologies from e-Business Solutions to the Semantic Web are relevant to AACR, the integration of audio and visual SP into SDR with suitable cognition architectures remains both a research challenge and a series of increasingly interesting radio systems designs opportunities. A CRA that is broadly supported by industry could accelerate such an evolution.

References

[1] J. Mitola III, *Software Radio Architecture*, Wiley Interscience, New York, September 2000.

[2] P. Mahonen, *Cognitive Wireless Networks*, RWTH Aachen, Aachen, DE, September 2004.

[3] www.sdrforum.org

[4] www.omg.org

[5] Eriksson and Penker, *UML Toolkit*, John Wiley & Sons, Inc., New York, 1998.

[6] T. Mowbray and R. Malveau, *CORBA Design Patterns*, John Wiley & Sons, Inc., New York, 1997.

[7] www.jtrs.mil

[8] Wireless World Research Forum (www.wwrf.com) 2004.

[9] http://www.wikipedia.org/

[10] J. Mitola III, "Software Radio Architecture: A Mathematical Perspective," in *IEEE JSAC*, IEEE Press, New York, April 1998.

[11] R. Hennie, *Introduction to Computability*, Addison-Wesley, Reading, MA, 1997.

[12] J. Mitola III, *Cognitive Radio Architecture*, John Wiley & Sons, Ltd., New York, 2006.

[13] www.omg.org/UML

[14] J. Mitola III, *Cognitive Radio: An Integrated Agent Architecture for Software Defined Radio*, KTH, The Royal Institute of Technology, Stockholm, June 2000.

[15] L. Erman, F. Hayes-Roth, V. Lesser and R. Reddy, "The Hearsay-II Speech Understanding System: Integrating Knowledge to Resolve Uncertainty," *ACM Computing Surveys*, Vol. 12, No. 2, June 1980, pp. 213–253.

[16] SNePS (Internet: ftp.cs.buffalo.edu:/pub/sneps/) 1998.

[17] Koser, et al., *"read.me" www.cs.kun.nl*, University of Nijmegen, The Netherlands, March 1999.

[18] The XTAG Research Group, *A Lexicalized Tree Adjoining Grammar for English Institute for Research in Cognitive Science*, University of Pennsylvania, Philadelphia, PA, 1999.

[19] PC-KIMMO Version 1.0.8 for IBM PC, February 18, 1992.

[20] J. Mitola III "Software Radio: Survey, Critical Evaluation and Future Directions," in *Proceedings of the National Telesystems Conference*, IEEE Press, New York, May 1992.

[21] J. Mitola III, "Software Radio Architecture," in *IEEE Communications Magazine*, IEEE Press, New York, May 1995.

[22] W. Tuttlebee, *Software Defined Radio Enabling Technologies*, John Wiley & Sons, Ltd., West Sussex, UK.

[23] J.H. Reed, *Software Radio: A Modern Approach to Radio Engineering*, Prentice Hall, Upper Saddle River, NJ.

[24] J. Mitola III and Z. Zvonar (Ed.), *Software Radio Technologies*, IEEE Press, New York, 1999.

[25] F. Jondral, *Software Radio*, Universität Karlsruhe, Karlsruhe, Germany, 1999.

[26] P. Marshall, *Remarks to the SDR Forum*, September 2003.

[27] C. Phillips, "Optimal Time-Critical Scheduling," in *STOC 97* (www.acm.org: ACM), 1997.

[28] L. Esmahi, et al., "Mediating Conflicts in a Virtual Market Place for Telecommunications Network Services," in *Proceedings of the Fifth Baiona Workshop on Emerging Technologies in Telecommunications*, Universidade de Vigo, Vigo, Spain, 1999.

[29] S.K. Das, et al., "Decision Making and Plan Management by Intelligent Agents: Theory, Implementation, and Applications," in *Proceedings of Autonomous Agents 97* (www.acm.org: ACM), 1997.

[30] J. Pearl, *Causality: Models, Reasoning, and Inference*, Morgan-Kaufmann, San Francisco, CA, March 2000.

[31] R. Michalski, I. Bratko and M. Kubat, *Machine Learning and Data Mining*, John Wiley & Sons, Ltd., New York, 1998.

[32] K. Clark and S.A. Tarnlund, *Logic Programming*, London Academic Press, 1982.

[33] J. Mitola III, "Cognitive Radio for Flexible Mobile Multimedia Communications," in *Mobile Multimedia Communications (MoMUC 99)*, IEEE Press, New York, November 1999.

Cognitive Radio Performance Analysis

James O. Neel, Jeffrey H. Reed and Allen B. MacKenzie

*Wireless@Virginia Tech, Bradley Department of Electrical
and Computer Engineering, Virginia Tech,
Blackburg, VA, USA*

15.1 Introduction

The preceding chapters covered many of the challenges that must be addressed to implement a cognitive radio, including how to build a cognitive radio on a software radio platform, how to achieve network and positional awareness, how a radio can be designed to learn, and how to incorporate internal representations of the world and the device into the radio. Even though realizing cognitive radio requires significant work to apply and refine these techniques, it seems reasonable to assume that cognitive radios are in our near future.

But what happens when we deploy cognitive radios? Will they perform as we expect when faced with a realistic environment? And exactly what is a realistic environment?

To answer these questions, we need to take a slightly different view of how cognitive radios will operate. The preceding chapters frequently made use of the cognition cycle to understand the operation of cognitive radios. However, the cognition cycle presents a limited view of the *outside world* (the environment in which the cognitive radios are observing, learning, and reacting). A more realistic view of the operation of cognitive radios would depict an outside world whose state is jointly determined by the adaptations of several cognitive radios. Thus, for the purposes of understanding how cognitive radios will behave, it is more accurate to envision cognitive radios operating as shown in Figure 15.1 where cognitive radios react to an outside world determined by both "dumb" and other cognitive radios.

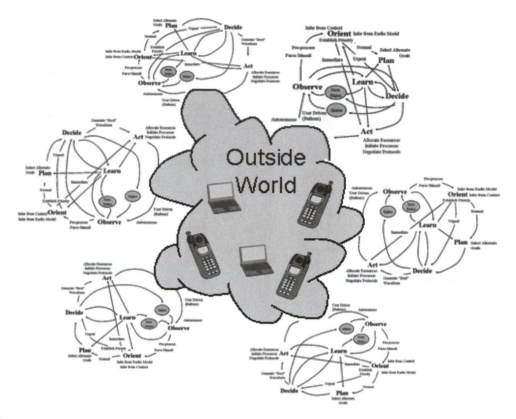

Figure 15.1: The interactive cognitive radio problem. In addition to the interference profile produced by the static devices in the outside world, cognitive radios will also have to respond to the adaptations of other cognitive radios. Modified from Mitola [1], Figure 4.2.

While we intuitively understand how a cognitive radio will react to a collection of "dumb" radios, the interaction of a collection of cognitive radios is less clear. In such a case, each cognitive radio waveform adaptation changes the state of the outside world and impacts the observations, performance, and adaptations of the other radios. Abstractly, this interaction can be viewed as a recursive interactive decision process in which each adaptation can spawn an infinite sequence of adaptations.

The existence of this recursive process poses several difficult questions that we should answer before deploying cognitive radios:

- Will the recursion have a steady state (or steady states) that we can identify so we can anticipate performance?

- Will that performance be desirable or will the adaptations result in a vicious cycle of deteriorating network performance?

- What conditions will be necessary to ensure these adaptations arrive at a desirable steady state?

- Will the steady states be stable or will the inherent variations of the wireless medium make the system unpredictable?

These are the sorts of questions that any regulatory body, any developer, and any end user of cognitive radios would want answered before deploying cognitive radios. However, cognitive radios with limited capabilities are already being deployed.

For example, a radio that implements the 802.11e protocol is for all intents and purposes a policy-based radio as its operation in terms of power levels and bands of operation are determined by the policy regime in which the radio operates. Staying within the WiFi® collection of standards, the Wireless World Research Forum (WWRF) [2] has noted that a key portion of the 802.11h protocol—dynamic frequency selection (DFS)—has been termed a "cognitive function." To see why an 802.11h wireless local area network (WLAN) might be considered a cognitive radio, consider that the 802.11h protocol requires that a WLAN be capable of the following tasks:

- *Observation*: 5.4.4.1 in IEEE Standard 802.11h [3] requires WLANs to estimate channel characteristics such as path loss and link margin, and 5.4.4.2 further requires the radios to estimate channel characteristics such as path loss and link margin.

- *Orientation*: Based on these observations, the WLAN has to determine if it is operating in the presence of a radar installation, in a bad channel, in a band with satellites, or in the presence of other WLANs.

- *Decision*: Based on the situation that it is encountering, it has to decide to change its frequency of operation (DFS), adjust the transmit power (*transmit power control* (TPC)), or both.

- *Action*: The WLAN has to then implement this decision.

Other than learning, each 802.11h WLAN is required to implement all of the major processes of the cognition cycle (observation, orientation, planning, decision, action, learning). Because definitions of cognitive radios vary from group to group, we should consider other formulations of a cognitive radio when evaluating whether cognitive radios are being deployed. The cognitive radio study group at Virginia Tech [4] defines a cognitive radio as:

> *an adaptive radio that is capable of the following:*
> *(a) awareness of its environment and its own capabilities,*

(b) *goal driven autonomous operation,*

(c) *understanding or learning how its actions impact its goal, and*

(d) *recalling and correlating past actions, environments, and performance.*

From this definition, it is apparent that only the capability of "recalling and corre-lating past actions, environments, and performance" is not required as part of the standard. However, if we move beyond the requirements of the standard to expected implementations, it seems reasonable that many vendors will include and leverage some memory of past observations (useful for detecting intermittent transmitters), which implies that the cognitive radio definition will be satisfied. With basic cognitive radios already experiencing limited deployment, and with more complex cognitive radios on the way, it is vital that we develop techniques for analyzing cognitive radio interactions sooner rather than later.

Frequently in engineering, we understand new complicated problems via sim-ulation and experimentation. However, this is generally a very time-consuming approach even when we are studying limited systems under limited scenarios. For example, Ginde et al. [5] presented a desktop simulation of an abstracted General Packet Radio Services (GPRS) network that incorporated power and rate adapta-tions that required days to fully simulate all possible combinations of powers and rates, and that system had just seven subscriber units operating from fixed posi-tions. Expanding this simulation to account for more units, different positions, and mobility would have required *months* of simulation time.

So, when possible, we would prefer to adopt an approach that is able answer our questions in just minutes by mathematically analyzing the structure and characteris-tics of the interactions of cognitive radio algorithms. As such, the goal of this chap-ter is to present a methodology suitable for quickly analyzing many cognitive radio networks with interactive and recursive decision processes with a particular focus on the kinds of adaptive and cognitive radio algorithms that are deployed today.

The remainder of this chapter is arranged as follows. Section 15.2 provides an overview and formalization of the cognitive radio network analysis problem. Section 15.3 addresses traditional engineering techniques that are useful for ana-lyzing distributed adaptive radio systems, including dynamical systems theory, contraction mappings, and Markov models. To introduce a foundation for address-ing a broader range of applications, Section 15.4 defines the basic elements of game theory and describes how this tool, originally used in the field of economics, can be applied to the analysis of cognitive radio networks. Building upon this foundation, Section 15.5 presents two important game models that facilitate rapid analysis of many cognitive radio network algorithms. Section 15.6 then presents a number of extended analyses of distributed algorithms that we can expect to see implemented in cognitive radio networks.

15.2 The Analysis Problem

This section develops a formal model of a network of cognitive radios that is used for the reminder of this chapter, defines the analysis objectives of this chapter, and provides a brief mathematical refresher.

15.2.1 Mathematical Preliminaries

To a large extent, this chapter is self-contained, which means that we believe everything in this chapter can be understood by anyone with an undergraduate exposure to engineering mathematics. However, to go beyond the discussion of this chapter, the reader would benefit from an understanding of real analysis, optimization theory, parallel processing, control theory, and game theory. Before continuing, however, we need to review some symbology and terminology used throughout the remainder of this chapter—a necessity for a chapter premised on mathematical analysis. For instance, throughout this chapter, the symbols listed in Table 15.1 express concepts in a more concise manner.

Table 15.1: Mathematical symbology.[1]

Symbol	Meaning	Symbol	Meaning
$\forall$	"for all"	$\rightarrow$	"from ... to" (functions)
$\in$	"contained in"	$\exists$	"there exists"
$\setminus$	"excluding"	$:$	"such that"
$\Rightarrow$	"implies"	N	Set of natural numbers
$\rightarrow$	"goes to" (sequences)	$\Re$	Set of real numbers
$\|\cdot\|$	"norm" "Euclidean distance"	$\cup$	Union of sets
$|\cdot|$	"magnitude of" [a variable] "number of elements in" [a set]	$\subset$	"is a subset of"
∞	"infinity"	$\times$	"Cartesian product"
∇	"gradient"	$\lesssim_j$	"j prefers"
$\equiv$	"is defined as"	∂	"partial derivative"

---

[1] Due to the considerable mathematical treatment in this chapter, we will provide prose explanations as required for readability.

Additionally, this chapter makes use of a few other terms whose meanings are important to review. A concept important to optimization and convergence, a set is said to be *closed* if it contains the *limit points* of all sequences in the set. Given a sequence $\{a^1, a^2, \ldots a^n, \ldots\}$, a limit point to the sequence is a point, a^*, such that $a^n \rightarrow a^*$ as $n \rightarrow \infty$. For example, the interval $[0,1]$ and the set $\{a \in \Re \cup \infty: a \geq 0\}^2$ are both closed, but the interval $[0,1)$ is not closed as all the points in the infinite sequence $1 - 1/n$ are in $[0,1)$, but its limit point, 1, is not in the set.

A set is said to be *open* if every point in the set has a neighborhood (the set of all points that are less than some distance away from a specified point) that is also in the set. A set, A, is said to be *convex* if given two points $a^1, a^2 \in A$, all points of the form $a^3 = \lambda a^1 + (1 - \lambda)a^2$ are also in A where $\lambda \in [0,1]$. In other words, a set, A, is convex if for every pair of points $a^1, a^2 \in A$, all points on the line connecting a^1 and a^2 are also in A.

Among other definitions for compactness, a set is *compact* if it is both closed and bounded. For example $[0,1]$ is compact but $\{a \in \Re \cup \infty: a \geq 0\}$ is not. If a cognitive radio operated over a set of 10 frequencies, perhaps by virtue of being a Velcro radio,[3] then this set of frequencies is compact, but not convex.

A function $f: A \rightarrow f(A)$, where A is some set, is said to be *continuous* if $f(a^k) \rightarrow f(a^\infty)$ as $k \rightarrow \infty$ for all $a^\infty \in A$ such that there is a sequence $\{a^k\} \in A: a^k \rightarrow a^\infty$. It should be noted that every function with a finite domain (A is finite, i.e., $|A| < \infty$) is continuous.

15.2.2 A Formal Model of a Cognitive Radio Network

While algorithm specific variables will be defined as they are used, this chapter assumes a basic model of cognitive radio interactions that uses of the following symbols and conventions:

N: The (finite) set of cognitive radios under study.

For convenience we say that n represents the number of elements in N (i.e., $n = |N|$).

i, j: Particular cognitive radios in N.

A_j: The set of actions available to cognitive radio j.

---

[2] This example means "the set of all real numbers a which are greater than or equal to zero union with infinity."

[3] Velcro radios are radios with multiple single-band radio frequency (RF) chip sets in a given handset; they are sometimes called slice radios.

Although these sets are quite limited for today's cognitive radios, they include all available adaptations to the radio. Because the adaptations can include a number of independent types of adaptations (e.g., power levels, modulations, channel and source coding schemes, encryption algorithms, medium access control (MAC) algorithms, center frequencies, bandwidths, and routing algorithms), A_j will generally be a multidimensional set. To simplify matters, throughout this chapter we assume we are analyzing adaptations only over a short time interval so that A_j will not be a function of time. However, for longer time intervals, A_j could be expected to grow as cognitive radio j learns new waveforms.

A: The *action space*, or the set of all possible combinations of actions by the cognitive radios.

Throughout this chapter we assume that A is formed by the Cartesian product of the radios' action sets, or $A = A_1 \times A_2 \times \cdots \times A_n$. For some algorithms, it is convenient to think of A as a vector space with orthogonal bases A_1 through A_n.

a: An *action tuple*, or a particular combination of actions in which each cognitive radio in N has implemented a particular action or waveform (equivalently, a is a point in A).

Cognitive radio j's contribution to a is written as a_j, and the choice of actions by all cognitive radios other than j is written as a_{-j}.

O: The *observed outcome space*, or the set of all possible realizations of the outside world as determined by the choice of actions available to each cognitive radio and each radio's operating environment and as observed by each radio.

o: An *observation tuple*, or a particular combination of observations in which each cognitive radio in N has made a particular observation (equivalently, o is a point in O).

Cognitive radio j's contribution to o is written as o_j, and the observations made by all cognitive radios other than j is written as o_{-j}. Frequently, we also refer to o as an *outcome* or observed outcome. The importance in distinguishing between actions and observations can be seen in the following example.

   Consider two cognitive radios, {1,2}, with actions (waveforms) $\{\omega_{1_a}, \omega_{1_b}\}$ and $\{\omega_{2_a}, \omega_{2_b}\}$, respectively, that are communicating with a common receiver that reports back to each cognitive radio its *signal-to-interference ratio* (SIR).

Assuming, for simplicity, fixed transmission powers, the actual outcomes the radios observe can be described by:

$$o_j = \gamma_j = \frac{g_j}{g_{-j} \, |\rho(\omega_{j_k}, \omega_{-j_m})|} \qquad (15.1)$$

where $j \in \{1,2\}$, γ_j is the observed SIR for radio j, g_j is the link budget gain of radio j to the common receiver, g_{-j} is the gain of the radio other than j to the common receiver, and $|\rho(\omega_{j_k}, \omega_{-j_m})|$ is the absolute value of the statistical correlation between the waveforms chosen by the radios. In this scenario, there are four different possible elements in A, which form the set $\{(\omega_{1_a}, \omega_{2_a}), (\omega_{1_a}, \omega_{2_b}), (\omega_{1_b}, \omega_{2_a}), (\omega_{1_b}, \omega_{2_b})\}$. However, there are an infinite number of possible observations due to the infinite number of possible channel realizations between the radios and their common receiver. As was the case in this example, we assume throughout this chapter that traditional communications have been used to supply an expression that relates the observed outcome with the implemented waveforms (action tuple).

T_j: The times at which cognitive radio j can update its decision (a radio may have a time allocated for updating, but choose not to update its decision).

Unless stated otherwise, we assume that each T_j is infinite, or $T_j = \{t_j^0, t_j^1, \ldots, t_j^m, \ldots\}$.

T: The set of all times where decision updates can occur, or $T = T_1 \cup T_2 \cup \ldots \cup T_n$, where $t \in T$ denotes a particular time at which an update can occur.

For notational convenience, we use t^k to represent the kth element of T arranged in chronological order.

$f_j(o)$: The *decision update rule*, which describes how radio j updates its decisions given the observation o, or $f_j: O \rightarrow A_j$.

When there is a clear and well-defined mapping between A and O, we treat f_j as a function mapping a to an a_j, or $f_j: A \rightarrow A_j$. When appropriate, we also use the notation f^t to denote the network update function at time t where in general f^t captures the adaptations of the subset $M \subset N$ of radios that update their decisions at time t, or $f^t = \underset{j \in M}{\times} f_j$. Although it is also possible that a radio bases its decisions on past observations and predictions about the future state of the network, for this chapter we assume that f_j^t is a function of only cognitive radio j's most recent observation.

    Additionally, we make use of the following conventions when describing the timing of the decision update process. If $\forall t \in T, f^t = \underset{j \in N}{\times} f_j$, then we say that the network is *synchronously* updating its decisions and we write $a(t^{k+1}) = f^{tk}(a(t^k))$. If $t_1^m < t_2^m < t_3^m < \cdots < t_n^m < t_1^{m+1}$, then we say that the network is updating its decisions in *round-robin* order; if $\forall t \in T\ f^t = \underset{j \in N}{rand}\{f_j\}$, then we say that the network is updating its decisions in *random* order; and if no structure can be inferred about the updating process, we say that the network is updating its decisions in an *asynchronous* manner.

For asynchronously updating networks, there may be some points in time where $t_i^m = t_j^k$ (the mth update of radio i occurs at the same time as the kth update of radio j) is satisfied for two or more radios. As shown in subsequent sections of this chapter, these different decision update timings—synchronous, round robin, random, and asynchronous—can have a significant impact on the behavior of a cognitive radio network.

Systems with synchronous timings are most frequently encountered in centralized systems and rarely will be encountered in an interactive cognitive radio decision process, which implies some degree of distributed decision timings. A round-robin scheme can occur in centralized systems with distributed decision-making with scheduling (as might occur in a hybrid automatic repeat request (ARQ) scheme). Without a synchronizing agent and assuming an arbitrary fineness in the time scale, every distributed cognitive radio algorithm is a randomly time-ordered process. However, due to the coarseness of the timing of observation processes, many systems with random timings can be expected to behave in a manner that is indistinguishable from an asynchronous system.

15.2.3 Analysis Objectives

Before formally presenting the analysis objectives, consider a network of three radios, {1,2,3}, where each radio, j, can choose an action, a_j, which is drawn from a convex action set according to a decision update rule, f_j. Starting at any initial action vector, repeated applications of the radios' decision update rules trace out paths in the action space. Sometimes these paths terminate in a stable point; under different conditions the paths may enter into an infinite loop. There may also be points in the action space that the decision update rules would not independently adapt away from, but after a small external perturbation (perhaps from noise or channel fluctuations) the decision rules drive the network away from these points.

Each of these concepts is illustrated in the example interaction diagram shown in Figure 15.2, in which paths are shown by arrows and steady states are labeled as "NE (Nash Equilibrium)" for reasons that will become clear later in this chapter.

This conceptual interaction diagram illustrates the four different analysis questions that we would like to answer when analyzing a network of cognitive radios:

- What is the expected behavior of the network?
- Does this behavior yield desirable performance?
- What conditions must be satisfied to ensure that adaptations converge to this behavior?
- Is the network stable?

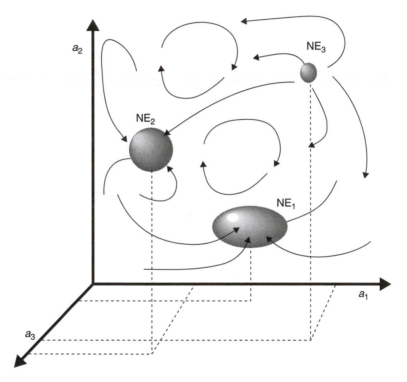

Figure 15.2: A three-radio interaction diagram. The diagram shows three steady states (NE$_1$, NE$_2$, and NE$_3$) and adaptation paths.

The following four subsections formalize each of these questions into specific analysis objectives.

Establishing Expected Behavior

As is the case for many systems, the analysis in this chapter assumes that the expected behavior of a cognitive radio network is equivalent to its steady-state behavior. Accordingly, to establish expected behavior, we analyze the following steady-state issues:

- *Existence*: Does the system have a steady state?
- *Identification*: What are the specific steady states for the system?

The following sections introduce a number of different techniques to address these issues, including showing that f^t is a variant of a *contraction mapping*, that the network can be modeled as an *absorbing Markov chain*, and that f^t has a fixed point, that is, an action vector $a*$ such that $a* = f^t(a*) \ \forall t \in T$.

Desirability of Expected Behavior

Of course, determining a cognitive radio network's steady states tells us little about the desirability of the algorithm under study. We also need to address whether those steady states are "good" steady states or "bad" steady states, and whether there are other action vectors that would be preferable from a network designer's perspective. Again, there are two specific issues to be addressed:

- *Desirability*: How "good" are the steady states of the algorithm?
- *Optimality*: Does an optimal action vector exist, and how close do the steady states come to achieving optimal performance?

There are many different ways of identifying whether an action vector is a "good" steady state, but we assume in this text that there is some global objective or cost function, $J: A \rightarrow \Re$ that we wish to maximize or minimize (perhaps total system goodput or total network interference, respectively).[4] Assuming we wish to maximize J, we would treat action vector a^2 as more desirable than a^1 if $J(a^2) > J(a^1)$. To determine if an optimal action vector exists and if our steady states are indeed optimal, we introduce gradient techniques and *Pareto optimality* criteria. These concepts are discussed in greater detail in sections *Establishing Optimality* and *Desirability*.

Convergence Conditions

Along with identifying that a cognitive radio algorithm has desirable steady states, it is important to also identify the conditions under which the decision update algorithms *converge* to a steady state. Formally, given a sequence of action

---

[4] J is a global objective or cost function that maps from the action space to the set of real numbers.

vectors, $\{a(t)\}$, formed from the decision update rules as $a(t^{k+1}) = f^t(a(t^k))$ and an initial action vector $a(0)$, we say that $\{a(t)\}$ *converges* to action vector a^* if for every $\varepsilon > 0$, there is some time $t^* \in T$ such that $t \geqslant t^*$ implies $\|a(t), a^*\| < \varepsilon$. In other words, $\{a(t)\}$ converges to a^* if for every arbitrarily small *neighborhood* of a^* (a neighborhood of a^* is the set $S \subset A$ such that $\|a, a^*\| < \varepsilon \; \forall a \in S$), there is a time after which $\{a(t)\}$ remains "trapped" in that neighborhood of a^*.

For convergence, we are interested in addressing the following issues:

- *Rate*: Given $\{a(t)\}$, a^*, f^t, and our neighborhood, what is the value or expected value of t^*?

- *Sensitivity*: Do changes in the value of $a(0)$ or a different realization of f^t (perhaps modeling an asynchronous system instead of a round-robin system) impact convergence, and if so, how?

For the various analysis techniques that we discuss in the following sections, we highlight how those techniques provide insight into convergence rate and how different decision update rules impact convergence.

Network Stability

The preceding analysis objectives assume the radios are behaving in a deterministic manner, but wireless networks are stochastic. Thus the cognitive radios are actually responding to estimates of their operating environment in terms of their desired signals and their interference profiles. Accordingly, the radios will frequently make mistakes in their adaptations due to the imperfect information with which the radios operate.

For this chapter, we assume the radios' estimates are unbiased and their errors random. However, this does lead to stability concerns because small perturbations could potentially lead to undesirable behavior. Accordingly, this chapter addresses the following analytical issues with respect to an algorithm's steady state(s):

- *Lyapunov stability*: After a small perturbation, will the system stay within a neighborhood of the steady state?

- *Attractivity*: After a small perturbation, will the network converge back to the steady state?

Section 15.3.3 defines Lyapunov stability and attractivity with greater rigor and introduces techniques for determining whether a network has Lyapunov stable and attractive fixed points.

15.3 Traditional Engineering Analysis Techniques

This section reviews some traditional engineering techniques from dynamical systems, optimization theory, parallel processing, and Markov chain theory that can be leveraged to analyze cognitive radio networks.

15.3.1 A Dynamical Systems Approach

Dynamical systems theory is concerned, as the name implies, with analyzing the behavior of dynamical systems, and also designing mechanisms so the systems act in a desirable manner. Typical analysis goals of dynamical systems theory are similar to the ones set out in Section 15.2.3: to determine the expected behavior, convergence, and stability of the system.

Formally, a dynamical system is a system whose change in state is determined by a function of the current state and time. In other words, a dynamical system is any system of the form given by:

$$\dot{a} = g(a,t) \tag{15.2}$$

which describes the change in the state of a system as a function of the system state, a, and time, t.[5] Implicitly, the system is assumed to be at state $a(0)$ at time $t = 0$.

When Eq. (15.2) is not directly dependent on t (i.e., $\dot{a} = g(a)$), the system is said to be *autonomous*. For our purposes, it makes sense to treat synchronous systems as autonomous, but for random and asynchronous systems, it is difficult to eliminate the time dependency.

The first goal of a dynamical systems analyst is to solve Eq. (15.2) to yield the *evolution function* that describes the state of the system as a function of time. This typically involves solving an ordinary differential equation—a task preferably not undertaken without knowing that a solution exists. In general, this solution would take the form we supposed existed for the decision update rule, f^t, in our model of Section 15.2.2.

Given a dynamical systems model, we can be assured that such a solution exists by the Picard–Lindelöf theorem [6], which states that given an open set $D \subset A \times T$, if g is continuous on D and *locally Lipschitz*[6] with respect to a for

---

[5] The change in system state is a function of the current state and time.
[6] A function is Lipschitz continuous if there exists a finite real K, such that for all action states a^1 and a^2 in the action space, the Euclidean distance between their next action state in time is less than K times the distance between their current action states.

every $a \in D$, then there is a unique solution, f^t, to the dynamical system for every $a(0)$ while f^t remains in D.

A function,[7] $f^t \colon A \times T \to A, A \subset \Re^n$ is said to be *Lipschitz continuous* at (a, t) if there exists a $K < \infty$ such that $\|f^t(a^1, t) - f^t(a^2, t)\| \leq K \|a^1 - a^2\|$ for all a^1, $a^2 \in A$; f^t would be *locally* Lipschitz continuous if this condition were satisfied only for some open set $D \subset A \times T$. Similarly the function f is *Lipschitz continuous* if it is Lipschitz continuous for all $(a, t) \in A \times T$. Any function that is Lipschitz continuous is also continuous.

Fixed Points and Solutions to Cognitive Radio Networks

A solution for the evolution function f^t may imply a system that is changing states over time, perhaps bounded within a certain region or wandering over the entire action space. For some systems, continual adaptations may not be an issue and may even be desirable. However, continual adaptations for a cognitive radio network implies that significant bandwidth is being consumed to support the signaling overhead required to support these adaptations.

For a cognitive radio network, we would prefer that the network settle down to a particular steady state and adapt only as the environment changes. Identifying these steady states also allows a cognitive radio designer to predict network performance. In the context of our state equation, such a steady state is a *fixed point* of f^t—a point $a^* \in A$ such that $a^* = f^t(a^*) \; \forall t \geq t^*$. For one-dimensional sets, it is convenient to envision a fixed point of a function as a point where the function intersects the line $x = f(x)$. Figure 15.3 illustrates a function, $f(x)$, that has three fixed points.

Figure 15.3: A function with three fixed points (circled). For functions on one-dimensional sets, the points at which the function intersects the line $f(x) = x$ (dashed) are fixed points.

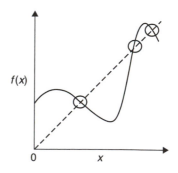

---

[7] f^t is a function that maps from the Cartesian product of the action space with the set of all update times to the action space, where the action space is a subset of all real n tuples; that is, given an initial action state, the function describes how the network state changes overtime.

Solving for fixed points can be tedious because it may involve a search over the entire action space (an impossibility for an infinite action space, and a considerable undertaking for most realistic finite action spaces), so we would like to know if a fixed point even exists before we begin our search. Fortunately, this can be readily established by the Leray–Schauder–Tychonoff fixed point theorem, given by Proposition 1.3 in Chapter 3 of Bertsekas and Tsitsiklis [7], which states that if $A \subset \mathfrak{R}^n$ is *nonempty*, *convex*, and *compact* (recall the definitions in Section 15.2.1), and if $f^t \colon A \to A$ is a *continuous* function, then there exists some $a* \in A$ such that $a* = f^t(a*)$.[8] Note that this definition is inappropriate for finite action sets, which, although compact, are not convex.

However, actually solving for a fixed point under such general conditions can be much more difficult. Subsequent sections refine these conditions to facilitate fixed point identification.

Establishing Optimality

Perhaps the easiest way to establish that a solution to a cognitive radio network is optimal is to show that it maximizes (or minimizes) some objective function $J \colon A \to \mathfrak{R}$. For networks with a finite action space, we can perform an exhaustive search and evaluate J at each point in A.

However, this approach is impractical for infinite action spaces. When J is differentiable and A is a compact interval of $\mathfrak{R}^n$, however, we can reduce the search space by noting that if a particular action vector, $a*$, is optimal, then $a*$ must either be a boundary point or $\nabla J(a*) = 0$. Recall that

$$\nabla J(a) = \frac{\partial J(a)}{\partial a_1} \hat{a}_1 + \frac{\partial J(a)}{\partial a_2} \hat{a}_2 + \cdots + \frac{\partial J(a)}{\partial a_n} \hat{a}_n,\text{[9]}$$

where each $\hat{a}_j$ is a dimension of A. So, in effect, this condition says that for $a*$ to optimize J, there must be no direction that can be followed from $a*$ that increases J. If J is *pseudoconcave*, we can change this to a sufficient condition—if there exists some point such that $\nabla J(a*) = 0$, then is optimal. J is said to be pseudoconcave if $\nabla J(a'')(a' - a'') \leq 0 \Rightarrow J(a') \leq J(a'')$ for all points $a', a'' \in A$ [8]. More familiarly, a

---

[8] Leray–Schauder–Tychonoff actually considers continuous *mappings* instead of continuous functions, but a continuous function is a continuous mapping.

[9] The gradient of the cost function J, is in general a vector-valued function that, when evaluated at a particular point, a, indicates the magnitude and direction of greatest increase for J at a. When J is a function of a *single* dimension, then the gradient of J is equivalent to the slope of J.

function that is *concave* is also pseudoconcave. Formally, a function, $J{:}A \rightarrow \mathfrak{R}$, is concave on the set A if for all $a_1, a_2 \in A$, $J(\lambda a_1 + (1 - \lambda)a_2) \geqslant \lambda J(a_1) + (1 - \lambda)$ $J(a_2)$ for all $\lambda \in [0,1]$. Equivalently, a function is concave if it is impossible to join two points in the function with a line that contains points above the function.

Figure 15.4 shows an example of a function that is pseudoconcave, but not concave. This function can be verified to not be concave by considering a line joining the points $(0,0)$ and $(1,1)$ (shown as a dashed line); except for the end-points, all of the points in this line lie above the function.

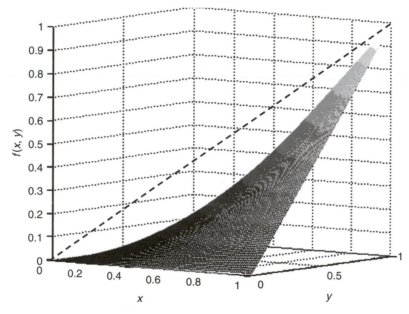

Figure 15.4: $f(x,y) = xy$, $x, y > 0$. A function that is pseudoconcave, but not concave.

Convergence and Stability

When discussing convergence and stability of a decision rule's fixed point, it is convenient to make use of two forms of stability: *Lyapunov stability* and *attractivity*.

Formally, we say that an action vector, a^*, is *Lyapunov stable* if for every $\varepsilon > 0$ there is a $\delta > 0$ such that for all $t \geqslant t^0$, $\|a(t^0), a^*\| < \delta \Rightarrow \|a(t), a^*\| < \varepsilon$.[10] Even though no particular relation between δ and ε can be inferred from this

---

[10] Equivalently, the action vector a^* is said to be Lyapunov stable if for every arbitrarily sized $\varepsilon > 0$, it is possible to identify a $\delta > 0$ such that after a perturbation to any point $a(t^0)$, all subsequent action vectors are no more than a Euclidean distance of ε away from a^*.

definition, an engineer may be more comfortable thinking of Lyapunov stability as akin to bounded-input, bounded-output stability, wherein after a bounded "stimulus" of δ is added to a system operating at $a*$, the system remains within a bounded distance ε of $a*$.

The action vector $a*$ is said to be *attractive* over the region $S \subseteq A$, $S = \{a \in A \mid \|a, a*\| < M\}$, if given any $a(t_0) \in S$, the sequence $\{a(t)\}$ converges to $a*$ for $t \geqslant t_0$. We say that $a*$ is *asymptotically stable* if it is both Lyapunov stable and attractive.

Note that Lyapunov stability does not imply attractivity nor does attractivity imply Lyapunov stability. For instance, the fixed point (0,0) in Figure 15.5 is Lyapunov stable, but not attractive; in contrast, the fixed point (0,0) in Figure 15.6 is attractive, but not Lyapunov stable. However, a method does exist for simultaneously establishing both stability and attractivity—Lyapunov's Direct Method.

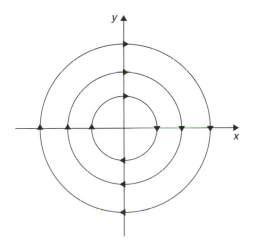

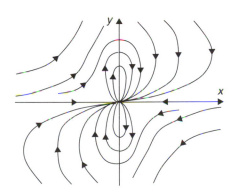

Figure 15.5: Paths for a system that is Lyapunov stable but not attractive (paths are formed by iterative application of f^t and shown arrows).

Figure 15.6: Paths for a fixed point that is attractive but not Lyapunov stable.

Lyapunov's Direct Method for Discrete Time Systems

Instead of attempting to directly apply the definitions of Lyapunov stability and asymptotic stability, we can use Lyapunov's direct method. The discrete time version of Lyapunov's direct method is given by Medio and Lines [9], Theorem 3.4, as follows:

Given a recursion $a(t^{k+1}) = f^t(a(t^k))$ with fixed point $a*$, we know that $a*$ is Lyapunov stable if there exists a continuous function (known as a Lyapunov function) *that maps a neighborhood of a* to the real numbers, or L:N(a*) → ℜ, such that the following three conditions are satisfied:*

1. $L(a^*) = 0$.
2. $L(a) > 0 \; \forall a \in N(a^*) \backslash a^*$.
3. $\Delta L(a(t)) \equiv L[f^t(a(t))] - L(a(t)) \leqslant 0 \; \forall a \in N(a^*) \backslash a^*$.

Further, if conditions 1–3 hold and:

(a) *$N(a^*) = A$, then a* is globally Lyapunov stable;*

(b) *$\Delta L(a(t)) < 0 \; \forall a \in N(a^*) \backslash a^*$, then a* is asymptotically stable;*

(c) *$N(a^*) = A$ and $\Delta L(a(t)) < 0 \; \forall a \in N(a^*) \backslash a^*$, then a* is globally asymptotically stable.*

Lyapunov's direct method says, in effect, that if we can find a function that strictly decreases along all paths created by the adaptations of a cognitive radio network, then that cognitive radio network is asymptotically stable.

Note that the existence of a Lyapunov function can be used to establish the existence and identify the network's steady states, namely, all points where $L(a^*) = 0$. Further, unlike the Picard–Lindelöf equation, Lyapunov's second method can be readily applied to both synchronous and asynchronous cognitive radio networks—the only requirement being that each adaptation must decrease the value of the Lyapunov function.

15.3.2 Contraction Mappings and the General Convergence Theorem

The preceding discussion assumed a closed-form expression for the next network state as a function of current network state. Now suppose that, after one recursion of the network update rule, we are unable to precisely predict the next network state. However, we are able to bound the network state within a particular set of states $A(t^1)$. Then suppose that armed with the knowledge that the network starts in $A(t^1)$, we could say that after the second iteration, the network state would have to be within another set $A(t^2)$, which is a subset of $A(t^1)$. Extending this concept, suppose that given any set of network states, $A(t^k)$, we know that the decision update rule always results in a network state in the set, $A(t^{k+1})$, which is a subset of $A(t^k)$.

In effect, this process is saying that as the recursion continues finer and finer approximations on the operating point of the network are possible, perhaps resulting in a prediction of a specific steady state for the network. Such a sequence of finer approximations might look as shown in Figure 15.7, in which the recursion of subsets, $A(t^k)$, converges to a single point. This iterative restriction on a recursion's possible points forms the basis of numerous valuable algorithms and is a characteristic of a special class of algorithms known as *contraction mappings*.

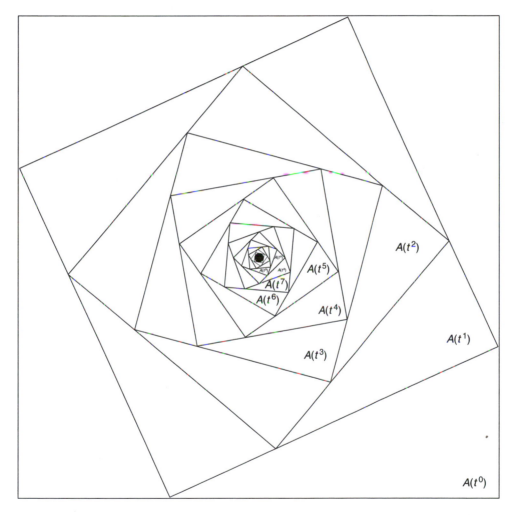

Figure 15.7: A sequence of contracting sets, $\cdots \subset A(t^2) \subset A(t^1) \subset A(t^0)$.

Contraction Mappings

Given a recursion $a(t^{k+1}) = f(a(t^k))$, f is said to be a *contraction mapping* with modulus α if there is an $\alpha \in [0,1)$ such that $\| f(a), f(b) \| \leq \alpha \| a, b \| \ \forall b, a \in A$.

While applying this definition to a decision rule can be difficult, we can show that an arbitrary recursion, f, is a contraction mapping if the following two conditions are satisfied [10]:[11]

1. *Monotonicity*: Given bounded functions $g_1, g_2: A \to \mathfrak{R}$ where $g_1(a) \leq g_2(a)$ $\forall a \in A$, f must satisfy $f(g_1(a)) \leq f(g_2(a))$ $\forall a \in A$.

2. *Discounting*: There exists a $\beta \in (0, 1)$ such that $f(g_1(a) + c) = f(g_1(a)) + \beta c$ for all bounded $g_1: A \to \mathfrak{R}$, $c \geq 0$, $a \in A$.

Analysis Insights

Knowing that our decision rule constitutes a contraction mapping immediately provides us with several valuable insights. From Banach's contraction mapping theorem [11], we know that f has a unique fixed point to which the recursion f converges from any starting point. After k iterations, a bound on the distance of the current state from the fixed point is given by:

$$\left\| a(t^k), a^* \right\| \leq \frac{\alpha^k}{1 - \alpha} \left\| a(t^1), a(t^0) \right\| \tag{15.3}$$

Eq. (15.3) is also useful for bounding the error in estimating f's fixed point by recursively evaluating f. Additionally, a Lyapunov function for any contraction mapping with fixed point a^* is given by:

$$L(a) = \left\| a, a^* \right\| \tag{15.4}$$

Thus, every contraction mapping, f, has a unique stable fixed point to which f converges at a predictable rate.

Pseudocontractions

A pseudocontraction eliminates the contraction mapping's requirement that all points move closer to each other after each iteration, but still requires that after each iteration all points move closer to a unique fixed point. Formally, given mapping $f: A \to A$ with fixed point, a^*, we say f is a pseudocontraction if there is an $\alpha \in [0, 1)$ such that $\| f(a), f(a^*) \| \leq \alpha \| a, a^* \| \forall a \in A$. By definition, f has a unique fixed point, a^*, and the distance to a^* at time t^k is given by:

$$\left\| a(t^k), a^* \right\| \leq \alpha^k \left\| a(0), a^* \right\| \tag{15.5}$$

---

[11] These conditions are known as *Blackwell's conditions*.

Note that evaluation of Eq. (15.5) requires knowledge of the fixed point, so, unlike Eq. (15.3), it is not appropriate for bounding the error on an estimate of the system's fixed point while iterating to solve for the fixed point.

General Convergence Theorem

Most contraction mappings assume that the updating process occurs synchronously (recall the discussion of decision timings in Section 15.2.2). We can relax this assumption by introducing the general convergence theorem presented by Bertsekas and Tsitsiklis [7] in Proposition 2.1 of Chapter 6:

> *Suppose we know that $\cdots \subset A(t^{k+1}) \subset A(t^k) \subset \cdots \subset A(t^0)$ where $A(t^k)$ represents the possible states of the network after k iterations and $A(t^0)$ represents all possible initial states for the network. Then if the following two conditions hold, f also converges asynchronously.*

1. Synchronous Convergence Condition
   (a) $f(a) \in A(t^{k+1}) \; \forall k, \, a \in A(t^k)$
   (b) *If* $\{a(t^k)\}$ *is a sequence such that* $a(t^k) \in A(t^k)$ *for every* k, *then every limit point of* $\{a(t^k)\}$ *is a fixed point of* f.

2. Box Condition
   For every k, there exist sets $A_j(t^k) \subset A_j$ such that $A(t^k) = A_1(t^k) \times \cdots \times A_n(t^k)$.

For our purposes, the general convergence theorem states that under our regular action sets assumption, any contraction or pseudocontraction mapping that converges synchronously also converges asynchronously. However, we can also apply the general convergence theorem to algorithms that are not obviously contraction mappings, as seen in the extended example presented in next section (*Standard Interference Function Model*).

Standard Interference Function Model

Many traditional analyses consider specific decision rules that model specific applications. The following discussion presents such an analysis that is also an example of a nonobvious contraction mapping. Yates [12] considers a power control algorithm operating on the uplink of a cellular system. For this algorithm, there is a set of N mobiles where each mobile, j, attempts to achieve a target received signal-to-interference and noise ratio (SINR), $\hat{\gamma}_j$. The development of this algorithm assumes that each mobile is capable of observing its received SINR (perhaps via feedback from a base station) which is generally given by:

$$\gamma_j = \frac{g_{jj}p_j}{\sum\limits_{k \in N} g_{kj}p_j + N_j} \tag{15.6}$$

where g_{kj} can be the link budget gain from mobile k to the base station mof mobile j, p_k is the transmit power of mobile k, and N_j is the noise power at the base station that is receiving mobile j's signal.

Based on observations of Eq. (15.6), the mobiles compute a scenario-dependent *interference function*, $I_j(\mathbf{p})$, which is formed as the ratio of the target SINR, $\hat{\gamma}_j$, and the effective SINR, γ_j, or $I_j(\mathbf{p}) = \hat{\gamma}_j / \gamma_j$ where $\mathbf{p}$ is the vector of transmit powers, $\mathbf{p} = (p_1, p_2, \dots, p_n)$, drawn from the power vector space $\mathbf{P}$.

Generalizing beyond this ratio formalization, Yates [12] defines any interference function to be *standard* if it satisfies the following three conditions.

1. *Positivity*: $I(\mathbf{p}) > 0$
2. *Monotonicity*: If $\mathbf{p}^1 \geqslant \mathbf{p}^2$ then $I(\mathbf{p}^1) \geqslant I(\mathbf{p}^2)$
3. *Scalability*: For all $\alpha > 1$, $\alpha I(\mathbf{p}) > I(\alpha \mathbf{p})$

Here we write $\mathbf{p}^1 \geqslant \mathbf{p}^2$ if $p_j^1 \geqslant p_j^2 \; \forall j \in N$ and $I(\mathbf{p})$ is the synchronous evaluation of all $I_j(\mathbf{p})$.

Assuming the existence of a standard interference function, Yates [12] defines a synchronous updating process of the form $\mathbf{p}(t^{k+1}) = f(\mathbf{p}(t^k))$, where $f(\mathbf{p}) = f_1(\mathbf{p}) \times \cdots \times f_n(\mathbf{p})$ and f_j is given by:

$$f_j(\mathbf{p}(t^k)) = p_j(t^k) I_j(\mathbf{p}(t^k)) \tag{15.7}$$

When the target SINR vector, $\hat{\boldsymbol{\gamma}}$, is *feasible* (there exists a $\mathbf{p} \in \mathbf{P}$ such that $\gamma_j \geqslant \hat{\gamma}_j \; \forall j \in N$), Yates [12] is able to show that an algorithm updating the power vector according to Eq. (15.7) has the following properties:

1. A fixed point exists, that is, there is some $\mathbf{p}^*$ such that $\mathbf{p}^* = f(\mathbf{p}^*)$.
2. This fixed point is unique.
3. Starting from any initial power vector, f converges to $\mathbf{p}^*$.

Whereas Yates [12] shows these results in an ad hoc manner, Berggren [13] shows that this updating process constitutes a pseudocontraction that could be used to establish these same results in a different manner.

The fact that f constitutes a pseudocontraction implies that $\cdots \subset \mathbf{P}(t^{k+1}) \subset \mathbf{P}(t^k)$ $\subset \cdots \subset \mathbf{P}(t^0)$, where $\mathbf{P}(t^k)$ is the power vector space for iteration k. Coupled with the just-established synchronous convergence of f and implicit satisfaction of the box condition, this means that f has satisfied the conditions for the general convergence theorem. Thus it is known that f converges both synchronously and asynchronously.

Further Standard Interference Function Analysis Insights

Assuming the SINR feasibility criterion is satisfied, Yates [12] also shows that the following arrangements of base stations and mobiles have standard interference functions and thus converge synchronously and asynchronously to a unique power vector when the decision update rule is given as in Eq. (15.7):

- *Fixed assignment*: Each mobile is assigned to a particular base station.
- *Minimum power assignment*: Each mobile is assigned to the base station in the network where the mobile's SINR is maximized.
- *Macro diversity*: All base stations in the network combine the signals of the mobiles.
- *Limited diversity*: A subset of the base stations combine the signals of the mobiles.
- *Multiple connection reception*: The target SINR must be maintained at a number of base stations.

Feasible SINR

Previously we defined a target SINR vector, $\hat{\gamma}$, as being feasible if there exists a $\mathbf{p} \in \mathbf{P}$ such that $\gamma_j \geqslant \hat{\gamma}_j \; \forall j \in N$. Rather than performing an exhaustive search over $\mathbf{P}$, Zander and Kim [14] present the following approach for determining whether $\hat{g}$ is feasible.

Consider a network with link gain matrix $\mathbf{G}$ formed as:

$$\mathbf{G} = \begin{bmatrix} g_{11} & g_{12} & \cdots & g_{1n} \\ g_{21} & \ddots & & \vdots \\ \vdots & & \ddots & \vdots \\ g_{n1} & g_{n2} & \cdots & g_{nn} \end{bmatrix}$$

where g_{jk} is the link gain as used in Eq. (15.6). Now form the normalized link matrix H with elements $h_{ij} = \gamma \dfrac{g_{ji}}{g_{ii}} \; i \neq j$ and $h_{ii} = 0$. Then Zander and Kim

[14, p. 155] state that the uniform target SINR vector $\hat{\gamma}_u = (\hat{\gamma}, \hat{\gamma}, \dots, \hat{\gamma})$ is achievable if the spectral radius (largest eigenvalue)[12] of **H** is less than or equal to 1. When the spectral radius is exactly 1, then $\hat{\gamma}$ is achievable only when there is no noise in the system.

A similar expression can be found for the nonuniform target SINR scenario, where $\hat{\gamma} = (\hat{\gamma}_1, \hat{\gamma}_2, \dots, \hat{\gamma}_n)$ as follows where the link matrix, H', is formed as

$$h_{ij} = -\frac{\hat{\gamma}_i}{\hat{\gamma}_{max}} \frac{g_{ji}}{g_{it}}, \; i \neq j$$

where $\hat{\gamma}_{max} = \max_{i \in N}\{\hat{\gamma}_i\}$.

Assuming the target SINRs are feasible, then the power vector corresponding to the unique fixed point specific can be found by solving the following equation:

$$\mathbf{Zp} = \vec{\gamma} \tag{15.8}$$

where

$$\mathbf{Z} = \begin{bmatrix} g_{11} & -\hat{\gamma}_1 g_{21} & \cdots & -\hat{\gamma}_1 g_{n1} \\ -\hat{\gamma}_2 g_{1n} & g_{22} & \cdots & -\hat{\gamma}_2 g_{n2} \\ \vdots & & \ddots & \vdots \\ -\hat{\gamma}_n g_{1n} & -\hat{\gamma}_n g_{2n} & \cdots & g_{nn} \end{bmatrix}$$

and

$$\vec{\gamma} = [\hat{\gamma}_1 N_1 \quad \hat{\gamma}_2 N_2 \quad \cdots \quad \hat{\gamma}_n N_n]^T$$

where the superscripted T denotes the transpose operation.

15.3.3 Markov Models

Perhaps due to uncertainty in the order of adaptation (as would be the case for a randomly or asynchronously timed process) or due to uncertainties in the decision rules, it may be impossible to derive a closed-form expression for an evolution equation or to even to bound the adaptations into sequential subsets. Instead, suppose we can model the changes of the cognitive radio network from one state to another as a sequence of probabilistic events conditioned on past states that the

---

[12] Due to the work of Hilbert, spectral theory refers to a set of theories relating to matrices, eigenvalues and eigenvectors. In spectral theory, the set of eigenvalues for a matrix is said to be its spectrum.

system may have passed through. When the probability distribution for the next state in time, $a(t^{k+1})$, is conditioned solely on the most recent state as shown in the following equation:

$$P\left(a(t^{k+1}) = a^k \mid a(0), \ldots, a(t^k)\right) = P\left(a(t^{k+1}) = a^k \mid a(t^k)\right) \qquad (15.9)$$

the random sequence of states, $\{a(t)\}$, is said to be a *Markov chain*. A model of a system whose states form a Markov chain is said to be a *Markov model*. Throughout the remainder of this section, we use these two terms interchangeably.

Formalizing our model, let us assume that our state space is finite. This is not a requirement for a Markov chain, but the assumption is useful for the subsequent discussion. Further, let us assume that if the network is in state $a^m \in A$ at time t^k, then at time t^{k+1}, the network transitions to state $a^n \in A$ with probability p_{mn}, where $p_{mn} \geq 0 \; \forall a^m, a^n \in A$ and $\sum_{j \leq |A|} p_{mj} = 1$. Of course, it is also permitted that the system remains in state a^m, which it does with probability p_{mm}. To simplify notation, we make use of a *transition matrix*, which we represent with the symbol $\mathbf{P}$. The transition matrix is formed by assigning p_{mn} to the entry corresponding to the mth row and nth column.

Markov Model Analysis Insights

From $\mathbf{P}$ we can then form $\mathbf{P}^2$ as the matrix product $\mathbf{PP}$. Now entry p_{mn}^2 in the mth row and nth column of $\mathbf{P}^2$ represents the probability that the system is in state a^n two iterations after being in state a^m. Similarly, if we consider the matrix $\mathbf{P}^k$ formed as $\mathbf{P}^k = \mathbf{PP}^{k-1}$ (an example of a Chapman–Kolmogorov equation for a Markov chain [15]), then entry p_{mn}^k in the mth row and nth column of $\mathbf{P}^k$ represents the probability that system is in state a^n k iterations after being in state a^m.

A similar relationship can be found when the initial state is specified by a random probability distribution arranged as a column vector $\boldsymbol{\pi}$, where $\pi_m \in [0, 1]$ and $\sum_{m=1}^{|A|} \pi_m = 1$, and π_m represents the probability of starting in state a^m. For such a situation, the state probability distribution after k iterations is given by $\boldsymbol{\pi}^T \mathbf{P}^k$.

There may also be some distribution $\boldsymbol{\pi}^*$ such that $\boldsymbol{\pi}^{*T} \mathbf{P} = \boldsymbol{\pi}^{*T}$. Such a distribution $\boldsymbol{\pi}^*$ is said to be a *stationary distribution* for the Markov chain defined by $\mathbf{P}$. Note that solving for a stationary distribution is equivalent to solving the eigenvector equation $\boldsymbol{\pi}^{*T} \mathbf{P} = \lambda \boldsymbol{\pi}^{*T}$, where $\lambda = 1$. A related concept is the *limiting distribution*, which is the distribution that results from evaluating $\lim_{k \to \infty} \boldsymbol{\pi}^T \mathbf{P}^k$. Although the network is not generally a steady state as considered in the previous discussion, showing that a Markov chain has a unique distribution that is both stationary and limiting would permit us to characterize the behavior of the network. Specifically, given the unique stationary limiting distribution π^*, we could predict that at a particular instance in time and after a sufficient number of iterations, the network would be in state a^m with probability π_m. Thus, it is desirable to be able to identify when such a unique stationary limiting distribution exists.

The ergodicity theorem [16] states that if a Markov chain is *ergodic*, then there exists a unique limiting and stationary distribution for all initial distributions π. A Markov chain is ergodic if it is: (a) *irreducible*, (b) *positive recurrent*, and (c) *aperiodic*. A Markov chain is *irreducible* if $\forall a \in A$, there exist sequences of state transitions with nonzero probability that lead to every state. A Markov chain is *positive recurrent* if the expected time to return to every state is finite. Finally, a Markov chain is *aperiodic* if for each state in the chain there is no integer, $m > 1$, such that once the system leaves the state, it can only return to the state in multiples of m iterations.

Identifying that these three conditions are satisfied can be a rather daunting task. However, Kemeny and Snell's [17] Theorem 4.1.2 shows that for finite Markov chains, $\mathbf{P}$ is ergodic if and only if there is some k such that $\mathbf{P}^k$ has no zero entries. Thus, by identifying this simple condition, we know that a unique stationary limiting distribution exists.

Example 15.1 _____

Consider a network consisting of two cognitive radios that each can choose between two actions. This network would have four possible states, which we could label $\{a^1, a^2, a^3, a^4\}$. Suppose that from experimental observation we observe the probability transition matrix shown in the following:

$$\mathbf{P} = \begin{array}{c} \\ a^1 \\ a^2 \\ a^3 \\ a^4 \end{array} \begin{array}{cccc} a^1 & a^2 & a^3 & a^4 \\ \begin{bmatrix} 0.1 & 0.3 & 0.1 & 0.5 \\ 0.4 & 0.0 & 0.3 & 0.3 \\ 0.4 & 0.1 & 0.3 & 0.2 \\ 0.1 & 0.4 & 0.3 & 0.2 \end{bmatrix} \end{array} \qquad (15.10)$$

We see that $\mathbf{P}$ gives the probability of transitioning from state a^2 to state a^3 as 0.3. After calculating $\mathbf{P}^2$ as shown below, we can immediately determine the probability of the system operating in state a^4 two iterations after starting in a^3 as $p_{34}^2 = 0.33$:

$$
\mathbf{P}^2 = \begin{array}{c} \\ a^1 \\ a^2 \\ a^3 \\ a^4 \end{array} \begin{array}{cccc} a^1 & a^2 & a^3 & a^4 \\ \begin{bmatrix} 0.22 & 0.24 & 0.28 & 0.26 \\ 0.19 & 0.27 & 0.22 & 0.32 \\ 0.22 & 0.23 & 0.22 & 0.33 \\ 0.31 & 0.14 & 0.28 & 0.27 \end{bmatrix} \end{array}
$$

Similarly, given an initial distribution of states $\boldsymbol{\pi} = [0.1\ 0.2\ 0.3\ 0.4]^T$, after two iterations, the probability of being in each state is given by $\boldsymbol{\pi}\mathbf{P}^2 = [0.25\ 0.203\ 0.25\ 0.297]^T$. Because all elements in $\mathbf{P}^2$ are positive, there exists a stationary distribution $\boldsymbol{\pi}^*$ that we can find by solving the eigenvector equation $\boldsymbol{\pi}^{*T}\mathbf{P} = \boldsymbol{\pi}^{*T}$ to yield $\boldsymbol{\pi}^{*T} = [0.2382\ 0.2352\ 0.2272\ 0.2938]$.

Absorbing Markov Chains

Unfortunately, the limiting distribution for an ergodic matrix is rather unsatisfying because all states will have nonzero probability of being occupied. However, this is not a problem for absorbing Markov chains. A state a^k is said to be an *absorbing state* if there are no paths that leave a^k, that is, $p_{km} = 0\ \forall k \neq m$ and $p_{kk} = 1$. A Markov chain is said to be an *absorbing Markov chain* if:

(a) it has at least one absorbing state;

(b) from every state in the Markov chain there exists a sequence of state transitions with nonzero probability that lead to an absorbing state. These nonabsorbing states are called *transient states*.

Eq. (15.11) is an example of a transition matrix for an absorbing Markov chain where a^4 is the absorbing state and a^1, a^2, and a^3 are the transient states:

$$
\mathbf{P} = \begin{array}{c} \\ a^1 \\ a^2 \\ a^3 \\ a^4 \end{array} \begin{array}{cccc} a^1 & a^2 & a^3 & a^4 \\ \begin{bmatrix} 0.1 & 0.3 & 0.1 & 0.5 \\ 0.4 & 0.0 & 0.3 & 0.3 \\ 0.4 & 0.1 & 0.3 & 0.2 \\ 0 & 0 & 0 & 1 \end{bmatrix} \end{array} \qquad (15.11)
$$

Note that when represented as a transition matrix, state a^m is an absorbing state if and only if $p_{mm} = 1$.

Absorbing Markov Chains Analysis Insights

Within the context of our analysis objectives, an absorbing state is a fixed point or steady state that, once reached, the system never leaves. Similarly, valuable convergence insights can also be gained when the system can be modeled as an absorbing Markov chain as follows.

First, form the modified transition matrix, $\mathbf{P}'$, as shown in the equation:

$$\mathbf{P}' = \left[\begin{array}{c|c} \mathbf{Q} & \mathbf{R} \\ \hline \mathbf{0} & \mathbf{I}^{ab} \end{array}\right] \tag{15.12}$$

where $\mathbf{I}^{ab}$ is the identity matrix corresponding to the state transitions between the absorbing states of the chain, $\mathbf{Q}$ represents the state transitions between the nonabsorbing states of the chain, $\mathbf{0}$ is a rectangular matrix filled with all zeros representing the probability of transition from absorbing states to nonabsorbing states, and $\mathbf{R}$ represents the rectangular matrix of state transition probabilities from nonabsorbing states to absorbing states.

Given $\mathbf{P}'$, Markov theory provides us with information on convergence and the expected frequency that the system visits a transitory state. First, $\lim_{k \to \infty} \mathbf{Q}^k \to \mathbf{0}$ implies that the probability of the system not being "absorbed" (i.e., not terminating in one of the absorbing states of the chain) goes to zero.

Beyond this basic result, more specific convergence results can be stated by introducing the *canonical form* for the absorbing chain. Given an absorbing chain with a modified transition matrix, as in Eq. (15.12), the *fundamental matrix* is given by:

$$\mathbf{N} = (\mathbf{I} - \mathbf{Q})^{-1} \tag{15.13}$$

Solving for the fundamental matrix $\mathbf{N}$ permits a number of valuable analytic insights. First, Kemeny and Snell's [17] Theorem 3.2.4 states that the entry n_{km} gives the expected number of times that the system will pass through state a_m given that the system starts in state a_k. Second, their Theorem 3.3.5 [17] states that if we evaluate $\mathbf{t} = \mathbf{N1}$, where $\mathbf{1}$ is a column vector of all ones, then t_k gives the expected number of iterations before the state is absorbed when the system starts in state a_k. Finally, their Theorem 3.3.7 [17] states that if we evaluate $\mathbf{B} = \mathbf{NR}$ where $\mathbf{R}$ is as given in Eq. (15.12), then entry b_{km} in $\mathbf{B}$ specifies the probability the system ends up in absorbing state a_m if the system starts in state a_k.

Thus, once we show that a Markov model for a network of cognitive radios with transition matrix **P** is an absorbing Markov chain, the following insights are readily gained:

- Steady states for the system can be identified by finding those states a^m for which $p_{mm} = 1$.
- Convergence to one of these steady states is assured, and the expected distribution of states can be found by solving for **B**.
- Given an initial state, a^k, convergence rate information is given by solving for **t**.

15.4 Applying Game Theory to the Analysis Problem

The techniques presented in Section 15.3 provide us with tools that allow us to answer many of our analysis questions, but there are still some noticeable limitations. Establishing the existence and uniqueness of a fixed point for an evolution function says little about convergence or stability. Finding an appropriate Lyapunov function can provide valuable convergence and stability information, but finding that Lyapunov function is largely a hit-or-miss affair. Contraction mappings provide many of the results that we desire, but are encountered infrequently. Although Markov models can do an excellent job of handling the nondeterministic nature of one of the most promising cognitive radio adaptation algorithms—genetic algorithms—solving for a network's transition matrix can be a daunting task. Finally, for all of these approaches, analyzing one decision rule says little about the performance of related decision rules. It would be nice if, given the application-specific goal of a cognitive radio, we were able to immediately predict what decision processes would be required to achieve the desired level of network performance.

Frequently, progress in analysis occurs by introducing additional information. In this case, introducing cognitive radios' goals allows us to apply techniques from game theory to gain additional insights. As we will see in the remainder of this chapter, game theory has a number of advantages over more traditional techniques. A number of readily identifiable game models allow us to simultaneously identify the existence of steady states, convergence criteria, and stability of cognitive radio algorithms. Further, these game models provide the capacity to analyze broad classes of decision rules, such as the random better response dynamic considered in section *Identification of Exact and Ordinal Potential Games*, which encompasses the genetic algorithms of Chapter 7. Further, game theory provides a means to analyze the interactions of the ontologically defined cognitive radios of

Chapter 13 for which no predefined decision rules exist—an impossible analysis problem for more conventional techniques.

It is assumed that game theory is a new concept to most readers, so this section provides an extended discussion of the basic elements of game theory, how game theory can be applied to cognitive radio networks, basic game models, and the analytic insights that can be gained by applying the game models. Section 15.5 goes into greater depth by describing two game models that permit rapid analysis of cognitive radio networks in terms of our analysis objectives.

15.4.1 Basic Elements of Game Theory

Game theory is a collection of models and analytic tools used to study *interactive decision-making processes*. In brief, an interactive decision process is a process whose outcome is a function of the inputs (actions) from several different decision-makers (players) who may have conflicting goals with regard to the outcome of the process.

The fundamental modeling tool of game theory is the *game*. Whether explicitly or implicitly, every game includes the following components:

- A set of *players*.
- *Actions* for each of the players.
- Some method for determining *outcomes* according to the actions chosen by the players.
- *Preferences* for each of the players defined over all the possible outcomes.
- *Rules* governing the order of play.

The following subsections provide brief descriptions of these components in light of the cognitive radio network model introduced in Section 15.2.

Players

The players in a game are the decision-making entities in the modeled interactive process. In our case, the players are the cognitive radios in the network. For notational continuity, we refer to the set of players (cognitive radios) as N and individual players as i or j. As a rule, games consider situations only of two or more players because a single-player game would by definition not be an interactive process.

Actions and Outcomes

For our purposes, we continue to use actions and outcomes (or observations of the outcomes) in the same manner as introduced in Section 15.2. The actions are the adaptations (waveforms) available to the radio, and the outcomes are the observations of the network.

Preferences and Utility Functions

Preferences were not considered in the model introduced in Section 15.2, so a longer discussion of preferences here is merited.

In a game, it is assumed that each player has a set of preference relations, $\{\lesssim_j\}$, that describe that player's preferences with respect to all the possible outcomes in the outcome space. We write $o^2 \lesssim_j o^1$ if player j prefers outcome o^1 at least as much as it prefers outcome o^2; $o^2 <_j o^1$ if j strictly prefers o^1 to o^2; and $o^2 \sim_j o^1$ if j is indifferent between o^1 and o^2.

Identifying the preferences of a radio may seem difficult, but this is generally not the case. For example, consider a radio with a goal of achieving a target SINR. Presumably, an intelligent radio would prefer any SINR that is closer to its target over an SINR that is farther away. Or assume the radio is attempting to minimize a particular parameter, such as in-band interference; then any outcome with lower in-band interference would be preferable to an outcome with higher in-band interference.

Even for algorithms for which such a clear objective is not available, if we know the radio's decision update rules, then we can infer some preference relations by examining adaptations.[13] Assuming other radios' adaptations are fixed, if an adaptation of radio j causes a change from o^1 to o^2, then it is reasonable to assume that $o^1 \lesssim_j o^2$, i.e., radio j "prefers" o^2 to o^1.

Particularly in light of inferring preferences from the behavior of algorithms, it can be readily seen that the concept of preferences can be applied to adaptive radios and not just to cognitive radios. Still, a game theorist may feel more comfortable describing the preferences of a device that is actually aware of what it is doing.

For a small game, we could list all of the preference relations for every player over all possible outcomes. However, as the size of the game grows, a complete listing quickly becomes unwieldy. For instance, an n-player game in which each player has m actions could reasonably have $m \times n$ different outcomes.

---

[13] For the game theorist, we are implicitly assuming our players exhibit perfect rationality—an assumption that seems reasonable in light of the fact that our players are programmable machines.

Accordingly, a full listing of all the preferences for a single player requires defining $(m \times n)(m \times n + 1)/2$ preference relations; and a listing for all players requires defining $n(m \times n)(m \times n + 1)/2$ preference relations.

To capture these preference relations in a more compact way, game theorists frequently employ *utility functions* (sometimes called *objective functions*), which assign a real number to each outcome for a particular player (for mathematical rigor, $u_i: O \rightarrow \Re$) in such a way that if $o^2 \precsim_i o^1$ then $u_i(o^2) < u_i(o^1)$, if $o^2 \sim_i o^1$ then $u_i(o^2) = u_i(o^1)$, and if $o^2 \precsim_i o^1$ then $u_i(o^2) \leq u_i(o^1)$. Of course, when a cognitive radio's goal is expressed numerically, finding the utility function is trivial (the goal is the utility function).

But utility functions are really just a stand-in for the preference relations in a compact manner, so the exact numbers assigned by utility functions are generally of secondary importance, assuming that the utility functions preserve the preference relations. For example, suppose player j prefers apples to oranges (yes, we are comparing apples and oranges). From the perspective of preference relationships, $u_j (apple) = 1$ and $u_j (orange) = 0.5$ is equivalent to writing $u_j (apple) = 1000$ and $u_j (orange) = -3$; faced with the choice between an apple and an orange, player j still prefers the apple and would be predicted to choose the apple. Or, in a cognitive radio example, a goal (utility function) of maximizing linear SINR is equivalent in terms of preference relationships to a goal of maximizing SINR expressed in dB.

Frequently, we write the utility functions as functions of the action vectors that yield the outcomes, $u_j: A \rightarrow \Re$, instead of as functions of the outcome space, $u_j: O \rightarrow \Re$. As was the case for the model of Section 15.2, this simplification is appropriate as long as the mapping between A and O is clear. Because different players generally ascribe different valuations to the same action vector (or outcome), we sometimes make use of a *payoff vector* that lists the utility that each player assigns to a particular action vector. For example, rather than writing $u_1(a) = 1$, $u_2(a) = -3$, and $u_3(a) = 4$, we could write $u(a) = (1, -3, 4)$. With this notation, it also sometimes makes sense to describe a single utility function that maps A into $\Re^n$ where n is the number of players in the game.

Rules Governing the Order of Play

Different game models assume different rules for when the players are allowed to "play" (choose an action). As with the cognitive radio network model introduced in Section 15.2, for our games, we concern ourselves with games that adopt synchronous, asynchronous, round robin, or randomly ordered timings for decisions.

15.4.2 Mapping the Basic Elements of a Game to the Cognition Cycle

Fundamentally, game theory can be applied to the analysis of the adaptations of any set of intelligent agents, and the cognition cycle represents the processes that go on in any intelligent being—including humans. So it is not surprising that we can establish connections between the components of a game and the cognition cycle.

A cognitive radio network described via interactive cognition cycles can be modeled by using the elements of a game as follows. First, the players of the game are all the nodes in a network adapting their waveforms, N. Each player's action set is formed from the various adaptations available to the radio, A_j, and the action space for the game is simply formed from the Cartesian product of the radios' available adaptations or A. A utility function for each player is provided by the cognitive radio's goal, and the arguments and valuation for this utility function are taken from the outputs of the cognitive radio's observation and orientation steps. Loosely, the observation step provides the player with the arguments to evaluate its utility function, and the orientation step determines the valuation of the utility function. Figure 15.8 shows an illustration of how the different components of the cognition cycle map into the player-specific components of a game.

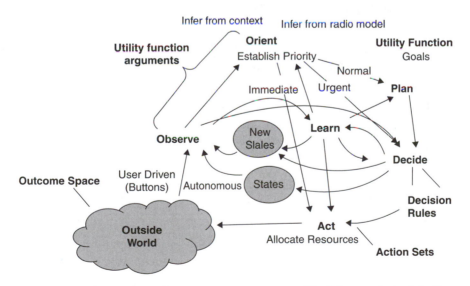

Figure 15.8: Cognition cycle and game components. Modified from Mitola [1], Figure 4.2.

Note that we have ignored the learning step of the cognition cycle. This is neither an oversight nor indicative of a limitation of game theory. Subsequent sections show that some simple models can also be used to handle many situations in which the cognitive radios must learn how their operating environment impacts their goals.

15.4.3 Basic Game Models

Different game models incorporate these basic game elements in different ways. Some game models, such as an extensive form game model, have complex iteration-varying rules governing the actions and order of play. Other game models, such as the normal form game model, are much simpler. For the purposes here, we are particularly interested in the normal form and the repeated game models.

Normal Form Game Model

The simplest and most frequently encountered game model used to describe an interactive decision-making process is the *normal form game*. In addition to the basic elements, a normal form game adds the following rules:

- *Synchronous single-shot play*: All players make their decisions simultaneously and make only a single decision. Thus T has a single element.
- *Perfect information*: The players know their own utility functions and the utility functions for all the other players in the game.
- *Perfect implementation*: All players exhibit perfect implementation[14] (i.e., no player accidentally implements action a_j^1 instead of a_j^2).

With these rules governing the timing and implementation of play, a normal form game is defined by the 3-tuple, $\Gamma = \langle N, A, \{u_j\}_{j \in N} \rangle$, where N is the set of players, A is the action space, and u_j is the utility function for player j (each player in N has its own utility function).

Particularly for two-player games, it is convenient to represent a normal form game in *matrix form*. In a matrix form representation of a two-player normal form game, all possible action vectors are arrayed in a matrix such that player 1's actions (the first component of the action vector) are given by the rows of the matrix and player 2's actions (the second component of the action vector) are given by the columns of the matrix. Each cell in this matrix is thus determined by a unique action vector (row, column) and is filled with the payoff vector associated with that action vector.

---

[14] The *trembling hand* model is an example where perfect implementation is not assumed.

Example 15.2 *Modeling a Game of Paper–Rock–Scissors* _____

For example, consider a game of paper–rock–scissors where winning the game is associated with a utility of 1, losing with -1, and a tie with 0.[15] This game could be expressed in matrix form as shown in Figure 15.9. For example, if player 1 played paper, p, and player 2 played rock, R, the action vector (p, R) indicates a payoff vector of $(1, -1)$.

Figure 15.9: Matrix form representation of a game of paper–rock–scissors. *P*, *R*, and *S* are the actions, or moves, available to player 2, and *p*, *r*, and *s* are the actions available to player 1.

Γ	P	R	S
p	(0, 0)	(1, −1)	(−1, 1)
r	(−1, 1)	(0, 0)	(1, −1)
s	(1, −1)	(−1, 1)	(0, 0)

Example 15.3 *The Cognitive Radios' Dilemma* _____

Two cognitive radios are operating in the same environment and are attempting to maximize their throughput. Each radio can implement two different waveforms—one a low-power narrowband waveform, the other a higher-power wideband waveform. If both radios choose to implement their narrowband waveforms [action vector (n, N)], the signals will be separated in frequency and each radio will achieve a throughput of 9.6 kbps. If one of the radios implements its wideband waveform while the other implements a narrowband waveform [action vectors (n, W) or (w, N)], then interference will occur. In this event, the narrowband signal will achieve a throughput of 3.2 kbps and the wideband signal will yield a throughput of 21 kbps. In the event that both radios choose to implement wideband waveforms, then each radio will achieve a throughput of 7 kbps.

These waveforms can be visualized in the frequency domain as shown in Figure 15.10 and represented in matrix form as shown in Figure 15.11. Without going into the analysis of this game (presented in Example 15.6), the insightful reader may already anticipate that the design of this algorithm will tend to lead to less than optimal performance.

---

[15]As the sum of the values for each payoff vector is zero, this game is an example of a special normal form game known as a *zero-sum* game.

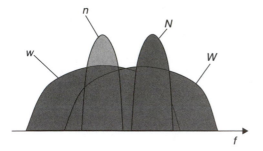

Γ	N	W
n	(9.6, 9.6)	(3.2, 21)
w	(21, 3.2)	(7, 7)

Figure 15.10: Frequency domain representation of waveforms in cognitive radios' dilemma.

Figure 15.11: The cognitive radios' dilemma in matrix form.

Repeated Game Model

A repeated game is sequence of "stage games," in which each stage game is the same normal form game. Based on their knowledge of the game (e.g., past actions, current observations, and future expectations), players choose *strategies* (choices of actions over subsequent stages). These strategies may be fixed (not responsive to choices by other players in later stages) or adapt in response to the actions of other players. Further, these strategies can be designed to punish players who deviate from agreed-upon behavior. When punishment occurs, players choose their actions to minimize the payoff of the offending player.

For our purposes, we consider repeated games defined by the 4-tuple, $\Gamma^R = \langle N, A, \{u_j\}_{j \in N}, \{T_j\}_{j \in N} \rangle$, where N, A, and $\{u_j\}_{j \in N}$ are specified by the normal form stage game $\Gamma = \langle N, A, \{u_j\}_{j \in N} \rangle$ and T_j represents the times (or with a simple notational change, the stages) at which player j can change its decisions. For modeling purposes, this is the same T_j that we used in the model of Section 15.2.2.

Example 15.4 *Paper–Rock–Scissors Repeated Game*

To illustrate the concept of a repeated game, consider a repeated game of paper–rock–scissors where it is assumed that $T_1 = T_2$ (making this a synchronous repeated game) and the paper–rock–scissors game of Example 15.2 forms the stage game. Figure 15.12 shows the result of a single iteration of the game where player 1 has chosen scissors and player 2 has simultaneously chosen paper. This dictates an outcome where player 1 wins (and player 2 loses), so player 1 accrues a utility of $+1$ and player 2 receives a utility of -1. As indicated by the loop, the players would then continue to adapt their decisions in subsequent stages guided by their decision rules, observations about previous actions/outcomes and their expectations for future play.

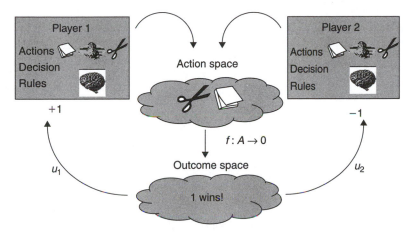

Figure 15.12: An instance of a repeated game of paper–rock–scissors. After playing a single iteration of the game where player 1 won by playing scissors to player 2's paper, the players incorporate this result into their decision process for the next iteration of the game.

Example 15.5 *FM–AM–Spread Spectrum Repeated Game*

Two cognitive radios are operating in the same environment and are attempting to achieve high voice quality and adapt synchronously. Each radio can implement three different waveforms—a frequency modulation (FM) waveform, an amplitude modulation (AM) waveform, and a spread spectrum waveform. Figure 15.13 illustrates the result of a stage of this game in which each radio has chosen a waveform. The combined choices specify an action vector, which in combination with the operating environment, determines an observable outcome (SINR, γ). These observations are then evaluated in the context of the cognitive radio goals, which are functions of SINR ($u_1(\gamma_1)$, $u_2(\gamma_2)$). For instance, these radios may both have the goal of independently attaining sufficient voice quality and intelligibility as part of a process of satisfying a telephony standard (e.g., mean opinion score (MOS) > 4.0). An MOS of 4.0 requires a bit error rate of less than 0.1 percent on 8000 bps vocoded speech and could be achieved when the SINR is at least 15 dB. Based on their observed SINR from this stage and inferences about the future, the radios' cognition cycles would determine their next actions in the repeated game.

For example, if both radios observe operating SINRs greater than 15 dB, then future adaptations would be unlikely; however, if a radio is not achieving its MOS, then that radio is likely to adapt its waveform in the next stage.

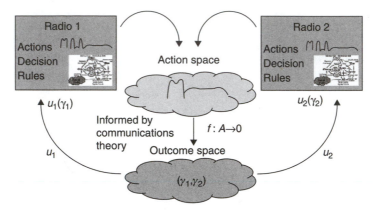

Figure 15.13: A repeated two-player cognitive radio game. After playing a single iteration of the game where player 1 achieved an SINR γ_1 and player 2 achieved an SINR γ_2 determined by their choice of waveforms and the operating environment, the players observe this result and update their decision process for the next iteration of the game.

15.4.4 Basic Game Theory Analysis Techniques

Unlike in Section 15.3, where we analyzed predefined decision rules, game theory permits us to predict the behavior of large classes of decision rules by considering the goals and preferences of the radios. The following discussion shows how game theory addresses the general issues of identifying steady states, measuring optimality, and determining convergence. Stability is not particularly well addressed for the general game models presented in Section 15.4.3, but stability is considered in the discussion of the game models presented in Section 15.5.

Steady States

In game theory, the typically discussed steady-state concept is the NE. An action vector, a^*, is said to be an NE if and only if Eq. (15.14) is satisfied $\forall i \in N, b_i \in A_i$:

$$u_i(a^*) \geq u_i(b_i, a^*_{-i}) \tag{15.14}$$

In other words, an action vector is an NE if no player (radio) can improve its performance by itself. When only one player changes its action, the adaptation is called a *unilateral deviation*. In a normal form game, where each player has knowledge of the other players' utility functions, an NE is predicted as the most likely action tuple that the players will choose; in a repeated game without complex punishment strategies, an NE is similarly predicted to be like a traditional steady state, in which once that action vector is reached, adaptations cease because no player can improve its performance.

Both of these assertions can be seen to hold as long as each player's decision rule results in adaptations that further the players' own self-interest (i.e., the player acts in a way that increases its own utility or goal). In game theory parlance, a player acting in its own interest is said to be *rational*. Now, consider a repeated game whose previous iteration resulted in an action vector that satisfies the conditions for an NE. Assuming no coordination between groups of players, then if the players are following decision rules that adapt in the direction of improving performance, there is no directional adaptation that increases any player's payoff, and thus the next iteration of the game will again be in the NE. If at each iteration each player chooses the action that maximizes its performance, then the locally optimal choice is still the NE, and play does not change in the next iteration. If possible new actions to be considered for implementation are generated randomly (perhaps via a genetic algorithm), then the NE will remain inescapable if each player's decision rule requires the new action to improve performance. So by analyzing the utility functions instead of a specific decision rule, an NE identifies the steady state for all decision rules that a game theorist would say are rational.

This rationality requirement readily extends to cognitive radio networks because every cognitive radio can be considered to be acting in its own interest, in its user's interests, or in the interests of its network. In all these cases, the cognitive radio is observing, orienting, deciding, learning, and acting in a way that maximizes an objective. And if the network arrives in a state from which no radio in the network can find a profitable adaptation that increases its objective, then a game theorist would identify this state as an NE, and for any rational decision rule the network would remain in the NE.

Example 15.6 *Identifying the NE of Cognitive Radios' Dilemma*

Consider the Cognitive Radios' Dilemma of Example 15.3 again. This game has a unique NE of (w, W), which is circled in Figure 15.14. Note that although (n, N) would actually yield superior performance for both radios, neither radio can unilaterally deviate from (w, W) and improve its performance ($3.2 < 7$).

Figure 15.14: The cognitive radios' dilemma.
This game has a unique NE at (w, W).

Γ	N	W
n	(9.6, 9.6)	(3.2, 21)
w	(21, 3.2)	(7, 7)

NE Existence

Now that we have seen the power of the NE concept, how can we know that our network of cognitive radios has an NE? As we did in the discussion of dynamical systems, we turn to fixed point theorems, but now as applied to the goals of the players, not the decision rules.

The most important fixed point theorem for normal form and repeated games is the Glicksberg–Fan fixed point theorem ([18], Theorem 1.2):

> *Given a normal form game* $\Gamma = \langle N, A, \{u_i\} \rangle$ *where all* A_i *are nonempty compact convex subsets of* $\mathfrak{R}^m$ $\forall i \in N$. *If* $\forall i \in N$ u_i *is continuous in a and* quasi-concave *in* a_i *then* Γ *has a pure strategy NE.*

Note that unlike the Leray–Schauder–Tychonoff fixed point theorem (see section *Fixed Points and Solutions to Cognitive Radio Networks*), which needed a specific decision rule, the Glicksberg–Fan theorem considers utility functions on which any self-interested decision rule could be implemented.

The previously undefined term from the Glicksberg–Fan theorem, *quasiconcave*, is used for a function if all of its *upper level sets* are convex. Given a point a^* and a function $f: A \rightarrow \mathfrak{R}$, the upper level set for a^* is given by $U(a^*) = \{a \in A: f(a) \geq a^*\}$. Contrasting quasiconcavity to the previously considered concepts of concavity and pseudoconcavity, Figure 15.15 provides an example of a function that is quasiconcave, but neither concave nor pseudoconcave. Lack of concavity can be verified by noting that a line between a^0 and a^2 contains points above $f(a)$. Lack of pseudoconcavity can be verified by noting that $f(a)$ is not differentiable at

Figure 15.15: A quasiconcave function that is neither concave nor pseudoconcave.

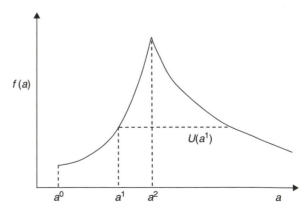

a^2. Relating these three concepts, all concave functions are pseudoconcave, and all pseudoconcave functions are quasiconcave.

It is also important to note that if A is finite, then Glicksberg-Fan cannot be applied to show that the system has a fixed point because no sets would be convex. For instance, the paper–rock–scissors game of Example 15.2 has no NE. However, if the radios are permitted to mix their strategies (i.e., if a radio is permitted to randomly alternate between playing actions a_i and b_i), then even games with nominally finite action spaces will have an NE. The existence of an NE under mixed strategies is a result of Nash's fixed point theorem.

NE Identification

As was the case for the dynamical systems approach, identifying the steady states of a general normal form or repeated game can be quite difficult. Because the only generally applicable approach is to perform an exhaustive search with repeated application of Eq. (15.14), NE identification for a game is an NP-complete problem [19]. When attempting to identify all of the NE in a game, some analysts are forced to turn to simulations—the very step we are intent on minimizing. For example, Ginde et al. [5] used an exhaustive simulation that ran for days to show that a GPRS network employing joint rate-power adaptations had four NE, even though the modeled system included only seven players. Fortunately, the potential game model presented in Section 15.5 provides further information that can be used to simplify the NE identification process.

Desirability

The most typically encountered criterion in the game theory literature that demonstrates that an NE is desirable is *Pareto optimality* [20–22]. Formally, an action vector, a^*, is said to be Pareto optimal if there exists no other action vector, $a \in A$, such that $u_i(a) \geqslant u_i(a^*) \ \forall i \in N$ with at least one player strictly greater.

Unfortunately, Pareto optimality is a very weak concept because a very large number of states may be Pareto optimal, and some Pareto optimal states are neither desirable nor fair, as Example 15.7 demonstrates. Accordingly, it is preferable to adopt the optimality approach of Section 15.3 wherein steady states are evaluated via some network objective function that is appropriate to the cognitive radio engineer's (and hopefully the user's) objectives such as Erlang capacity.[16]

---

[16]Erlang capacity refers to the call arrival rate that a network can accommodate without the network's quality of service (frequently measured via an inverse relationship with the probability of a call being blocked) dropping below some threshold.

Example 15.7 **SINR Maximizing Power Control** _____

Neel et al. [19] consider a single-cluster Direct Spread Spread Spectrum (DS-SS) network with a centralized receiver, in which all of the radios are running power control algorithms in an attempt to maximize their signals' SINR at the receiver.

A normal form game, $\Gamma = \langle N, A, \{u_i\} \rangle$, for this network can be formed with the cognitive radios as the players, the available power levels as the action sets and the utility functions given by:

$$u_i(\mathbf{p}) = h_i p_i / \left[(1/K) \sum_{k \in N \setminus i} h_k p_k + \sigma \right] \qquad (15.15)$$

where K is the statistical cross-correlation of the signals.

As might be expected, the unique NE for this game is the power vector where all radios transmit at maximum power. This outcome can be verified to be Pareto optimal because any more equitable power allocation will reduce the utility of the radio closest to the receiver, and any less equitable allocation will reduce the utility of the disadvantaged nodes.

However, this is not a network we would want to implement because:

1. This state greatly reduces capacity from its potential maximum due to near-far problems (unless our network is in the unlikely configuration of having all radios at the same radius from the receiver).

2. The resulting SINRs are unfairly distributed (the closest node will have a far superior SINR to the farthest node).

3. Battery life is greatly shortened.

Convergence

It makes little sense to speak of convergence of a normal form game because it is defined as having only a single iteration. Accordingly, convergence is more frequently discussed in the context of repeated games. However, the properties of repeated games are largely defined by their stage game, which we are assuming to be a normal form game. So, to analyze the convergence of repeated games, we must identify properties of normal form games that lead to convergent behavior when the normal form game is played repeatedly.

In particular, there are two normal form game properties for games with finite action spaces that can be used for establishing some conditions for convergence: (1) the *finite improvement path* (FIP) property and (2) the *weak finite improvement path* (weak FIP) property.

A normal form game is said to have the FIP property if every sequence of profitable unilateral deviations (called an *improvement path*) is finite. Implicit in this property are the results that every game with FIP has at least one NE, that there are no improvement cycles, and that play converges to an NE [23]. Accordingly, any network that can be modeled as a repeated game with a stage game that has FIP will converge if the following two conditions are met:

1. The radios implement rational decision rules.
2. The network has either random or round-robin decision timings.

A normal form game is said to have the weak FIP property if, from every action vector, there exists at least one FIP that leads to an NE. By this definition, every game that has the FIP property also has weak FIP. Weak FIP implies the existence of at least one NE, but permits the existence of improvement cycles.

There are two different scenarios for which play converges when the normal form game has weak FIP. First, if the decision rules are designed such that the adaptations always follow the FIP, then play converges. Although this may seem like a difficult condition to achieve, supermodular games, which are discussed in Section 15.5.2, have a well-defined algorithm for achieving this result.

Second, if we consider decision rules by which adaptations are chosen randomly from the set of possible actions that improve performance, where all elements in the set have nonzero probability of being selected, such as would be the case for a genetic algorithm, then a repeated game with weak FIP converges to an NE for any decision update timing. The convergence of this process can be assured by noting that the process can be modeled as an absorbing Markov chain, where the game's NE form the absorbing states for the game.

An example of a normal form game with FIP is shown in the Cognitive Radios' Dilemma of Example 15.3. An example of a game with weak FIP but not FIP is shown in Figure 15.16. Here an improvement cycle [(*a*, *A*), (*a*, *B*), (*b*, *B*), (*b*, *A*), (*a*, *A*)] is labeled with the circular arrow, and an NE is circled at (*c*, *C*). However, starting from any action vector, it is possible to find an FIP that terminates in the NE. An example of a normal form game that has neither FIP nor weak FIP is shown in the paper–rock–scissors game of Example 15.2.

For an arbitrary normal form game, there is no generalizable technique for establishing FIP or weak FIP outside of an exhaustive search of a game's improvement paths. Due to this difficulty, many authors have considered

Figure 15.16: A game with the weak FIP property. The game has an improvement cycle (shown by the circular arrow) and an NE (circled). Reproduced from Neel et al. [19], Figure 2.

Γ	A	B	C
a	(1, −1)	(−1, 1)	(0, 2)
b	(−1, 1)	(1, −1)	(1, 2)
c	(2, 0)	(2, 1)	(2, 2)

convergence separately from the game theoretic analysis [22] or through an exhaustive simulation [5].

Fortunately, more powerful game models exist that not only permit establishment of FIP and weak FIP, but also perform NE identification and establish stability criteria. Two such game models—the potential game model and the supermodular game model—are discussed in Sections 15.5.1 and 15.5.2, respectively.

15.5 Relevant Game Models

This section presents two readily identified normal form game models—*potential games* and *supermodular games*—which will enable us to immediately establish the following results for decision update rules:

- existence and sometimes identification of an NE/steady state;
- convergence conditions;
- stability conditions.

15.5.1 Potential Games

As formalized by Monderer and Shapley [24], a potential game is a normal form game that has the property that there exists a function known as the *potential function*, $V: A \rightarrow \Re$, that reflects the change in value accrued by a unilaterally deviating player. In many ways, the concept of a potential function is identical to that of a Lyapunov function—a topic discussed in section *Convergence and Stability* and return to in section *Stability*.

Given an arbitrary unilateral deviation by player j from a_j to b_j (the actions of the other players (a_{-j}), remain fixed), five different types of potential games are defined by the relationship between the value accrued by the unilateral deviation,

$\Delta u_i(a, b_i) = u_i(b_i, a_{-i}) - u_i(a_i, a_{-i})$, and the change in value of the potential function, $\Delta V(a, b_i) = V(b_i, a_{-i}) - V(a_i, a_{-i})$:

- If there exists a function, V, such that $\Delta u_i(a, b_i)$ and $\Delta V(a, b_i)$ are *exactly* equal for all unilateral deviations, then the game is said to be an *exact potential game*.
- If every $\Delta u_i(a, b_i)$ is equal to a *weighted* version of $\Delta V(a, b_i)$, that is, $\Delta u_i(a, b_i) = \alpha_i \Delta V(a, b_i)$, then the game is said to be a *weighted potential game*.
- If V preserves the *ordinal* relations of u_i for all unilateral deviations and for all $i \in N$, that is, $u_i(b_i, a_{-i}) > u_i(a_i, a_{-i}) \Leftrightarrow V(b_i, a_{-i}) > V(a_i, a_{-i})$, then the game is said to be an *ordinal potential game*.
- If the ordinal relationships for u_i are reflected by V, but not necessarily in the reverse direction, then the game is said to be a *generalized ordinal potential game*.
- If there exists a function V such that every $\varepsilon_i > 0$ increase in u_i from a unilateral deviation results in a $\delta_i > 0$ increase in V, then the game is said to be a *generalized ε-potential game*.

These relationships are summarized in Table 15.2. For purposes of shorthand notation, we refer to the potential function of an exact potential game as an *exact potential function* and the potential function of a weighted potential game as a *weighted potential function* and so on.

Table 15.2: Potential game definition summary.

Potential game	Relationship ($\forall i \in N, \forall a \in A$)
Exact (EPG)	$u_i(b_i, a_{-i}) - u_i(a_i, a_{-i}) = V(b_i, a_{-i}) - V(a_i, a_{-i})$
Weighted (WPG)	$u_i(b_i, a_{-i}) - u_i(a_i, a_{-i}) = \alpha_i[V(b_i, a_{-i}) - V(a_i, a_{-i})]$
Ordinal (OPG)	$u_i(b_i, a_{-i}) - u_i(a_i, a_{-i}) > 0 \Leftrightarrow V(b_i, a_{-i}) - V(a_i, a_{-i}) > 0$
Generalized ordinal (GOPG)	$u_i(b_i, a_{-i}) - u_i(a_i, a_{-i}) > 0 \Rightarrow V(b_i, a_{-i}) - V(a_i, a_{-i}) > 0$
Generalized ε (GεPG)	$u_i(b_i, a_{-i}) > u(a_i, a_{-i}) + \varepsilon_i \Rightarrow V(b_i, a_{-i}) > V(a_i, a_{-i}) + \delta_i$

EPG: exact potential game; WPG: weighted potential game; OPG: ordinal potential game; GOPG: generalized ordinal potential game; GεPG: generalized ε potential game.

By these definitions, every exact potential game is a weighted potential game (where $\alpha_i = 1 \; \forall i \in N$); every weighted potential game is an ordinal potential game; and every ordinal potential game is a generalized ordinal potential game.

Further every weighted potential game and every finite generalized ordinal potential game are also generalized ε-potential games. The relationships between these game models can be visualized as a Venn diagram, as shown in Figure 15.17. Implicit to these relationships, each subset preserves the properties of its larger parent set.

Figure 15.17: Relationships between forms of potential games. An EPG is a WPG is an OPG is a GOPG. Further, a WPG is a GεPG [23].

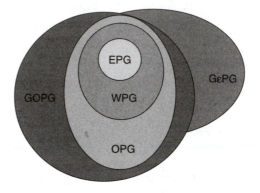

Example 15.8 *Potential Game Examples*

To illustrate these games in a more concrete manner, we present a series of examples of exact, weighted, ordinal, and generalized ordinal potential games taken from Neel [23]. Throughout this example, we make use of a two-player normal form game, Γ, with action sets $\{a, b\}$ and $\{A, B\}$ portrayed in matrix representation. Each game is accompanied by an associated potential function, also presented in matrix representation.

Figure 15.18 depicts an exact potential game Γ and its associated exact potential V. Note that $u_1(b, A) - u_1(a, A) = V(b, A) - V(a, A) = 2$, and that $u_2(b, A) - u_2(b, B) = V(b, A) - V(b, B) = 1$. Similar relationships hold for the other two possible unilateral deviations.

Γ	A	B
a	(3, 3)	(0, 5)
b	(5, 0)	(1, 1)

$V(\cdot)$	A	B
a	0	2
b	2	3

Figure 15.18: An exact potential game and its associated exact potential function.

Figure 15.19 shows an example of a weighted potential game Γ and its associated weighted potential function V. Note that this is the same potential function as in Figure 15.18, but now weighted by the factors $\alpha_1 = 2$ and $\alpha_2 = 1$.

Γ	A	B
a	(3, 3)	(0 ,5)
b	(7, 0)	(2, 1)

V(·)	A	B
a	0	2
b	2	3

Figure 15.19: A weighted potential game and its associated weighted potential function.

Figure 15.20 shows an ordinal potential game Γ and its associated ordinal potential function *V*. Note that this potential function is neither a weighted nor an exact potential function.

Γ	A	B
a	(1, −1)	(2, 0)
b	(2, 0)	(0, 1)

V(·)	A	B
a	0	2
b	2	3

Figure 15.20: An ordinal potential game and its associated ordinal potential function.

Figure 15.21 shows a generalized ordinal potential game Γ and its associated generalized ordinal potential function *V*. A careful application of the definition of an ordinal potential game will reveal that although Figure 15.21 is a generalized ordinal potential game, it is not also an ordinal potential game.

Γ	A	B
a	(1, 0)	(2, 0)
b	(2, 0)	(0, 1)

V(·)	A	B
a	0	3
b	1	2

Figure 15.21: A generalized ordinal potential game and its associated generalized ordinal potential function.

For purposes of brevity, the remaining subsections focus on exact and ordinal potential games. The interested reader can find a more extensive discussion of potential games in Neel [23].

Identification of Exact and Ordinal Potential Games

At this point, potential games may appear as difficult to identify as Lyapunov functions. Fortunately, some techniques have been developed for identifying potential games.

Exact Potential Games

First, Monderer and Shapley [24] state that when all $u_k \in \{u_i\}$ are everywhere differentiable and V is an exact potential function, Eq. (15.16) must hold $\forall i, j \in N$, $\forall a \in A$. They also state that when all $u_k \in \{u_i\}$ are everywhere twice differentiable, Eq. (15.17) is a sufficient condition for the existence of an exact potential function:

$$\frac{\partial^2 u_i(a)}{\partial a_i \partial a_j} = \frac{\partial^2 u_j(a)}{\partial a_j \partial a_i} = \frac{\partial^2 V(a)}{\partial a_i \partial a_j} \tag{15.16}$$

$$\frac{\partial^2 u_i(a)}{\partial a_i \partial a_j} = \frac{\partial^2 u_j(a)}{\partial a_j \partial a_i} \tag{15.17}$$

When these equations are satisfied (Monderer) gives Eq. (15.18) for finding the potential function:

$$V(a) = \sum_{i \in N} \int_0^1 \frac{\partial u_i}{\partial a} (x(t)) x_i'(t) dt \tag{15.18}$$

where x is a piecewise continuously differentiable path that connects some fixed action tuple b to some other action tuple a such that $x: [0, 1] \to A$ ($x(0) = b$, $x(1) = a$).

Even when the utility functions are twice differentiable, the evaluation of Eq. (15.18) can be quite tedious. Instead, the solution of a potential function is more readily accomplished by demonstrating that the game satisfies the conditions of one of a handful of common exact potential game forms and then applying its associated equation to find its exact potential function. Neel [23] provides a listing of types of exact potential games and their associated exact potential functions (Table 15.3). For our purposes, when the goals or utility functions of all radios in the network take the form shown in the first column of Table 15.3, then the network has an exact potential function given by the corresponding entry in the second column and is said to be the type of game listed in the third column.

When reading this table, note that a function that is not subscripted (e.g., $C(a)$) is an arbitrary function that all players use as part of their utility function. When a function is subscripted with i (e.g., $D_i(a_{-i})$), then each player has its own arbitrary function using the same kind of argument. Specifically, when $D_i(a_{-i})$ is present, each player has its own "dummy" function whose valuation is only a function of the actions of the other players. For an entry that includes $S_i(a_i)$, the player has a

Table 15.3: Common exact potential game forms (from Neel [23]).

Utility function form	Potential function	Game
$u_i(a) = C(a)$	$V(a) = C(a)$	Coordination game
$u_i(a) = D_i(a_{-i})$	$V(a) = c, c \in \mathfrak{R}$	Dummy game
$u_i(a) = C(a) + D_i(a_{-i})$	$V(a) = C(a)$	Coordination–dummy game
$u_i(a) = S_i(a_i)$	$V(a) = \sum_{i \in N} S_i(a_i)$	Self-motivated game
$u_i(a) = \sum_{j \in N \setminus \{i\}} w_{ij}(a_i, a_j) - S_i(a_i)$ where $w_{ij}(a_i, a_j) = w_{ji}(a_j, a_i)$	$V(a) = \sum_{i \in N} \sum_{j=1}^{i-1} w_{ij}(a_i, a_j) - \sum_{i \in N} S_i(a_i)$	BSI game
$u_i(a) = \sum_{\{S \in 2^N ; i \in S\}} w_{s,i}(a_s) + D_i(a_{-i})$ where $w_{S,i}(a_S) = w_{S,j}(a_S) \, \forall i, j \in S$	$V(a) = \sum_{S \in 2^N} w_s(a_s)$	MSI game

MSI: multilateral symmetric interaction.

function whose valuation is only a function of its own action. These sort of utility function components can be encountered when interference ($D_i(a_{-i})$) can be separated from the signal component ($S_i(a_i)$) of a player's goal (e.g., an SINR goal measured in dB). When a function is subscripted with multiple players (e.g., ij or S), then all players listed receive the same payoff. For example, assuming that the players' goals are functions of the correlations between waveforms (as occurs when measuring interference), then given choices of waveforms a_i and a_j, both players would measure the same correlation between a_i and a_j, w_{ij}.

Example 15.9 *A Bilateral Symmetric Interaction Interference Avoidance Game*

Consider a network with a frequency reuse scheme such that cross-cluster interference is negligible. Each cluster is power controlled so that received power at the cluster head for all radios is constant. However, each radio that is communicating with the cluster head is also attempting to minimize the interference its signal experiences at the receiver by adapting its waveform.

We can consider this as a repeated game with the normal form stage game modeled as follows. Each adaptive radio in a cluster is a player, the actions for each radio are its available waveforms, and utility functions are given as Eq. (15.19), where $\rho(a_j, a_k)$ is the statistical correlation between waveforms a_j and a_k with the assumption that $\rho(a_j, a_k) = \rho(a_k, a_j)$:

$$u_j(a) = -\sum_{k \in N \setminus j} \rho(a_j, a_k) \tag{15.19}$$

From Table 15.3, we can see that Eq. (15.19) satisfies the conditions for a bilateral symmetric interaction (BSI) game, where $S_j(a_j) = 0 \; \forall j \in N$. Table 15.3 indicates that Eq. (15.20) is an exact potential function for this game:

$$V(a) = \sum_{i \in N} \sum_{j=1}^{i-1} \rho(a_i, a_j) \tag{15.20}$$

Ordinal Potential Games

Monderer and Shapley [24] and Voorneveld and Norde [25], have introduced techniques by which an ordinal potential game can be identified. Monderer and Shapley [24] state that if a game has FIP and if $\forall a_{-i} \in A_{-i}$ and all $i \in N$, $u_i(a_i, a_{-i}) \neq u_i(b_i, a_{-i}) \; \forall a_i, b_i \in A_i$, then the game is an ordinal potential game. Now consider the generalized ordinal potential game in Figure 15.21, which we claimed was not also an ordinal potential game. Even though the game does have FIP, $u_2(a, A) = u_2(a, B)$ and thus the game fails the condition set out in Monderer and Shapley.

Voorneveld and Norde [25] introduce a different characterization of ordinal potential games wherein the game lacks *weak improvement cycles*. A sequence of unilateral deviations $\{a^1, \ldots, a^n\}$ is said to be a *cycle* if $a^1 = a^n$. A cycle is said to be a *weak improvement cycle* if $u_j(a^{k+1}) \geq u_j(a^k)$ for all $a^k \in \{a^1, \ldots, a^n\}$, where j is the unique deviator at step $k + 1$ and for at least one $a^m \in \{a^1, \ldots, a^n\} \; u_j(a^{m+1}) \geq u_j(a^m)$.

Both of these approaches to identification require an exhaustive characterization of the game, which can be quite time consuming. A less systematic but frequently quicker approach is to apply the concept of *better response equivalence*. A game $\Gamma = \langle N, A, \{u_i\} \rangle$ is said to be a *better response equivalent* to game

$\Gamma' = \langle N, A, \{v_i\} \rangle$ if $\forall i \in N$, $\forall a \in A$, $u_i(a_i, a_{-i}) \geqslant u_i(b_i, a_{-i}) \Leftrightarrow v_i(a_i, a_{-i}) \geqslant v_i(b_i, a_{-i})$. For notational simplicity, if Γ is a better response equivalent to Γ', we write $\Gamma \sim \Gamma'$.

An example of better response equivalence is given in the first three games presented in Example 15.8. These games also imply a few important properties that are preserved by better response equivalence, namely NE, improvement path properties, and ordinal potential functions.

These preservation properties imply a different approach to identifying ordinal potential games—identifying a *better response transformation* (a reformulation of a game's utility functions in a way that the resulting game is a better response equivalent to the original game) that yields an identifiable exact potential game. Examples of better response transformations include scalar multiplication and logarithmic transformations.

Example 15.10 *An Ordinal Potential Interference Avoidance Game*

Suppose we modify the network of Example 15.9 so that instead of minimizing interference, the radios adapt their choice of operating frequencies to maximize their throughput. In this context, throughput is a monotonic function of interference in that any decrease in interference results in an increase in throughput. Accordingly these two games are a better response equivalent and we can conclude that this modified game is an ordinal potential game.

Fixed Points and Steady States for Potential Games

It is relatively easy to show that every potential game for which a maximum value of V exists (e.g., games with a finite action space and games where V is bounded and A is compact) has an NE. Specifically, if a^* is a global maximizer of V, then there can be no a' such that $u_j(a') > u_j(a^*)$, where a' and a^* differ only in the jth component. Otherwise, $V(a') > V(a^*)$, and a^* is not a global maximizer of V.

Thus, by showing that each iteration of a cognitive radio algorithm can be modeled by the same potential game and that the associated potential function has a maximum, we also show that a steady state exists for the algorithm. Further, the preceding discussion also shows that we can identify the steady states for the algorithm by solving for the global maximizers of V. Although other NEs may exist for a potential game, as we show in section *Stability*, only those NEs that are maximizers of V are stable.

Desirability

In general, little can be said about the optimality or desirability of the steady states of a cognitive potential game. They need not be Pareto efficient, and they are not generally maximizers of a design objective function. However, when the potential function is also the network objective function, that is, when $V = J$, then if V admits a global maximum, there exists an NE that is optimal. Further, because deterministic play increases the value of V with each iteration, it is safe to say that the stable steady state of the network will give better performance than the initial state of the network.

As an example, consider the adaptive interference avoidance game of Example 15.9. This game necessarily has numerous steady states, but if the network's designer is attempting to minimize total network interference, then the steady states will generally be desirable and performance will improve with each adaptation.

Convergence

Up to this point we have considered only the goals of the radios. Let us now consider the set of possible decision rules, f^t, that could be implemented to achieve convergence. Specifically, let us consider rational adaptations, in which, each radio, j, chooses to change its action from a_j to b_j if and only if $u_j(a_j, a_{-j}) < u_j(b_j, a_{-j})$. Further, let us restrict ourselves to those algorithms with either round robin or random decision timings. With these assumptions, the sequence $\{V(a(t))\}$, where $a(t^{k+1}) = f^{t^k}(a(t^k))$ is monotonically increasing.

If we consider the situation in which the potential functions are bounded, both $\{V(a(t))\}$ and the sequence $\{a(t)\}$ must converge. For finite games implementing random or round-robin self-interested algorithms, $\{a^t\}$ converges to an NE. However, this need not be the case for games with infinite action spaces because infinitesimally small steps could be taken to cause convergence to an action vector other than an NE.

However, for games that are generalized ε-potential games (recall that all exact potential games are generalized ε-potential games), we can introduce a different class of algorithms that can be shown to converge, called ε-self-interested updates. For an ε-self-interested update, each radio, j, chooses to change its action from a_j to b_j if and only if $u_j(a_j, a_{-j}) + \varepsilon < u_j(b_j, a_{-j})$. For a generalized ε-potential game with a bounded potential function, this implies that there is a $\varepsilon_V > 0$ such that for every round-robin or random ε-self-interested adaptation, V increases by at least an ε_V. In this case, the adaptations are guaranteed to converge to an ε-NE. An action vector a^* is said to be an ε-NE if $\forall j \in N$ when there is no $b_j \in A_j$ such that $u_j(a_j^*, a_{-j}^*) + \varepsilon < u_j(b_j, a_{-j}^*)$.

For synchronous algorithms, note that little can be said about the convergence of a traditional potential game. For instance, consider the exact potential game shown in Figure 15.22. Starting from (a, A), it is possible to enter into the oscillation $(a, A) \rightarrow (b, B)$ for a deterministic self-interested algorithm with synchronous timing. However, it is readily observed that such an oscillation is broken when random decision timing is assumed.

Figure 15.22: An exact potential game. This game could oscillate between (a, A) and (b, B) for synchronous decision timings.

Γ	A	B
a	(0, 0)	(2, 2)
b	(2, 2)	(0, 0)

Convergence Rate

An infinite generalized ε-potential game with a potential function bound by $|V| \leq K$ and step size of ε_V with round-robin decision timing cannot have more than $2|N|K/\varepsilon_V$ iterations. This particular bound on the number of iterations comes from an assumption of an initial action vector such that $V(a(0)) = -K$ as well as the requirement that for every $|N|$ iterations, at least one player must be able to improve its payoff by at least an ε (otherwise an ε-NE has been reached).

Stability

As alluded to earlier in Section 15.5.1, the definition of a potential function is similar to that of a Lyapunov function. Recall that Lyapunov's Direct Method (see section *Convergence and Stability*) stated that a decision rule f^t is Lyapunov stable if there exists a continuous function L that satisfies the following conditions:

1. $L(a^*) = 0$
2. $L(a^*) > 0 \ \forall a \in N(a^*)\backslash a^*$
3. $\Delta L(a^{t_k}) \equiv L[f^{t_k}(a^{t_k})] - L(a^{t_k}) \leq 0 \ \forall a \in N(a^*)\backslash a^*$

Now consider the function given by:

$$L^V(a) = -V(a) + V(a^*) \tag{15.21}$$

where V is a bounded continuous potential function with isolated potential maximizer a^*. In the neighborhood of a^*, Eq. (15.21) satisfies Lyapunov function conditions 1 and 2. Now consider any rational decision rule with random or round-robin timing with iterations that can be modeled as a potential game with potential function V. Because this is a potential game, $V(a^{t_{k+1}}) \geqslant V(a^{t_k})$ so $V[f^{t_k}(a^{t_k})] - V(a^{t_k}) \geqslant 0$ and $\Delta L(a^{t_k}) \leqslant 0 \; \forall a \in N(a^*) \backslash a^*$, satisfying condition 3.

Thus L^V is a Lyapunov function for any self-interested decision rule with round-robin or random timing that can be modeled as a potential game with a bounded continuous potential function. Further, all maximizers of V (our method for finding NE in a potential game) are also Lyapunov stable.

Also note that if the game has a unique NE, then a^* is globally Lyapunov stable. And if the decision rule fits into one of the classes of algorithms that deterministically converge discussed in section *Convergence*, then the algorithm is asymptotically stable as well. Note that before, Lyapunov stability had to be considered for a specific decision rule, but for potential games, we can show that an entire class of decision rules is stable by examining the goals of the radios (the radios are the players in a game model of a cognitive radio network).

15.5.2 Supermodular Games

Supermodular games are encountered frequently in algorithms in which an increase in a_{-i} results in a corresponding increase in a_i—a concept known as *increasing differences*. Beginning with an initial discussion of some mathematical concepts related to supermodular games—such as increasing differences—this section defines the concept of the supermodular game and describes what analytical insights can be gained by demonstrating the iterations of a cognitive radio algorithm can be modeled as a supermodular game.

Supermodular Game Mathematical Preliminaries

Before beginning our discussion of supermodular games, we need to first define two concepts that are not explicitly part of other game models, namely *lattices* and *supermodularity*. Our discussion of lattices requires us to first discuss and define partially ordered sets and the join and meet operations.

Partially Ordered Sets

Consider a set X and some relation R (some operator that compares two elements of a set and returns a Boolean true or false), termed the order of X, such that the following three properties are satisfied for all $x, y, z \in X$:

- *Reflexivity* – $x \, R \, x$
- *Transitivity* – $x \, R \, y \, R \, z \Rightarrow x \, R \, z$
- *Antisymmetry* – $x \, R \, y \, R \, x \Rightarrow x = y$

Given a set X with a relation R that satisfies the above three properties, X is said to be a *partially ordered set* (or *poset*) if there are some $x, y \in X$ such that neither $x \, R \, y$ nor $y \, R \, x$. If there are no such elements, then X is said to be *totally ordered* on R.

We state, without proof, that the relation $\geqslant$ (greater than or equal to) induces a partial order on the set $X = \mathfrak{R}^n$. Because we make use of this result later in this chapter, we further define here the relation $\geqslant$ as applied to $\mathfrak{R}^n$. By convention, we write $x \geqslant y$ if $x_k \geqslant y_k \, \forall k = 1, \ldots, n$. Similarly we would write $x > y$ if $x \geqslant y$ and for some $k \, x_k > y_k$. This is the same meaning for $\geqslant$ that we used when relating two different power vectors in section *Standard Interference Function Model*.

As examples, suppose $n = 4$ and $x = (1, 0, 0, 3)$ and $y = (1, 2, 2, 3)$. Then we could accurately write both $y \geqslant x$ and $y > x$. If $x = (1, 0, 4, 3)$ and $y = (1, 2, 2, 3)$, then by the rules of $\geqslant$ no relation between x and y can be established.

Join and Meet

Given two vectors, $x, y \in \mathfrak{R}^n$, we define the *meet* of x and y, $x \wedge y$, as $(\min\{x_1, y_1\}, \min\{x_2, y_2\}, \ldots, \min\{x_n, y_n\})$. The meet operation is equivalent to evaluating the *infimum* (greatest lower bound) of x and y on an element-by-element basis. The *join* of x and y, $x \vee y$, is defined as $(\max\{x_1, y_1\}, \max\{x_2, y_2\}, \ldots, \max\{x_n, y_n\})$. The join operation is equivalent to evaluating the *supremum* (least upper bound) of x and y on an element-by-element basis. Note that if $y \geqslant x$, then $x \wedge y = x$ and $x \vee y = y$.

As examples, if $x = (1, 0, 0, 3)$ and $y = (1, 2, 2, 3)$, then $x \wedge y = \{1, 0, 0, 3\}$ and $x \vee y = \{1, 2, 2, 3\}$. If $x = (1, 0, 4, 3)$ and $y = (1, 2, 2, 3)$, then $x \wedge y = \{1, 0, 2, 3\}$ and $x \vee y = \{1, 2, 4, 3\}$.

Lattices

A set X is termed a *lattice* if the following two conditions are satisfied:

1. X is a partially ordered set,
2. $x \wedge y \in X$ and $x \vee y \in X \, \forall x, y \in X$.

While applying this definition may be difficult, the following are several commonly encountered lattices taken from Topkis [26], Examples 2.23 and 2.25:

(1) $\mathfrak{R}$ with the normal ordering relationship, $\geq$.
(2) $X \subset \mathfrak{R}$.
(3) A totally ordered set (called a *chain* by Topkis [26]).[17]
(4) $\mathfrak{R}^n$, where for $x', x'' \in \mathfrak{R}^n$ $x' \vee x'' = (x'_1 \vee x''_1, \ldots, x'_n \vee x''_n)$ and $x' \wedge x'' = (x'_1 \wedge x''_1, \ldots, x'_n \wedge x''_n)$.
(5) $X = x_{\alpha \in A} X_\alpha$, where each X_α is a lattice, $x' \vee x'' = (x'_\alpha \vee x''_\alpha \colon \alpha \in A)$ and $x' \wedge x'' = (x'_\alpha \wedge x''_\alpha \colon \alpha \in A)$.

If X is a lattice and every nonempty subset has a supremum and an infimum, then X is a *complete lattice*. We consider two particularly useful examples of complete lattices:

- All compact lattices.
- The Cartesian product of compact subsets of $\mathfrak{R}$.

We also define a *sublattice* as a subset of a lattice that is itself a lattice.

Supermodular Functions

Suppose $f \colon X \to \mathfrak{R}$, where X is a lattice; if Eq. (15.22) is satisfied, then f is said to be *supermodular* on X:

$$f(x') + f(x'') \leq f(x' \vee x'') + f(x' \wedge x'') \ \forall x', x'' \in X \qquad \textbf{(15.22)}$$

For single-dimensional X, every f is supermodular. For multidimensional X, supermodularity can be difficult to determine. Instead, a more tractable way to determine whether f is supermodular is given by Topkis's Characterization Theorem [27] which states that if X is a subset of $\mathfrak{R}^n$ and a complete lattice, and if f is twice continuously differentiable on X, then f is supermodular if and only if Eq. (15.23) is satisfied:

$$\partial^2 f(x) \ / \ \partial x_i \partial x_j \geq 0 \forall x \in X, \forall i \neq j \qquad \textbf{(15.23)}$$

---

[17]Note that $\mathfrak{R}$ is a totally ordered set.

Increasing Differences

Definition 12.4 in Fudenberg and Tirole [18] states that a function, f, has *increasing differences* in (x_i, x_{-i}) if for all pairs of points $x_i, \overline{x}_i \in X_i$, such that $x_i \geqslant \overline{x}_i$, and all pairs of points $x_{-i}, \overline{x}_{-i} \in X_{-i}$ such that $x_{-i} \geqslant \overline{x}_{-i}$, Eq. (15.24) holds:

$$f(x_i, x_{-i}) - f(\overline{x}_i, x_{-i}) \geqslant f(x_i, \overline{x}_{-i}) - f(\overline{x}_i, \overline{x}_{-i}) \qquad \textbf{(15.24)}$$

Similarly, f is said to be *increasing* in x_i if $f(x_i) \geqslant f(\overline{x}_i) \; \forall x_i \geqslant \overline{x}_i$.

For utility functions, increasing differences has an intuitive meaning. Any increase in x_{-i} makes an increase in x_i more desirable. For example, consider any utility function appropriate to being applied to a cognitive radio algorithm in which the radios are attempting to achieve a target SINR at their receivers by altering their power levels, perhaps as shown in the following equation:

$$u_j(\mathbf{p}) = -\left(\overline{\gamma}_j - 10 \log_{10}(g_{jj} p_j) + 10 \log_{10} \left(\sum_{k \in N \setminus j} g_{kj} p_k + N_j \right) \right)^2, \qquad \textbf{(15.25)}$$

where the symbols are the same as used in section *Standard Interference Function Model*, Eq. (15.6).

By examining the increasing differences property at a particular point, if the system is operating at an NE and the other radios in the network increase their transmit power levels, p_{-j}, cognitive radio j will see a benefit from also increasing its transmit power. Further, u_j is supermodular in its own transmit power, p_j, because the set of transmit powers is a single-dimensional set. As we see in section *Supermodular Games*, such a target SINR power control algorithm would satisfy the conditions of a *supermodular game*.

Supermodular Games

According to Topkis [26], a normal form game $\Gamma = \langle N, A, \{u_i\} \rangle$ is said to be a supermodular game if $\forall i \in N$ the following three conditions are satisfied:

(1) A_i is a complete lattice.

(2) u_i is supermodular in a_i.

(3) u_i has increasing differences in (a_i, a_{-i}).

The preceding discussion gave an example of a cognitive radio game that satisfies these conditions, but identifying a supermodular game via its definition can be difficult. A more readily evaluated special case of these conditions is given in Theorem 4 of Milgrom and Roberts [27], which gives the following conditions for a *smooth supermodular game*:

1. A_i is a closed interval in $\mathfrak{R}^{k_i}$, that is, $A_i = [\underline{y}_i, \overline{y}_i] = \{a | \underline{y}_i \leq a \leq \overline{y}_i\}$.[18]
2. u_i is twice continuously differentiable on A_i.
3. $\partial^2 u_i / \partial a_{ik} \partial a_{im} \geq 0$ for all $i \in N$ and all $1 \leq k \leq m \leq k_i$.
4. $\partial^2 u_i / \partial a_{ik} \partial a_{jm} \geq 0$ for all $i \neq j \in N$, $1 \leq k \leq k_i$ and $1 \leq m \leq k_j$.

Note that this definition permits each player to have k_i dimensions to its action set. For example, a two-dimensional set might include adaptations of both power and frequency.

Example 15.11 *Supermodular Game Example*

Let us define a supermodular game with the following components, player set $N = \{1, 2, \ldots, n\}$, action sets given by $A_i = [0, 1] \; \forall i \in N$, and utilities given by

$$u_i(a) = \sum_{j \in N} a_i a_j.$$

This game can be shown to be a smooth supermodular game. Condition (1) is satisfied as A_i is a compact subset of $\mathfrak{R}$ ($k_i = 1 \; \forall i \in N$). Condition (2) holds as u_i is twice differentiable. Condition (3) does not apply as the action sets are single dimensions. Condition (4) can be verified as $\partial^2 u_i(a) / \partial a_i \partial a_{-i} = 1$. Note that this also satisfies the conditions for a BSI potential game.

It is also interesting to note that this game is also a BSI game and thus an exact potential game.

Properties of Supermodular Games

This section describes some of properties of supermodular games that are relevant to the analysis of cognitive radio networks. In particular, supermodular games are useful for cognitive radios that implement decision rules that perform local optimizations, or in game theory parlance, best response decision rules.

---

[18]In this notation, each i may have k_i free dimensions from which to choose its waveform. For example, an 802.11h radio is free to choose both power and center frequency (both real numbers), so the radio would have $k_i = 2$.

Given a_{-j}, player j's *best response* to a_{-j} is given by the set of action tuples for which no higher utility can be found. This set can be formally specified as in:

$$\text{BestResponse} = \{b_j \in A_j : u_j(b_j, a_{-j}) \geq u_j(a_j, a_{-j}) \forall a_j \in A_j\} \quad \text{(15.26)}$$

We can define a best response function for each player $j \in N$, $\hat{B}_j(a)$ where $\hat{B}_j(a)$ returns an action that satisfies Eq. (15.26). We can also define a joint best response to a as $\hat{B}(a)$, where $\hat{B}(a)$ returns an action vector formed by the synchronous application of $\hat{B}_j(a) \ \forall j \in N$.

For supermodular games where u_j is upper semicontinuous in $a_j \ \forall j \in N$, $\forall a_{-j} \in A_{-j}$ and each A_j is an interval of $\Re^{k_j}$, Lemma 4.2.2 in Topkis [26] gives the following properties for $\hat{B}_j(a)$ and $\hat{B}(a)$:

1. $\hat{B}_j(a_{-j})$ is a nonempty compact sublattice of $\Re^{k_j} \ \forall a_{-j} \in A_{-j}$.

2. $\hat{B}(a)$ is a nonempty compact sublattice of $\Re^m \ \forall a \in A$ ($m = \sum_{j \in N} k_j$).

3. There exists a greatest and a least element of $\hat{B}_j(a_{-j}) \ j \in N, \ \forall a_{-j} \in A_{-j}$.

4. There exists a greatest and a least best joint response $\hat{B}(a) \ \forall a \in A$.

5. $\hat{B}_j(a_{-j})$ is increasing in a_{-j} on $A_{-j} \ \forall j \in N$.

6. $\hat{B}(a)$ is increasing in a on A.

7. $\max_{a_j \in A_j} \hat{B}_j(a_{-j})$ is an increasing function from A_{-j} into $A_j \ \forall j \in N$.

8. $\max_{a \in A} \hat{B}(a)$ is an increasing function from A into A.

A function $u_i : A \to \Re$, is upper semicontinuous if for all $a^0 \in A$, $\lim_{a \to q^0} \sup (u_i(a)) \leq u_i(a^0)$. Note that because differentiability implies continuity, and recalling the definition of a continuous function from Section 15.2, it is clear that a real-valued continuous function is also upper semicontinuous. Thus, smooth supermodular games necessarily have upper semicontinuous $u_j \ \forall j \in N$.

Fixed Points in Supermodular Games

Before considering NE and fixed points in supermodular games, we need to introduce Tarski's fixed point theorem:

> Let A be a nonempty compact sublattice of $\Re^n$. Let $f: A \to A$ be an increasing function. Then f has a fixed point.

Unlike the previous fixed point theorems considered in this chapter, there is no requirement that f be continuous, nor is there a requirement that A be a convex set. The only requirement is that A is a lattice and f is an increasing function.

Now consider a cognitive radio network whose iterations can be modeled as a repeated smooth supermodular game. Then assume that the decision rule is synchronous and of the form $f^t(a) = \hat{B}(a)$. By property (6) of Lemma 4.2.2 of Topkis [26], $f^t(a)$ is then an increasing function, and by virtue of being smooth and by Tarski's fixed point theorem, $f^t(a)$ must have at least one fixed point (NE).

For supermodular games, identifying NE is not as easy as it is for potential games. However, by leveraging property (2) of Lemma 4.2.2 of Topkis [26], we see that the set of NE must form a complete lattice. Thus, once we identify a pair of NE, a^*, a^{**}, we know that their joins and meets must also be NE.

Example 15.12 *Visualizing Tarski's Fixed Point Theorem for Supermodular Games*

Although we have been referring to multidimensional action spaces and a multidimensional f^t, we can visualize why $f^t(a) = \hat{B}(a)$ must have at least one fixed point for a single-dimensional action space, as shown in Figure 15.23.

Figure 15.23: Visualization of Tarski's fixed point theorem for an increasing (and upper semicontinuous) function $\hat{B}(a)$ defined on a. In this example, $\hat{B}(a)$ has two fixed points (circled).

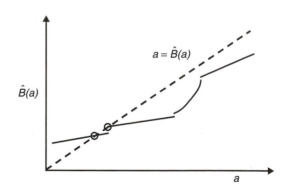

The solid line in Figure 15.23 shows an increasing and upper semicontinuous function $\hat{B}(a)$, and the dashed line plots the line $f(a) = a$. Whenever $\hat{B}(a)$ intersects $f(a)$, we have $a = \hat{B}(a)$—a fixed point. Assuming $\hat{B}(a)$: $A \rightarrow A$, then if $\hat{B}(a)$ starts above $f(a)$, $\hat{B}(a)$ must eventually intersect $f(a) = a$, indicating a fixed point. Note that even though $\hat{B}(a)$ is not continuous, as was necessary for the Leray–Schauder–Tychonoff fixed point (see secton *Fixed Points and Solutions to Cognitive Radio Networks*), $\hat{B}(a)$, still has two fixed points.

Desirability

Little can be said about the desirability of the fixed points in a supermodular game in general. As is the case for many systems, the fixed points of the game should be evaluated via some network-wide cost function to determine desirability.

Convergence

Consider a synchronous locally optimal decision rule, $f^t(a) = \hat{B}(a)$, a round-robin locally optimal decision rule, $f^t_k(a) = \hat{B}_{mod(k,|N|)}(a)$, or a randomly timed locally optimal decision rule, $f^{t_k}(a) = \hat{B}_{rand(k)}(a)$. Because $\hat{B}(a)$ and $\hat{B}_j(a)$ are increasing functions, for all $a^0 \leqslant \inf(\{a \in A: a \in B(a)\})$, the sequence formed by f^t is nondecreasing. Similarly, for all $a^0 \geqslant \sup(\{a \in A: a \in \hat{B}(a)\})$, the sequence formed by f^t is nonincreasing. As these two sequences squeeze together, the recursive calculation of f^t must converge to a region bounded by the greatest and least NE. Of course, if the NE lattice has only one unique element, then these adaptations must converge to this unique element.

Theorem 4.3.1 in Topkis [26] states that when the round-robin best response algorithm is modified so that the least element in $\hat{B}(a)$ is selected at each stage and the algorithm is started at the least element in A, then the number of iterations this algorithm will take to reach an NE (in particular, the least NE in the NE lattice) is bounded by:

$$\text{\# Iterations} = \left(|N| - 1\right)\left(\sum_{i \in N} |A_j|\right) - |N|^2 + |N| + 1 \qquad (15.27)$$

Adaptive Dynamic Process

A variation on the simultaneous best response algorithm is presented by Milgrom and Roberts [27], wherein the players follow what is termed an *adaptive dynamic process*. In an adaptive dynamic process, all players play a best response to some arbitrary weighting of actions played by other players in the recent past (i.e., not just the most recently observed actions).

Formally, a process is defined as an adaptive dynamic process ([27], (A6)) if $\forall T \; \exists \; T'$ such that $\forall t \geqslant T'$, $a^t \in \bar{U} ([\inf(P(T, t)), \sup(P(T, t))])$, where $P(T, t)$ denotes the actions played between times T and t, $U(a)$ be the list of undominated responses to a for each player, and $\bar{U}(a) = [\inf(U(a)), \sup(U(a))]$.

The corollaries to Theorem 8 in Milgrom and Roberts [27] show that a smooth supermodular game following an adaptive dynamic process converges to a region bounded by the NE lattice and that iterative elimination of dominated strategies converges to a region defined by the NE lattice. Note that when the NE is unique, the adaptive dynamic process converges to NE.

Random Sampling

For supermodular games, there exists an improvement path that terminates in an NE starting from every action tuple. Accordingly, all smooth supermodular games have the weak FIP property described in *Convergence,* a subsection of 15.4.4.

Friedman and Mezzetti [28] present a modified better response dynamic, here called a *random better response dynamic*, that assumes the normal form game has weak FIP. In the random better response dynamic, the player that adapts its behavior is chosen at random, and the choice of adaptation by that player is chosen randomly and implemented if the adaptation would improve performance. Specifically, when deciding upon its action, player i randomly samples action a_i' from $A_i\backslash\{a_i\}$ where all actions in $A_i\backslash\{a_i\}$ have equal probability of being chosen. If $u_i(a_i', a_{-i}) > u_i(a_i, a_{-i})$, then player i will implement a_i'. Friedman and Mezzetti proceed to show in that any finite supermodular game, (finite A) converges according to random better response dynamics. Based on the discussion in this chapter, we can see that such a process on a game with weak FIP forms an absorbing chain and thus must end in an absorbing state, in this case, an NE.

In the context of cognitive radios, a cognitive radio network following random sampling would permit the radios to "try out" a particular waveform and then continue using that waveform if it improved its performance. In effect, algorithms that converge via random sampling are algorithms that permit cognitive radios to learn about their operating environment, specifically learning how the operating environment impacts performance even when they start with no knowledge of the environment.

Because this result requires only that the stage game has weak FIP, all potential games with finite action spaces (which have FIP and thus also weak FIP) also converge under random sampling.

Stability

For any finite best response convergent normal form game following any one of the best response algorithms discussed in the preceding section, a Lyapunov function exists and is given by:

$$V(a) = |\sigma(a)| + \sum_{a' \in \sigma(a)} V(a')$$

(15.28)

where $\sigma(a) = \{a' \in A: a \in \hat{B}(a')\}$.

For supermodular games with infinite action sets (for which the concept of stability is better defined), additional information must be introduced to determine stability, perhaps by leveraging whatever information can be gleaned from traditional analysis techniques. An example analysis for which additional information can be used to determine the stability of a supermodular game is given in section *Analysis*.

15.6 Case Studies

This section presents several examples that demonstrate the analysis of interactions of cognitive radio networks. Specifically, we consider and analyze the (distributed) scenarios of ad hoc power control, DFS, and general waveform adaptation for adaptive interference avoidance. For each scenario, we briefly overview the algorithm under study, present a game model for the algorithm, and analyze the algorithm for steady states, optimality, convergence, and stability.

Many of these examples make extensive use of game theoretic techniques to provide the most generalizable results, but the analyses are not limited to only game theoretic techniques. When appropriate, traditional analysis techniques refine the results of our analysis.

15.6.1 Distributed Power Control

Neel et al. [29] present an analysis of a distributed power control algorithm on an ad hoc network in which each link, j, varies its transmit power in an attempt to achieve a target SINR, γ_j, measured in dB at the receiving end of the link. This scenario can be thought of as analogous to the fixed assignment scenario presented by Yates [12] and presented in section *Standard Interference Function Model*. Indeed, this analysis can be considered an extension of that scenario to ad hoc networks with additional consideration given to stability.

Using the notation presented in section *Standard Interference Function Model*, in a network, N, of cognitive radios the SINR of the signal transmitted by j and received by its node of interest measured in dB is given by:

$$\gamma_j = 10 \log_{10}(g_{jj} p_j) - 10 \log_{10}\left(\sum_{k \in N \backslash j} g_{kj} p_k + N_j\right) (\text{dB}) \qquad (15.29)$$

where g_{kj} is the effective fraction of power transmitted by node k that is received at j's node of interest (receiving end of j's link) and N_j is the noise at the receiving end of link j.

Stage Game Model

Based on the preceding discussion, a normal form stage game can be formulated as follows:

- Player set, N: Set of decision-making links.

- Player action set, A_j: The real convex, compact set of powers, $[0, p_j^{\text{max}}]$, where p_j^{max} is the maximum transmit power of cognitive radio j. The action space, $A \subset \Re^n$, is given by $A = A_1 \times A_2 \times \cdots \times A_n$.

- Utility: An appropriate utility function for a target SINR (dB) algorithm given by:

$$u_j(p) = -\left(\hat{\gamma}_j - 10 \log_{10} (g_{jj} p_j) + 10 \log_{10}\left(\sum_{k \in N \setminus j} g_{kj} p_k + N_j\right)\right)^2 \quad \text{(15.30)}$$

where $\hat{\gamma}_j$ is the SINR target of cognitive radio j.

Analysis

Altman and Altman [30] claim that the cellular fixed assignment scenario of Yates [12] on which this ad hoc network model is based on supermodular. The following analysis parallels that given by Neel et al. [29], which showed that this stage game constitutes a smooth supermodular game for an ad hoc network.

A stage game can be shown to be a smooth supermodular game by applying the second-order conditions presented in section *Supermodular Games*. First, notice that the action space forms a complete lattice because it is a compact interval of Euclidean space. Then evaluating the second derivative with respect to p_j and p_k, where k is any cognitive radio $k \in N \setminus j$, yields:

$$\frac{\partial^2 u_j(p)}{\partial p_j \partial p_k} = \frac{200 g_{kj}}{p_j \left(\sum_{k \in N \setminus j} g_{kj} p_k + N_j\right) \ln(20)} \quad \text{(15.31)}$$

Because Eq. (15.31) is strictly positive, the last condition for a smooth supermodular game is satisfied. Accordingly, we know the following about the network:

- It has at least one steady state.

- The network converges for synchronous and asynchronous best response algorithms (local optimization), even with adaptation errors to a region bound by the greatest and least NE. Thus, convergence to a bounded region is assured even as the radios learn their operating environment.

Further, because the best response algorithm given by:

$$p_j^{t_{k+1}} = p_j^{t_k} \frac{\overline{\gamma}_j}{\gamma_j} \tag{15.32}$$

is a known standard interference function, we know the following:

- The network adaptation process constitutes a pseudocontraction and thus has a unique fixed point. Thus the greatest and least NE are the same point and play must converge to a unique point.
- The network achieves the target SINR vector with the smallest possible power vector (when the SINR vector is feasible) implying the algorithm is optimal in terms of minimizing power consumption.
- A Lyapunov function is given by the distance between the current power vector and the fixed point power vector.

Finally, for a feasible SINR target vector (see section *Standard Interference Function Model*), the unique steady state for this game can be found by solving the linear program $\mathbf{Z}\overline{\mathbf{p}} = \mathbf{Y}$:

where

$$\mathbf{Z} = \begin{bmatrix} h_{1v_1} & -\hat{\gamma}_1 h_{1v_2} & \cdots & -\hat{\gamma}_1 h_{1v_n} \\ -\hat{\gamma}_2 h_{2v_1} & h_{2v_2} & \cdots & -\hat{\gamma}_2 h_{2v_n} \\ \vdots & \vdots & \ddots & \vdots \\ -\hat{\gamma}_n h_{nv_1} & -\hat{\gamma}_n h_{nv_2} & \cdots & h_{nv_n} \end{bmatrix}$$

$$\mathbf{Y} = [\hat{\gamma}_1 N_1 \quad \hat{\gamma}_2 N_n \quad \cdots \quad \hat{\gamma}_n N_n]^T$$

and

$$\mathbf{p} = [p_1 \quad p_2 \quad \cdots \quad p_n]^T$$

where the *h* variables as are in section *Standard Interference Function Model*.

Validation

Consider the ad hoc network shown in Figure 15.24, which is operating at a single frequency, where each terminal is attempting to maintain a target SINR at a cluster head and where each cluster head is maintaining a target SINR at the gateway node. The signals employed by the radios have a statistical spreading factor of K. Here, $h_{jk} = g_{jk}/K$ for $j \neq k$ and $h_{jj} = g_{jj}$.

Figure 15.24: Simulation scenario for ad hoc power control example.

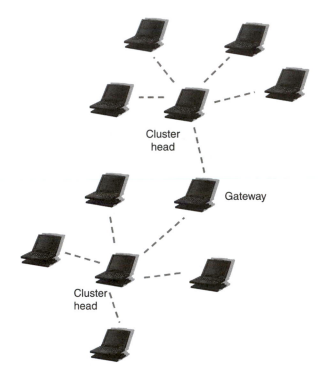

Assuming these devices implement decision rules that are locally optimal, the network is implementing a decision rule that we know converges to the unique steady state. Accordingly, we would expect that any initial power vector would converge to the unique power vector and that even when corrupted by noise, the system would remain in a region near this steady state as the analysis predicts Lyapunov stability.

To confirm these predictions, a synchronous simulation was constructed for deterministic and stochastic simulation scenarios. The simulation results for these scenarios are shown in Figures 15.25 and 15.26, respectively, where the upper graphs plot objective function values as a function of iteration, and the lower graphs plot power level as a function of iteration. Note that the locally optimal algorithm rapidly converges to the steady state in both scenarios and that even in the presence of random noise–induced perturbations, the network remains in a region around the deterministic steady state.

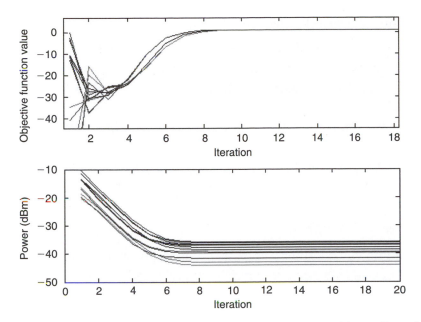

Figure 15.25: Deterministic simulation of an ad hoc network of cognitive radios with synchronous adaptations and utility functions given by Eq. (15.30). The top graph shows the value of each radio's utility function versus the iteration; the lower graph shows the power levels for each radio versus iteration.

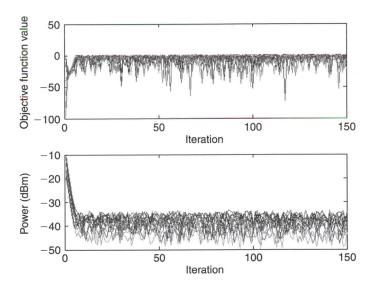

Figure 15.26: Simulation of an ad hoc network of cognitive radios with synchronous adaptations, utility functions given by Eq. (15.30), and stochastic channel models. The top graph shows the value of each radio's utility function versus the iteration; the lower graph shows the power levels for each radio versus iteration.

15.6.2 Dynamic Frequency Selection

Suppose each cognitive radio in a network of cognitive radios is adjusting the center frequency of its waveforms to minimize in-band interference. Frequently, this process is called DFS such and is implemented by WiFi radios that support the 802.11h MAC[3].

Although the 802.11h MAC is concerned with avoiding radar installations and minimizing interference between access points, we can generalize this problem to one of minimizing in-band interference by selecting a center frequency from a set F while avoiding "off-limit" frequency bands (presumably excluded from F).

Specific Network Under Study

For this example, we consider a cognitive radio network in which each radio is utilizing a signal with a brick wall bandwidth of B and can operate over the set of frequencies F. Although it is important that F is compact, it is not as important that F is convex for the subsequent analysis. Thus F may be a discrete set, F may be a closed interval of frequencies, or F may be a collection of closed intervals of frequencies with gaps in the location of the off-limit frequency bands.

Stage Game Model

A single iteration of this game can be expressed as a normal form game as follows:

- Players, N: Set of cognitive radios.
- Action space, A: $A = F_1 \times F_2 \times \cdots \times F_n$, where each F_j is a compact set.
- Utility function, u_j : A utility function for any player, j, is given by:

$$u_j(f) = \sum_{k \in N \setminus j} \sigma(f_j, f_k), \tag{15.33}$$

where $\sigma(f_j, f_k) = \min\{|f_j - f_k|, B\}$.

Analysis

Neel et al. [29] showed that this stage game is an exact potential game with an exact potential function given by:

$$V(f) = \sum_{j=1}^{|N|} \sum_{k=j+1}^{|N|} \sigma(f_j, f_k) \tag{15.34}$$

Specifically, this game can be readily identified as an example of a BSI game with BSI terms given by $\sigma(f_j, f_k)$.

Because the system can be modeled as an exact potential game, we know the following about this network:

- There are numerous steady states for this network (i.e., numerous maximizers for Eq. (15.34) exist).
- Round-robin and randomly ordered best response algorithms converge.
- Round-robin and randomly ordered ε-better response algorithms converge to ε-NEs.
- Assuming finite A_j random sampling algorithms converge so convergence is assured even if the radios must first learn their operating environment.
- The attractive states of this network can be identified via the maximizers of Eq. (15.34).
- As with most desirable potential games, this game provides an example of self-ish behavior leading to a "common good," in this case, increasing sum capacity. Specifically, the steady states of the network are optimal (or near optimal for ε-Nash equilibria) if we are interested in maximizing sum capacity.

Validation

Figure 15.27 is the output of a simulation of this system in which there are 10 links, $B = 1\,\text{MHz}$, $F = [0,10]\,\text{MHz}$, and each master node chooses the frequency that maximizes its utility. Figure 15.28 shows another realization of this simulation that starts from a different initial frequency vector. In this second scenario, a different, sub-optimal, steady state is reached. A key point to remember from potential game analysis is that although the analysis guarantees existence of a steady state, convergence of self-interested algorithms to a steady state, and local stability of any potential maximizer steady state, as in this case, there may be numerous steady states in the network, not all of which result in the same cost or objective function valuation. Also note that because there are numerous fixed points, no fixed point is globally stable.

15.6.3 Adaptive Interference Avoidance

Rather than the single-dimensional action sets considered so far, cognitive radios have numerous parameters under their control. So when minimizing interference, the cognitive radios will not be limited to adapting only their frequencies. Instead,

Figure 15.27: Simulation results for frequency selection. The top plot depicts the frequencies chosen by each radio and the lower plot depicts the utilities for each radio.

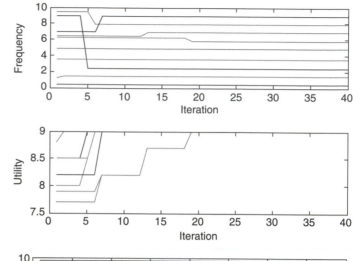

Figure 15.28: Frequency selection simulation results for the same network as in Figure 15.27, but with a different starting point. Note that while network performance is fairly good, it is not optimal.

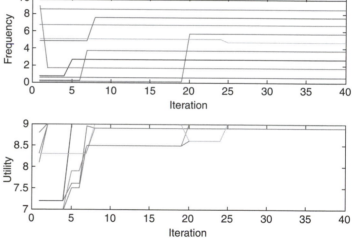

they may also adapt transmission times, spreading codes, and spatial channels. Further, the cognitive radios may also have other parameters under their control that are uniquely determined by the choice of other parameters (e.g., coding rate is uniquely determined by power in a GPRS system [5]).

Specific Network Under Study

For this example, we consider a cognitive radio network where each radio, j, has available for adaptation a set of frequencies, F_j, a set of spreading codes C_j, and a set of transmission times, T_j. Using these available adaptations, each radio attempts to minimize the effective interference it sees from the remaining radios in the network.

Stage Game Model

A single iteration of this game can be expressed as a normal form game as follows:

- Players, N: set of cognitive radios.
- Action space, A: $A = A_1 \times A_2 \times \cdots \times A_n$ where $A_j = F_j \times C_j \times T_j$.
- Utility function, u : A utility function for any player, j, is given by:

$$u_j(f) = \sum_{k \in N \setminus j} \rho(a_j, a_k) \tag{15.35}$$

where $\rho(a_j, a_k)$ is the statistical correlation between the waveforms determined by a_j and a_k.

Analysis

As in Example 15.9, this utility function is in the form of a BSI game, which means that it has a potential function of the form of Eq. (15.20), which is repeated here:

$$V(a) = \sum_{i \in N} \sum_{j=1}^{i-1} \rho(a_i, a_j)$$

Because this is an exact potential game, we know the following about the system:

- Steady states for this system exist.
- Round-robin and randomly ordered best response algorithms converge.
- Round-robin and randomly ordered ε-better response algorithms converge to ε-NEs.
- Random sampling algorithms converge assuming finite A_j, so convergence is assured even if the radios must first learn their operating environment.
- The asymptotically stable states of this network can be identified via the maximizers of V.
- As with most desirable potential games, this game provides an example of selfish behavior leading to a "common good," in this case, minimizing total network interference, with each adaptation decreasing the interference.

Note that an operating environment that would result in utility functions of the form in Eq. (15.35) would rarely be encountered—typically, only when all radios are communicating with the same receiver and all waveforms are power controlled. A more generally encountered system could be described wherein the waveform adaptation of a radio has a greater impact on its own receiver than on the receivers of the other cognitive radios in the network, as expressed in the following equation:

$$u_j(b_j, a_{-j}) - u_j(a_j, a_{-j}) > \sum_{k \in N \setminus j} [u_k(b_j, a_{-j}) - u_k(a_j, a_{-j})] \qquad \textbf{(15.36)}$$

Such a scenario is considered by Sung and Leung [31] for spreading code adaptation in a multicell cellular network. Any system that satisfies Eq. (15.36) can be shown to be an ordinal potential game with an ordinal potential given by:

$$V(a) = \sum_{k \in N} u_k(a) \qquad \textbf{(15.37)}$$

Fortunately, this game is also an ordinal potential game and thus preserves all of the relevant analytic properties of the less realistic scenario for finite A_j.

15.7 Summary and Conclusions

This chapter has presented several modeling approaches for describing the interactive decision processes that occur in a network of cognitive radios, specifically covering dynamical systems, contraction mappings, standard interference functions, Markov models, games, potential games, and supermodular games. These models and the techniques for establishing whether a cognitive radio network satisfies the conditions of the model are summarized in Table 15.4.

For these game models, this chapter has presented analysis insights that can be gleaned by demonstrating that a cognitive radio network satisfies the modeling conditions for one of the models listed in Table 15.4. The steady-state properties, the convergence properties, and the stability properties for each of these models are summarized in Tables 15.5, 15.6, and 15.7, respectively. As Section 15.6.2 showed, sometimes cognitive radio networks satisfy the conditions of multiple models. In these cases, the analytic insights from each of the applicable multiple models are available.

This chapter also presented two different model-independent approaches to determining the desirability of network behavior: Pareto optimality and evaluation of a network objective function. Demonstrating that a network state is Pareto optimal was shown to be of less value than demonstrating that the state maximized a network objective function.

Table 15.4: Presented models.

Model (section number)	Basic model	Identification
Dynamical systems (15.3.1)	$\dot{a} = g(a, t)$, evolution equation $a(t^{k+1}) = f^t(a(t^k))$	$\dot{a} = g(a, t)$ always exists. Solve g for f^t. f^t exists if g satisfies Picard–Lindelöf theorem
Contraction mappings (in 15.3.2)	$\|f(a), f(b)\| \leq \alpha\|a, b\|$ $\forall b, a \in A$	Blackwell's conditions
Standard interference function power control (in 15.3.2)	$f_j(\mathbf{p}(t^k)) = p_j(t^k)I_j(\mathbf{p}(t^k))$	$I(\mathbf{p})$ satisfies positivity, montonicity, and scalability
Finite ergodic Markov chain (in 15.3.3)	$P(a(t^{k+1}) = a^k|a(0), \ldots, a(t)) = P(a(t^{k+1}) = a^k|a(t^k))$	$\exists\, k$ such that $\mathbf{P}^k$ has all positive entries
Absorbing Markov chain (in 15.3.3)	$\mathbf{P'} = \begin{array}{\|c\|c\|} \hline \mathbf{Q} & \mathbf{R} \\ \hline \mathbf{0} & \mathbf{I}^{ab} \\ \hline \end{array}$	Apply model definition
Game (15.4.3)	$\Gamma = \langle N, A, \{u_j\}\rangle$	Map from cognition cycle
Potential game (15.5.1)	$\Delta u_i(a, b_i)$ everywhere related to $\Delta V(a, b_i)$	$\dfrac{\partial^2 u_i(a)}{\partial a_i \partial a_j} = \dfrac{\partial^2 u_j(a)}{\partial a_j \partial a_i}$ (others in Section 15.5.1)
Supermodular game (in 15.5.2)	(1) A_i is a complete lattice (2) u_i is supermodular in a_i (3) u_i has increasing differences in (a_i, a_{-i})	(1) A_i is a closed interval in $\mathfrak{R}^{k_i}$ (2) u_i is twice continuously differentiable on A_i (3) $\partial^2 u_i/\partial a_{ik}\partial a_{im} \geq 0$ for all $i \in N$ and all $1 \leq k < m \leq k_j$ (4) $\partial^2 u_i/\partial a_{ik}\partial a_{jm} \geq 0$ for all $i \neq j \in N$, $1 \leq k \leq k_i$ and $1 \leq m \leq k_j$

This chapter has presented a significant number of useful analytic results, but note that it was able to include only a brief treatment of these extensive models. In fact, many of these models have entire disciplines dedicated to their analyses and applications. Accordingly, the interested reader is encouraged to explore the texts listed in the references for further study.

Table 15.5: Steady-state properties by model.

Model (section number)	Existence	Identification
Dynamical systems (15.3.1)	Maybe, evaluate Leray–Schauder–Tychonoff theorem on evolution equation	Exhaustive search
Contraction mappings (in 15.3.2)	Yes (Banach's theorem)	Recursion (unique steady state)
Standard interference function power control (in 15.3.2)	Yes [12]	Recursion (unique steady state), $\mathbf{Zp} = \overline{\gamma}$
Finite ergodic Markov chain (in 15.3.3)	Yes (ergodocity theorem)	Recursion (unique distribution), Solve $\boldsymbol{\pi}^{*T}\mathbf{P} = \boldsymbol{\pi}^{*T}$
Absorbing Markov chain (in 15.3.3)	Yes (definition)	$p_{mm} = 1$
Game (15.4.3)	Maybe, evaluate Glicksberg–Fan theorem on cognitive radio goals	Exhaustive search
Potential game (15.5.1)	Yes, if A is compact and V bounded	$\arg \max_{a \in A} V(a)$
Supermodular game (in 15.5.2)	Yes	Exhaustive search (must lie in a lattice)

Note also that certain analytical difficulties arise when information is limited—a situation that will become increasingly prevalent as the technologies of the preceding chapters are realized. We may not be able to precisely describe a radio's future decision update rule as the decision processes and goals evolve to better reflect a user's preferences. A radio's available actions may also evolve in time to incorporate new waveforms that could not be anticipated ahead of time. From an analysis perspective, this situation is analogous to attempting to solve a system of equations of unknown order with unknown coefficients and an unknown number of variables. Perhaps it will be possible to broadly classify the decision update processes and action sets, in which case a game theoretic preference approach should be able to address this situation. Perhaps different models will be needed or perhaps a completely different approach should be taken.

Table 15.6 Convergence properties by model.

Model (section number)	Sensitivity	Rate
Dynamical systems (15.3.1)	Apply Lyapunov's direct method (when possible)	No general technique
Contraction mappings (in 15.3.2)	Everywhere convergent	$\left\|a(t^k), a*\right\| \leq \dfrac{\alpha^k}{1-\alpha}\left\|a(t^1), a(t^0)\right\|$
Standard interference function power control (in 15.3.2)	Everywhere convergent	$\left\|\mathbf{p}(t^k), \mathbf{p}*\right\| \leq \alpha^k \left\|\mathbf{p}(0), \mathbf{p}*\right\|$
Finite ergodic Markov chain (in 15.3.3)	Converges to distribution from all starting distributions	Transition matrix dependent
Absorbing Markov chain (in 15.3.3)	$\mathbf{B} = \mathbf{NR}$	$\mathbf{t} = \mathbf{N1}$
Game (15.4.3)	Apply FIP, weak FIP	Length of longest improvement path
Potential game (15.5.1)	All rational decision rules converge	Length of longest improvement path
Supermodular game (in 15.5.2)	All locally optimal decision rules converge	Length of longest optimal improvement path

Although this chapter describes methodologies suitable for analyzing the cognitive radios of today, analyzing the cognitive radios of the future will continue to be an active area of research and a "remaining hard problem."

15.8 Questions

1. Prove that every contraction mapping satisfies the Synchronous Convergence Condition of the Asynchronous Convergence Theorem.

2. Given that the Standard Interference Function is a pseudocontraction, prove the following:

    (a) That it has a unique fixed point.

    (b) Synchronous adaptations converge to the fixed point.

    (c) Synchronous adaptations are stable.

Table 15.7: Stability properties by model.

Model (section number)	Lyapunov stability	Attractivity
Dynamical systems (15.3.1)	Apply Lyapunov's direct method (when possible)	Apply Lyapunov's direct method (when possible)
Contraction mappings (in 15.3.2)	Global	Global
Standard interference function power control (in 15.3.2)	Global	Global
Finite ergodic Markov chain (in 15.3.3)	No	No
Absorbing Markov chain (in 15.3.3)	Only if unique steady state	Only if unique steady state
Game (15.4.3)	Not generally	Not generally
Potential game (15.5.1)	Isolated potential maximizers are Lyapunov stable for all rational decision rules	Attractive to potential maximizers if finite action space or finite step size
Supermodular game (in 15.5.2)	Stable for all locally optimal decision rules for finite spaces	Attractive to NE lattice for locally optimal decision rules

3. Consider a network consisting of three terminals and an access node with noise power of -80 dBm implementing TPC with a statistical spreading gain of 64. Suppose gains to the access node 1, access node 2, and access node 3 are -10, -15 and -20 dB, respectively, and each node would like to achieve a target SINR of 8 dB. Assume each radio's set of transmit power levels is convex:

   (a) Determine whether these target SINRs are feasible.

   (b) If these SINRs are feasible, solve for the operating power vector.

   (c) Based on the discussion in this chapter, what conditions are necessary to ensure convergence?

   (d) Is this operating power vector stable? How do you know?

4. Repeat question 3, assuming target SINR for the access node 1, access node 2, and access node 3 are 6, 8, and 10 dB, respectively.

5. Repeat question 3, assuming the radios operate with discrete power levels.

6. Consider a pair of cognitive radio local area networks (LANs) where each LAN is attempting to maximize its network capacity. Each LAN must make a one-time choice of frequencies $\{f_1, f_2, f_3\}$. If the two LANs choose the same frequency, the capacity of both LANs is 1/2 (reflecting a probability of collision of 1/2); if they choose different frequencies, each LAN has a capacity of 1 (reflecting a probability of collision of 1/2). Model this situation as a normal form game in matrix representation and solve for any NE.

7. For the scenario in question 6, suppose the LANs are allowed to repeatedly adapt their decisions. Model this situation as a repeated game and describe what conditions need to be met to ensure convergence. How might a random timer like the one used in a random back-off scheme be useful here?

8. Suppose a cognitive radio network consists of a set of ontologically defined radios where each radio is attempting to minimize its observed interference by adapting its waveform. Model this network as a game and identify conditions for which this network could be modeled as a potential game. (Hint: Suppose the radios have access to a radio environment map.)

9. Consider a set of 802.11 devices implementing DFS and TPC in Europe. Model this network as a normal form game by specifying the players, action sets, and appropriate goals. Identify an appropriate network objective function. Under what conditions could we expect uncoordinated cognitive radios implementing genetic algorithms to converge to a desirable steady state for the network?

References

[1] J. Mitola, *Cognitive Radio: An Integrated Agent Architecture for Software Defined Radio*, PhD Dissertation, Royal Institute of Technology, Stockholm, Sweden, May 2000.

[2] *Cognitive Radio, Spectrum and Radio Resource Management*, Wireless World Research Forum, Working Group 6 White Paper, December 2004.

[3] Std. 802.11h™-2003 Part 11: Wireless LAN Medium Access Control (MAC) and Physical Layer (PHY) Specifications Amendment 5: Spectrum and Transmit Power Management Extensions in the 5 GHz band in Europe. *IEEE* New York, NY, October 2003.

[4] "Cognitive Radio Definition," Virginia Tech Cognitive Radio Work Group Wiki. Available Online: http://support.mprg.org/dokuwiki/doku.php?id=cognitive_radio: definition

[5] S. Ginde, R.M. Buehrer and J. Neel, "A Game Theoretic Analysis of the Joint Link Adaptation and Distributed Power Control in GPRS," *Fall VTC 2003*, Vol. 2, pp. 732–736.

[6] J. Walker, *Dynamical Systems and Evolution Equations: Theory and Applications*, Plenum Press, New York, 1980.

[7] D. Bertsekas and J. Tsitsiklis, *Parallel and Distributed Computation: Numerical Methods*, Athena Scientific, Belmont MA, 1997.

[8] W. Zangwill, *Nonlinear Programming: A Unified Approach*, Prentice Hall, Englewood Cliffs NJ, 1969.

[9] A. Medio and M. Lines, *Nonlinear Dynamics: A Primer*, Cambridge University Press, Cambridge, UK, 2001.

[10] D. Blackwell, "Discounted Dynamic Programming," *The Annals of Mathematical Statistics*, Vol. 36, No. 1, February 1965, pp. 226–235.

[11] R. Sundaram, *A First Course in Optimization*, Cambridge University Press, New York, 1999.

[12] R. Yates, "A Framework for Uplink Power Control in Cellular Radio Systems," *IEEE Journal on Selected Areas in Communications*, Vol. 13, No. 7, September 1995, pp. 1341–1347.

[13] F. Berggren, *Power Control, Transmission Rate Control and Scheduling in Cellular Radio Systems*, PhD Dissertation, Royal Institute of Technology, Stockholm, Sweden, May, 2001.

[14] J. Zander and S. Kim, *Radio Resource Management for Wireless Networks*, Artech House, Norwood, MA 2001.

[15] William J. Stewart, *Introduction to the Numerical Solution of Markov Chains*, Princeton University Press, Princeton, NJ, 1994.

[16] W. Turin, *Digital Transmission Systems: Performance Analysis and Modeling*, McGraw-Hill, 1999.

[17] J. Kemeny and J. Snell, *Finite Markov Chains*, D. Van Nostrand Company, Princeton, NJ, 1960.

[18] D. Fudenberg and J. Tirole, *Game Theory*, MIT Press, Cambridge, MA, 1991.

[19] J. Neel, J. Reed and R. Gilles, "Game Models for Cognitive Radio Analysis," *SDR Forum 2004 Technical Conference*, November 2004.

[20] C. Sung and W. Wong, "A Noncooperative Power Control Game for Multirate CDMA Data Networks," *IEEE Transactions on Wireless Communications*, Vol. 2, No. 1, January 2003, pp. 186–19.

[21] D. Krishnaswamy, "Game Theoretic Formulations for Network-assisted Resource Management in Wireless Networks", *VTC Fall* 2002, pp. 1312–1316.

[22] M. Hayajneh and C. Abdallah, "Distributed Joint Rate and Power Control Game-Theoretic Algorithms for Wireless Data," *IEEE Communications Letters*, Vol. 8, No. 8, August 2004, pp. 511–513.

[23] J. Neel, "Potential Games," *MPRG Technical Report*, May 2005. Available Online: www.mprg.org/people/gametheory/publications.shtml

[24] D. Monderer and L. Shapley, "Potential Games," *Games and Economic Behavior*, 14, 1996, pp. 124–143.

[25] M. Voorneveld and H. Norde, "A Characterization of Ordinal Potential Games," *Games and Economic Behavior*, 19, 1997, pp. 235–242.

[26] Topkis and M. Donald, *Supermodularity and Complementarity*, Princeton University Press, Princeton, NJ, 1998.

[27] Milgrom, Paul and John Roberts, "Rationalizability, Learning, and Equilibrium in Games with Strategic Complementarities," *Econometrica*, Vol. 58, No. 6, November 1990, pp. 1255–1277.

[28] J. Friedman and C. Mezzetti, "Learning in Games by Random Sampling," *Journal of Economic Theory*, 98, 2001, pp. 55–84.

[29] J. Neel, R. Menon, A. MacKenzie and J. Reed, "Using Game Theory to Aid the Design of Physical Layer Cognitive Radio Algorithms," *Conference on Economics, Technology and Policy of Unlicensed Spectrum*, Lansing, Michigan, May 16–17, 2005.

[30] E. Altman and Z. Altman, "S-Modular Games and Power Control in Wireless Networks," *IEEE Transactions on Automatic Control*, Vol. 48, May 2003, pp. 839–842.

[31] C.W. Sung and K.K. Leung, "On the Stability of Distributed Sequence Adaptation for Cellular Asynchronous DS-CDMA Systems," *IEEE Transactions on Information Theory*, Vol. 49, No. 7, July 2003, pp. 1828–1831.

CHAPTER 16

The Really Hard Problems

Bruce A. Fette

Chief Scientist, Communications Networks Division
General Dynamics C4 Systems, Scottsdale, AZ, USA

"The telecommunications market is nearly 1 trillion dollars per year. With a market of that size, even a few percent is large enough to be an interesting business." [1]

16.1 Introduction

The ability to improve spectral efficiency, enhance network efficiency, and serve the telecommunication user—the purview of cognitive radio (CR)—is an application that adds significant value to the telecommunication market. This book has provided a detailed review of the major technologies that are well understood and stand ready to enable CR. This concluding chapter recaps those technologies, and reviews the current state of the art. This chapter stitches together the pieces, showing how they intermesh to build CR systems, and then presents remaining problems that must be solved.

16.2 Review of the Book

As stated in Chapter 1, we are taking the perspective that the CR is a radio that is sufficiently intelligent to: (1) aid spectrum efficiency, (2) aid the radio networks and network infrastructures, and (3) aid the user. At this time, no "cognitive" radios are in production, although several are in development, prototyping in several laboratories is an ongoing activity, and several international testbeds have begun.

Chapter 2 presented the regulatory perspective regarding CR. The regulatory climate is positive and is encouraging the demonstration of CR technologies. US research and development (R&D) organizations are providing support for such demonstrations. In the United States, 10 MHz has been provided for experimental evaluation testbeds of cognitive technologies, and in Australia, 25 MHz has just been allocated to CR. Several testbed programs are under way, funded by the National Science Foundation (NSF) and the Defense Advanced Research Projects agency (DARPA), as well as by certain corporations, with experiments at several leading universities.

Chapter 3 reviewed the hardware and software architectures of software defined radio (SDR). These architectures are real, and products are in production. Furthermore, it is clear that the cognitive applications can be added to these SDR architectures as one or more additional applications. Some of the SDRs that are in production already have sufficient general purpose processor (GPP) computational power as well as enough memory to support the additional tasks of a cognitive engine running alongside the waveform and protocol processing.

Chapter 4 presented the technologies required to build a CR. Assuming an SDR as a platform, it showed that the CR must have an ability to sense its environment (spectrum activity, locally available networks, position, orientation, time, biometrics); a reflection capability to assess its own radio link performance behaviors; a policy engine; and a library of applicable policies, protocols, and waveforms that are useful in its current context. This chapter demonstrated that SDRs exist and are in production. Even though this chapter did not address policy engines as products, it showed that they exist.

In fact, they have been integrated with SDR radios in General Dynamics Laboratories, and in at least one university research setting, and have been demonstrated at the IEEE Milcom 05 conference [2]. Furthermore, General Dynamics digital modular radios (DMRs) have also demonstrated that by simply adding the proper software applications, SDR radios were able to perform a spectral sensing function and geopositional awareness. Under a demonstration contract with the Air Force Rome Labs (AFRL), General Dynamics teamed with SCA Technica to demonstrate that an SDR could perform biometric analysis and integrate the biometric reports into the radio's waveform [3].

Chapter 5 presented the details of making efficient use of spectrum. Spectrum efficiency is not simply a matter of choosing waveforms with many bits per symbol. Rather, it is a combination of understanding the interaction of a waveform within the context of legacy communications systems and cognitive systems, as well as understanding the antenna beam management, adaptive power control, and member network topology of the existing users of the spectrum. The DARPA NeXt Generation (XG) program has demonstrated that radios can network together, and they can share sufficient information to utilize spectral time and frequency opportunities. The XG program is now evolving protocols and etiquettes to standardize spectral sharing techniques.

Chapter 6 discussed the design properties of a policy engine. The policy engine includes a current environmental status component, a policy decision/analysis component, and a policy enforcement component. The policy enforcement component acts as a final check that proposed waveforms, frequencies, or other actions to aid the network or the user are compliant with the radio architecture capabilities, network capabilities, and local regulatory policy. The details of such an architecture and the corresponding application programming interfaces (APIs) are now under a standardization study and development effort at the SDR Forum.

Chapter 7 explored how the genetic algorithm (GA) can be used to evaluate the performance of many waveforms and assess which waveform properties lead to a waveform that works well for not just the current link conditions of one network member but for the spectral efficiency of the whole network. Other optimization criteria, such as direct current (DC) power drain on the batteries of the local radio, or even network overhead traffic, may also be included. The GA may be performed in real time on the current link by the radios currently closing the link, or non-real time by a computer server that has been provided with a channel and environment model. By moving the GA to a non-real-time server, the analysis results can be one of many choices made available to other network members when similar conditions are encountered. In addition, while the analysis is being performed, the actual spectrum and radio traffic is not encumbered by the GA experiments. The currently active radios can be updated to use the most appropriate waveforms by a network server as soon as an acceptable solution is found, whether that acceptable solution comes from over-the-air GA experiments, from prior experience discovered on a server, or from GA analysis performed on a remote compute server. Although the GA is well known and demonstrated, it is not available at this time as a network service, or to offload link and network optimization, even though there is little or no economic barrier to preclude developing and offering such a service. The immediate next step to accomplishing this is developing and publishing a standardized API for such an interface and the corresponding standardized API for the radio to download the optimized waveform recommendation, and install it.

Chapter 8 reviewed many of the techniques available and in use to perform global, regional, or local position analysis, as well as time awareness. It covered the US global positioning system (GPS), the European Galileo program, and the Russian GLObal NAvigation Satellite System (GLONASS), as well as the concepts of very high frequency (VHF) omnidirectional ranging (VOR), LOng RAnge Navigation (LORAN), time of arrival (ToA), time difference of arrival (TDoA), and the television (TV) ghost canceling reference (GCR) signal used with regional signal databases. This chapter also lightly touched on the topic of precision time and its availability from GPS, LORAN, TV, and WWV sources. GPS functionality is quite prevalent and is now available in single-chip implementations. Therefore, it is not necessary to utilize an SDR channel for this function. In addition, precision time is available from high-precision, low-power watch electronics and precision oscillators. Time, with sufficient precision for human users, includes WWV, Loran, and TV, where available. In locations where GPS may be unavailable, such as urban areas and the interiors of buildings, GPS may be augmented by inertial navigation techniques. Thus, these technologies are cost effective and available.

Chapter 9 discussed how wireless network protocols must be different from wired network protocols, and how a cognitive system can identify protocols that improve network performance under the current conditions. Such protocol improvements can also be made available from a server to network members, much as the waveform protocols discussed in Chapter 7. We note that considerable research has been performed in wireless ad hoc networking, and that these systems frequently define their own networking protocols. However, wireless networks often terminate into wired networks. At the gateway between the wireless and wired world, a need generally exists for protocol conversion, and it is precisely at these points where intelligent protocols provide critical functionality. It seems likely that gateway wireless servers will keep a collection of protocol applets and apply the most appropriate applets as a function of best choice for the applications under available conditions. They may also serve the corresponding applets to SDRs, to a radio network, or eventually just call them out by number. Gateways have the opportunity to become a very interesting business. Section 16.3 discusses this further.

In the United States, DARPA programs have been designed to demonstrate the substantial value of the components of a CR. The XG program has demonstrated improvements in spectral efficiency, and the Situation Aware Protocols in Edge Network Technologies (SAPIENT) program has demonstrated improvements in radio networking protocols. The recently begun Adaptive Cognition-Enhanced Radio Teams (ACERT) and Brood of Spectrum Supremacy (BOSS) programs have the objectives of demonstrating the value of CR networking to the user. In addition, the DARPA Phraselator program has demonstrated the value of having language translation immediately available to the user.

Chapter 10 explored speech input and output. Because speech is analyzed with a signal analysis process that extracts vocal tract shape and excitation properties, these properties can be encoded for voice communication or can be statistically analyzed to extract speech vocal tract behaviors that map into word recognition, language recognition, or speaker identification. The radio's ability to be aware of its user is very significant to providing intelligent services that the user can appreciate and value. These technologies are commercially real. Voice coders (vocoders) are available as software products. Speech (word) recognition is commercially available in productized format. Language recognition has been demonstrated in very useful forms, but it is not generally available as a product. General Dynamics and others have demonstrated such systems. All of these technologies are readily integrated into smart radio systems and into commercial infrastructure to support CR to the user. This is discussed in Section 16.3.4.

Chapter 11 discussed how services can be provided to a radio user from an infrastructure server. This chapter specifically focused on providing a server, called a Radio Environment Map (REM), that tracks local radio activity and makes that information available to other radios. As mentioned previously, in addition to radio communication activity, other types of relevant information may also be provided as services to wireless users. How this information can be delivered to various types of radio networks is further detailed in Section 16.3.

Chapter 12 described knowledge representation, reasoning, and learning algorithms, ranging from neural networks to case-based reasoning (CBR). Reasoning and learning are essential parts of exhibiting cognitive behavior. If the user is to expect the radio to provide intelligent help, it must exhibit the ability to reason and to learn. Humans reason and learn without being taught how they do so, but this remains a daunting computational task to a radio, regardless of whether it has large computational power or the very small computational power of a handheld radio. This task is further daunting in that the human performs such tasks very quickly. The radio may take longer than a human to reach the same conclusions, depending on the complexity of the reasoning and the learning algorithms employed. There are, however, other things that the human will not learn quickly—or maybe not learn at all—that may more readily be learned by the radio. The timeliness of the cognitive processes will be quite important to the user. The means for tying together these timeliness goals and the complexity of reasoning required is still very much a developing field, with current exploration in the artificial intelligence (AI) community.

There are other things that the radio may readily have all the information to assess, such as the choices of local radio networks, their corresponding performance (e.g., data rates), and their cost effectiveness. Analysis of radio network selection can be expected to be learned from the user's behaviors and choices, and eventually become fully automatic activity by the radio. We expect this type of behavior in the next generation personal digital assistant (PDA)-type wireless devices.

Chapter 13 described ontology (knowledge representation) in great detail, and gave some simple examples of knowledge about radio protocols in the Web Ontology Language (OWL). In order for a PDA radio to deliver the level of intelligence of a "Radar O'Reilly," the radio must learn a great deal about many specialties. It is probably impractical for a radio to learn everything to help its user on its own. Rather, specialized areas of expertise may be methodically captured and transcribed, making the radio a "savant" about one topic at a time. Gradually, other economically interesting areas will be added to the domain of radio expertise. Consider, as an example, that making an airline reservation has been automated to the point of making it possible for a computer to capture the necessary keywords from speech and then perform the computerized transactions. (One DARPA contractor may actually be using this as a standard computerized service for its employees.) In a similar fashion, we can expect other important areas of specialization, in which the learning process can be presented with adequate training (including non-routine conditions), monitored and optimized for the purpose, and then rolled out to the larger set of radios and radio users. The list of references for Chapter 13 shows that this area is an incredibly active field of AI, but commercial deployment is still limited to a few unique applications.

Chapter 14 presented the software architectures of adaptive, aware, and CRs, with the major focus of attention on the Radio Knowledge Representation Language (RKRL) and AI functions. Considerable attention is paid to "computability," to assure that the reasoning engines can analyze and reach a conclusion in a practical finite time and not be stuck in an endless loop of contingencies. This chapter defines five perspectives of CR Architecture, and defines basic APIs to interface the various CR functions. It also addresses how to integrate the CR Architecture into the SCA. Finally, this chapter provides examples of what a user might expect his radio to learn, setting goals for the industry.

Chapter 15 provides a deep mathematical treatment of game theory with application to spectrum efficiency and stability analysis. As CRs adapt their waveforms, they must avoid interference with each other. It is important that the result is not a continuous chain of dominoes, by which each radio's change of behavior causes a change to yet another radio, resulting in unbounded cycles of adaptation and re-adaptation. This chapter provides an analysis of the stability criteria and fairness issues related to this problem.

16.3 Services Offered to Wireless Networks through Infrastructure

In this final chapter, we specifically recognize that there are several classes of radios. To begin with, there is the radio designed and built for defense applications. When used in peacekeeping missions, such radios will be expected to comply with the regulatory requirements of the regions where they are used. They will need to have details of the radio networks of other allies and coalition partners, including waveforms, protocols, frequencies, and what conditions should be communicated with the partners. These radios will gateway many local users to many global networks and have access to numerous computer servers and database servers through those gateways. Most such radios will be designed to have considerable built-in intelligence; in addition, the access to intelligent pull and push services through their networks will be very significant. The information available to such a radio is limited by the bandwidth of the most restrictive wireless link providing the access, by the battery power available, and by the total time the mission must be operative on that battery power. Consequently, distributed caching throughout the network, so that information can be found nearby, will become standard in such networked radios.

Although similar to the defense radio in that much of its capability must be embedded, the public safety and emergency response radio will have significantly different applications than the defense radio. Three-dimensional positioning inside buildings is critical, as is the ability to bridge and gateway among multiple networks, depending on which support organizations arrive to the emergency and what their radio communications capabilities are. In this application, the CR's ability to track which responders need what information is essential, and we will expect CRs to network the proper information to the correct emergency responders. The ability to synthesize multiple waveforms in multiple frequency bands, sometimes in the absence of infrastructure support, is a major issue requiring both SDR and CR functionality.

The second class of radio involves the cellular telephone subscriber unit, the corresponding base station, and its corresponding infrastructure. Due to the very high volume of subscriber units that are manufactured, the manufacturers go to great effort to push any complexity from the subscriber unit over to the base station. Two main drivers for this are that the power dissipation and complexity of the base station and the infrastructure behind the base station are less limiting than for the subscriber units, and costs can be amortized across thousands or even millions of subscribers.

The third class of radio is those embedded into computing devices. This includes laptop computers, PDAs, and similar devices, for which the primary access is a wireless personal area network (WPAN, such as Bluetooth or ZigBee); wireless local area network (WLAN, which is usually 802.11 waveform); or wireless metropolitan area network (WMAN, which will probably be 802.16 WiMAX). These devices begin with the assumption that there is ample bandwidth in the network and that there are gateway devices providing connectivity to the Internet.

We also predict a fourth class of radio: automobiles that will soon have an add-on business of transmitting and receiving useful services to the driver and passengers by using wireless regional area network (WRAN) services. The automotive industry refers to this as "telematics." General Motors has been successfully marketing its OnStar product in luxury vehicles for nearly a decade. This is the forerunner of very intelligent services provided to the driver. We can easily imagine that many of the following services would be useful and can be enabled by broadband wireless service: directions to specified locations; specific types of services locally available (gas stations, restaurants, banks, hotels, businesses); specific types of products locally available (tickets to a play, a copy of this book, etc.); interaction with dispatch services (nearest taxi, pickup and delivery services, etc.); traffic avoidance (only practical as a radio network service); entertainment (personally selected music or movies for the kids); drive-through shopping and pickup. As these specialized services finely tune their services to the users, they will become cognitive to specialize and optimize their services for each particular user.

16.3.1 Stand-Alone Radios with Cognition

SDR radios of manpack size and larger are reasonably capable of built-in cognitive capabilities. This can include reasoning and learning about spectral activity, node locations, local networks, network protocols, data caches, standard and nonstandard activities, and functions that local users expect the radio to perform automatically and routinely.

Such a radio should also be sufficiently sophisticated to assess how its own communication affects the static and dynamic performance of the network and spectral utilization density, the behavior of local legacy systems, and so forth.

Advanced versions of defense radios, will likely organize and maintain themselves by using ad hoc protocols, and gateways will provide connectivity to a wide variety of wireless networks of varying data rate, varying range, and network membership. Such a collection of networks will offer robustness of message forwarding, although optimization is truly a high-dimensionality problem. Defense radios can also expect to have access to wideband data networks, which can provide higher cognitive functions for authentication, non-repudiation, speech understanding, translation, schedule keeping, and prioritized objectives. Much of this functionality is feasible through system integration of existing functionality.

16.3.2 Cellular Infrastructure Support to Cognition

The economic model of cellular telephony strongly encourages two principles: (1) shift as much complexity as possible to the base station, and (2) keep the subscriber locked to his or her service provider.

Sophisticated cognitive functionality in handheld devices will exist because it is provided by the base stations and the infrastructure made available by the base stations. Subscriber devices may become more versatile and more fully capable, but it is likely that their intelligence will be activated or downloaded only as necessary to enable selected features. The actual reasoning, spectral evaluation, and network management will be managed by the base station. The base station will convert that logic into controls and software applets to control the subscriber units.

Higher application-level cognitive functions will probably first be integrated by cellular service providers. By putting highly polished special purpose applications somewhere in the data infrastructure, time scheduling, location tracking, event awareness, location-based opportunity awareness, and business-specific knowledge niches can readily be served with incremental investment and nearly negligible cost to the position-aware subscriber unit. Furthermore, each new capability can be developed, polished, test-marketed, and rolled out to the customer base, resulting in the perception of gradually increasing cognitive capability in the subscriber unit.

To enable this, the subscriber units will need to be SDRs. The ability to shift waveform and channel protocol may be limited to the current properties of the telecommunication network, but sufficient computational resources must be provided to serve the application and the application interface to the user, and to download the application servelets upon request. Thus, we will expect to see some maturing of the user interface, the memory, and computational support in subscriber devices.

However, we expect to see more significant change to the base stations and infrastructure, as cognitive services begin to be integrated. We can reasonably expect heavier data services, to bring significant change to the base station. Voice Over IP (VoIP) will become a more common voice-coding choice, especially if Internet Protocol (IP) is already being provided for data services. The ability to authenticate a subscriber, download a servelet to the subscriber, authenticate his or her service requests, and then perform the corresponding network transactions suggests that base stations will have significantly more server functionality and wider bandwidth data networks.

Telephony services experience high variability of traffic demand, from very few calls late at night, to incredible demand when airplanes land or when the football game is over. Servicing high-demand variability has a large cost impact. The network neither wants to deny calls at peak time, nor have idle infrastructure when there is far less demand. There are significant economic opportunities associated with spectrum sharing. If there are other services willing to briefly sublease their spectrum to serve these conditions, the radio technology will need to be able to shift to new frequencies under control channel handover. This implies additional changes to the handsets and to the base stations to accommodate this degree of flexibility, and it remains to be seen if the industry considers this to be an economically cost-effective move.

16.3.3 Data Radios

Unlike telephony devices, PDAs and portable computers are first a computer with data network capabilities (possibly including WPAN (Bluetooth, ZigBee or ultra-wideband (UWB)), WLAN (802.11), WMAN (WiMAX), etc.), and cellular telecommunication services are an additional feature. Therefore, the PDA software can, in principle, make economic trade-offs on behalf of the user as to which functions ought to be performed through which networks by considering cost effectiveness, timeliness, and specialized advanced capabilities of various networks.

The choice between VoIP delivered through the WLAN or the WMAN networks or voice telephony through cellular could easily be made as a function of the user's or subscriber's habits (e.g., walking a short distance or driving a long distance). Whichever network will experience the fewest disruptions over the predicted performance of the next 5 minutes may be the preferred initial access, and thereby minimize disruption associated with a service handover.

It is also likely that more memory and computational resources are available than currently in a standard cellular subscriber device. Therefore, it is likely that such devices will choose which cognitive functions are performed locally and which functions are performed in the infrastructure. Because these devices are primarily data devices, they may inherently offer a broader portfolio of cognitive capabilities, matched appropriately to their form factor. For example, a PDA may act on scheduled activities and timeliness issues, operating from the user's pocket, whereas a laptop may operate on longer-range time windows with more complex tasks. Both may find it convenient to draw upon database servers and distributed computational servers to perform cognitive selections, trade-offs, and priorities. Although these data services may be similar to the telecommunication services, the economic model supporting deployment is likely to be quite different. Perhaps we will see academic prototypes rapidly spin out as cognitive server support businesses (CSSBs) based on drag-along advertising rather than the pay-per-month or pay-per-user-minute models of telecommunications infrastructure cost.

16.3.4 Cognitive Services Offered through Infrastructure

Chapter 11 introduced the topic of cognitive services offered through infrastructure, which is reviewed here to more fully list its relevant opportunities. Populating these opportunities will rapidly proliferate CR functionality.

At the physical layer, the REM is a powerful way to keep track of node location, waveforms, networks, time slots, traffic volume, motion prediction, the local interference noise level, hidden nodes, silent emergency services, telecom, frequency modulation (FM), TV, and locally relevant spectral considerations. Access to the local server in support of PDA or telecommunication base stations seeking access to extra spectrum are quite feasible. Furthermore, it is likely that the REM can readily be expanded to perform the spectrum access requests between primary and secondary spectrum users, to carry out any required financial transactions, and to manage access and priority override.

In addition, the same infrastructure server can readily serve the local regulatory policy within the local regulatory boundary. This can be handled in a fashion similar to a cellular handover between base stations, with the infrastructure recognizing a local boundary crossing and performing a handover message to the appropriate adjacent server. If financial transactions are required for spectrum access, the policy server and the spectrum server will likely serve as the banker for all parties.

The schedule keeper and the task priority manager can utilize the node position information transacted with the REM to keep track of how closely the user is following his or her activity plan. Significant deviations from the plan can be used to notify subsequent interaction spots of progress or schedule updates, and even schedule "meet me" locations and times.

Perhaps one of the most significant servers we all want is the extension of the GPS-based tour guide that tells you how to get from location to location, but adds the current traffic flow into consideration. If we all knew about roads with excessive congestion, accidents, or road maintenance activities, and the best current alternate choices, perhaps we would all get to work and get home more expeditiously. This function can be derived from reports to the local REM database from mobile PDA devices. By observing velocity and location reports, the average velocity and traffic light wait times along various road choices leading to the destination can be compared, and recommended alternate paths can be identified to the motorist.

Likely, there will be a server for local service advertisements. For food or services of special interest to the user, the location, the schedule pressure, the agenda of activities, and the experience of the user's common choices provide all the information necessary for local service advertisements to selectively provide welcome suggestions and avoid the nuisance factor of excessive advertisement.

Such services do not require breakthroughs of science or technology. They can be implemented from a rigorous functionality specification. The industry must prepare the functionality specifications, the interface message requirements, and the transaction protocol sequences. This work can be brought forward by a standards organization or by a de facto standard product that experiences widespread adoption.

The primary issue is the business model. How will the cost of the equipment and its maintenance be supported and be profitable? And how many subscribers must already exist before it makes sense to deploy the infrastructure? These questions are not unlike those for the rollout of cellular infrastructure. Small infrastructure to serve large regions is gradually reduced in radius as subscriber density grows and the business grows in cost effectiveness. Perhaps the most compelling business-case arguments are that gasoline consumption and pollution can be lowered and communities can become more desirable places to live if the cognitive services are made available as local services. It is therefore likely that funding for some cognitive services startups will be progressive municipal investments.

Other services will be invented to serve the users. Gradually many niches will be populated, providing the aggregated learning experience of academy, industry, and regulator communities; the creativity of the user community; and further validation of new marketing opportunities. New and clever services that can be provided by adding a new database server or a new computational server to the existing infrastructure can appear to make the CR grow smarter, but with no hardware change and minimal software change.

Even so, there will always be specialized services that do not exist, or are not directly supported by infrastructure servers. Some of these needs may be met as a script of many basic services woven together to meet a specific objective. But in the limit, the radio must learn to implement cognitive functions not supported by well-tuned infrastructure-based servers. These functions will truly follow the ontology-based reasoning and learning now in development in AI laboratories as described by Bostian, Kovarik, Kokar, and Mitola in earlier chapters of this book.

16.3.5 The Remaining Hard Problems

Many of the problems to bring the vision of CRs forward to products are technical implementation issues, and business case issues. However, the truly hard problems are going to be the regulatory problems. These non-technical problems are frequently decided on criteria that cannot be anticipated by technologists.

Each country has a body of regulators who specify how citizens and non-citizens are allowed to use spectrum. In the United States, the Federal Communications Commission (FCC) specifies this policy, and currently, WPAN, WLAN, and WMAN (or WRAN) devices must meet manufacturers' requirements to operate in a certain band (say 2.4 GHz) and at a certain effective isotropic radiated power (EIRP), with a specified spectral shape (spectral mask) on the out-of-band transmissions. Most countries have similar organizations, each with different rules. These rules can be provided to the CR, which can enforce its compliance to those rules if they are effectively transcribed. Rules will include necessary extensions to accommodate spectral cognition (e.g., etiquettes or protocols to request access to the spectrum from primary owners), and to perform the corresponding transactions. The regulators will need assurance that the protocols are robust under all network-loading conditions, that they cannot be tampered with (hacked), that radios are guaranteed to follow the policy in a regulatory region, and that the networks of the primary owners will remain stable under a wide range of CR-loading requests.

The specification and analysis of the protocols and etiquettes is currently a very hard problem at the boundary between the technical community and the regulatory community. The regulatory organization needs to be able to specify the local rules. They need to be assured that those rules are flawlessly translated. They need to be assured that the primary spectrum owners will retain performance and network stability under all conceivable (and inconceivable) conditions. However, flawless translation into machine-readable form and protocol analysis under a wide regimen of usage conditions are hard to assure the regulator. System performance analysis can be done offline by computer simulation. Formal methods have some promise to allow us to make assertions about the machine-readable policies. It remains unclear, however, whether we can assure the regulators that no machine failure can cause grave impact to the legacy radio networks. This appears to be a new domain, with hard problems, requiring new science.

This book should create sufficient interest in the R&D community to find solutions to these remaining hard problems.

References

[1] Pete Cook and Stephen Hope, "Technology and Application Considerations for 3G Profitability," *SDR Forum Technology Conference*, Phoenix, AZ, November 2004.

[2] D. Cohlman, "Demonstration or DMR Networking and Policy Engine," at *IEEE Milcom 2005 Conference*, Atlantic City, 17–20, October 2005.

[3] J. Kleider, S. Gifford, S. Chuprun and B. Fette, "Radio Frequency Watermarking for OFDM Wireless Networks," in *IEEE ICASSP Conference*, Montreal, Canada, 17–21, May 2004.

Glossary

802.11	An IEEE standard for Physical layer for Wireless LANs
802.11h	Radio resource management—utilizing adaptive power control
802.22	A standard in development for regional wireless networks, intended to work in rural areas over long distances, must be cognitive, must sense what TV channels are in use and must avoid active channels
3G	3rd generation wireless cellular telephony
A/D	Analog to Digital
A3V	Advanced Amphibious Assault Vehicle (Marine Corps)
AA	Aware–Adaptive
AAAI	American Association for Artificial Intelligence
AACR	Aware, Adaptive, and Cognitive Radio
AB	Asynchronous Balanced
ABCS	Army Battle Command System
ACERT	Adaptive Cognition-Enhanced Radio Teams
ACIP	Architectures for Cognitive Information Processing
ACL	Access Control List
ACR	Adjacent Channel Rejection
ADC	Analog to Digital Converter
ad hoc	A routing technique in which all radios in a network participate in forwarding messages from the source to the destination, along the path of available wireless connectivity—generally intended to apply to a mobile radio network
AE	Array Processing Element
AES	Advanced Encryption Standard
AFRL	Air Force Research Laboratories
AGC	Automatic Gain Control
AGFL	Affix Grammars over a Finite Lattice
AI	Artificial Intelligence
ALE	Adaptive Link Establishment
AM	Amplitude Modulation
AME	Amplitude Modulation Equivalent

AML	Autonomous Machine Learning
AMPS	Advanced Mobile Phone System or Advanced Mobile Phone Service
AOA	Angle of Arrival
AOP	Aspect-Oriented Programming
API	Application Programming Interface
AR	Asynchronous Response
ARM	<u>A</u>dvanced <u>RISC</u> <u>M</u>achine, a 32 bit microprocessor architecture
ARQ	Acknowledge Receipt of Message
ARQ	Automatic Repeat reQuest
ASIC	Application-Specific Integrated Circuits
ASIP	Advanced System Improvement Program
ATM	Asynchronous Transfer Mode
AWGN	Average White Gaussian Noise
B3G/4G	Beyond 3G—the 4th generation of cellular, currently planned to support high-speed data applications
BAA	Broad Agency Announcement
BCV	Battle Command Vehicle (Army)
BER	Bit Error Rate
BIT	Built-In-Test
BLAST	Bell Labs Layered Space Time Transmission (an MIMO Technique for high information Spectral Density)
Bluetooth	A moderate data rate, very short range, full duplex, hopping wireless link, commonly used for headset to handset connection
bps	bits per second
BPSK	Binary Phase Shift Keying
BREW	Binary Runtime Environment for Wireless
BSI	Bilateral Symmetric Interaction
BSU	
BTS	Cellular telephony base station
BW	Bandwidth
C++	An Object-Oriented Programming Language
C2V	Command and Control Vehicle
CASA	Computational Auditory Scene Analysis
CASE	Computer-Aided Software Engineering—Software Development Tools
CAT5	Computer cable
CB	Citizen Band
CBDT	Case-Based Decision Theory

CBP	Component-Based Programming
CBR	Case-Based Reasoning
CCM	Custom Computing Machines
CDMA	Code Division Multiple Access
CDPD	Cellular Digital Packet Data
CEP	Circular Error Probability
CISC	Complex Instruction Set Computer
CLDC	Connected Limited Device Configuration
CLI	Command Line Interface
COMSEC	Communications Security
COPS	Common Open Policy Service
CORBA	Common Object Request Broker Architecture
COTS	Commercial Off the Shelf
CPDA	Cognitive Personal Data Assistant
CPE	Customer Premises Equipment
CR	Cognitive Radio
CRA	Cognitive Radio Architecture
CRA I–V	Cognitive Radio Architecture—Building Phases
CRA I	Functions, Components and Design Rules
CRA II	The Cognition Cycle
CRA III	The Inference Hierarchy
CRA IV	Architecture Maps
CRA V	Building CRA on an SDR
CRC	Communications Research Center of Canada
CRC	Cyclic Redundancy Check
CSEL	Combat Survivor/Evader Locator
CSIS	Center for Strategic and International Studies
CSM	Cognitive System Module
CSMA	Carrier-Sense Multiple-Access
CTS	Clear to Send
CVSD	Continuously Variable Slope Delta modulation
CW	Continuous Wave
CWN	Cognitive Wireless Network
CWPDA	Cognitive Wireless Personal Data Assistant
D/A or DAC	Digital to Analog Converter
DAMA	Demand Assign Multiple Access
DAML	DARPA Agent Markup Language
DARPA	Defense Advanced Research Projects Agency

DBS	Direct Broadcast Satellite
DC	Direct Current
DCD	Device Configuration Descriptor
DCOM	Distributed Component Object Model, Microsoft Proprietary
DECT	Digital European Cordless Telephone
DES	Digital Encryption Standard
DH3	Data—High rate, 3 slots
DL	Data Link
DM1	Data—Medium rate, 1 slot
DMD	DomainManager Configuration Descriptor
DMR	Digital Modular Radio
DoC	Department of Commerce
DoD	Department of Defense
DPD	Device Package Descriptor
DPSK	Differential Phase Shift keying
DSB	Defense Science Board
DSP	Digital Signal Processor
DSR	Distributed Speech Recognition
DSSS	Direct Sequence Spread Spectrum
DTE	Digital Terminal Equipment
DTN	Disruption Tolerant Networking
EARS	Effective Affordable Reusable Speech Recognition
ECA	Event–Condition–Action
ECCM	Electronic Counter CounterMeasure
ECF	Earth Centered Fixed
EER	Equal-Error Rate
EGG	ElectroGlottoGraph (electronic sensing of the vibration of the vocal cords)
EIA/TIA	Electronic/Telecommunications Industries Association
EIRP	Effective Isotropic Radiated Power
EPG	Exact Potential Game
EPLRS	Enhanced Position Location Reporting System
ERFC	Complementary Error Function
ETSI	European Telecommunications Standards Institute
FA	False Alarm
FAA	Federal Aviation Administration
FCC	Federal Communications Commission
FDMA	Frequency Division Multiple Access

FDoA	Frequency Difference of Arrival
FEC	Forward Error Correction
FFT	Fast Fourier Transform
FIFO	First In First Out memory
FIP	Finite Improvement Path
FIR	Finite Impulse Response
FM	Frequency Modulation
FOPC	First-Order Predicate Calculus
FPGA	Field Programmable Gate Array
FRS	Family Radio Service
FSK	Frequency Shift Keying
FTP	File Transfer Protocol
GA	Genetic Algorithm
GAO	Government Accountability Office
GεPG	Generalized Epsilon Potential Game
GHz	GigaHertz, 10^9 cycles per second
GIS	Graphical Information System (a map for example)
GNU	GNU's Not Unix, an open operating system with a copy left license structure
GNU Radio	An SDR software and hardware toolkit offered by GNU
GOPG	Generalized Ordinal Potential Game
GPP	General Purpose Processor
GPRS	Generalized Packet Radio Service
GPS	Global Positioning System
GRACE	Graduate Robot Attending Conference (a functional robot)
GSM	Groupe Special Mobile (a cellular protocol stack)
HAL	Hardware Abstraction Layer (creates a standard way for GPP to interface to FPGA)
HCI	Human–Computer Interface
HF	High Frequency
HMI	Human–Machine Interface
HTML	Hyper-Text Markup Language
HTTP	Hyper-Text Transfer Protocol
HW	Hardware
I/O	Input/Output
IA	Information Assurance
IBL	Instance-Based Learning
IBS	Integrated Broadcast Service

ICNIA	Integrated Communication Navigation IFF and Avionics
iCR	ideal Cognitive Radio
IDL	Interface Description Language
IEEE	Institute of Electrical and Electronics Engineers
IETF	Internet Engineering Task Force
IF	Intermediate Frequency
IIR	Infinite Impulse Response
IJCAI	International Joint Conference on Artificial Intelligence
INFOSEC	Information Security
IP	Internet Protocol
ips	instructions per second
IRDA	InfraRed Data Association
ISA	Instruction Set Architecture
ISAPI	Information Services APIs
ISM	Industrial, Scientific, and Medical (Frequency Band around 2.4 GHz)
ISP	Internet Service Provider
IT	Information Technology
ITU	International Telecommunications Union
J2EE	Java 2 Enterprise Edition
J2ME	Java 2 Micro Edition
J2SE	Java 2 Standard Edition
Java	An Object-Oriented Programming Language
JINI/JADE	JAVA based Programming Languages and Environments
JPO	Joint Program Office
JTP	Java Theorem Prover
JTRS	Joint Tactical Radio System
JTT	Joint Tactical Terminal
JVM	Java Virtual Machine
KDD	Knowledge Discovery and Data Mining
KQML	Knowledge Query and Manipulation Language
KS	Knowledge Source
LAN	Local Area Network
LDAP	Lightweight Directory Access Protocol
LE	Logical Elements
LF	Low Frequency
LID	Language Identification
LISP	List Processing Language (an artificial Intelligence Language)

LLC	Logical Link Control
LM	Logic Modules
LNA	Low-Noise Amplifier
LOS	Line of Sight
LP	Linear Programming
LTM	Long-Term Memory
LWSR	Land Warrior Squad Radio
MOS	Mean Opinion Score (a voice quality measure)
M3	Multi-band, Multi-mode, Multi-media
MAC	Media Access Control (the ISO stack layer controlling how a radio finds time and frequency to minimize interference)
MANET	Mobile Ad hoc Networking
MBITR	Multi-Band Intra/Inter Team Radio
MBMMR	Multi-Band, Multi-Mode Radio
MELP	Mixed Excitation Linear Predictive (voice coding)
MEMS	Micro Electro-Mechanical Systems
mesh	Similar to an ad hoc network but applied to a stationary network
MHz	MegaHertz, 10^6 cycles per second
MIB	Management Information Base
MIMO	Multiple Input Multiple Output—a technique in which both transmitter and receiver use multiple antennas to optimize the spectral efficiency of a link
MIPS	Million Instructions Per Second
ML	Machine Learning
MLE	Maximum Likelihood Estimation
MMITS	Modular Multifunctional Information Transfer System
MMSE	Minimum Mean Square Error
MMU	Memory Management Unit
MODM	Multi-Objective Decision Making
MPRG	Mobile and Portable Radio Research Group
MPSK	*M*-ary Phase Shift Keying
MSE	Mean Square Error
MSI	Multilateral Symmetric Interaction
MT	Machine Translation
NAQ	Negative Acknowledgement of Message
NASA	National Aeronautics and Space Administration
NE	Nash Equilibrium
NEC	National Economic Council

NET	Network Layer
NGI	Next Generation Internet
NIST	National Institute of Standards and Technology
NL	Natural Language
NLP	Natural Language Processing
NOAA	National Oceanic Atmospheric Administration
NP	Description of the hardness of a computational problem, as in NP complete
NR	Normal Response
NSA	National Security Agency
NSC	National Security Council
NSF	National Science Foundation
NTDR	Near-Term Digital Radio
NTIA	National Telecommunications and Information Administration
OBR	Ontological or Ontology Based Radio
ODP	Open Distributed Processing
OFDM	Orthogonal Frequency Division Modulation
OIL	Ontology Interface Layer
OMAP	Open Multimedia Application Platform—a TI Proprietary Chip
OMEGA	An end-point architecture
OMG	Object Management Group
OODA	Observe, Orient, Decide, Act
OOK	On–Off Keyed
OOP	Object-Oriented Programming
OOPDAL loop	Observe (sense, perceive), Orient, Plan, Decide, Act, Learn
OPG	Ordinal Potential Game
ops	Operations per second
OS	Operating System
OSI	Open System Interconnect (defines a seven layer protocol stack for networking)
OSSIE	Open Source SCA Implementation: Embedded, a low power SCA demonstration by MPRG at Virginia Tech
OTAR	Over-The-Air Rekeying
OWL	Web Ontology Language
P2P/PMP	Point to Point, and Point to Multi-point
PA	Power Amplifier
PBNM	Policy-Based Network Management
PC	Personal Computer; Portable Computer

PCIM	Policy Core Information Model
PCS	Personal Communications Systems
PDA	Personal Digital Assistant
PDF	Probability Density Function
PDL	Policy Definition Language
PDP	Policy Decision Point
PDR	Programmable Digital Radios
PEP	Policy Enforcement Point
PEP	Performance-Enhancing Proxies
PER	Packet Error Rate
PHY	Physical layer of the ISO stack—usually meaning the waveform used to send information bits (the modem)
PKI	Public key infrastructure
PLL	Phase Lock Loop
PLRS	Position Location Reporting System
POEMA	Policy Enabled Mobile Applications
POLYSURV	DARPA Policy-Based Survivable
POSIX	Portable Operating System Interface, a DoD Standard Interface to an Operating System
PPS	Pulses per Second
PRF	Properties Descriptor File
PROLOGUE	An Artificial Intelligence Programming Language
PSF	Pulse Shape Filter
PSK	Phase Shift Keying
PTT	Push To Talk
PU	Primary User
PVM	Parallel Virtual Machines
QAM	Quadrature Amplitude (and phase) Modulation
QoI	Quality of Information
QoS	Quality of Service (implies importance of bit errors, latency, flow rate)
QPSK	Quadrature Phase Shift Keying
R&O	Rule and Order
RA	Regulatory Authorities
RAKE	Filter used to track multi-path components of spread-spectrum signals
RAM	Random Access Memory
RBAC	Role-Based Access Control

RDF	Resource Description Framework
RDFS	RDF Schema language
REAL	Real World Reasoning
REM	Radio Environment Map
RF	Radio Frequency—usually implying the energy emitted from the antenna
RFFE	RF Front End
RFI	Request for Information
RF-LAN	Radio Frequency Local Area Network
RFOPL	Restricted First Order Predicate Logic
RISC	Reduced Instruction Set Architecture
RKRL	Radio Knowledge Resource Language
RL	Reinforcement Learning
RMI	Remote Method Invocation
RPC	Remote Procedure Calls
RSS	RDF Site Summary
RSSI	Received Signal Strength Indicator
RSVP	Resource reSerVation Protocol
RTS	Request to Send
RTT	Round-Trip Times
Rx	Receiver
RXML	Radio XML
SAD	Software Assembly Descriptor File
SAPIENT	Situation Aware Protocols In Edge Network Technologies
SATCOM	Satellite Communication
SCA	Software Communication Architecture
SCARI	SCA Reference Implementation
SCD	Software Component Descriptor File
SDCR	Software Defined Cognitive Radio
SDR	Software Defined Radio
SDRF	Software Defined Radio Forum
serModel	Stimulus Experience Response Model
SINC GARS	Single Channel Ground and Airborne Radio System
SINR	Signal to Interference and Noise Ratio
SIP	System Improvement Program
SIR	Signal-to-Interference Ratio
SKAT	Semantic Knowledge Articulation Tool
SMS	Short Message System (text paging)

SMT	Spectrum Management Tools
SNePS	A Natural Language Research System
SNMP	Simple Network Management Protocol
SNR	Signal-to-Noise Ratio
SoC	System-on-a-Chip
SOF	Special Operations Forces
SONET	Synchronous Optical Network
SP	Sensory and Perception
SPD	Software Package Descriptor
SPEAKeasy	SPEAKeasy software defined radio development program
SPI	Spectrum Policy Initiative
sps	samples per second
SPS	Standard Positioning Service
SPTF	Spectrum Policy Task Force
SR	Software Radio
SRA	Software Radio Architecture
SS	Skill Set
SSB	Single Side Band
SSIT	Spectrum Sharing Innovation Testbed
SSP	Subset-Sum Problem
STM	Short-Term Memory
STT	Speech-to-Text
SU	Secondary User
SUO SAS	Small Unit Operations Situational Awareness Systems
SUSAS	Speech Under Stress And Strain speech recognition database
SVM	Support Vector Machines
SW	Software
SWR	Software Radio
SWRL	Semantic Web Rule Language
TC	Turing-Capable
TCP	Transmission Control Protocol
TCP/IP	TCP with Internet Protocol
TD	Temporal Difference
TDD	Time Division Duplex
TDMA	Time Division Multiple Access
TDoA	Time Difference of Arrival
TI	Tactical Internet
TM	Trademark

ToA	Time of Arrival
TOC	Tactical Operations Center
TTS	Text-to-Speech
TV	Television
Tx	Transmitter
UDP	User Datagram Protocol
UHF	Ultra High Frequency
UML	Unified Modeling Language
UNIX	Uniplexed Information and Computing System (OS)
URI	Universal resource indicator
US	United States
USRP	Universal Software Radio Peripheral
UW	Ultra Wide
UWB	Ultra WideBand
VHDL	<u>V</u>HSIC <u>H</u>igh level <u>D</u>esign <u>L</u>anguage (usually a specification for the register transfer functionality of a chip)
VHF	Very High Frequency
VHSICD	Very High Speed Integrated Circuit Design
VM	Virtual Machine
VoCoder	Voice Coder
VoIP	Voice over Internet Protocol
VOR	VHF Omni-directional Range (radio navigation system for aircraft)
VSWR	Voltage Standing Wave Ratio ($>$1:1 implies that the antenna does not perfectly radiate all energy)
VT	Virginia Tech
VTC	Video Teleconference
VT-CWT	Virginia Tech Center for Wireless Technologies
W3C	World Wide Web Consortium
WAN	Wide Area Network
W-CDMA	Wideband Code Division Multiple Access
WF	Waveform
WiFi	Wireless Fidelity (an 802.11 based Wireless Local Area Network)
WIN-T	Warfighter's Internet
WLAN	Wireless Local Area Network—usually 802.11
WPG	Weighted Potential Game
WRAN	Wireless Regional Area Network
WRC	World Radiocommunication Conferences
WSGA	Wireless System Genetic Algorithm

WWRF	World Wireless Research Forum
WWV	NIST radio station transmitting time of day
XG	Darpa's neXt Generation radio development program
XML	eXtensible Markup Language
XTAG	UPenn system for Tree Adjoining Grammar TAG
XTM	XML Topic Maps
Zigbee	A low data rate ad hoc wireless network meant for in home connectivity

Index

Note: Italicized page numbers refer to figures and tables and page numbers with "n" refer to footnotes.